Structure and Being

Structure and Being

A Theoretical Framework for a Systematic Philosophy

Lorenz B. Puntel
Translated by and in collaboration with Alan White

The Pennsylvania State University Press
University Park • Pennsylvania

Copyright © 2008 The Pennsylvania State University

Published by The Pennsylvania State University Press,
University Park, PA 16802-1003

It is the policy of The Pennsylvania State University Press to use acid-free paper. Publications on uncoated stock satisfy the minimum requirements of American National Standard for Information Sciences—Permanence of Paper for Printed Library Material, ANSI Z39.48–1992.

Library of Congress Cataloging-in-Publication Data

Puntel, Lorenz B. (Lorenz Bruno)
[Struktur und Sein. English]

Structure and being : a theoretical framework for a systematic
philosophy / by Lorenz B. Puntel ; translated by and in collaboration
with Alan White.
      p.   cm.
   Includes bibliographical references and index.
Summary: "Presents, and in part develops, a systematic philosophy as the universal science, or the theorization of the unrestricted universe of discourse, explicitly including being as such and as a whole. Argues that complete exploration of the theoretical domain requires such a science"—Provided by publisher.

   ISBN-13: 978-0-271-03374-7 (pbk : alk. paper)
   1. Philosophy.   I. White, Alan, 1951–   II. Title.
   B53.P8613   2008
   193—dc22

2008007415

To my students and to the participants in my graduate courses
at the University of Munich

Lorenz B. Puntel, born 1935, studied philosophy, psychology, classical philology, and Catholic theology in Munich, Vienna, Paris, Rome, and Innsbruck. He received a doctorate in philosophy in 1968 and one in Catholic theology in 1969. He qualified as a university lecturer in philosophy in 1972 and became Professor of Philosophy at the University of Munich in 1978. In 2001, he became Professor Emeritus.

Alan White received his BA from Tulane University in 1972 and his PhD from the Pennsylvania State University in 1980. In 2000, he became Mark Hopkins Professor of Philosophy at Williams College.

Overview of Contents

Preface to *Struktur und Sein* xviii

Preface to *Structure and Being* xx

Key to Abbreviations and Logical/Mathematical Symbols xxii

Introduction 1

1. **Global Systematics: Determination of the Standpoint of the
   Structural-Systematic Philosophy** 22
   1.1 A Theoretical Framework for a Systematic Philosophy: The Complexity
       of the Concept and of Its Presentation 22
   1.2 A First Determination of Systematic Philosophy 26
   1.3 Structure and Being: A First Characterization of the Basic Idea
       Behind the Structural-Systematic Philosophy 36
   1.4 The Idealized Four-Stage Philosophical Method. 41
   1.5 (Self-)Grounding of Systematic Philosophy?. 52

2. **Systematics of Theoreticity: The Dimension of Philosophical
   Presentation** 74
   2.1 Theoreticity as a Dimension of Presentation. 74
   2.2 Language as the Medium of Presentation for Theoreticity 76
   2.3 The Epistemic Dimension as the Domain of the Accomplishment of
       Theoreticity. 99
   2.4 The Dimension of Theory in the Narrower Sense. 121
   2.5 Fully Determined Theoreticity: First Approach to a Theory
       of Truth 141

3. **Systematics of Structure: The Fundamental Structures** 155
   3.1 What Is the Systematics of Structure?. 155
   3.2 The Three Levels of Fundamental Structures 172
   3.3 Theory of Truth as Explication (Articulation) of the Fully
       Determinate Connections among Fundamental Structures 222

4. **World-Systematics: Theory of the Dimensions of
    the World** **246**

    4.1 The Concept of World 247
    4.2 The "Natural World" 250
    4.3 The Human World 263
    4.4 The Aesthetic World 305
    4.5 The World as a Whole 324

5. **Comprehensive Systematics: The Theory of the Interconnection of All
    Structures and Dimensions of Being as Theory of Being as Such and
    as a Whole** **357**

    5.1 The Philosophical Status of Comprehensive Systematics 357
    5.2 Basic Features of a Theory of Being as Such and as a Whole 413
    5.3 Starting Points for a Theory of Absolute Being 441

6. **Metasystematics: Theory of the Relatively Maximal Self-Determination
    of Systematic Philosophy.** **461**

    6.1 The Status of Metasystematics. .461
    6.2 Immanent Metasystematics 467
    6.3 External Metasystematics 469
    6.4 Self-Determination, Metasystematics, and the Self-Grounding of the
         Structural-Systematic Philosophy. 481

Works Cited. 485

Index. 499

Detailed Table of Contents

Preface to *Struktur und Sein* xviii

Preface to *Structure and Being* xx

Key to Abbreviations and Logical/Mathematical Symbols xxii

Introduction 1

**1. Global Systematics: Determination of the Standpoint of the
Structural-Systematic Philosophy** 22

  1.1 A Theoretical Framework for a Systematic Philosophy:
The Complexity of the Concept and of Its Presentation 22

  1.2 A First Determination of Systematic Philosophy 26

    1.2.1 A Quasi-Definition of the Structural-Systematic Philosophy . . . 26

    1.2.2 "Theory" 27

    1.2.3 "Structure" 27

    1.2.4 "Unrestricted Universe of Discourse". 30

    1.2.5 "Most General or Universal Structures" 33

    1.2.6 "Systematic Philosophy" and "Philosophical System". 35

  1.3 Structure and Being: A First Characterization of the Basic Idea
Behind the Structural-Systematic Philosophy 36

  1.4 The Idealized Four-Stage Philosophical Method 41

    1.4.1 The Problem of Method. 41

    1.4.2 First Methodic Stage: Identification of Structures and
Construction of Minimal and Informal Formulations
of Theories. 42

    1.4.3 Second Methodic Stage: Constitution of Theories 44

    1.4.4 Third Methodic Stage: Systematization of Theories. 50

    1.4.5 Fourth Methodic Stage: Evaluation of the Comprehensive
System or Network with Respect to Theoretical Adequacy and
Truth Status 51

  1.5 (Self-)Grounding of Systematic Philosophy?. 52

    1.5.1 On the Concept of Grounding in General 52

    1.5.2 The Problem of Grounding in Philosophy. 54

      1.5.2.1 On the Nonsystematic Concept of Grounding 55

1.5.2.2   The Systematic Process of Grounding and Its Forms, Stages, or Levels 64

1.5.2.3   The Process of Systematic Grounding as an Idealized Form of the Practice of Systematic Grounding 71

2. **Systematics of Theoreticity: The Dimension of Philosophical Presentation** **74**

2.1 Theoreticity as a Dimension of Presentation 74

2.2 Language as the Medium of Presentation for Theoreticity 76

2.2.1 Language, Communication, and Dimension of Presentation . 77

2.2.2 Normal or Ordinary Language and Philosophical Language . . 78

2.2.3 Philosophical Language as Theoretical Language 89

2.2.3.1   The Linguistic Criterion for Theoreticity 89

2.2.3.2   Basic Features of a Program for the Development of a Systematic Philosophical Language. 94

2.2.4 The Centrality of Language to Philosophy 96

2.2.5 From the Dimension of Language to the Dimension of Knowledge: The Roles of Speaker and Subject 97

2.3 The Epistemic Dimension as the Domain of the Accomplishment of Theoreticity 99

2.3.1 On the Problem of the Epistemic Subject (or of Epistemic Subjectivity) 99

2.3.2 On the Systematic Status of the Epistemic Dimension— The Dimension of Knowledge. 101

2.3.2.1   The Ambiguity of "Knowledge". 101

2.3.2.2   Knowledge as a Philosophical Problem 102

2.3.2.3   Knowledge and Cognition in Kant 108

2.3.2.4   Subjectivity and Knowledge with Respect to Systematicity 110

2.3.2.5   A Reversal of Perspective: The Indispensable but Secondary Theoretical Status of the Epistemic Dimension 117

2.4 The Dimension of Theory in the Narrower Sense 121

2.4.1 On the Proper Concept of Theory in General 121

2.4.2 The Theory-Concept in Metalogic/Metamathematics and in Philosophy of Science 121

2.4.2.1   The "Logical" Theory-Concept 121

2.4.2.2   The "Scientific" Theory-Concept I: The "Received View" 122

2.4.2.3   The "Scientific" Theory-Concept II: "Semantic Approaches". 124

2.4.2.3.1   Bas van Fraassen's Constructive-Empiricist Position. 124

2.4.2.3.2   The Structuralistic Conception of Theory. . 127

2.4.3 A Structural Theory-Concept for (Systematic) Philosophy . . 130

2.4.3.1   The Problematic 130

    2.4.3.2 The Essential Components of a Structural
             Theory-Concept for Systematic Philosophy 136
    2.4.3.3 The Structural Theory-Concept as Regulative,
             and Its Approximative-Partial Realization 138
2.5 Fully Determined Theoreticity: First Approach to a Theory
     of Truth 141
    2.5.1 Preliminary Questions 141
        2.5.1.1 The Word "Truth" and the Problem of the Concept
                 of Truth 141
        2.5.1.2 Substantialism and Deflationism 144
        2.5.1.3 "Truth" as Predicate and as Operator 145
        2.5.1.4 Comprehensive Theory of Truth and Subtheories
                 of Truth 147
    2.5.2 The Basic Idea of Truth. 148
        2.5.2.1 The Fundamental Fact About Language: Linguistic
                 Items Require Determination 148
        2.5.2.2 The Three Levels of Semantic Determination 149
        2.5.2.3 The Interconnection of the Three Levels: The
                 Explicitly Semantic Dimension as Fundamental 152
        2.5.2.4 Informal-Intuitive Formulation of the
                 Fundamental Idea of Truth 153

3. Systematics of Structure: The Fundamental Structures 155
3.1 What Is the Systematics of Structure? 155
    3.1.1 The Basic Idea 155
    3.1.2 Preliminary Clarifications of Terms and Concepts. 158
        3.1.2.1 "Concept," "Meaning," "Sense," "Bedeutung,"
                 "Semantic Value," "Thought," "Proposition,"
                 "State of Affairs" 158
        3.1.2.2 "Object," "Property," "Relation," "Fact," and
                 Other Entities 163
        3.1.2.3 "Category" 164
    3.1.3 The Systematic-Architectural Status in Philosophy of the
           Expanded Concept of Structure 167
    3.1.4 The Program of a Philosophical Systematics of Structure. . . 168
    3.1.5 The Status of Language and Semantics Within
           the Systematics of Structure 170
3.2 The Three Levels of Fundamental Structures 172
    3.2.1 Formal Structures 172
        3.2.1.1 Logic, Mathematics, and Philosophy 172
        3.2.1.2 Mathematical Structures 175
        3.2.1.3 Logical Structures 178
    3.2.2 Semantic Primary Structures 183
        3.2.2.1 General Characterization 183

3.2.2.2 The Decisive Option: Ontologically Oriented
Semantics for Philosophical Language 185
3.2.2.3 Critique of the Semantics and Ontology That Are Based
on the Principle of Compositionality 186
   3.2.2.3.1   Basic Features of Compositional Semantics:
Compositional Semantic Structures 186
   3.2.2.3.2   Critique of Compositional Semantics and
Ontology: The Unacceptability of Substance
as Fundamental Ontological Category 190
      3.2.2.3.2.1 Substance Ontology and Its
Alternatives in Contemporary
Philosophy. 190
      3.2.2.3.2.2 The Root Problem with All
Conceptions of Substance 193
      3.2.2.3.2.3 Quine's Procedure for the
Elimination of Singular Terms:
An Insufficient Means for Accomp-
lishing a Philosophical Revolution . . 195
3.2.2.4 Basic Features of a Semantics Based on a Strong
Version of the Context Principle. 199
   3.2.2.4.1   A Strong Version of the Semantic Context
Principle. 199
      3.2.2.4.1.1 Incompatibility of the Context
Principle and the Principle of
Compositionality 200
      3.2.2.4.1.2 Basic Features of and Requirements
for a Strong Version of the
Context Principle 201
      3.2.2.4.1.3 The Problem of Identity Conditions for
Primary Propositions
(and Primary Facts). 203
   3.2.2.4.2   The Concept of Contextual Semantic
Structure: Primary Propositions as Semantic
Primary Structures 207
3.2.3 Ontological Structures. 208
3.2.3.1 Definition of Ontological Primary Structures
(Primary Facts) 208
3.2.3.2 Simple Primary Facts as Simple Ontological
Primary Structures 209
3.2.3.3 Forms of Configuration as Ontological Structures 214
   3.2.3.3.1   On the Relation Between Logical/
Mathematical Structures and
Ontological Structures 215
   3.2.3.3.2   Configurations and Propositional Logic 216
   3.2.3.3.3   Configurations and First-Order
Predicate Logic 218

3.2.3.3.4 Forms of Configurations: Expansions of
Classical Logic and the Multiplicity of
Mathematical Structures 222

3.3 Theory of Truth as Explication (Articulation) of the Fully
Determinate Connections Among Fundamental Structures 222

3.3.1 A More Precise Characterization of the Basic Idea of Truth 223

3.3.2 The So-Called "Truth-Bearers" and the Fundamental
Structures 226

3.3.3 Truth as Composition of Three Functions: The Tristructural
Syntactic-Semantic-Ontological Connection 227

3.3.3.1 The Syntactic-Semantic Dimension: A "Cataphoric"
Theory 227

3.3.3.2 The Semantic-Ontological Dimension:
The Identity Thesis 231

3.3.3.2.1 The Fully Determined Semantic Status of
Language and the Ontological Dimension 231

3.3.3.2.2 The Ontological Import of Truth as Identity of
Proposition and Fact (the Identity Thesis) 232

3.3.3.2.3 The Ontology of Primary Facts as the
Ontology Appropriate to the Structural
Truth-Theory .235

3.3.4 Three Concluding Questions .236

3.3.4.1 Starting Points for a Theory of Falsity 237

3.3.4.2 On the Ontological Import of the Truth of Formal
(Logical and Mathematical) Propositions or Structures . .239

3.3.4.3 A Moderate Relativism with Respect to Truth 241

4. **World-Systematics: Theory of the Dimensions of the World** **246**

4.1 The Concept of World 247

4.1.1 World, Universe of Discourse, and Being as a Whole 247

4.1.2 The Most Important Domains or Subdimensions of the
Actual World 249

4.2 The "Natural World" 250

4.2.1 Is a Philosophy of the Natural World at all Possible? 251

4.2.1.1 An Instructive Example: The Philosophical Incoherence
of Quine's Attempted Reconciliation of "Naturalism"
and "Global Ontological Structuralism". 251

4.2.1.2 The Interdependence of Philosophy and the
Natural Sciences 257

4.2.2 Major Tasks and Global Theses of a Philosophy of the
Natural World Connected to the Natural Sciences 260

4.2.2.1 The Categorial-Structural Constitution of the
Natural World 261

4.2.2.2 The Natural World and the Plurality of Domains
of Being(s): The "Ontological Difference". 262

4.3 The Human World 263

4.3.1 Philosophical Anthropology, or Philosophy of Mind 263

4.3.1.1 What is an Individual, Categorially/Structurally
Considered? 263

4.3.1.2 The Individual Human Being as Person 264

4.3.1.2.1 On the Problematic of the Adequate
Formal Articulation of the Concept of
Configuration 265

4.3.1.2.2 Is "Configuration" the Adequate
Ontological Structure of the Individual
Human Being as Person? 269

4.3.1.2.2.1 A Fundamental Systematic-
Methodological Consideration 270

4.3.1.2.2.2 The Elements of the Configuration
Constituting the
Human Individual 273

4.3.1.2.2.3 The Unifying Point as the Factor
Configuring the Configuration 275

4.3.1.2.2.4 Intentionality and
Self-Consciousness 278

4.3.1.2.3 Is the Human Individual or
Person Explicable Materialistically/
Physicalistically? 282

4.3.1.2.3.1 On the Current Discussion. 282

4.3.1.2.3.2 An Argument Against Physicalism. . .287

4.3.2 Moral Action and Moral Values (Ethics). 290

4.3.2.1 On the Theoretical Character of Ethical Sentences 291

4.3.2.1.1 The Ambiguity of "Practical Philosophy"
and of "Normative Ethics" 291

4.3.2.1.2 Primarily Practical, Theoretical-Deontic,
and Theoretical-Valuative Sentences 293

4.3.2.2 The Ontological Dimension of Ethical Truth:
Ontological Values. 296

4.3.2.3 The Distinction Between Basal-Ontological Values
and Moral-Ontological Values 298

4.3.2.4 The Ontological Status of Basal-Ontological Values 300

4.3.2.4.1 The General-Metaphysical Perspective 300

4.3.2.4.2 The Metaphysical-Anthropological
Perspective .301

4.3.2.5 The Ontological Status of Moral-Ontological Values 303

4.4 The Aesthetic World 305

4.4.1 The Three Central Logical-Semantic Forms of
Aesthetic Sentences. 306

4.4.2 The Universal Aesthetic Dimension: Beauty as
Fundamental Concept 314

4.4.3 The Specific Dimension of Art 318

4.4.4 Two Objections. 322

4.5 The World as a Whole 324

4.5.1 Natural-Scientific Cosmology. 324

4.5.2 The Phenomenon of the Religious and the Plurality
of Religions: The Necessity of a Philosophical Interpretation . . .329

4.5.3 World History 332

    4.5.3.1 Philosophy of World History and the Science
of History. 333

    4.5.3.2 The Ontology of World History. 334

    4.5.3.3 Does World History Have an Inner Structure? 340

    4.5.3.4 Does World History Have a Meaning? 342

        4.5.3.4.1 Preliminary Clarifications. 342

        4.5.3.4.2 Reasons for the Necessity of a Comprehensively
Systematic Theory of World History. 345

        4.5.3.4.3 Presuppositions for a Comprehensively
Systematic Theory That Clarifies the
Meaning of World History 350

**5. Comprehensive Systematics: The Theory of the Interconnection
of All Structures and Dimensions of Being as Theory of Being
as Such and as a Whole 357**

5.1 The Philosophical Status of Comprehensive Systematics 357

5.1.1 Comprehensive Systematics as Structural Metaphysics 357

5.1.2 The Primary Obstacle to the Development of a Comprehensive
Systematics as Structural Metaphysics 359

    5.1.2.1 The Problem of the Gap Putatively Separating the
Theorist from Reality as It Is "In Itself" 360

    5.1.2.2 Examples of Failed Attempts to Solve the Problem
of the Putative Gap .361

5.1.3 Comprehensive Clarification of the Problem of the
Putative Gap as Starting Point for a Theory of
Comprehensive Systematics: Four Fundamental Theses 369

    5.1.3.1 Thesis One: The Appropriate Form of Presentation
for the Structural-Systematic Philosophy Requires
Sentences in the Purely Theoretical Form 369

    5.1.3.2 Thesis Two: Semantics and Theories of Beings and
of Being are Fundamentally Interrelated 370

    5.1.3.3 Thesis Three: Expressibility Is a Factor Fundamental
to the Structurality of Beings and of Being 371

    5.1.3.4 Thesis Four: Philosophical Languages Are Languages
of Presentation. 371

5.1.4 The Adequate Concept of Theoretical-Philosophical Language . .371

    5.1.4.1 Language, Communication, and Presentation 372

5.1.4.2 The Fundamental Criterion for the Determination
of the Basic Structures of an Adequately Clarified
Philosophical Language .373

5.1.4.3 Philosophical Language as Semiotic System with
Uncountably Many Expressions 374

5.1.4.3.1 The Realism/Anti-Realism Debate as a
Dead End: Reasons and Consequences 374

5.1.4.3.2 An Essential Presupposition for the Universal
Expressibility of the World (of Being):
Theoretical Languages with Uncountably
Many Expressions 378

5.1.4.3.2.1 The Possibility in Principle of
Semiotic Systems with Uncountably
Many Signs/Expressions 378

5.1.4.3.2.2 A Fundamental Problem: Language
and "Tokening System" (the Position
of Hugly and Sayward) 380

5.1.4.3.2.3 The Status of Tokening Systems for
Theoretical Languages 384

5.1.4.3.3 The Segmental Character of Effective
Theoretical Languages 387

5.1.4.4 Are There Uncountably Many Entities? 392

5.1.4.5 Is Philosophical or Scientific Language a Purely Human
Production? Or What, Ultimately, Is (a) Language? 394

5.1.5 The Plurality of Languages: Its Ontological Interpretation, and
Several Consequences 397

5.1.5.1 In What Sense and on What Basis Is There a Plurality of
(Theoretical) Languages? 397

5.1.5.2 The Ontological Ramifications of the Plurality of
Theoretical Languages. .398

5.1.5.2.1 On Various Approaches to the Problem 398

5.1.5.2.2 A Suggested Three-Step Solution to
the Problem 401

5.1.5.2.2.1 First Step: The Ontologization of the
Theoretical Sphere 402

5.1.5.2.2.2 Second Step: Changing the Focus of
the (Philosophical/Scientific)
Perspective from Subjectivity to Being
(Nature, the World) 402

5.1.5.2.2.3 Third Step: Three Pairs of Concepts as
Criteria for Judging the Strength or
Weakness of the Ontological Adequacy
of Theoretical Frameworks 405

5.1.6 Summary: Comprehensive Systematics as Universal Theory. . . . 411

5.2 Basic Features of a Theory of Being as Such and as a Whole 413

   5.2.1 What is Being as Such and as a Whole? 413

   5.2.2 Talk of "the Whole (the Totality)": Semantics, Logic/Mathematics, and Philosophy 421

   5.2.3 The Primordial Dimension of Being, the Actual World, and the Plurality of Possible Worlds. 431

   5.2.4 The Inner Structurality of the Primordial Dimension of Being: The Most Universal Immanent Characteristics. 436

5.3 Starting Points for a Theory of Absolute Being. 441

   5.3.1 Preliminary Clarifications 441

   5.3.2 The Decisive Step: The Primordial Difference with Respect to Being as the Difference Between the Absolutely Necessary and the Contingent Dimensions of Being 443

   5.3.3 Additional Remarks and Clarifications 446

   5.3.4 Additional Steps in the Explication of the Absolutely Necessary Dimension of Being 451

6. Metasystematics: Theory of the Relatively Maximal Self-Determination of Systematic Philosophy 461

6.1 The Status of Metasystematics. 461

   6.1.1 Metasystematics and Metaphilosophy. 461

   6.1.2 The Metasystematic Self-Determination of the Structural-Systematic Philosophy and the Criterion of Relatively Maximal Intelligibility and Coherence 463

6.2 Immanent Metasystematics 467

   6.2.1 What is Immanent Metasystematics? 467

   6.2.2 Three Aspects of Immanent Metasystematics 467

6.3 External Metasystematics. 469

   6.3.1 What Is External Metasystematics? 469

   6.3.2 External Intratheoretical Metasystematics 470

      6.3.2.1 External Intratheoretical Interphilosophical Metasystematics 470

      6.3.2.2 External Intratheoretical Philosophical-Nonphilosophical Metasystematics 477

   6.3.3 Extratheoretical Metasystematics 480

6.4 Self-Determination, Metasystematics, and the Self-Grounding of the Structural-Systematic Philosophy 481

Works Cited. 485

Index. 499

Preface to *Struktur und Sein*

This book is the realization of a project that I undertook in the first years of my teaching and research activity at the University of Munich more than two decades ago. In my lectures and seminars on systematic philosophy, I wrote, with the supportive collaboration of my then-student Geo Siegwart, a manuscript titled "*Systematic Philosophy—Outline of a Program*" [*Eine Programmschrift*] intended to serve as a text for the participants in my courses. Before long, hundreds of copies of this text circulated among students and colleagues. Thereafter, a noted German press offered to publish the text as a book. The offer was attractive, but I turned it down. I was convinced that serious philosophical labor has its price, and that this was particularly the case for a book with the comprehensive scope and the ambitious intent of this particular work; such a book could be a significant contribution, I concluded at the time, only if it was the result of a long process of ripening.

From my current perspective, I deem my earlier conviction to have been correct. My conception of that time, which had developed chiefly under the influence of the pre-Kantian metaphysical tradition and the classical German tradition in philosophy, has changed significantly in the ensuing years. There remain some basic issues, theses, and intuitions characteristic of those traditions, but all have been subjected to radical transformations, particularly with respect to methodology and in light of developments in the philosophy of science. The transformations arose from my intensive engagement with contemporary formal logic, philosophy of science, and analytic philosophy. The kind of systematic philosophy that I now advocate and attempt to develop can be viewed as a form of analytic philosophy, but it differs from analytic philosophy in its traditional and in most of its contemporary forms in two respects. First, it resolutely takes up central elements of the tradition of *philosophia perennis*. Second, it opposes the fragmentary character of the understanding and of the treatment of philosophical issues characteristic of the majority of analytic philosophers in that it strives to reconfer upon philosophy its traditional systematic status.

Although this book is the expression of my own philosophical path and of the conception to which that path led me, it could not have taken its current form without the stimulation that I received over the course of many years from many of my colleagues, my assistants, my doctoral students, the participants in my regularly occurring graduate courses at the University of Munich, and not least from many with whom I engaged in lively intellectual exchanges during my in some cases lengthy stays at various foreign universities, particularly in the United States. To this large group of people I owe my primary thanks.

I must refrain from personally thanking all of these individuals by name. I must, however, make exceptions for two American colleagues.

Nicholas Rescher (University of Pittsburgh) gave me the decisive impulse to overcome the megalomaniacal idea, championed by the German Idealists, of developing an absolute system of philosophy, but without leading me to embrace the equally senseless alternative of philosophical fragmentation. From Rescher's books—especially *The Coherence Theory of Truth*, which appeared in 1973—I learned how one can do systematic philosophy in a sober, clear, and rigorous manner. Rescher's path later led to his monumental three-volume work, *A System of Pragmatic Idealism* (1992–94). Although my own philosophical path diverges from his in significant ways, I am grateful for the fundamental inspirations I have gained from the conversations we have had in our many encounters.

Alan White (Williams College, Williamstown, MA) merits wholly unique acknowledgment and recognition. After he had read some of my work, he contacted me via the Internet in September 2003, posing several interesting philosophical questions. From that arose a lively discussion that led to a remarkable collaboration: despite the distance between Germany and the United States, current electronic means of communication made possible nearly daily contact. Soon, White offered to translate into English the parts of this book that were already in first-draft form (something more than half of what became the entire book), and indeed promptly set himself to the task. He then translated the other parts as soon as they were available in first drafts. In this manner, an English version of this text emerged almost simultaneously with the German version. White's enthusiasm for the philosophical conception presented in the book, his encouragement, and his suggestions contributed decisively to my decision to commit all of my time and energy to the completion of the book. During a visit of several weeks' duration that I made to beautiful Williamstown in fall 2004, I was able not only to enjoy the hospitality of White and of his wife, Jane Nicholls, but also, by working on the book together with him, to bring the project a good step further. This constructive collaboration continued during White's stay in Germany in June 2005. The significance of his contributions, particularly with respect to the development of the English edition, cannot be overestimated.

I owe deep thanks to a number of philosophical friends who have accompanied the development of my philosophical conception with great interest and intensive discussions. In particular, I thank Constanze Peres, Christina Schneider, and Karl-Heinz Uthemann for their in various ways critical but always constructive readings of the final draft of the manuscript of this book; they contributed much with respect to clarifying it linguistically as well as contentually. I would also like to thank Sylvia Noss for her scrupulous corrections to the manuscript and the galley proofs.

Finally, I would like to thank Dr. Georg Siebeck of the Mohr Siebeck Verlag for his stimulating interest in the publication of the book, his energetic engagement, and his fairness.

Lorenz B. Puntel

Munich-Augsburg, February 2006

Preface to *Structure and Being*

As Lorenz B. Puntel reports in the Preface to *Struktur und Sein*, translated above, my involvement with him on the development of this book, and of its German counterpart, began in fall 2003. I was initially motivated to contact him by my interest in reviving the enterprise of systematic philosophy. My readings of some of his articles on truth-theories suggested to me not only that he shared that interest but also that he had made significant progress in developing his own conception of a systematic philosophy. Equally clear to me, once our correspondence had begun, were his intellectual generosity and his openness to and indeed eagerness for conversations about his work and suggestions concerning how best to present it, particularly in English. I was delighted when he agreed to the unusual undertaking of our producing the German and English versions in tandem. It has been an honor and a privilege to work so closely with him as he brought his systematic philosophy to the point of its presentation in this book, and I am deeply grateful to him for his having allowed me to do so. An added bonus is my friendship with him and his wife, Christina Schneider.

The title page of this book identifies me as both translator and collaborator. Puntel partially describes the nature of my collaboration in the Preface to *Struktur und Sein*, but readers of *Structure and Being* need additional information because there are passages in this book that are not direct translations from *Struktur und Sein*. In some cases (e.g., Section 1.5.1), the divergences from direct translation are motivated by richly consequential differences between German and English terminology; in some other cases (e.g., Section 1.2.5), they stem from the availability of examples more familiar to (especially) British and American readers. In yet other cases, they correct errors in *Struktur und Sein* identified only after its publication; these will be corrected in the second edition of that book. Most of those modifications are relatively minor, but some are sufficiently significant to be worth noting here. These are changes to pp. 10–12 (*S.u.S.* 13–14), p. 113 (*S.u.S.* 151), Section 2.4.2.3.2, Section 2.4.3, Chapter 2 note 32 (p. 140, *S.u.S.* p. 187), p. 468 (*S.u.S.* p. 622), and Chapter 6 note 7 (p. 478, *S.u.S.* p. 637). In all significant cases (save inadvertent errors on my part, to be listed, as I become aware of them, on a web page accessible from www.structureandbeing.com), the English text differs from the German only in places where Puntel and I agreed that our allowing them to do so would result in making the English version clearer and more intelligible than it would have been had it more closely mirrored the German. To facilitate comparisons of the English with the German, page numbers from *Struktur und Sein* are indicated in the margins of the pages of *Structure and Being*.

I am delighted to have the opportunity publicly to express my gratitude to Boston University (BU) Professors of Philosophy Charles Griswold and David Roochnik, who were instrumental in providing me with the opportunity, when I served as BU's John Findlay Visiting Professor of Philosophy in fall 2005, to focus two seminars on an earlier draft of this book. Additional thanks are due Roochnik and his wife Gina Crandell for the warm hospitality they provided me during that semester.

This book has been much improved by comments on earlier drafts from participants in my seminars at Boston University and at Williams College. I cannot thank all of those participants by name, but I would be remiss in not once again mentioning Roochnik, along with Williams philosophy majors Ben Roth ('04) and Jon Zeppieri ('99). These three, although not enrolled in my graduate seminar at BU, found the time, despite their full-time jobs, to read the translation-in-progress, to attend class sessions, and to make valuable comments and suggestions.

Puntel and I are both deeply grateful to two sources for subsidizing the initial production costs of this book. One of those sources is Williams College; the other is a family that we would love to thank publicly but that prefers to remain anonymous. The generous support provided by these sources enabled Penn State Press to make the book available at prices far lower than would otherwise have been possible.

Finally, Puntel and I much appreciate the support of this book provided by Sandy Thatcher, Director and Philosophy Editor at Penn State Press.

Alan White
Williamstown, MA, February 2008

Key to Abbreviations and Logical/Mathematical Symbols[1]

$B_{S,P}$	operator, "Subject S particularistically believes that" (112)
CPP	principle of compositionality (186)
CPP_{PC}	propositional-conditional principle of compositionality (188)
CPP_S	principle of compositionality for sentences (186)
CPP_{TC}	truth-conditional principle of compositionality (188)
CTC	complete truth-concept (234)
CTP	context principle (199)
CTP_F	context principle in Frege's formulation (199)
D_I	domain of arguments for a function (47)
D_{II}	range of values for a function (47)
DN	deductive-nomological explanation (123)
FMSO	factive mental state operator (Williamson 2000) (107)
G	designates a schema for lawlike, justificatory formulations (57)
G_S	designates a schema for lawlike, systematic formulations (63)
H-O	Hempel-Oppenheim schema for scientific explanation (123)
I_K	idealized knowledge (120)
(K)	knowledge as defined in this book (106)
(K_G)	Gettier-type definition of knowledge (103)
L	(1) language; (2) "links": a class of intertheoretical connections according to the structuralist theory-concept (129)
O(A)	operator "It is obligatory that A" (294)

---

[1] For abbreviations and unusual symbols, the page numbers on which they first appear are provided; the abbreviations and symbols are explained in those places. The text provides no additional explanations of the normal logical/mathematical symbols listed here.

Throughout this book, all footnotes designated by numerals have counterparts in *Struktur und Sein*; footnotes designated by letters do not.

Perproposition	proposition to be determined (or qualified) (228)
Persentence	sentence to be determined (or qualified) (228)
PL	particularistic-lifeworldly (as index for the subject-operator $\circledS$) (406)
PL1	first-order predicate logic (218)
Q-Def	quasi-definition (26)
S	(1) dimension of structure (136); (2) sentence (generally)
$S_{Æ}$	aesthetic sentence (306)
S^f	Legenhausen's non-standard semantics for PL1 based on the concept *fact* (220)
S^P	Legenhausen's non-standard semantics for PL1 based on the concept *property* (219)
TS	truth-schema (360)
T-schema	truth-schema (225)
U	universe of discourse (136)
$\mathfrak{a}$	semantic function (interpretation function for compositional semantics: denotation and designation) (184)
$\mathfrak{a}^*$	union of the functions $\mathfrak{a}$ and η ($\mathfrak{a}^* = \mathfrak{a} \cup \eta$) (184)
$\mathfrak{b}$	semantic function (interpretation function for contextual semantics: the expression of a primary proposition by means of a primary sentence) (207)
$\mathfrak{b}^*$	union of the functions $\mathfrak{b}$ and μ ($\mathfrak{b}^* = \mathfrak{b} \cup \mu$) (207)
$\mathfrak{A}$	(1) structure (generally) (28); (2) semantic structure (184)
$\mathfrak{B}$	semantic structure (207)
$\mathfrak{C}$	ontological structure (208)
$\mathfrak{U}$	universal class (following Kelly-Morse) (425)
$|\mathfrak{a}|$	domain or universe of the structure $\mathfrak{a}$ (380)
η	function (assigning to variables values from a set A of sentences of the subject-predicate form) (184)
μ	function (assigning values to variables from a set B of primary sentences) (207)
$\varphi, \psi, \chi \ldots$	sentence letters (sentence constants or parameters or variables)
Φs	operator "knows that" (following Williamson) (107)

A	(1) non-empty domain or universe for the compositional-semantic structure $\mathfrak{A}$: the totality of objects (184); (2) formal system (380)
B	(1) non-empty domain or universe for the compositional-semantic structure $\mathfrak{B}$: the totality of primary propositions (207); (2) set of basic signs of a formal system (380)
f	primary fact (234)
f*	simple primary fact (210)
F	non-empty domain or universe for the contextual-ontological structure $\mathfrak{C}$ (208); the totality of primary facts (234)
I	(1) set of intended applications of a theory according to the structuralist conception of theory (129); (2) interpretation function (188)
L_O	observation language (75)
L_T	theoretical language (75)
M	(actual) model (128)
M_p	potential model (128)
M_{pp}	partial-potential model (129)
T (T', T'' ..., T_1, T_2...)	true, truth, truth-value, truth operator, theory (depending on context) (224)
T_R	reconstructed Tarskian truth-schema (229)
T^*	first truth-theoretical function (229, 231)
T^+	second truth-theoretical function (229, 231)
$T^\times$	third truth-theoretical function (229, 234)
V_O	observation vocabulary (75)
V_T	theoretical vocabulary (75)
$Ⓔ\chi$	aesthetic operator ("There is an aesthetic presentation such that χ") (91)
$Ⓟ\psi$	practical operator ("It is ethically obligatory (or forbidden or permissible) that ψ") (91)
$Ⓢ$	subject operator ("It is the case from the perspective of subject S that...") (111)
$Ⓢ_P$	particularistic subject operator (111)
$Ⓢ_{PL}$	particularistic-lifeworldly subject operator (406)
$Ⓢ_U$	universal subject operator (111)

$\mathbb{T}\varphi$	theoretical operator ("It is the case that φ") (91)
$\mathbb{T}_D$	theoretical-deontic operator (295)
$\mathbb{T}_{DE}$	theoretical-deontic-empirical operator (295)
$\mathbb{T}_{DE/O}$	theoretical-deontic-empirical operator in the mode of obligation (295)
$\mathbb{T}_{DU}$	theoretical-deontic-universal operator (295)
$\mathbb{T}_{DU/O}$	theoretical-deontic-universal operator in the mode of obligation (295)
$\mathbb{T}_V$	theoretical-valuative operator (295)
$\mathbb{T}_{VE}$	theoretical-valuative-empirical operator (296)
$\mathbb{T}_{VU}$	theoretical-valuative-universal operator (296)
$p, q, r \ldots \varphi, \psi, \chi$	sentence letters (sentence constants or parameters or variables)
$x, y, z \ldots$	variables for individuals or sets (depending on context)
$\neg, \sim$	connective "negation"
$\wedge, \&$	connective "conjunction"
$\vee$	connective "disjunction"
$\rightarrow, \supset$	connective "conditional" (or "(material) implication")
$\leftrightarrow$	connective "biconditional" (or "equivalence")
$\nrightarrow$	negation of the connective "conditional" (or "(material) implication")
$\vdash$	logical derivability (180n14)
$\models$	(1) model relation: a relation between sentences and structures (180n14); (2) logical consequence: a relation between (sets of) semantically interpreted sentences and sentences following from them (180n14)
$\Vdash$	logical consequence or validity (180n14)
$\Rightarrow$	relation of implication (Koslow) (181)
$\exists$	existence quantifier
$\forall$	universal quantifier
ι	iota operator (ιx : the x such that) (198n20)
$[]$	semantic value of the enclosed expression (187)
$\ulcorner \urcorner$	quasi-quotation marks ("Quine corners") (146)
$E!$	existence predicate

$\langle p \rangle$	proposition p (following Horwich) (146)
p	fully determined primary sentence or the fully determined proposition expressed by it (302)
$\Diamond$	possibility operator (443)
$\Box$	necessity operator (443)
∇	contingency operator (443)
$\in$	relation of being-an-element
$\notin$	relation of not-being-an-element
$\{x \mid x \dots\}$	the set of all x such that...
$\{a_1, \dots, a_n\}$	the set of elements $a_1, \dots, a_n$
$\langle a_1, \dots, a_n \rangle$	ordered n-tuple
$\varnothing$	null set
$\subseteq$	relation is-a-subset-of
$\subset$	relation is-a-proper-subset-of
$\cup$	union
$\cap$	intersection
$\cup X$	the *union* of X: the class of all members of the members of X
$\cap X$	the *intersection* of X: the class of all common members of all members of X
$\times$	Cartesian product (of the sets X and Y: $X \times Y := \{z \mid z = \langle x, y \rangle\}$)
$\circ$	(1) composition of functions (e.g., $f \circ g$); (2) in mereology: overlapping (268)
$\approx$	the same as (in mereology) (268)
$\wp$	power set
$\mapsto$	specification of a function with respect to its assignments (e.g., $x \mapsto f(x)$)
$f\mid_M$	restriction of the function f to M (184)
$\mathbb{Z}^+$	the set of positive integers
$\leftrightharpoons$	relation between the dimension of structure S and the universe of discourse U in the triple $\langle S, \leftrightharpoons, U \rangle$ (137)
$\langle\!\!-\!\!\rangle$	symbol for a rule governing the application of (mathematical) structures to the world (following Ludwig) (137n31)
$\otimes$	combination function (in the definition of the mathematical structure "group") (47)

The systematic philosophy presented in this book has arisen from two insights, formulable as two theses, resulting from a long and intensive occupation with the fundamental philosophical conceptions from history and of the present. The first thesis is that, in terms of its intention, self-understanding, and accomplishments, the theoretical enterprise that for over two thousand years has been designated "philosophy" is fundamentally a form of knowledge with a comprehensive or universal character. The second thesis is that contemporary philosophy—and quite particularly so-called analytic philosophy—today does scarcely any justice to this universal character of philosophy, in that it exhibits, virtually exclusively, a fragmentary character that is conditioned by various distinct factors.

[1] To designate the comprehensive character of philosophy, modernity introduces the term "system," which then develops a significant history. For reasons presented at the end of this Introduction, this term is used in this book, if at all, only marginally, and certainly not as the proper designation of the philosophy here presented. That designation is instead "systematic philosophy" (and, more specifically, "the structural-systematic philosophy").

To be emphasized at the outset is that contemporary philosophy uses the term "systematic" in two distinct senses—or, more precisely, that the term currently has both a central signification and a secondary one. In its central philosophical signification, "systematic" designates a conception of philosophy distinguished by two characteristics: the completeness of its scope, in terms of its subject matter, and its concern with articulating the interconnections among all its various thematic components. Neither this completeness nor this interconnectedness is, as a rule, taken in an absolute sense. Thus, it is not meant that all the details relevant to a philosophical subject matter or domain and all of the interconnections among those details are explicitly presented. What is meant is instead that what this book calls the *unrestricted universe of discourse* is understood and articulated at least in its global structuration.

According to the secondary signification of "systematic" in contemporary philosophy, the term is the counterpart to "(purely) historical": a "systematic" treatment of a topic, a "systematic" view, etc., is one that is not historically oriented.

This secondary signification is *not* of primary importance for this book; here, the chief signification is intended except in cases where either the context or explicit notation indicates the relevance of the secondary signification.

Throughout most of its long history, philosophy has attributed to itself a comprehensive character, even if that character has taken various distinct forms. In the golden age of antiquity, for example, philosophy is more or less identified with scientific knowledge as a whole,[1] in the Middle Ages it is primarily understood as taking the form of a *Summa*, and in modernity it develops, increasingly, as a *system*; this development leads to the duality of Rationalism and Empiricism, which itself then leads to Kant's historical attempt to overcome the duality between these two schools of thought by developing a new form of philosophical system, albeit a radically limited one. Kant's critical enterprise has, as a consequence that only appears to be paradoxical, the development of the highest and most daring variants of philosophy as comprehensive; these are the philosophical systems that come to be grouped under the designation "German Idealism." It is not a historical accident that the collapse of these systems, particularly Hegel's, coincides, in the second half of the nineteenth century, with the impressively self-conscious rise, in the arenas both of theory and of experimentation, of the natural sciences and with the beginnings both of contemporary mathematical logic and of what later becomes known as analytic philosophy.

An additional, later line of separation is to be noted; this is between analytic philosophy and various other schools of thought that have developed, some of which persist, with varying degrees of vivacity, into the present. Those other schools of thought include Husserlian phenomenology, the philosophy of life, hermeneutics, and Heidegger's philosophy of being. The comprehensive character of philosophy—earlier brought into question only rarely and never fundamentally—remains present in these schools of thought, albeit only in a somewhat paradoxical manner. It is present explicitly in a manner that is virtually exclusively negative (i.e., as rejection), but implicitly in one that is astonishingly positive: the attempt has been and continues to be made to relativize precisely this (traditionally) comprehensive character in various ways, by means of the development of some kind of metaconception of it. This is exemplarily the case in the hermeneutic philosophy developed especially by Hans-Georg Gadamer: of central importance to this school of thought is the comprehensive context of the history of interpretation, within which attempts are made to situate the various philosophies that claim to be comprehensive. Heidegger, above all, presses such a metaconception to the greatest extreme in that he attempts to develop a thinking that understands itself as explicitly superior to all preceding philosophies, and thereby claims to have a character yet more radically comprehensive than any of those others.[2]

---

[1] More precisely, the borders between "philosophy" and what is currently termed "empirical science" were, in antiquity, largely undetermined. Aristotle's *Physics* (more precisely: *lectures* on physics, ΦΨΣΙΚΗ ΑΚΡΟΑΣΙΣ) serves as a characteristic example. Throughout the history of philosophy, this work is understood and interpreted as a work of philosophy. On the basis of an understanding of the relationship between philosophy and the empirical or natural sciences that is clarified by modern and contemporary insights, this historical classification can scarcely be maintained.

[2] For a presentation and critique of Heidegger's position, see Puntel (1997).

Philosophy cannot simply ignore or abstract from the tradition because that would be tantamount to a kind of self-denial and thus to self-destruction. But attendance to its own history can be and in fact is concretized in various ways. Thus, philosophy can simply restrict its concern to the history of philosophy or indeed identify itself with this concern. But it can also go to the opposite extreme; it does so if it turns completely and explicitly *against* the entire history of philosophy. Even a simple ignoring of the history of philosophy is a particular way of denying that history any positive significance, and indeed, in a certain respect, the most radical way of doing so. The spectrum of possibilities between these two extremes is quite extensive. It can, however, be established that the most productive new initiatives in philosophy are those that develop on the basis of appropriately balanced attention to the history of philosophy.

In opposition to the schools of thought just introduced, analytic philosophy develops along significantly more modest lines. Fundamentally (and almost exclusively), it has always been *systematic* in the *secondary* sense; as is shown below, it continues to be so. Whether it has been or is systematic in the chief sense is a completely separate question that is addressed shortly below. The "systematic"—in the sense of "not (purely) historical"—character of analytic philosophy, starting from its beginnings, has as one of its consequences the fact that it has neglected and often indeed simply ignored the grand philosophical tradition. Much could be said about this neglect, but a general remark suffices here: analytic philosophers are at present increasingly concerned not only with the history of analytic philosophy, but also with the entire history of philosophy.

[2] The question whether contemporary philosophy is systematic in the chief sense is answered in the negative by the second thesis articulated in the opening paragraph of this Introduction. This thesis has a global character and cannot be defended in detail here; nevertheless, some further specifications are possible and also requisite. For this, it is necessary to distinguish between non-analytic (so-called "continental") and analytic philosophy. As far as non-analytic philosophy since the end of World War II is concerned, the following may be noted globally: to the extent that this philosophy has a distinctly theoretical character, it is concerned essentially with ever new interpretations and reinterpretations of traditional philosophical texts, and not with systematic philosophy in the second of the senses introduced above ("systematic" as "non-historically oriented").[3] Works that are systematic in the chief sense of "systematic" and thus in continuity with the continental tradition of philosophy are scarcely to be found.

The thesis introduced above that analytic philosophy has a solely fragmentary character requires more extensive explanation and specification. In a lecture presented in 1975 (1977/1978), Michael Dummett treats the question posed in his title: "Can Analytic Philosophy Be Systematic, and Ought It to Be?" His answer is illuminating in some respects but not in all. Dummett does not directly pose the question whether analytic philosophy up to and including 1975 is systematic; he does, however, treat this

[3] With respect to German philosophy (since 1945), this thesis is formulated and defended in Puntel (1994). To be emphasized however, is that the situation in German philosophy has changed significantly since 1994.

question indirectly, although even then not comprehensively. He distinguishes between two meanings of "systematic":

> In one sense, a philosophical investigation is systematic if it is intended to issue in an articulated theory, such as is constituted by any of the great philosophical 'systems' advanced in the past by philosophers like Spinoza or Kant. In the other sense, a philosophical investigation is systematic if it proceeds according to generally agreed methods of enquiry, and its results are generally accepted or rejected according to commonly agreed criteria. These two senses . . . are independent of one another. (455)

5   Dummett contends that to the extent that the philosophy of the past—pre-Fregean philosophy—is systematic, it is systematic only in the first sense, not in the second. As far as analytic philosophy is concerned, Dummett appears to hold that to the extent that it is systematic up to 1975, it is so only in the second sense. Dummett restricts this "to the extent" in two ways. He deems such philosophers as Gilbert Ryle, John Austin, and the later Wittgenstein to be explicitly non-systematic in both of his senses. With respect to other analytic philosophers, above all Rudolf Carnap, W. V. O. Quine, and Nelson Goodman, he maintains that it would be absurd to pose to them the question whether analytic philosophy can be systematic; he appears to consider these thinkers to be systematic in both of his senses.

Dummett defends the thesis that "at least in the philosophy of language, philosophy ought henceforward to be systematic in both senses" (455). In part for this reason, he deems Frege to be "the fountain-head of analytical philosophy" (440) and to be the central figure in the entire history of this now-dominant philosophical movement. He maintains "that philosophy failed, throughout most of its long history, to achieve a systematic methodology" (456–57). An explanation is required, he contends, for "how it comes about that philosophy, although as ancient as any other subject and a great deal more ancient than most, should have remained for so long 'in its early stages'" (457), but he provides no such explanation in the essay under consideration. Instead of offering one, he reasons as follows: "The 'early stages' of any discipline are, presumably, to be characterised as those in which its practitioners have not yet attained a clear view of its subject-matter and its goals." He adds that philosophy has "only just very recently struggled out of its early stage into maturity: the turning-point was the work of Frege, but the widespread realisation of the significance of that work has had to wait for half a century after his death, and, at that, is still confined only to the analytical school."

Dummett takes an additional step by contending, "Only with Frege was the proper object of philosophy finally established" (458); to explain this development, he introduces three factors. First, the goal of philosophy is the analysis of the structure of thought, second, this thought is to be distinguished strictly from the *thinking* studied by psychology, and third, the only correct method for the analysis of thought is that of the analysis of language.

6   On this basis, Dummett provides his clearest determination of analytic philosophy: "We may characterise analytical philosophy as that which follows Frege in accepting that the philosophy of language is the foundation of the rest of the subject" (441).

Dummett's reflections well reveal the difficulty encountered with any attempt to describe, generally, what specifically characterizes analytic philosophy; the difficulty is yet clearer if one attempts to answer the question whether analytic philosophy is systematic. As accurate as Dummett's remarks are in mutual isolation, viewed as a whole they are quite one-sided, short-sighted, and in part even incorrect. His distinction between his two senses in which a given philosophical investigation can be "systematic"—on the one hand, "if it is intended to issue in an articulated theory," and, on the other, "if it proceeds according to generally accepted methods of enquiry, and its results are generally accepted or rejected according to commonly agreed criteria" (455)—is both one-sided and artificial. As indicated above, however, Dummett identifies such a method in the philosophical legacy of Frege; he deems this method, which involves the analysis of language, the "only proper" one (458).

These contentions are problematic in several respects. A method determined by the sociology of knowledge ("generally accepted, . . . commonly agreed . . ." [455]) cannot raise the claim of being the "only proper" one; factors of the sociology of knowledge are subject to a volatility far too great to qualify them as a firm basis for evaluating systematic philosophical methods. It cannot, for example, (or can no longer) be said that the method of the analysis of language is currently widely accepted. Dummett says that it is "amazing that, in all its long history, [philosophy] should not yet have established a generally accepted methodology, generally accepted criteria of success, and, therefore, a body of definitively achieved results" (455), and it follows from various of his own theses that *his* method, the analysis of language, should not only be generally accepted but should (or would) also establish a "body of definitively achieved results." Talk in philosophy of "definitive results" is, however, extraordinarily problematic. In any case, Dummett's method has not produced any such results, and again, it cannot be said that his philosophical methodology is generally accepted.

Does it then follow that Dummett's philosophy lacks a "systematic method"? That would be strange, but then it is likewise strange and even incoherent to ascribe to thinkers of the past "articulated theories" (and in this sense *systematicity*) while simultaneously denying that they had systematic philosophical methodologies. In addition, if one attributes to the criteria of general agreement and acceptance as central a significance as does Dummett, then it would be only consequent to apply the criteria not only to systematicity as requiring a universal methodology but also to systematicity as "intending to issue in articulated theor[ies]." But then one could no longer contend, as does Dummett, that Spinoza, Kant, and other philosophers develop "articulated theories" and are in this sense "systematic philosophers," because it is simply a fact that there are no "commonly agreed criteria" in accordance with which their results are "generally accepted or rejected."

From this arises the more general question: to which philosophies and/or philosophers could one, on the basis of Dummett's criteria, ascribe *systematicity*? Dummett appears not to have been aware of this problem that emerges from his thesis. At the end of his essay, he maintains that many philosophers have suffered from the illusion that they have succeeded in overcoming the scandal caused by the lack of a systematic philosophical methodology, explicitly naming such philosophers as Descartes, Spinoza, Kant, and Husserl. He also maintains that the era of systematic philosophy (in both of his senses) begins with Frege. But then he writes (458),

> I have mentioned only a few of many examples of this illusion; for any outsider to philosophy, by far the safest bet would be that I was suffering from a similar illusion in making the claim for Frege. To this I can offer only the banal reply which any prophet has to make to any sceptic: time will tell.

One should perhaps instead say that the philosopher does well to avoid acting like a prophet. This of course presupposes that the philosopher develops a conception of systematic philosophy that does not simply do away with the history of philosophy and that wholly and coherently makes possible an open future for philosophy.

The systematic conception presented in this book shares the view that philosophy must ascribe to language a role that is not only important but even indeed fundamental. This view remains, however, relatively uninformative until the senses of "language," "analysis of language," and "philosophy of language" are clarified. The two great deficiencies in Dummett's philosophy of language (which he understands as a "theory of meaning") are the following: first, he does not consider the question of which language is adequate and therefore requisite for the development of philosophical (or scientific) theories. He contends that the philosophy of language is concerned "with the fundamental outlines of an account of how language functions" (442). But which language? Ordinary (natural) language, or a philosophical language, perhaps yet to be developed? The primary matter at hand is not pure "functionality," important though that is; of primary importance is instead clarification of the implications of a given language for the treatment of complexes of philosophical problems. Second, Dummett considers the fundamental domain of ontology, if at all, only quite inadequately. Among the most important implications of language however, are its *ontological* implications.

The conception presented in this book avoids or overcomes these two deficiencies in that it explicitly develops both the concept of a philosophical language and of its basic features and an innovative ontology fundamentally in relation to its semantics. These developments reveal that the semantics and the ontology of philosophical language are fundamentally two sides of the same coin. As far as the method of systematic philosophy is concerned, it is in no way reduced to the "analysis of language" or to anything that could be formulated so simply. Instead, it presents a completely thorough philosophical method consisting of four methodological stages (or, for sake of simplicity, four methods). These are the identification of structures and constitution of minimal or informal theories, the constitution of genuine theories (theories presented in the form appropriate to them as theories), the systematization of the component theories, and the evaluation of the theories with respect to theoretical adequacy and truth status. In philosophical practice, the four methods are virtually never applied *comprehensively*; they therefore represent an ideal case of a philosophical theory, one that is not an insignificant abstraction, but instead serves as an important regulative idea with respect to the development of philosophical theories. Taking the complexity of a completely developed philosophical method into consideration, it is possible to gain clarity about the current status of philosophical theories that are either under development or already available.

As far as the fragmentary character of analytic philosophy is concerned, Dummett himself makes clear that from Frege's "fundamental achievement"—that he managed to "alter our perspective in philosophy" (441)—no developed theory has yet emerged. Frege's thus remains what can be termed a fragmentary philosophy. The fragmentary character of contemporary analytic philosophy mentioned at the beginning of this Introduction is, however, a different sort of fragmentarity, and one that is far more radical and therefore significantly more important. This is now to be shown with respect to the "analytic method" and to analytically "articulated theories."

Even the philosophical method known as "generally analytic" can adequately be described only as a fragmentary method, not as a systematic one, because the factors

introduced to characterize it are at most necessary, and certainly not sufficient for the   9
systematic characterization of a method. These factors include the following: logical
correctness, conceptual clarity, intelligibility, argumentative strength, etc. The listing of
such factors in no way provides a systematic understanding and articulation of the factors
required by a complete or integrally determined method. In this sense, analytic philosophy
on the whole is, as far as methodology is concerned, fragmentary. Only in isolated cases can
one find attempts to identify a comprehensive and thus systematic method for philosophy.

An incomparably more important fragmentarity concerns what Dummett terms
"articulated theories." Beyond question, analytic philosophy contains such theories
in significant numbers. As a rule, however, these theories treat quite specific topics;
articulated, *comprehensive* theories are not developed, so the relations between the individual theories remain unthematized. A few examples well illustrate this phenomenon.
Works on topics in the domain of the philosophy of mind have directly ontological
components and implications, but what ontology is presupposed or used by a given
theory in the philosophy of mind remains, as a rule, unsaid. If ontological concepts
such as "object," "properties," etc., are used, it remains wholly unexplained how the
corresponding ontology is more precisely to be understood, and there is no consideration of whether that ontology is intelligible and thus acceptable. Something wholly
analogous happens with most works concerning theories of truth. Theories of truth
that are developed or defended virtually always have implications or presuppositions
with respect to "the world," to "things," to "facts," etc., but these ontological factors, at
least in the majority of cases, remain utterly unexplained. As a rule, these theories simply presuppose some form of the substance ontology that dates to Aristotle; according
to such ontologies, "the world" is the totality of substances (for which analytic works
almost always use the term "objects") that have properties and stand in relations to one
another. If a sentence qualifies as true and if thereby some form of "correspondence"
to something in the world is assumed, how is this "something" understood? Analytic
works do not pose this question and therefore do not answer it. That they do not makes
questionable the coherence of the conceptions they present.

By far the most important evidence of the theoretical fragmentarity of analytic
philosophy is the lack of comprehensive theories concerning actuality as a whole—in
the terminology of this book, theories of being. For the most part, some comprehensive
conception of actuality (of the world, of the universe) is presupposed; in the
overwhelming majority of cases, this is a diffusely materialistic view of the whole, but
this view is scarcely explained, much less subjected to serious theoretical examination.   10
To be sure, there are some moves in the direction of the development of comprehensive
theories, but those theories themselves are nowhere to be found.[4] In sum: the systematic

---

[4] This is the case, for example, with David Lewis, above all in his *On the Plurality of Worlds* (1986).
His position is treated and criticized extensively in Section 5.2.3.

Two other contemporary philosophers must be mentioned, ones who are significant
exceptions in the domain of analytic philosophy in that both have produced systematic
philosophical works. Nicholas Rescher, an extraordinarily productive philosopher, has
collected in systematic form the philosophical conception developed in many individual works
over the course of many years; the result is the imposing, three-volume·*A System of Pragmatic
Idealism* (1992–94). In its goals and many of its central methodological aspects, Rescher's

conception presented in this book arises from the insight that the deficiencies in contemporary philosophy just described ought to be overcome, and that they can effectively be overcome. Only if they are can philosophy do justice to its primordial task and fully develop its potential.

[3] Along with the preceding critical remarks on Dummett's position, some of the central thoughts and theses presented in this book are introduced and preliminarily explained. In what follows, the comprehensive architectonic of the book is briefly presented and preliminarily clarified. The presentation is of course quite general and summary; for more precise orientation with respect to details, the quite detailed Table of Contents is available.

In this book, philosophy is understood uncompromisingly and consequently as *theory*. For this reason, wholly *excluded* are such conceptions as philosophy as therapy or therapeutics, particularly as therapeutic criticism of language, all forms of philosophy that have practical aims (philosophy as wisdom, as practical reflection, as educational technique, as a way of life, as a way of shaping one's life or orienting oneself

---

systematic work is similar to what is presented in this book. Distinctions consist particularly with respect to three points. First, the interconnection ("systematic interrelatedness," according to his Preface) presented by Rescher in the domain of philosophical topics and theories is only quite general and loose. Second, the generally *pragmatic-idealistic* perspective (in this book's terminology, the pragmatic-idealistic theoretical framework) is far too narrow to be appropriate for the immense task of systematic philosophy. Third, Rescher's theory lacks central components of a comprehensive theory of actuality as a whole, quite particularly an ontology and a metaphysics. Nevertheless, the significance of his works can scarcely be valued sufficiently highly.

The second exception is the German philosopher Franz von Kutschera, who has published an impressive number of treatments in many philosophical disciplines (philosophy of language, epistemology, ethics, aesthetics, etc.). The first and third points of difference between this book and Rescher's position hold as well, in analogous fashion, between this book and the works of Kutschera. Above all, the utter absence of a comprehensive theory is all too evident in the book designated by its title as treating just this topic, *The Parts of Philosophy and the Whole of Actuality* (1998). In all brevity: according to Kutschera, "the entirety of actuality" is treated in the distinct parts of philosophy, among which he includes neither ontology nor metaphysics as a comprehensive theory. He writes,

> One can . . . well say that, at the center of Aristotle's and of later conceptions, there stands a conception of metaphysics that concerns the totality of actuality in its most general and fundamental features—its ontological structures along with their effective interconnections, be they causal or teleological. Within our contemporary understanding, a so-understood metaphysics is not a subdiscipline of philosophy, because its themes appear in all disciplines. Formal ontology is often ascribed nowadays to logic, the problem of universals is treated in the philosophy of mathematics, rational theology in the philosophy of religion, the mind-body problem in the philosophy of mind. The entirety of actuality is thus a topic for philosophy as a whole. (15–16)

Despite the closing sentence in this quotation, that the topics of ontology/metaphysics, in the sense of a comprehensive theory of actuality, *appear* in all philosophical disciplines does not in any way entail that these topics are or can be also *treated* in these disciplines in any manner that is at all appropriate. To the contrary, they are presupposed by these other disciplines and, therefore, if they are not explicitly treated, they are a background that is left in the dark.

with respect to life, as educational, etc.). A significant amount of the book is devoted to the clarification of the dimension of theoreticity in general and of the concept *philosophical theory* in particular.

Central to that clarification is the concept of the *theoretical framework*, which is presented in connection with and as a modification of the concept, introduced by Rudolf Carnap, of the linguistic framework. The account proceeds from the fundamental insight that every theoretical questioning, every theoretical sentence, argument, every theory, etc., is intelligible and evaluable only if understood as situated within a theoretical framework. If this presupposition is not made, then everything remains undetermined: the meaning of a given sentence, its evaluation, etc. To every theoretical framework belong, among other things, the following constitutive moments: a language (with its syntax and its semantics), a logic, and a conceptuality, along with all of the components that constitute a theoretical apparatus. Failure to attend to this fundamental fact—or, as is most common, failure even to recognize it—is the source of countless catastrophic mistakes from which philosophy has suffered throughout its history and into the present.

It suffices here to introduce a single example: the question raised in modernity and particularly in classical German philosophy concerning the grounding or self-grounding, and indeed the ultimate grounding of philosophy, is one that for the most part has floated in empty space, that is, utterly independently of any theoretical framework. Without the explication of a language, a logic, a conceptuality, fundamental assumptions, etc., the procedure has been one of immediately requesting and indeed demanding that any contention or thesis put forth be grounded (or, often, "justified") immediately. The presuppositions for meaningful questions concerning grounding are not clarified to the slightest degree. In opposition to this way of proceeding, this book treats philosophical grounding in a manner that stringently 12 attends to the insight, introduced above, concerning the central importance of the theoretical framework.

As its subtitle indicates, this book develops a theoretical framework—which it defends as the best currently available—for a systematic philosophy. The basic thesis that theories require theoretical frameworks, which provides the fundamental architectonic for the systematic philosophy presented here, is made more precise by the additional thesis that a *plurality* of theoretical frameworks is potentially and indeed even actually available.

This second thesis brings with it a cluster of serious problems, such as the following: How are these various theoretical frameworks to be evaluated? Can philosophical sentences be true only in *one* theoretical framework, the "absolute" one? Are all theoretical sentences that do not arise within this absolute theoretical framework false? But is there such an absolute theoretical framework, and if so, is it at all accessible to us human beings? The conception defended in this book is a systematically well-balanced one: true sentences emerge within every theoretical framework, but not all the true sentences are on the same level. Sentences are true only relative to their theoretical frameworks. This relativity is a specific form of a moderate, non-contradictory relativism.

Any philosophical theoretical framework is highly complex; taken as a whole, each consists of numerous particular theoretical frameworks that are to be understood as stages in the process of the development of the complete systematic theoretical framework. At the outset, the philosophical theoretical framework is only quite globally determined, as including quite general elements (concepts, etc.). In the course of the systematic determination and concretization of the theoretical framework, new elements are added in such a way that, step by step, broader, more determinate, more powerful subframeworks emerge *as* more concrete forms of the general theoretical framework. The comprehensive presentation in this book traces this process of the increasingly precise determination and concretization of the (general) systematic-theoretical framework; this matter is explained more precisely and in more detail in Chapter 1.

On the basis of the concept of the theoretical framework, *systematic philosophy*—specifically, the *structural-systematic philosophy* developed in this book—is understood as the universal science or—more precisely, with the aid of a preliminary quasi-definition—as a *theory of the most general and universal structures of the unrestricted universe of discourse*. This is an ambitious formulation whose worth is determined only by the degree of success achieved in clarifying the concepts on which it relies and demonstrating its relevance for philosophy. A better preliminary evaluation of this quasi-definition is provided by its comparison with a similar and well-known formulation of a philosopher who undertakes a strikingly similar philosophical project: Alfred North Whitehead. He calls the systematic philosophy presented in his monumental work *Process and Reality* "speculative philosophy," and characterizes it as follows:

> Speculative philosophy is the endeavour to frame a coherent, logical, necessary system of general ideas in terms of which every element of our experience can be interpreted. By this notion of 'interpretation' I mean that everything of which we are conscious, as enjoyed, perceived, willed, or thought, shall have the character of a particular instance of the general scheme. (1928/1978: 3)

This "definition" (Whitehead's term!) contains a number of concepts that are quite problematic because they are ambiguous; these include "general ideas," "interpretation," "experience," "particular instance of the general scheme," etc. Nevertheless, the "definition" does provide a generally intuitive insight into the project termed "speculative philosophy." Magnificent though Whitehead's comprehensive presentation of that philosophy is, this book proceeds quite differently: methodically—indeed, strictly methodically—rather than intuitively. It proceeds patiently and step-by-step rather than immediately holistically (in the sense of somehow communicating a great deal at once), introducing strict and detailed distinctions.

The two most important concepts in the quasi-definition presented above are *structure* and *unrestricted universe of discourse*. Methodically, the latter term or concept is utterly neutral in that it contains no more precise contentual determinations; it designates that "dimension" (this too an intentionally chosen neutral term and concept) that represents the subject matter of systematic philosophy (Heidegger speaks, famously, of the "subject matter [*Sache*] of thinking"). The dimension of the universe of discourse is the *comprehensive datum* in the literal sense: what is *given to philosophy to be conceptualized and/or explained* (i.e., everything with which philosophical theorization can and must be concerned). The term "datum" is thus here a kind of technical term that must be strictly distinguished from the various alternative notions of data to be found in philosophy, including sense data, what is given by the senses, etc. In addition, the topic much discussed at present of the "myth of the given"[5] is related only indirectly to the datum in the sense intended here.

"Datum" here can be understood as a *candidate* for inclusion in a theory or for truth.[6] The dimension of the so-understood datum is not simply empty; the datum, thus the particular data, is/are available as prestructured, at the fundamental or zero-level of theorization, within everyday theoretical frameworks relying on ordinary language, and on higher levels of theorization within the theoretical frameworks of the various sciences, including philosophy. They include all the "somethings" that emerge as articulated theoretically in the universe of ordinary discourse when there is talk of "things," "the world," "the universe," etc. Systematic philosophy must attend   14 to these and to relevant higher-level articulations and attempt to bring all these data into a comprehensive theory. Doing so does not involve accepting such data as in any important sense "ready-made" components of the theory; quite to the contrary, they are precisely candidates for restructuration within the theory, items that must be conceptualized and explained, a process that involves radical corrections and transformations.

This state of affairs is visible in the relation between ordinary language and the philosophical language briefly described above. The latter connects to ordinary language and indeed begins from it, but then fundamentally corrects it, semantically if not necessarily syntactically. On the basis of the criterion of intelligibility, this book develops an alternative semantics that has, as an implication, an alternative ontology.

In the course of the presentation, the dimension termed the *universe of discourse* is determined step by step in that additional designations are introduced: "world," "universe," ultimately "being" (at first in the sense of the objective counterpole to "structure"). Up to the beginning of Chapter 4, these terms are used more or less synonymously, because differentiating among them is not important before that point. In Chapter 4 and thereafter, however, "world" is used in a sense that is there delimited and explained. The term/concept that emerges in Chapter 5 as the most adequate counterpole to "structure" is "being" (in the sense explained there).

The other crucial concept in the quasi-definition and in the main title of this book is *structure*. In brief, this concept designates everything a theory makes explicit. Conceptualizing and explaining are characterized most concisely as the discovery and presentation, respectively, of the structure(s) of what is conceptualized or explained (i.e., of the data). The term "structure" is attached to a concept central to this book not because of but despite the fact that the term has become popular. Its use in this book is justified by the fact that here, "structure" is scrupulously introduced, defined, and applied. Because of the centrality of this concept, the systematic philosophy presented here is termed the *structural-systematic philosophy*. How the dimension of structure and the dimension of the universe of discourse or of being fit together is articulated in detail in Chapter 1; moreover, the entire book is nothing other than the thematization

---

[5] The term is used by Wilfrid Sellars (1956) to designate a philosophical error he strongly criticizes.

[6] Nicholas Rescher uses "datum" as a technical term for "truth-candidate" (1973, esp. 53ff).

of this fitting together, developed step by step. Central to the endeavor are three sorts of fundamental structures that are introduced and investigated separately and in their interrelationships: formal, semantic, and ontological structures. These form the heart of the theoretical framework of the structural-systematic philosophy.

15   [4] At this point, the question presses concerning the relationship between the structural-systematic philosophy and the sciences.[7] Careful clarification of this question, so central precisely at present, is a task undertaken in this book in various places. To evaluate accurately the precise sense and significance of this question, one must consider a significant phenomenon in the history of philosophy. As indicated above, at the beginning of the history of philosophy, in Greece, the word "philosophy" designated a corpus of knowledge that was quite comprehensive, one that indeed was, in a certain respect, virtually coextensive with scientific knowledge as a whole. In the course of the history of philosophy, many branches of knowledge have developed, ones that earlier had been, in one way or another, parts of the philosophical corpus, but then came no longer to be understood as such parts. On the whole, one can speak of the gradual development of the sciences as we know them today as a process of their emancipation from philosophy.

Many authors interpret this process—a historical one in the truest sense of the word—as an utterly negative development for philosophy, maintaining that philosophy is, increasingly, deprived of its subject matter. Some go so far as to contend that by now philosophy no longer has any subject matter of its own. This book maintains the opposing thesis that this process can have an eminently positive effect in that it can clarify the theoretical undertaking that, from its very beginning, has borne the name "philosophy," making possible the identification of that undertaking's specific status. In light of this thesis, the history of philosophy, viewed as a whole, appears as philosophy's theoretical self-explication. This process has now reached the point at which, more than ever before, philosophy has the possibility of avoiding confusions, unclarities, hypertrophies of its status and its tasks, etc. Recognition of this process makes clear that it is a waste of time to speak about or to discuss philosophy, its subject matter, its tasks, etc., purely abstractly or *a priori*; only the concrete demonstration that philosophy does have its own subject matter, distinct from the subject matters of any of the sciences, can be meaningful and persuasive, and this demonstration can be provided only by the identification of that subject matter. This book provides that identification and with it the demonstration.

16   The relation between philosophy and the sciences with respect to subject matter comes to expression in the quasi-definition of philosophy introduced above: "the most general and universal structures of the unrestricted universe of discourse." To be sure, it must be precisely determined *both* what distinguishes the most general and universal structures from the particular structures that constitute the subject matters of the sciences *and* why the (nonphilosophical) sciences, even in conjunction, cannot investigate the *unrestricted* universe of discourse. One of the theses of this book relevant to these

---

[7] As is indicated above and explained more fully in Chapter 1, systematic philosophy, as understood in this book, is itself a genuine science. Nevertheless, throughout this section, and in various contexts in the book, "science" (or, usually, "sciences") is used in a narrower sense: to refer only to the empirical or natural sciences. Contexts make clear which signification is intended.

determinations is that certain structures have an indisputably universal character, with the consequence that they are not and cannot be thematized in the sciences. These are, most importantly, the structures that are treated in the theory of being presented in Sections 5.2 and 5.3.

Another thesis that, understandably, is highly topical and quite controversial concerns the structures in those domains that, taken globally and without differentiation, are thematized both by philosophy and by the sciences. Among these are several of the domains that are treated in part and quite summarily in Chapter 4 under the title "World-Systematics." With respect to the issue under current discussion, the most important and interesting of these domains is presumably that of the "human world," one aspect of which contemporary philosophy studies under the designation "philosophy of mind." In the cases of this and of similar domains, this book defends the thesis that the borders between philosophy and the sciences cannot be determined at the outset or once and for all; instead, the borders are flexible. The precise determination of the status of structures lying on or near such borders can be articulated only at specific stages of the historical developments of the sciences and of philosophy.

Methodologically, the criterion for the clarification of the relationships between philosophy and the sciences both in general and in concrete cases is the concept, introduced and fundamentally explained above, of the *theoretical framework*. It is utterly nonproductive and therefore senseless to discuss these relationships without making clear just what theoretical frameworks philosophy and the sciences presuppose and employ. Whether one should ascribe a specific question to philosophy or to the sciences can be rationally decided only on the basis of what the question asks about, what concepts are present in it or are presupposed or implied by it, what possibilities are available or requisite for its clarification, etc. A quite illustrative example is treated extensively in Section 4.5.1: when natural-scientific (physical) cosmology speaks 17 of the "beginning" of the world (or the cosmos), making scientific claims about it, it presupposes a specific natural-scientific framework, within which the concept "beginning (of the cosmos)" has a wholly determinate signification. Philosophy cannot question those natural-scientific theses that appear in models arising within such a theoretical framework. But an example of a question that does arise for philosophy is whether the concept *beginning* that appears within the physical-cosmological theoretical framework is identical to the *philosophical* (more specifically, *metaphysical*) concept *beginning*. As the considerations in that section reveal, the two concepts are quite different, so it is deeply regrettable that both are associated with a single term: the physical-cosmological and the philosophical concepts of beginning are fundamentally different concepts, which shows that there is a fundamental difference between the two theoretical frameworks. The tasks that result for philosophy are to explain carefully *its* concept of beginning—the genuinely metaphysical one—and to distinguish this concept clearly from the natural-scientific concept.

[5] A few introductory clarifications of the book's individual chapters are appropriate at this point. The six chapters present the stages of development of the complete theoretical framework of the structural-systematic philosophy; differently stated, each articulates a more determinate form of the theoretical framework, in that each adds significant new components.

Under the title "Global Systematics: Determination of the Standpoint of the Structural-Systematic Philosophy," Chapter 1 thematizes the factors or perspectives that distinguish the structural-systematic philosophy, both from non-theoretical and non-philosophical undertakings and from other philosophical ones, by articulating its initial, global determinations. This involves the formulation of the quasi-definition of this philosophy and the detailed explanations of the concepts found in it, as well as extensive treatment of the four-staged philosophical method and finally of the complex question of the grounding and self-grounding of the structural-philosophical theory (or theories). The most general form of the theoretical framework of this philosophy is thereby presented. In essence, these aspects are introductorily considered in [2] and [3] above.

Chapter 2 is devoted to the Systematics of Theoreticity; it thematizes the dimension of theoreticity as the philosophical dimension of presentation. The most important topics here are philosophical language, the domain of knowledge, the concept of theory in the narrower sense, and finally an initial account of the concept of truth based on the thesis that this concept articulates the fully determined status of the dimension of theoreticity. This chapter shows that and how philosophy must develop its own language, a language that is connected to ordinary language but then must diverge decisively from it. It also thematizes the linguistic criterion for theoreticity, which identifies as theoretical only sentences of a specific form that Wittgenstein makes explicit, in his *Tractatus*, in a different context; these are sentences beginning with the operator, "It is the case that . . ."[8] The domain of knowledge, or the epistemic dimension, is analyzed as a dimension that must be taken into consideration, but its analysis shows that—and why—the decisive status accorded to it by modern philosophy is to be denied. The standpoint of the knowing subject is in no way adequate for the development of theories. The necessity of freeing theories from the standpoint of the subject is one of this book's most important theses. Genuinely theoretical sentences do not have the (explicit or implicit) form, "Subject S believes/knows that p"; they have instead the form, "It is the case that p." Edmund Gettier's famous definition of knowledge is subjected to critical analysis and rejected; a different definition of knowledge is then provided.

Chapter 2 thoroughly treats the dimension of theory in the narrower sense by examining the most important theory-concepts defended at present. On the basis of this examination, a theory-concept suitable for philosophical purposes is developed. Finally, at the end of this chapter, the truth-concept is clarified on the basis of the thesis that it articulates the fully determined status of every theoretical sentence and of every theory, and thus of the entire dimension of theoreticity. Precise clarification of this understanding of truth is undertaken only at the end of Chapter 3, because fully unfolding the truth-theory presupposes the three sorts of fundamental structures.

---

[8] In *Struktur und Sein*, this footnote makes a point about the book's usage of commas that is not relevant to the English edition; this note is added only to make the footnote numbers of the editions agree in a way that avoids any mystery about the lack of an English footnote with the number "8." As is noted in the Preface, footnotes designated by letters have no counterparts in the German version.

The latter task is undertaken in Chapter 3 under the title "Systematics of Structure: The Fundamental Structures." This chapter presents the core of the structural-systematic philosophy. Beginning with its initial, basal mathematical definition, the concept of structure is expanded and made fully applicable philosophically. It is shown that on the basis of this concept, as it is understood and applied in this book, both an 19 enormous simplification of philosophical terminology as a whole and clarifications of philosophical conceptuality and philosophical entities can be attained: such terms as "concept," "meaning," "semantic value," "category," "proposition," "state of affairs," "object," "fact," "(logical) rule," etc., are reduced to and/or clarified as structures.

The fundamental formal structures are logical and mathematical structures, and this book must adequately characterize these structures. At the same time, it is of course not a work in the discipline either of logic or of mathematics; its concern is therefore with philosophically clarifying the kinds of entities with which logic and mathematics are concerned, and showing their significance for philosophical theories.

The section on semantic structures, opposing the "compositional" semantics based upon the principle of compositionality, develops an alternative semantics that is based upon a strong version of the Fregean context principle: "Only in the context of a sentence do words have meanings." One of its central theses is that sentences of the subject-predicate form are not acceptable for any philosophical language equipped with an appropriate semantics; what makes them unacceptable are their ontological consequences (if, as in this book, sentences with the subject-predicate syntactic form are nevertheless used, they must—as is explained particularly in Sections 2.5.1.3 and 3.2.2.4.1.3—be semantically interpreted and understood as convenient abbreviations of sentences without subjects and predicates.) The ontology that corresponds to subject-predicate sentences is one that this book calls "substance ontology"; the book shows this ontology to be unintelligible and therefore unacceptable. Sentences without subjects and predicates, like "It's raining," are termed "primary sentences"; they express "primary propositions" that are more precisely interpreted as "primary semantic structures." The qualifier "primary" is not a counterpart to anything like "secondary," and is not to be understood as synonymous with "simple" (or "atomistic," as in "atomistic sentence"). The term "primary" is instead employed, given the lack of any more appropriate alternative, to designate sentences that do not have the subject-predicate form. It is therefore wholly consequent to speak of "simple primary sentences and propositions" and of "complex primary sentences and propositions" (i.e., sentences or propositions that consist of more than one and indeed often of a great many simple primary sentences or propositions).

The ontological structures emerge directly from the semantic ones in that, as is noted above, semantics and ontology are two sides of the same coin. The fundamental ontological "category" (according to traditional terminology) is the "primary fact"; all "things" (in philosophical terms, all "beings" or "entities") are configurations of primary facts. The term "fact" is taken in a comprehensive sense, corresponding to the way this term is normally used at present (e.g., "semantic fact," "logical fact," etc.). It therefore does not necessarily connote, as it does in ordinary terminology, the perspective of empiricism. What is said above concerning the qualifier "primary" holds 20 correspondingly for the term as used in "primary facts." The concept *configuration of primary facts* or *complex primary facts* (thus also, correspondingly, *configurations of primary sentences/propositions* or *complex primary sentences/propositions*) emerges as one that is central within the structural-systematic philosophy.

As noted above, Chapter 3 completes the development of the theory of truth that begins at the end of Chapter 2. *Truth* is understood more precisely as the concept that articulates the interconnections among the three types of fundamental structures. Formally, it is explained as a composite function that consists of three individual functions. The third function articulates the connection between a true primary proposition (or primary semantic structure) and a primary fact (or primary ontological structure). The connection is simply an identity: the true primary proposition *is* (in the sense of identity) a primary fact. This identity thesis traces back to a famous passage from Frege's essay "The Thought," which reads, "What is a fact? A fact is a thought [at present, one would generally say: a proposition] that is true" (1918: 343). On the basis of this thesis, the ontology briefly sketched above proves to be completely and thoroughly consistent with contextual semantics. Its briefest characterization may be found in the second sentence of Wittgenstein's *Tractatus*: "The world is the totality of facts [for Wittgenstein: of existing states of affairs], not of things" (*Tractatus* 1.1).[9]

Chapter 4, "World-Systematics," opens a decisively distinct phase in the presentation of the structural-systematic theoretical framework. Chapters 1–3 present all the essential elements of this theoretical framework. Chapter 4 begins the application or the specification of this theoretical framework. From a globally architectonic perspective, one can say that this specification is the explicit thematization of the grand datum (i.e., of being). This thematization requires that the world (the datum, being) be determined more precisely. Its more precise determination in this book involves the introduction of a distinction fundamental with respect to these concepts; the distinction is between a restricted and an unrestricted dimension. Chapter 4 terms the former "the (actual) world," the latter, "the dimension of being." Not until Chapter 5 is it possible to provide more precise determination of these two dimensions. That chapter presents the restricted dimension as the totality of *contingent entities* and the unrestricted dimension as the *absolute dimension of being.*

Simply put, the world treated in Chapter 4 is *actuality* as the totality of the things and domains of things with which we are familiar and to which we relate in various ways. These are, globally viewed, (inorganic) nature, the domain of life, the human world—with all that belongs to it in one way or another, including human beings as minded persons, the domain of action (ethics), the social domain, etc.—the world of aesthetics, and finally the world as a whole: the cosmos, religion, and world history. From a book that intends to present only a theoretical framework for a systematic philosophy, one should neither expect nor demand that all these domains be treated in detail, because that would be the *comprehensive* presentation of the *fully developed* structural-systematic philosophy. The goal of Chapter 4 can be described as follows: in

---

[9] To be sure, Wittgenstein's understanding of this sentence differs fundamentally from the interpretation the sentence attains within the contextual semantics and ontology developed here. But the formulation as such, as a succinct formula, is appropriate as characterization of this semantics and ontology. Moreover, it is doubtful whether Wittgenstein's own formulation can be brought into harmony, without misunderstanding, with other passages found at the beginning of the *Tractatus*.

Chapters 1 through 3, the grand dimension of *structure* (or *structurality*) is developed in the form of the complete but still *abstract* theoretical framework for the structural-systematic philosophy; Chapter 4 begins to "apply," to concretize, or—to use a Fregean term—to "saturate" this abstract theoretical framework with respect to the central aspects of the grand datum. This can, however, be done in this book only incompletely, by means of treating some of the central questions from the grand domain of the world from the perspective of this philosophy. Other aspects, no matter how important they may be within the relevant philosophical domains, are not relevant to attaining this goal. Chapter 4 thus serves as an extensive *example* for the concretization or saturation of the theoretical framework presented in Chapters 1 through 3. For the most part, its account remains general, although in some cases important paradigmatic questions are treated in detail.

Chapter 5 is devoted to *comprehensive systematics*. As a *theory of the interconnection of all of the structures and all of the dimensions of being*, it is appropriately characterized as a *theory of being as such and as a whole*. In traditional terms, one would say that this chapter treats (general) metaphysics. But this designation must be used with care because this terminology is often connected to misunderstandings and prejudices of many sorts.

The extensive Section 5.1 clarifies the status of comprehensive systematics. That section analyzes the problem that is the root of all of the important critiques of the possibility of metaphysics to be found in the history of philosophy and into the present; Section 5.1 articulates this problem in a new form. The problem is the one that Hilary Putnam, with specific reference to Kant, locates at the center of philosophical inquiry; it is based in the thesis that there is a gap or cut between subject(ivity), thinking, mind, language, theories, etc., on the one hand, and the "system" (actuality, the world, the universe, being, etc.), on the other. The Kantian tradition takes this gap to be absolutely unbridgeable. In Putnam's words, "what it means to have a cut between the observer and the system is . . . that a great dream is given up—the dream of a description of physical reality as it is apart from observers, a description which is objective in the sense of being 'from no particular point of view'" (1990: 11). This passage describes the cut or gap in the domain of the physical world (of physics), but according to Putnam the problem also—and indeed especially—arises in the domain of philosophy as the putatively universal science. Instead of "observer," therefore, it would be better to say "theoretician," and instead of "physical reality," "actuality" or "being" (in the sense of the counterpole to "theoretician").

In opposition to the Kantian tradition and to all similar philosophical positions, this book establishes the thesis that the putative gap is one that is not only bridgeable, but indeed must be presupposed already to have been bridged by every serious and sensible science and philosophy. The central insight grounding this thesis is that science and philosophy, even on a minimal level, can be sensible (or, speaking loosely, can function) only on the basis of the presupposition that the segments of actuality with which they are concerned, and ultimately, thought through to the end, actuality or being as a whole, are *expressible*. In this book, "expressibility" is used as a technical term to designate the entire palette of our "accesses" to actuality or to being, or the modes of articulating (conceiving, understanding, explaining, etc.) actuality or being as a whole. What sense would it make to produce a scientific or philosophical statement about something if that something or indeed the whole were not expressible (in this sense)? That would be complete nonsense. If, however, absolutely everything—the entire universe of discourse—is expressible, then every form of fundamental gap in Putnam's sense must be viewed as already bridged, because both "poles" or sides of the gap or cut are only secondary or relative levels of a relationship in that each refers to the other, and in that the two are always already united. All the "gaps" that have appeared within the history of philosophy are based on the distinction, to be recognized but not to be interpreted as a dichotomy, between the dimension of structure and the dimension of being (understood as "objective" counterpole). But they are intelligible only as two different poles within one domain, i.e., only as within a primordial relationship; this primordial relationship appears for its part as the primordial dimension that first makes possible and therefore at once suspends the distinction between structure and being. This book terms this primordial and comprehensive dimension the *dimension of being*, and thematizes it in Sections 2 and 3 of Chapter 5.

This view, which literally *encompasses* both described dimensions, is expressed by means of sentences satisfying the linguistic criterion for theoreticity, i.e., sentences that (implicitly or explicitly) begin, "It is the case that. . . ." This phrase is thematized in this book as the *theoretical operator*. In a daring but philosophically well-grounded interpretation of the particle "it" in this formulation, the "it" can ultimately be understood as referring to what is here termed the primordial dimension of being. From this it follows that every theoretical sentence is a kind of *self-articulation* of this primordial dimension of being. Indeed, the result of the dispute with those who affirm the existence of a gap in Putnam's sense can in part be formulated as follows: every sort of exclusive restriction to one side of such a gap or dichotomy is excluded from the theoretical domain. Quite particularly excluded is any form of relativization of science and philosophy to the subject (or to subjectivity). Also excluded are explicitly formulated or even implicitly presupposed forms of presentation such as "From the transcendental perspective of the subject it is the case that . . ." Such forms express a restriction to one side of the rejected dichotomy. The alternative is the "absolute" form of presentation, "It is the case that . . .," which precedes sentences expressing the just-named self-articulation of the primordial dimension of being.

The relativization of science and philosophy to factors such as the subject is excluded, but not every form of relativization. As indicated above, all scientific and philosophical sentences presuppose the theoretical frameworks within which they arise, and within which alone they attain their determinate form or their determinate status. But it is also indicated above that there is a plurality of theoretical frameworks; the consequence is that every theoretical (scientific or philosophical) sentence has its determinate status only *relative* to its theoretical framework. But this relativity has nothing to do with any relativity to *one side* of the rejected gap or dichotomy, e.g., to a subject, to a time, to a social situation, or to any such factor. The relativity that holds here is only this: it designates a determinate *degree* of the self-articulation that is manifest in the form of expression, "It is the case that . . ." How this degree of the self-articulation of the entirety of the primordial dimension of being is to be interpreted presents one of the deepest and most difficult problems that the structural-systematic philosophy must consider.

To develop coherently the conception briefly sketched here, the thematic of philosophical language must be considered anew. To this topic is devoted a significant part of Chapter 5. From the semantics developed in outline in Chapter 3 and from various additional assumptions there results the necessity of developing a concept of philosophical language that is quite unusual. A philosophical language as a *semiotic system with uncountably infinitely many expressions* must be postulated in order to do justice to the basic thesis, formulated above, of universal expressibility. It is obvious that such a semiotic system does not correspond to the normal conception of language. The reasons for its postulation are strictly philosophical. Moreover, a plurality of such languages must be assumed, because of the plurality of theoretical frameworks. The many logical, semantic, and ontological aspects of this complex problematic are treated thoroughly in Chapter 5.

Sections 5.2 and 5.3 present a genuinely *comprehensive systematics*, which consists of the explication of the primordial dimension of being. Section 5.2 presents the basic features of a theory of being as such and as a whole. Here is clarified for the first time the difficult and highly timely semantic, logical, and mathematical problematic of talk about "the whole" or "(the) totality"; the account is developed in opposition to that presented by Patrick Grim in his book *The Incomplete Universe*. There follows an attempt to clarify the currently popular theory of the plurality of possible worlds in their relation to the actual world. Finally, the core of a structural-systematic theory of being is presented: under the title "the inner structurality of the dimension of being," and in fundamental harmony with the basic insights of the grand metaphysical tradition, the immanent characteristics of being and beings are presented: the universal intelligibility, universal coherence, universal expressibility, universal goodness, and universal beauty of the dimension of being.

Section 5.3, the last one in the chapter, presents the starting points for a theory of absolutely necessary being. This involves the extension and expansion of the theoretical framework applied here by means of the ontologically interpreted *modalities*. The result is that the primordial dimension of being is to be conceived of as two-dimensional, consisting of both an absolutely necessary and a contingent dimension. The task of determining more precisely how these dimensions relate to each other leads to determining the absolutely necessary dimension as free, minded, absolutely necessary being.

Chapter 6, the last chapter, treats *metasystematics* as the *theory of the relatively maximal self-determination of systematic philosophy*. This brings the presentation of the theoretical framework of the structural-systematic philosophy to its conclusion. This last topic is of ultimately decisive importance for the understanding and self-understanding of the conception presented in the book. As universal science, philosophy cannot rely upon any metascience that could determine its status. This fact brings with it a difficult and fundamental problem. Chapter 6 introduces various considerations that are indispensable to the solution of this problem, particularly the distinctions between immanent and external metasystematics, between external intratheoretical and external extratheoretical metasystematics, and between external intratheoretical interphilosophical and external intratheoretical philosophical-non-philosophical metasystematics.

Immanent metasystematics is what can be termed, to use a Kantian expression, the "architectonic" of the structural-systematic philosophy. In the complex expression "immanent (or internal) metasystematics," the term "systematics" designates the individual, specific systematics that are the components of the comprehensive philosophical conception: global systematics, systematics of theoreticity, systematics of structure, world-systematics, and comprehensive systematics.

The basic insight or thesis concerning *external metasystematics* results from two fundamental assumptions: the assumption introduced above of a plurality of theoretical frameworks, and the assumption that even if there is an ultimate or absolute theoretical framework, it is not one that is attainable by human beings. This means, among other things, that the structural-systematic philosophy is an open system (i.e., that it is essentially *incomplete*). One can think here of Gödel's famous incompleteness theorem, which is considered in this book in various passages. The situating or self-situating or self-determination of the structural-systematic philosophy always develops on a level of consideration that presupposes a more extensive and higher theoretical framework. This higher theoretical framework is, however, itself always a philosophical theoretical framework.

[6] Is the structural-systematic philosophy presented in this book a philosophical system? The answer to this question depends upon how one understands the formulation "philosophical system." The problem presses because this is a formulation that is heavily burdened by its history. One thinks of the "philosophical systems" that were regularly superseded by newer ones, particularly in the second half of the eighteenth and the first half of the nineteenth centuries; all of these systems, despite the extravagant claims made for them by their authors, have been and are now judged by most philosophers to be untenable, and are therefore largely abandoned. Such systems, still admired by some philosophers and studied and commented upon in never-ending chains of interpretations and reinterpretations, have not only benefited philosophy, they have also damaged it. The excessiveness of their claims and the poverty of their results have brought the term "philosophical system" into presumably irremediable discredit. For this reason, this formulation is avoided in this book or at most used only marginally; the term used instead is "systematic philosophy." That the conception of the systematic philosophy briefly sketched in this Introduction is far from those of the "philosophical systems" of the past should be obvious.

Worth emphasizing once again is that this book attempts *only* to present the *theoretical framework for a systematic philosophy*. Even this task is an extremely challenging one, but the completion of the structural-systematic philosophy itself would be immensely more so. It can be seriously undertaken only as a communal enterprise to which many philosophers must contribute. At the same time, however, one should not undervalue the significance of the development of the theoretical framework, because only insight into the necessity of treating every single philosophical question not in splendid isolation, but within a systematic framework, can overcome the fragmentation that is one of the chief defects of contemporary philosophy.

[7] In conclusion, it is appropriate to mention some specific aspects of the presentation that follows.

The book contains numerous cross-references to parts, chapters, sections, passages, etc., of the book; this could be irritating. It is, however, unavoidable because of the network character of the conception and consequently also of the presentation. Also, certain passages that are in part repetitious are so for just this reason.

Many topics are treated in this book in quite different ways at different places: some such treatments may appear quite brief, others disproportionally long. An example of the latter is the extensive treatment given to the topic of language in Chapter 5, 27 particularly to the problematic of language as a semiotic system with uncountably many expressions (Sections 5.1.4 and especially 5.1.4.3). There are two reasons for these inequities in treatments: on the one hand, different topics or subject matters vary greatly in complexity; on the other, some are more central and some peripheral to the structural-systematic philosophy. The topic just mentioned is both highly complex and quite central. It is central because it concerns the basic thesis of the universal expressibility of being, which, without recourse to a language with uncountably many expressions, can be neither made intelligible nor grounded.[a]

---

[a] At this point in *Struktur und Sein*, there are four paragraphs concerning the book's divergence, in specific cases, from normal German usage of commas and quotation marks; these are not relevant to the English edition. As is noted in footnote 1 to the Key to Abbreviations and Logical/Mathematical Symbols, footnotes designated by letters appear only in this book; they have no counterparts in *Struktur und Sein*.

Global Systematics: Determination of the Standpoint of the Structural-Systematic Philosophy

1.1 A Theoretical Framework for a Systematic Philosophy: The Complexity of the Concept and of Its Presentation

This book presents, according to its subtitle, a theoretical framework for a systematic philosophy. The task of this section is to clarify, in a preliminary fashion, what such a presentation involves. This clarification necessarily relies upon theses and concepts that are merely introduced in this section, but that are developed and defended in detail in later sections and chapters.

In a minimal but fundamental determination that this chapter expands on, philosophy is understood in this book as a *theoretical* activity, that is, as an activity aiming at the development and presentation of one or more theories (the latter themselves conveniently called "philosophies"). The development and presentation of any theory require, however, the satisfaction of various specific presuppositions. The factors satisfying these presuppositions jointly constitute what can be termed a *framework* or, better, a *theoretical framework*. Thus, clarification of the concept associated with the term "theoretical framework" requires clarification of those various factors.

[1] Although related notions are to be found throughout the history of philosophy, the term *framework* attains a central significance in philosophy with Rudolf Carnap's emphasis, in his essay "Empiricism, Semantics, and Ontology," on what he calls (1950/1956: 206) linguistic frameworks: "If someone wishes to speak in his language about a new kind of entities, he has to introduce a system of new ways of speaking, subject to new rules; we shall call this procedure the construction of a *linguistic framework* for the new entities in question." Carnap then distinguishes between "two kinds of questions of existence: first, questions of the existence of certain entities of the new kind *within the framework*; we call them *internal questions*; and second, questions concerning the existence or reality *of the system of entities as a whole*, called *external*

questions." Precisely how Carnap understands the distinction between the two kinds of   30
questions concerning existence is not fully clear. He continues,

> Internal questions and possible answers to them are formulated with the help of the new forms
> of expressions. The answers may be found either by purely logical methods or by empirical
> methods, depending upon whether the framework is a logical or a factual one. An external
> question is of a problematic character which is in need of closer examination. (205–06)

This is not the place for a detailed consideration of Carnap's position,[1] but the cited
passages reveal a basic difficulty: it is not clear why a framework that can be either "a
logical or a factual one" should be characterized as *linguistic*. Perhaps aware of this
lurking problem, Carnap adds a footnote to the second (and modified) version of the
essay: "I have made here some minor changes in the formulations to the effect that the
term 'framework' is now used only for the system of linguistic expressions, and not for
the system of the entities in question" (205n). Despite this terminological note, however, Carnap's various examples make clear that the frameworks he develops involve, in
addition to linguistic components, logical and conceptual ones; the designation of the
frameworks as *linguistic* is therefore misleading.

[2] This book uses "framework" in an expanded—indeed, in a comprehensive—
sense, that of *theoretical framework*. This term and concept encompass the totality
of the specific frameworks (chiefly the linguistic, semantic, conceptual, and ontological frameworks) that are, in one way or another, components of the framework
presupposed by a given theory.

In its various uses, the term "framework" connotes a specific difference or
distinction between two sides or levels or aspects of a determinate item; the difference
is between the framework and what the framework contains, presents, or enframes.
Framework and content can be interrelated in either of two ways; there are therefore
two partially distinct concepts of framework. The *abstract* or *underdetermined* concept
is that of the framework taken independently of that for which it is a or the framework   31
(e.g., a picture frame considered independently of any painting that it does or might
enframe). The *concrete* or *fully determinate* concept is that of the framework as explicitly related to its content (i.e., to that for which it is a or the framework).

[3] The theoretical framework presented and used in this book, in its *genuine or
complete form*, is a concrete, fully determined framework—a framework in the second
of the two senses just introduced. But the first sense is not ignored, it is explicitly
considered and presented. Its presentation does not cease with the abstract framework,
but instead shows how that abstract framework can and indeed must develop into a
concrete or fully determined framework. This complex way of proceeding is required
by systematic philosophy, as that discipline is understood in this book. The theoretical
framework appropriate for such a philosophy is such that the relationship between
framework and content (or between abstract framework and concrete or concretized
framework) must be quite subtly differentiated. Thus, for example, the "framework"
that is a *theoretical* framework for the systematic philosophy developed here
cannot be understood as an uninterpreted formal system; philosophical (and scientific)

---

[1] For additional details, see Section 1.5.2.2. Puntel (1999a) presents a detailed critique of Carnap's
position with respect to ontology.

theoretical frameworks are instead instruments that make possible the articulation, conceptualization, and explanation of theoretical contents or subject matters. Within or by means of theoretical frameworks, contents are available for theorization, but this presupposes that the theoretical frameworks include elements that are not purely formal, but instead contain interpretations (i.e., relations to contents). For this reason, every philosophical or scientific theoretical framework contains, in addition to purely formal elements and concepts, also material or contentual ones.

Of decisive importance here is distinguishing among different levels or stages of both the theoretical framework and its contents. The most comprehensive and complete theoretical framework for the structural-systematic philosophy developed here is highly complex and itself consists of or contains various individual theoretical frameworks. At the beginning of this presentation, as this chapter shows, the only theoretical framework available is a purely global one whose initial articulation is provided in section 1.2 by the quasi-definition of the structural-systematic philosophy. Section 1.2 introduces and preliminarily explains only the most comprehensive of the concepts relied on by this philosophy. The specification or concretization of this theoretical framework proceeds by means of the development of additional, increasingly specific or concrete theoretical frameworks *and* of the various theories developed within them. These specific or concrete theoretical frameworks are themselves different stages or levels of the comprehensive philosophical theory under development.

32   With respect to this process of concretization of the comprehensively systematic theoretical framework, three central factors must be kept in mind.

[i] During the process of concretization, the meanings of "framework" and "content" change. Every theoretical framework is a framework *for* a specific topic, content, subject matter, or universe of discourse. Thus, for example, the development of a semantics results from the concretization of a specific theoretical framework with respect to a language that is (or is to be) interpreted. In the process of the concretization or specification of the comprehensively systematic theoretical framework, this content or subject matter is itself embedded, at or in a broader or more concrete stage or level of the comprehensive philosophical theory, within a broader or more concrete theoretical framework; there, it has the status of being an integral component of the broader theoretical framework. To extend the example of semantics: there is a determinate theoretical framework that has semantics as its content or subject matter; this is the framework within which a semantic theory can emerge. There are also, however, more comprehensive theoretical frameworks that explicitly include the elements of the semantic theory among their components.

[ii] This process of the transformation of a content or subject matter into a component of a more comprehensive theoretical framework extends only as far as the end of Chapter 3. "World-Systematics" (Chapter 4) treats contents or subject matters that are purely objective or pure "data" in the precise sense of being only contents for theorization—of *not also* being additional *instruments* for theorization—and thus foci of the appropriate theoretical frameworks. This decisive difference provides the basis for the distinction fundamental to the structural-systematic philosophy, the distinction defining the comprehensive systematic architectonic; this is the distinction between the dimension of structure(s) and the dimension of data. It must be emphasized here that the term/concept "datum" has two significations, one that is general

and one that is specific. In the general sense, any content or subject matter for theorization—or, as is explained below in Section 1.2.4[2], any *truth-candidate*—qualifies as a datum. Data in this general sense are found in every chapter in this book, because each of its chapters presents theories. In the specific sense, "datum" (or "dimension of data") names the dimension correlative to that of structure(s). Considered architectonically, the data in this specific sense are the subject matter(s) of this book in the chapters following the presentation of the dimension of structure (thus, beginning with Chapter 4). Data in this specific sense are, however, also found in Chapters 1–3, but only in a restricted form: these are the structures that are the subject matters of the theories developed in those chapters. The data, in this restrictedly specific sense, are the correlates not of the entire dimension of structure, but only of the structures that are not subject matters but are instead components of the (not yet complete) theoretical framework used in these chapters. Precisely which senses the terms "datum" and "data" have is always made clear by the contexts wherein they appear.

Chapters 5 and 6 thematize the relations between fundamental structures and the data in the specific sense. In these chapters, therefore, the fundamental structures are themselves neither the objects of theories, as they are in Chapters 1–3, nor are they unthematized structures that are applied to data. Instead, *together with the data, now as restructured by them*, they constitute the subject matter of a broader theory. Chapter 5 thematizes this comprehensive complex, designating it as *being as such and as a whole*.[2]

[iii] The third factor to be considered is more methodological or metatheoretical: the process of presentation of the structural-systematic philosophy, in part described in the preceding paragraphs, is not to be understood as a completely noncircular construction of a comprehensive philosophical theory; the theory it presents is instead a theoretical network. Indeed, already at the beginning of the process of presentation of the comprehensive theory various factors and elements are presupposed and relied on; only further on in the course of systematic presentation are these factors and elements adequately explained. Thus, for example, a specific semantics is presupposed and used from the outset, but it is not fully and explicitly presented until Chapter 3.

[4] The considerations introduced in this section should make clear that the presentation of the theoretical framework for any systematic philosophy is an extremely complex undertaking. The reason for this is found in the complex structure of systematic philosophy itself, which is such that it cannot be presented by means of a noncircular construction, but only as a web or network. To be intelligible, the undertaking must develop by means of complex and at times redundant and pedantic presentations.

---

[2] The presentation, as it develops, introduces increasingly precise designations for what here, initially, is termed the "dimension of data." For the sake of simplicity and concision, this book's title uses the term "being" to name this dimension, but the term and concept "being" must be made scrupulously precise. This chapter begins the process of making them precise: Section 1.2.4 explains the concept "unrestricted universe of discourse," and Section 1.3 clarifies the book's title and the fundamental idea underlying the conception the book presents.

34   1.2 A First Determination of Systematic Philosophy

The title of this section introduces the concept *systematic philosophy*, and that of the next subsection, *the structural-systematic philosophy*. The latter is a specification of the former. The distinction between the two is meant to clarify two points. First, the goal of this book is to develop a theory that qualifies as a *systematic philosophy*; second, attainment of this goal requires the presentation of a specific form of systematic philosophy, one most appropriately designated *the structural-systematic philosophy*. It would, however, be impossible to define *systematic philosophy* in some completely neutral manner (i.e., in utter independence from a quite determinate understanding of how a specific systematic philosophy might be developed). The quite determinate conception of systematic philosophy defended in this book is precisely the one articulated by the designation *the structural-systematic philosophy*.

1.2.1 *A Quasi-Definition of the Structural-Systematic Philosophy*

This chapter introduces a quasi-definition because, at this point in the account, the various factors that a genuine definition would require are not yet available. A genuine definition—one satisfying the criteria identified in a rigorous theory of definition[3]—is a conceptual determination within a precisely specified theoretical framework, but, as is indicated above, no such framework has yet been presented. Nevertheless, particularly because "philosophy" has been and continues to be used in such an immense and in many cases mutually inconsistent variety of ways, the "systematic philosophy" that this book develops must be adequately defined, that is, delimited in such a way that activities, expressions, concepts, and notions of various sorts that constitute an immense and not precisely specifiable conglomeration are clearly excluded from it. A quasi-definition (Q-Def) of the structural-systematic philosophy is an initial delimitation of just this sort. The quasi-definition is global in that it introduces only the most general concepts that delimit this systematic philosophy. In addition, it relies upon concepts that themselves can be genuinely defined in specific parts of the system, but that, at the beginning, must simply be presupposed. This holds particularly for the concept *theory*. For these various reasons, only a quasi-definition is presented here, not a definition having a form that is, in strictly formal terms, correct. Its sole function is to provide adequate initial clarification of what this book develops under the rubric "the structural-systematic philosophy."

35

(Q-Def)   The structural-systematic philosophy is the theory of the most general or universal structures of the unrestricted universe of discourse.

The following sections clarify the elements of the quasi-definition: *theory* (1.2.2), *structures* (1.2.3), *unrestricted universe of discourse* (1.2.4), and *most general or universal* (structures) (1.2.5). These sections also articulate some of the implications of the quasi-definition.

---

[3] Here presupposed is a conception of definition that is also precisely characterized formally; see Belnap (1993). For a precise theory of definition accounting for the possibility and indeed the unavoidability of circular definitions, see Belnap and Gupta (1993, esp. Chapter 5).

1.2.2 *"Theory"*

Specification of *theory* requires the determination of various factors of various sorts; the broadest distinction is that between *external* and *immanent* factors. External factors distinguish theories from nontheories; immanent factors are those required for the proper definition of theory. This section considers only external factors; immanent factors are presented in Section 2.4.

The *purely external factors* relate theories to those domains that are not designated as "theoretical." In this respect (i.e., in relation to those domains), *theory* is understood more adequately as (the domain of) *theoreticity*. As indicated above, to be emphasized is the following: the systematic philosophy envisaged here is emphatically understood as a *theory*; it thus involves a determinate kind of activity or engagement, and also of presentation and accomplishment.

As is explained in Chapter 4, the theoretical, or the dimension of theoreticity, is one of three fundamental modes of activity, engagement, and presentation, (i.e., three dimensions none of which can be reduced to either of the others or to anything else); the three are theoreticity, practicity, and aestheticity. To undertake a sensible philosophical program, the most basic condition that must be satisfied is that philosophy be pursued as a strictly theoretical undertaking and that it thus not be confused with any undertakings emerging from either of the other two fundamental domains. At this point in the presentation, it is sufficient to specify that philosophy, as theoretical, aims exclusively at *truth*; it aims neither to produce or bring about a good life or lives nor to create works of art.

1.2.3 *"Structure"*

The concept of *structure* is used here only after careful consideration, in significant part because the term "structure" is in widespread use as a buzzword. It is nevertheless used in the quasi-definition for two reasons. First, it retains rich intuitive potential; its specification builds upon that potential by making the term/concept more precise and by holding it far from misleading or problematic uses and/or connotations.[4] Second, there is available a precise logical and mathematical determination of this concept, on which philosophy can rely.[5] To be sure, it must not be assumed that explicit mathematical definitions can have applications in philosophy that are rigorous in every respect (and particularly in formal respects), but they may nevertheless serve as fundamental guidelines of a sort or, in another respect, as regulative ideas for philosophy.

Intuitively, the concept of structure can be clarified as follows: a structure is a differentiated and ordered interconnection or interrelation of elements or parts or aspects of an entity, a domain, a process, etc. Structuration in this sense involves the negation of both the simple and the unconnected. Structure, in this intuitive sense, is a basic concept or basic factor within any theoretical endeavor.

---

[4] As is indicated in the Introduction, deep misunderstandings *will* arise if "structure" is associated with any form of structuralism or poststructuralism.

[5] It must, however, be acknowledged that in mathematics there is no complete consensus concerning the proper meaning and the precise status of the concept of structure. See, e.g., Corry (1992) and Section 3.2 of this book.

In contemporary mathematics and logic, "structure" is central, despite the fact that there is no single, universally accepted terminology interrelating, above all, "structure," "system," and "model." This section introduces the information, explanations, and (terminological) stipulations that are most important to the goals of this book.

In the broadest mathematical sense, a structure is a collection or an n-tuple consisting of elements (objects, entities of whatever sort) and relations (in a broad sense that includes functions and operations) among these elements. Formally, "structure" is usually defined as a triple, whereby relations are understood narrowly, as not including functions and operations.

A structure $\mathfrak{A}$ is a triple $\langle A, (R_i^{\mathfrak{A}})_{i \in I}, (F_j^{\mathfrak{A}})_{j \in J} \rangle$ such that

   i. A is a non-empty set (of so-called objects or individuals or entities of whatever sort), the so-called domain of individuals or simply the universe (of the structure),
   ii. $(R_i^{\mathfrak{A}})_{i \in I}$ is a (possibly empty) family of finitary relations in A, and
   iii. $(F_j^{\mathfrak{A}})_{j \in J}$ is a (possibly empty) family of finitary functions (operations) on A.

One can also list a fourth (possibly empty) component: a constant, a family $(C_k^{\mathfrak{A}})_{k \in K}$ of elements in A. When this fourth component is included, structures are quadruples. Simpler alternatives to $(R_i^{\mathfrak{A}})_{i \in I}$ include $(R_i^{\mathfrak{A}}, \ldots, R_n^{\mathfrak{A}})$ and even R, with corresponding forms of the other components.

Careful attention must be paid to linguistic usage. In the formal sciences, some authors use the term "structure" in a narrower sense. An example is the logician Stephen C. Kleene, who writes (1952/1974: 24–25):

> By a system S of objects we mean a (non-empty) set or class or domain D (or possibly several such sets) of objects among which are established certain relationships . . .
>
> When the objects of the system are known only through the relationships of the system, the system is *abstract*. What is established in this case is the structure of the system, and what the objects are, in any respects other than how they fit into the structure, is left unspecified.
>
> Then any further specification of what the objects are gives a *representation* (or *model*) of the abstract system, i.e., a system of objects which satisfy the relationships of the abstract system and have some further status as well. These objects are not necessarily more concrete, as they may be chosen from some other abstract system (or even from the same one under a reinterpretation of the relationships).[6]

---

[6] See also the excellent essay by Marchal (1975). Marchal (462–63) maintains "that there is a unique and interesting concept that underlies the expressed interests of general systems researchers and that it can be given a satisfactory explanation." Marchal suggests the following definition:

S is a system only if $S = \langle E, R \rangle$, where

(i) E is an element set and

(ii) $R = \langle R_1, \ldots, R_n \rangle$ is a relation set, i.e., $R_1, \ldots R_n$ are relations holding among the elements of E.

If one avoids the word "system" and instead—as is more usual—uses the word "structure," then a distinction is made between "pure (abstract) structure" and "specific (concrete) structure." What this book terms "pure structure" is Kleene's "structure of the system"; this book's "determinate or concrete structure" is Kleene's "system." A "pure structure" abstracts from what Kleene terms the "objects" that are structured by the pure structure.

The literature exhibits additional variants, both terminological and substantive; for example, some authors permit structures with empty sets (empty universes or domains).[7] With respect to the concept of mathematical structure, one must also note recent developments, particularly the (mathematical) theory of categories (theory of topoi; see Goldblatt 1979/1984 and Corry 1992).

Structures in the mathematical sense are mathematical rather than linguistic entities or configurations. Such structures can, however, be put into relation to languages. This book considers so relating them to be not a mere possibility, but instead a necessity. The results of such relatings are structures (usually) termed *models* of the languages (or of "theories" in the strictly formal sense, [i.e., uninterpreted formal systems]). Because this usage is not universal, this book adopts the following terminology to avoid misunderstandings: structures that are models (of languages or of theories in the relevant sense) are termed *model-structures*.

A question central to philosophy and particularly to the structural-systematic philosophy concerns just what the items or entities are that are structured by the pure (or abstract) structures. To anticipate: a central and fundamental thesis of the philosophy developed in this book is that the basic or primordial items that are structured by pure philosophical structures are, semantically, *primary propositions* (or *primary states of affairs*) and, ontologically, *primary facts*.[8] For these, the general designation "primary entities" may be used. These primordial entities are for their part structures of determinate sorts. As shown in detail in Section 3.2.3, two different kinds of primary entities or structures can be distinguished: simple and complex. Simple primary propositions or primary states of affairs are individual primary entities that, as explained below, contain only themselves as elements and therefore structure only themselves. They can be designated as primary structures on the zero or null level. Complex primary entities are *configurations* of primary entities.

Here, a further distinction must be introduced between two sorts of configurations: configurations of simple primary entities and configurations of complex primary entities. The latter are configurations of configurations of primary entities, whereby the scale of levels of complexity is in principle infinitely large. Because even the non-complex primary entities are structures (precisely simple primary structures), the primordial entities that are structured by abstract structures in the sense relevant here are also

---

[7] See, e.g., Ebbinghaus (1977/2003: 54): "In opposition to usual linguistic practice, we allow for structures with empty universes." A structure with an empty universe would presumably correspond in a determinate respect to what Kleene calls "structure of the system" and thus to what this book terms "pure (abstract) structure."

[8] As is indicated above in the Introduction, "primary," used to qualify sentences, propositions, and facts, is used *not* in any ordinary sense, but instead in a technical sense that is precisely explained in Chapter 3. That there are *primary* sentences, propositions, and facts does *not* mean that there are also secondary, tertiary, etc., sentences, propositions, and facts.

structures in the sense that has been explained. This makes it clear that the concept of structure has an unrestricted, comprehensive status.

This concept, particularly with respect to its application in philosophical domains, is treated in detail in Chapter 3, "Systematics of Structure." Through this application, this concept is extended in a significant manner and in part newly interpreted. The strictly formal (mathematical) definition serves only as a first, and still purely abstract, characterization of the concept of structure that is understood and applied philosophically.

1.2.4 "Unrestricted Universe of Discourse"

[1] The term "universe of discourse" appears to be the coinage of the logician Augustus de Morgan.[9] In this book, the term is used to designate the domain of inquiry (the subject matter) specific to systematic philosophy. What that means, precisely, is initially clarified within this section.

Within the structural-systematic undertaking, "universe of discourse," if not further qualified, would not serve non-problematically to designate the domain of inquiry specific to systematic philosophy. That is because (as is seen in the passage cited in Footnote 9) mathematicians and logicians usually deal with *restricted* universes of discourse, such that the universe of discourse for one account might be the set of rational numbers, for another that of mammals, etc. Hence, the qualifier "unrestricted" appears in the quasi-definition. If the quasi-definition used "universe of discourse" without qualification, that might lead some who are familiar with the term to ask *what* universe of discourse was meant.

The complete clarification of "unrestricted universe of discourse" would coincide with the entire presentation of the structural-systematic philosophy; at this point, it suffices to note one respect in which this book's use of the term presupposes a potentially controversial thesis [i], and one respect in which the term is helpfully neutral [ii].

[i] The potentially controversial thesis is that any subject matter for philosophical consideration or explanation is, or can be, articulated linguistically or, more colloquially, talked about. But what, more precisely, *is* meant by linguistic articulation or being talked about? To answer this question is one of the most important tasks of philosophy.

---

[9] Kneale and Kneale (1962: 408) write, "In practice, Boole takes his sign 1 to signify what de Morgan called the *universe of discourse*, that is to say, not the totality of all conceivable objects of any kind whatsoever, but rather the whole of some definite category of things which are under discussion." Kneale and Kneale refer to p. 55 of de Morgan's book *Formal Logic: or, The Calculus of Inference, Necessary, and Probable* (1847). There, however, de Morgan writes not "universe of discourse," but "universe of a proposition." Ignoring this historical detail, it is to be noted that in the passage quoted above, Kneale and Kneale restrict the notion of "universe of discourse" by distinguishing it from "the totality of all conceivable objects of any kind whatsoever." But the formulation itself shows how problematic this putative distinction is: How could the concept of "the totality of all conceivable objects of any kind whatsoever" make any sense at all if it were beyond any and every universe of discourse? Anyone introducing the distinction accomplishes, precisely by its means, the introduction of the concept into the universe of his or her discourse. This is not to say that "conceivable totality" is an element or subset of the universe of discourse; instead, the unrestricted universe of discourse *coincides* with this totality of conceivable objects—that totality is what the unrestricted universe of discourse *is*.

At this point, this task can be addressed only provisionally. Taken literally, to say that a subject matter is linguistically articulable is to say that there must be a language that relates to the subject matter's domain. Or, from the other direction: if a domain, no matter how designated, is taken into theoretical view, then language will always already have been, one way or another, at work.

Here arises, however, what might appear to be a weighty objection: does one not thereby restrict what can be thought about or theorized to what is within the "limits" of language (or of linguisticality as a whole)? This objection must be taken seriously, but so taking it requires considering extremely complex matters that can be treated at this point only quite generally and in broad outlines.

To make the problem clear, it is helpful to introduce a famous formulation from Wittgenstein's *Tractatus*: "The limits of my language are [*bedeuten*] the limits of my world" (*Tractatus* 5.6; see 5.62). Regardless of what Wittgenstein may actually mean here, if his formulation is taken at face value (as it normally is and indeed must be), then it is deeply misleading, because to *speak* of a limit drawn by language (or by *a* language) is to be, linguistically, beyond the putative limit. A limit is a limit only if there is something beyond it, so to identify a limit to language is also to enter the linguistic space making it possible to speak of what is beyond the limit, and thus to negate the identification of the "limit" as a limit of the language that bespeaks it *as* a limit.[10]

Unqualified talk of "limits" of language should therefore be abandoned. The issue of present concern is better characterized in different terms. What is crucial is how one understands "language." If one speaks of a specific language (e.g., Wittgenstein's "my language") and distinguishes this language from other languages, then in a specific sense speaking of "the limits of my language" (i.e., this specific language) is or can be correct. But this means only that one is beyond the limits of this determinate language ("my language"); precisely in surpassing (or again, in speaking of) the limits of *that* language, one relies upon a *more comprehensive* language that makes possible sensible and coherent talk of the limits of the first language. This procedure of speaking about a specific language by using a more comprehensive language can be repeated arbitrarily. For now, the question whether this leads to a problematically infinite regress may be ignored. What is crucial at this point is to note that the procedure and its iterability reveal the impossibility of getting outside of language on the whole, of getting outside of linguisticality. But it is not as though there is some "outside" to which language bars human access; on the contrary, its inescapability reveals language to be the universal dimension within which human beings as theoreticians always already find themselves and within which they proceed whenever they are involved with the theoretical undertaking. Instead of *limiting* theorization, language *makes comprehensive theorization possible*.

41

---

[10] Wittgenstein's distinction between saying and showing is not a solution, but an articulation of the misunderstanding contained in his formulation. Wittgenstein speaks of "my language" (*Tractatus* 5.6) and of the "limits of language (of the language that I alone understand)" (*Tractatus* 5.62). But what is a language "that I alone understand?" Wittgenstein's equation of "my language," "language," and "the language that I alone understand" reveals his position to be not only confused but also incoherent. This has not prevented others from citing his formulations as more or less indisputable theses.

For similar considerations undermining limitations placed on knowledge (rather than language), see Section 2.3.2.4.

Also worth noting at this point is that, in the context of theoretical undertakings, "language" need not be understood as natural or ordinary language; on the contrary, the languages of philosophy and the sciences should be understood as semiotic systems, in an unrestricted sense, with all the capacities such systems have to offer. This point is developed in detail in Chapter 5.

[ii] For the initial (quasi-)definition of the structural-systematic philosophy the term "unrestricted universe of discourse" is quite appropriate in that it is both comprehensive and broadly neutral: it in no way delimits or predetermines what may fall within it. That this is an advantage becomes clear when it is compared to the five alternative ways of initially determining the subject matter of systematic philosophy that are the most important both historically and at present (although this point is not argued for here.) Each of these, viewed from the perspective of the structural-systematic philosophy, thematizes only a restricted universe of discourse; this is of course not recognized by advocates of these alternatives. The first restricts the universe of philosophical discourse to what in the Introduction is termed the domain of *being* (in the objective sense—also termed *actuality, reality, the world,* etc.). This restriction is to be found throughout the long tradition of the metaphysics of being, although in widely differing, diverging, and often mutually exclusive variants including materialist-physicalistic philosophies of being, spiritualist philosophies of being, etc. The second restriction identifies the universe of philosophical discourse with the domain of subjectivity (thinking, self-consciousness, knowledge, etc.). This determination characterizes not only the entire philosophical program of German Idealism but also Kant's epistemology and the entire tradition of transcendental philosophy. Characteristic is one of Kant's formulations: "I understand by a system ... the unity of the manifold cognitions under one idea" (*Critique of Pure Reason* (CPR): B860). The third specification of the universe of philosophical discourse uses such terms as "intersubjectivity," "dialogue," "communication," and "historical context"; it considers the task of philosophy to be the development of the structures of this dimension, which are here named only globally (additional differentiations are required). The fourth option focuses on the dimension of linguisticality, such that the task of philosophy is seen to be that of articulating linguistic structures. The fifth restriction, finally, conceives of the universe of philosophical discourse from the perspective of the formal dimension (i.e., logic and mathematics, leaving open at this point the question whether the two are distinct, and if so, how).

The universe of philosophical discourse, as it is articulated within this book, proves not to be identified with or restricted or reducible to any single one of the five determinations just introduced, taken in isolation from the others, but is related most closely to the first, fourth, and fifth determinations. More precisely, the attempt is to conceive of the ontological, the linguistic, and the formal domains of the unrestricted universe of discourse as a unity in the strictest possible sense. The second and third dimensions (subjectivity and intersubjectivity) play no primary roles in determining the universe of philosophical discourse, but from this it does not of course follow either that they are inexistent or unimportant or that they are excluded. Instead, their status within the framework of the structural-systematic philosophy is derivative rather than primary. For this reason, the structural-systematic project is designated most precisely as a formal-semantic-ontological enterprise. Section 1.3 explains this project to the degree possible at that point in the book.

[2] At this point in the exposition, the unrestricted universe of discourse can be understood as the totality of the given, of the data, of the phenomena, etc., provided

that these terms are understood purely programmatically. They are thus not to be equated with "sense data" or what Kant terms the "raw material of sensible impressions" (*CPR*: B1).[11] On the contrary, "datum" and "given" are used simply because it is convenient to have technical terms to designate any item of any kind that is to be found in the universe of philosophical discourse: "datum" is the starting point for the theoretical undertaking. Nicholas Rescher (1973: passim) provides a helpful technical clarification of the datum, in this wholly general sense, as a "truth-candidate," nothing more and nothing less. Any theses concerning how such data are accessible, etc., cannot be introduced here, for such theses are constituents of theories, and at this point in the exposition, no theory is available.[12]

The just-described fundamentally undetermined status of the totality of the   44
data concerns the *theoretical* perspective; it does not exclude the possibility—which indeed is always an actuality—that this totality does present itself, *pretheoretically*, as largely determined. Within their everyday lives, human beings distinguish quite successfully and often quite precisely among things, domains, and totalities, including within their distinctions even the totality of all things and domains. But these pretheoretical distinctions are unstable and always subject to revision. If they were not, there would be no need for theoretical investigation: theory would be redundant, indeed literally pointless, in that its aim would always already have been attained. But this is obviously not the case; instead, the pretheoretically determined and articulated data are what theorization aims to conceive, to explain, to articulate *adequately* and *groundedly*.

1.2.5 *"Most General or Universal Structures"*

The component of the quasi-definition of the structural-systematic philosophy that is perhaps the most difficult and problematic is "most general or universal structures."

---

[11] When Kant further says that intuition is that "through which" cognition "relates immediately" to objects, and that this takes place only "insofar as the object is given to us," he proceeds on the basis of a powerfully contentual assumption (thesis) that he does not support either at the outset or during the course of the *Critique of Pure Reason* . The assumption or thesis is that the givenness of an object "is only possible if it affects the mind in a certain way" (A19/B33). In addition to being unsupported, the thesis involves various concepts and presuppositions that are insufficiently clarified.

[12] As is noted already in the Introduction, this book uses, in addition to the term "datum" (often in the form, "the grand datum"), the terms "objectivity," "world," "universe," and "being (as such and as a whole)." These terms are *not* synonyms. Their precise significations are provided at the appropriate systematic locations. Before the provision of these specifications, they are used *not as synonyms* , but in an *undifferentiated* manner, all designating the dimension named by the title of this book: that of *being*. The reason for this way of proceeding is that these various terms are used, in the philosophical literature, in utterly diverse contexts and in widely varying manners. Taking this literature into consideration—at times explicitly but often only implicitly—this book uses the terminology most common in discussions of the specific topics being considered. This holds especially for the term "world." But, as has just been indicated, when the presentation gets to a point at which one of the terms requires precise specification, that specification is provided; after that point, the term is used exclusively in its clearly specified sense.

The term itself is chosen not for expressive strength, which it lacks, but for want of anything better. Obviously, it requires clarification.

In an initial characterization, the most general or universal structures are those that are accessible within the universal perspective, (i.e., here, the perspective of the unrestricted universe of discourse as such and as a whole). In objectual terms, these are the structures of an entity or of a domain that characterize the entity or domain *insofar as it belongs within the unrestricted universe of discourse as such and as a whole.* This formulation presupposes that there are different levels of structures—thus, structures on different levels. Considered more closely, the most general or universal structures are those that structure the unrestricted universe of discourse as a whole, the universe of philosophical discourse. These structures provide the basis for restricting the universe of discourse in ways that make possible scientific inquiry in particular disciplines; further restrictions within disciplines open paths for subdisciplines.

45   As the distinction between disciplines and subdisciplines indicates, the universe of discourse as a whole encompasses different structural levels. The ways the levels interrelate may be clarified by means of an example.[a] Universities are typically divided into specific divisions and departments, with the divisions bearing such titles as "Humanities," "Sciences," and "Social Sciences," and departments, "Music," "Biology," and "Economics."[b] These divisions and departments *structure* the university as an academic institution, but the divisions and departments are themselves structured: divisions must and departments as a rule generally do at least implicitly contain subdivisions, and departments also contain individual courses and faculty members. Individual courses and class meetings are of course themselves structured. Typically (or at least quite often), the structures of individual class meetings are determined by instructors rather than by departments, but in any case not (in any remotely typical instance) by the university.

The structure of a given university, including its substructures, viewed synchronically, might well appear well defined, with, among other things, specific identifications of divisions, departments, faculty members, and courses. Viewed diachronically, of course, the structures are far more fluid, with fluidity increasing as one moves from the level of the universal or most general structures (structuration via divisions and departments) to those of substructures (faculty members, courses, class meetings). In principle, of course, the structuration of an individual class meeting can become a matter to be considered at the highest level, the level of the university as a whole; currently,
46   this might well happen if a class meeting were devoted to advocating racism.

In terms of this example, philosophy corresponds to the level of the university, whereas the other sciences are the counterparts of departments, disciplines, etc. Yet, as in the university, the boundaries are themselves flexible. Whereas—as is clarified below—there are issues that are clearly philosophical rather than scientific, and vice versa, there are also issues whose theoretical status or level must depend on specific developments within the various theoretical disciplines (just as, in the example given

---

[a] *Struktur und Sein* uses the example of the political structures and substructures of the Federal Republic of Germany. *Structure and Being* relies instead on an example likely to be more immediately familiar to many Anglophone readers.

[b] Universities also of course have administrations, but adding those and other complexities would needlessly confuse the example.

above, even a given class session can become a matter to be considered at the highest structural levels of the university). Thus, the concept "most general or universal structures" is in part *relative* in that it is a function of the relation of philosophy and the sciences as it actually exists at a given time.[13]

Structures can be interrelated in various ways: via horizontal connections, through being embedded in broader, more comprehensive structures (thus being substructures), etc. Correspondingly, any structure can stand in relation to other structures. Most important is the relation of structure(s) and substructure(s) or system(s) and subsystem(s). Formal instruments for precisely specifying such relationships are available. The third methodic stage (see Section 1.5), the *system-constitutive* stage, focuses on such relations. The method used for the task of specifying such relationships is here termed *structuring holistic nets*. Ultimately, there is a single holistic net in that there is, at the highest level, a comprehensive structure.

1.2.6 *"Systematic Philosophy" and "Philosophical System"*

The terms in the quasi-definition of the structural-systematic philosophy having now been clarified, the terms "systematic philosophy" and "(philosophical) system" require additional specification in two important respects. First, in regard to terminology, the titles both of this book and of its sixth chapter use the term "systematic philosophy," not "philosophical system." The only reason for avoiding the use of "system" in these titles is historical: because of how it has been used in the past, the term is easily misunderstood. The history of philosophy contains many examples of "philosophical systems," most relevantly (with respect to the question of terminology) those developed by the German idealists Fichte, Schelling, and Hegel. Particularly in the nineteenth century, many philosophers and philosophies reacted strongly against such systems of idealism. At present, in broad parts of contemporary philosophy and especially in Germany, the term "philosophical system" is usually associated with the idea or form of a philosophy modeled on these grand systems from the nineteenth century. This book rejects that idea and form of philosophy, taking it to be—despite the fact that it has fascinated and continues to fascinate many philosophers—fundamentally erroneous. That form of philosophy pays no attention even to the elementary factors that are absolutely indispensable for the development of any serious philosophical theory. These factors include the precise specification of concepts, clarification of the status of philosophy as theory, consequential argumentativity, and the like. This issue cannot be treated in detail here (but see Puntel 1996). To emphasize that the conception developed here differs fundamentally from that guiding the "philosophical systems" of the past, no title in this book uses the term "system."

The second respect in which the term "(philosophical) system" requires clarification arises from the fact that there is no *substantive* reason for avoiding the term "system"—there is *only* the historical reason just introduced. Because there is no substantive reason for avoiding it, this book—unlike its titles—does at times use that term. More specifically, it regularly uses the term "system" to designate two quite different types of entities: (1) nonlinguistic entities such as organisms, the solar system, etc., and (2) theories, where "theories" are generally understood as and taken to be linguistic

---

[13] This issue is explicitly considered in Chapters 4 (esp. Sections 4.2.1 and 4.2.1.1) and 6 (esp. Section 6.3.2.2).

complexes (sets of sentences; see Marchal 1975: 450ff). This ambiguity has caused complications throughout the history of philosophy; if for example one speaks of the "system of philosophy" of Fichte, Schelling, or Hegel, it is unclear whether one is referring to the set of sentences formulated by the philosopher in question or the whole (of being, of spirit, of the subject, etc.) expressed or articulated by this set of sentences.

If "theories" are understood—as, for example, by advocates of the "non-statement view" (discussed in Section 2.4.2.3)—not as linguistic, but instead as mathematical structures (more precisely, as classes of models), then "systems," as identified with such "theories," acquire an additional (third) signification: no such "system" is linguistic, and neither is that to which any such theory, as a mathematical structure, relates. In this sense, a "system" is a mathematical representation or articulation of something (of the objects of the theory or of the domain of those objects), whereby, however, as a rule it remains wholly unclear whether or not this domain of objects is identified with reality (the world, the universe, being).

When this book uses the term "system," the term always has one of the three significations just introduced; which one it has is always made clear by the context. But the three significations are closely interrelated, such that when one of the significations is intended primarily or in the first place, the other two are indirectly co-intended. However, for the third signification, this holds only if it is understood in a manner distinct from that in which it is understood by the advocates of the non-statement view of theories. Chapter 2 (especially Sections 2.4.2.3–2.4.3) clarifies this point.

Summarily and in part anticipatorily: the philosophy developed and presented in this book is systematic or *is a system*, first in that it is a theory in the sense of a system of sentences, second in that it is a theory in the sense of a system of propositions (i.e., a system of those entities that are expressed by the sentences of the theory in the first sense), and third in that it is a system in the sense of a comprehensive ontological whole with which the world (the universe, being) is identified. Philosophy as system in the second sense is the corrected or reinterpreted signification of "system," when this word is identified with a theory in the sense of the non-statement view of theories. The system of propositions is not a system of linguistic entities and also not (yet!) the system that is identical with the world (the universe, being). The system of propositions is a determinate whole, consisting of elements of the first two types of fundamental structures (formal and semantic structures), whereas the system in the ontological sense is the entirety of the ontological structures. At this point, again, much of this is anticipatory. The account that follows clarifies and grounds it step by step.

1.3 Structure and Being: A First Characterization of the Basic Idea Behind the Structural-Systematic Philosophy

[1] As is indicated above, the title *Structure and Being* is the maximally concise designation of the central contentual thought behind the structural-systematic philosophy. The title is also meant both to mirror and to oppose that of Heidegger's *Being and Time*, which—in contrast to Heidegger's self-interpretation—is the only one of his works that can in any reasonable way be classified as systematic. The title *Structure and Being* is also meant to echo other comparably programmatic book titles such as *Word and Object* (Quine 1960) and *Mind and World* (McDowell 1994/1996).

With the project of which *Being and Time* was initially envisaged as a part,[14] Heidegger aimed at "the explication of time as the transcendental horizon of the question concerning being" (subtitle of Part One of *Being and Time*). Appropriating Heidegger's language and conceptuality—if one can even speak of a Heideggerian "conceptuality"—the project undertaken in this book could be described as follows: *being* (understood initially as the unrestricted universe of discourse, nothing more and nothing less) is to be apprehended and presented within the horizon of the dimension of structure (in a comprehensive sense). An initial conceptual clarification of what is here meant by "structure" is provided above.

What more precisely does it mean to develop a systematic philosophy from the starting point of the title, *Structure and Being*? A beginning point from which to address this question is the fundamental observation that every large-scale philosophy deals with a *basic dichotomy* between two dimensions that are conceived of in a specific fashion (or, more often, simply presupposed); the task is to bridge the gap putatively revealed by the basic dichotomy, to close it, to accomplish its suspension or *Aufhebung*, etc. The best-known (and presumably most important) forms of this basic dichotomy include the following:

being and beings

being and appearance

thinking and being, in various forms or variants including subject(ivity) and object(ivity), whereby object(ivity) can take various forms, including being, actuality, world, etc.

humanity and nature (world, cosmos)

humanity and history

individual and society

humanity/world and God

language and world (being, actuality)

concept and world (being, actuality)

the formal dimension (logic, mathematics) and the world (being, actuality. . .)

As is presumably obvious, the basic dichotomies just listed are not all equally weighty, philosophically. There are therefore sufficient grounds for identifying six significant forms of the dichotomy as essential; these correspond to the determinations of the universe of discourse briefly introduced above. The first is the purely ontological basic dichotomy: the dichotomy between being and beings, between actuality as a whole and individual actualities or, briefly, between the ontological totality and ontological individualities. The second basic dichotomy, like the first, has been and continues to

50

---

[14] Heidegger published, as elements of this project, only the first two of three projected sections of the first of two envisioned parts. The title of the third section of the first part was to be "Time and Being" (*SZ*: 8), and in 1962 Heidegger presented a lecture bearing that title; the lecture was published in German and French in 1968 in *L'endurance de la pensée. Pour saluer Jean Beaufret* (Beaufret 1968: 12–68). The protocol from a seminar on this lecture appears, along with the lecture, in Heidegger (1969/1972).

be broadly influential; this is the basic dichotomy between subjectivity and objectivity, with the variants thinking/being, self-consciousness/world, knowledge/objectivity, etc. The third basic dichotomy focuses on intersubjectivity in relation to the individual subject or to individual subjects; different schools of thought in which this dichotomy is central are distinguished in part by how they designate the pole of intersubjectivity (e.g., as the intellectual community, the communication community, the domain of cultural traditions, etc). The fourth basic dichotomy is easily named as that of language and world, but it is anything but easy to show just how the relevant "language" is to be understood philosophically. The fifth basic dichotomy is classically termed that between concept(uality) and world. Equally classic is the formulation of the sixth basic dichotomy, that between the abstract sciences (logic, mathematics) and the world (actuality, being).

[2] According to a fundamental thesis of this book, all of the basic dichotomies named above (along with others that could also be named) are inadequate and in specific respects derivative forms of an absolutely fundamental basic dichotomy, that between the dimension of structure and the dimension of being. The term "being" is used here in a specific sense. The concept connected with it *initially* encompasses *both* poles of the first basic dichotomy (being and beings) and the second poles of the fourth, fifth, and sixth basic dichotomies (actuality, world, universe, objectivity, etc.). The term "structure" initially designates the first poles of the fourth, fifth, and sixth basic dichotomies (language, conceptuality, and logic/mathematics). The second and third basic dichotomies are genuine and weighty ones, but this book accords them a secondary significance.

Now briefly to be clarified is why the first poles of the fourth, fifth, and sixth basic dichotomies—respectively, language, conceptuality, and the formal dimension (logic and mathematics)—are designated and understood as structure(s). The first question posed by this task is how language, conceptuality, and the formal dimension are to be understood here.

In this book, language is understood as a dimension that plays a central role. Language can be either ordinary (natural, spoken) language or scientific (artificial) language. In direct opposition to the majority of contemporary philosophical schools of thought, this book denies both that ordinary language is an appropriate instrument for philosophy and science and that it provides a standard or measure appropriately applicable to philosophy and science. Detailed reasons for this denial are given below; at this point, what is important is that the artificial or technical language this book increasingly relies upon is a semiotic system for whose complete definition the three classic aspects of syntax, semantics, and pragmatics are essential. To these aspects correspond syntactic, semantic, and pragmatic structures. Semantic structures prove to be of particular significance, because they accomplish what may be termed the bridging or the suspension (*Aufhebung*) of the fundamental difference between language and the world (being).

In connection with the consideration of semantic structures, *ontological structures* must also be investigated. Ontological structures are indeed *structures* in the strict sense, as is demonstrated in detail below. This demonstration involves conceiving of the second pole of the basic dichotomy between structure and being as *structured being* or as *being as structurality*.

The term "formal dimension" here designates not only formal logic but also other formal instruments—thus, all of mathematics in all of its forms and domains. At this point, it suffices to note that, without formal structures, the formation of theories is simply unthinkable.

[3] The grand dimension of structure, as just described, has as its initial counterpole the second grand dimension, that of the unrestricted universe of discourse with its initial, pretheoretically given ontological determination. For the sake of brevity, this book's title designates this dimension "being." To simplify the presentation, the text also uses various other terms that are, save in noted exceptions, equivalent; these terms include, most importantly, "world," "universe," "actuality," and "reality." Generally, only one term is used in any given instance; most frequently used are "world" and "universe."

Given the approach taken here, clearly the task is to articulate the structures and structuration of the *world*, of *being*, thus of the *unrestricted universe of discourse*. The genitives "of the world," etc., are to be understood as cases of *genitivus subiectivus* in the strictest possible sense: the structures are *the world's own structures*, immanent to it and in no way imposed or forced on it from something somehow outside the world. This point is easily distorted and misunderstood if one or more of the fundamental dichotomies introduced above are taken in isolation or taken to be intelligible in isolation. To say that there is a structuration immanent to the world—that the structures to be articulated are structures *of the world itself*—is not to say that the world either is or must be conceived of as, in Hilary Putnam's phrase, a "ready-made world." It is not as though a "ready-made" or "prestructured" world were available for theoretical investigation in such a way that consideration of the domain of structuration (of formal, semantic, and ontological structures) would be not only not necessary but even detrimental. To think that the world was available in this manner would be to make (again in Putnam's terms) the mistake of metaphysical realism. If human beings were confronted with a world that was ready-made *in this metaphysically realistic sense*, then it would be unintelligible why philosophy and the sciences should not concentrate on and indeed restrict themselves to the simple description or—in some manner or other—mirroring of this world. The situation is far more complex than one of such a direct confrontation; its clarification shows that one can wholly correctly speak (using terms corresponding to Putnam's) of a ready-structured world and indeed of a prestructured world, albeit in senses *other* than the metaphysically realistic ones.

The decisive thesis guiding the development of the structural-systematic philosophy is that the truth of the universe, the actual universe, the true (theoretical) configuration of the universe[15]—thus, all with which philosophy and the sciences, as theoretical endeavors, are concerned—is neither such that it can be revealed, by any simple describing or copying or mirroring, as structured once and for all, nor such that it is in any way an *accomplishment* or *product* of any sort of subjective or intersubjective activity (e.g., that of a community of language users). Instead, what may anticipatorily be designated as the true configuration of the universe and its constituents veridically

52

---

[15] This book must rely on such formulations at this pretheoretical or presystematic stage of its presentation, but they are, obviously, precarious. At this point, (more) precise formulations cannot (yet) be provided.

manifests itself, within the theoretical framework of the structural-systematic philosophy, as the intersection of or agreement between the unrestricted universe of discourse
(and of various domains within it), understood as the comprehensive datum, and
the dimension of structure, characterized as the three dimensions of the formal, the
semantic, and the ontological; this could also be termed the manifestation of the coincidence of the comprehensive datum with the dimension of structure.[16]

The situation for systematic philosophy at the outset thus has two fundamental
aspects: first, the unrestricted universe of discourse as pretheoretically available, in a
determinate form, as structured, and second, the dimensions, already available for theorization (or readily accessible to theorization), of the formal, semantic, and ontological
structures. The entire theoretical undertaking of the structural-systematic philosophy
consists of bringing these two grand domains or dimensions into agreement. Viewed
from the perspective of the pretheoretically available data that initially constitute
the unrestricted universe of discourse, the task is to subject these pretheoretical data
(truth candidates) to theoretical examination and thereby clarify them (i.e., articulate
them semantically, ontologically, and formally). Viewed from the perspective of the
formal, semantic, ontological structures, the task is, as it is often put, to concretize these
structures, to project or discover models for them, etc.

As is indicated in Section 1.3[1], many philosophers simply assume, often as obvious
beyond question, that the pretheoretically available unrestricted universe of discourse,
on the one hand, and the domain of formal, semantic, and ontological structures, on
the other, constitute an irreducible dichotomy or are separated by a gap that is not
subject to explanation or clarification. This assumption is shared by a great number of
accounts, ranging from Kant's transcendental idealism to Putnam's "internal realism,"
that are variants of one and the same basic position. In direct opposition to all such
accounts, the structural-systematic position emphatically rejects this assumption. If the
two domains or dimensions can coincide or agree, then they cannot be essentially alien
to each other; if they were essentially alien to each other, any coincidence or agreement
would be miraculous and hence immune to theoretical investigation or clarification.

The structural-systematic counterthesis—further clarified and supported
particularly in Section 5.1—is this: there is an original or primordial unity between
being/world/universe, on the one hand, and the formal, linguistic/semantic, and ontological structures, on the other, such that being/world/universe is not (or no longer)
to be taken as it is pretheoretically given and the three kinds of structures no longer
as *abstract* in any ultimate or definitive sense. One therefore can and indeed must
say that the world understood as *originally* structured in the sense just explained is a
ready-structured world—a world whose structure is immanent to it—in the following
sense: at the outset of the theoretical undertaking it has the status of a world in general
to be discovered or developed or articulated, whereas at the end of the undertaking its
status is that of a discovered and articulated *ready-structured world*. But at this end the
tristructural dimension is seen to have been always *already integrated* into the world.

---

[16] As is shown in Section 1.5 and, in detail, in Chapters 5 and 6, a central thesis of the
metasystematics of the structural-systematic philosophy is that being (the world, the
universe, actuality) veridically manifests itself in different ways within different theoretical
frameworks, with differing degrees of theoretical adequacy (i.e., of intelligibility, coherence,
and comprehensiveness).

Without the integrating or integration, the putative thought of a ready-structured world is nonsense. Differently stated: the structuration of the world (*genitivus subiectivus*) *is* (the "is" of identity) the structuration of the dimension that at first, at the outset of the theoretical undertaking, appears as the "other side" with respect to the universe of discourse. Conversely, what first appears as or is assumed to be a dimension of structure *that is different and distinct from* the universe of discourse *is* (in the sense of identity) the structuration of the world. In *this* sense, the world can and indeed must be acknowledged to be an *initially or primordially structured dimension*.

[4] As is clarified in various passages above, the articulation of the world as a dimension that is structured formally, semantically, and ontologically can unfold only within one or another determinate theoretical framework. Any determinate framework is, however, one among many, and the claim that any such framework is simply the right one, the best one in an absolute sense, would presuppose the accomplishment of the impossible task of identifying and comparing all such frameworks, both actual and possible. Despite various philosophical projects (that of Hegel is paradigmatic) that have sought utterly conclusive results, such absolutely complete and potentially conclusive undertakings cannot even be begun, much less completed. The consequences of their impossibility are considered more fully in Section 1.5 and in Chapter 6.

1.4 The Idealized Four-Stage Philosophical Method

1.4.1 *The Problem of Method*

Can the basic systematic program just sketched actually be developed? The first challenge arising from this question is that of method. To be sure, "method" is a broadly vague and comprehensive word that designates many, often quite disparate things. Despite this vagueness, the term is here used to designate a four-staged procedure.

Among contemporary philosophers, there is not even minimal consensus concerning *a* or *the* philosophical method, nor throughout the history of philosophy has there ever been any such agreement that would qualify as a consensus. At the most, the practice of philosophizing exhibits various rules, customs, and standards that are accepted and applied by many philosophers, although not in a unified fashion.

As is well known, philosophers rely on many so-called philosophical methods; these include, among others, the Socratic, the phenomenological, the hermeneutical, the dialectical, the analytical, the axiomatic, and the network methods. This is not the place to examine or to evaluate these various methods individually. It suffices here to note that fundamentally each of these methods includes elements that can be used philosophically; claims that any one method is the *only* method for philosophy are, however, unacceptable.

The structural-systematic philosophy presented in this book is based on a complex method. The reason for the complexity is that the development of philosophical theories involves numerous tasks of distinct kinds and that accomplishing these tasks requires reliance on a variety of procedures. Viewed systematically, the multiplicity of procedures may be reduced to four methodic stages. This matter is now to be considered in detail.

The philosophical method characterized here is a *complete* method that can be understood, more precisely, as four-staged. For the sake of brevity, somewhat undifferentiated use is made of the notions of the four stages of the philosophical method and, more simply, of four philosophical methods. However, these methods are always to be understood as four indispensable steps in the comprehensive procedure of philosophizing. A presentation that was complete in every respect would have to apply this philosophical method as a whole and in a detailed manner. Because, however, this book considers and presents only the theoretical framework for such a theory, different components of which are concretized to differing degrees, it does not follow the method in such complete and comprehensive detail.

Summarily, the four stages are as follows:

Stage 1: Inventory: Identification of structures and their preliminary collection, seeking maximal coherence or structuration, into informal theories

Stage 2: Interrelation of (informal) theories initially and informally articulated at the first stage into holistic networks and (in particular cases) axiomatized theories

Stage 3: Interrelation of component (informal and/or axiomatized) theories into increasingly comprehensive theories (holistic networks)

Stage 4: Evaluation of the comprehensive system or network with respect to its theoretical adequacy and truth

The guiding idea behind the four-stage method is simple: with respect to the *data*—be they individual phenomena or events, entire domains, or even the comprehensive datum (i.e., the unrestricted universe of discourse as a whole)—the task of theorization is the following: first, the theoretician must seek structures for the data (and thus structures of the data), thereby both acquiring the material for theorization and formulating initial or informal theories. Next, these informal or elementary theories must be put into strictly theoretical form; that is, theories in the genuine or strict sense are to be formulated. Third, the thus-presented individual theories must be brought into systematic form, which requires the development of a network of theories, thus a system (more precisely, a system of systems). Fourth and finally, it must be determined whether the individual theories and the theoretical network within which they are integrated, thus the comprehensive theory, are theoretically adequate, (i.e., whether they satisfy the criteria for theoreticity, which include, most importantly, that of truth).

In normal philosophical practice, these steps are scarcely ever even recognized, much less taken in this order. The second and third steps or stages are usually wholly ignored. Typically, only incidental aspects of the first and fourth stages are applied, and the fourth stage is usually taken to involve "justification" of an only vaguely determinate sort. More ambitious philosophical presentations ignore only the second step. In such cases, the informal or minimal theories that result from the application of the first stage of the method are directly—in this book's terms, without reliance upon the second stage of the method—integrated into a network-theory that is itself only informally articulated.

1.4.2 *First Methodic Stage: Identification of Structures and Construction of Minimal and Informal Formulations of Theories*

A modified form of some basic features of the coherence methodology developed by Nicholas Rescher characterizes this first stage. It essentially clarifies the procedure to be followed with respect to philosophical treatment of the data from the starting point of identifying the relevant structures up to the point of developing initial, minimal formulations of theories. To be noted at the outset is that Rescher's procedure has a defect, but also that, as is shown below, the procedure can be freed of this defect.

Rescher (1992–1994 I:157) presents the basic idea of his coherence methodology as follows:

> Acceptance-as-true is in general not the starting point of inquiry but its terminus. To begin with, all that we generally have is a body of prima facie truths, that is, propositions that qualify as potential—perhaps even as promising—candidates for acceptance. The epistemic realities being as they are, these candidate truths will, in general, form a mutually inconsistent set, and so exclude one another as to destroy the prospects of their being accorded in toto recognition as truths pure and simple. The best that can be done in such circumstances is to endorse those as truths that best cohere with the others so as to "make the most" of the data as a whole in the epistemic circumstances at issue. Systemic coherence thus affords the criterial validation of the qualifications of truth candidates for being classed as genuine truths. Systematicity becomes not just the organizer but the test of truth.—A coherentist epistemology thus views the extraction of knowledge from the data by means of an analysis of best-fit considerations. 58

"Datum" is for Rescher a technical term meaning simply "truth-candidate" (thus *not* "established truth," in any sense whatsoever). Data in this sense are not "sense data" or anything of the sort; instead, they are the linguistically/logically given starting points for any scientific (philosophical) undertaking. At the outset, the theoretician is confronted with a set of such data $S = \{P_1, P_2, P_3, \ldots\}$. The task is to introduce order into the set by means of considerations of coherence. Such considerations involve various procedures, both logical and extralogical. The extralogical procedures can be grouped under the designation "considerations of plausibility" (1973, esp. Ch. IV, V).

Rescher (1992–1994: I:159) identifies three steps or tasks required by his coherence methodology:

1. To gather in all of the data (in the present technical sense of this term).

2. To lay out all the available conflict-resolving options that represent the alternative possibilities that are cognitively at hand.

3. To choose among these alternatives by using the guidance of plausibility considerations, invoking (in our present context) the various parameters of systematicity—regularity, uniformity, simplicity, and the rest—to serve as indices of plausibility.

Rescher describes in detail the formal mechanism of this procedure, which he appropriately terms (1992–1994: I:159) "inference to the best systematization."[17]

Further clarification is provided by a brief description of the step with which the systematization of the data begins. That step is a purely logical one that must be completed prior to and independently of the application of considerations of plausibility;

---

[17] Rescher's coherence methodology can be viewed as a more precise and detailed form of the procedure known as inference to the best explanation. See esp. Rescher (1973) and (1979).

it involves analyzing a set of data with respect to consistency. Here is a brief example (adapted from Rescher 1973: 55f.):

Data ("truth candidates")	Truths yielded by analysis for consistency
1. p, q	p, q
2. $p, \neg p$	—
3. $p, q, \neg q$	p
4. $p, q, r, \neg p \vee \neg q, p \vee \neg p$	$r, p \vee \neg p$

59   More complex data sets must be thoroughly analyzed so that their maximally consistent subsets can be exhibited. An extremely simple example: let there be the inconsistent set of propositions

$$S = \{p, p \rightarrow q, \neg q\}$$

This inconsistent set has three maximally consistent subsets:

$S_1 = \{p, p \rightarrow q\}$ (equivalent to $p \wedge q$)

$S_2 = \{p \rightarrow q, \neg q\}$ (equivalent to $\neg p \wedge \neg q$)

$S_3 = \{p, \neg q\}$ (equivalent to $p \wedge \neg q$)

This example makes clear not only that logic alone can proceed no further but also that logic is indispensable. Only extralogical considerations could make possible the acceptance of one of the maximally consistent subsets (in the example, only one of the three) and the rejection of the others.

At this point, two critical remarks on Rescher's method are appropriate. [i] Rescher ultimately fails to clarify the status of the datum. He takes this term as a technical one in a purely methodological sense: for him, "datum" is simply that X that can and/or should be brought into a coherence nexus. He speaks of the datum as a "proposition," but he does not make explicit whether he means that in the usual technical sense (i.e., as the *expressum* of a sentence) or in some other possible sense. He also fails to specify its semantic or ontological status. [ii] At the other extreme, Rescher never specifies just what he means by the "coherence" or "whole" to which he ultimately aspires.

Rescher's coherence methodology involves various other aspects and procedures. This brief description of the general idea behind this method sufficiently characterizes the first stage of the comprehensive philosophical method used by this book.

1.4.3 *Second Methodic Stage: Constitution of Theories*

Completion of the first methodic stage yields a "coherence framework" or a "structural framework"; thereby, one attains sentences that can be characterized as general and universal sentences, as formulations of laws, or, technically, as *structural* sentences. With these, however, the goal of the theoretical enterprise is not yet reached. The material

for theorization has been gathered and informally collected, but only a first, informal, and minimal formulation of theories has been achieved; the form that is adequate for theories is not yet explicit. The second stage of the method, the theory-constitutive stage, completes precisely this task: the transfer of all the theoretical material presented within the framework of the first stage of the method in an initial, informal, minimal   60
formulation into genuine theoretical form.

[1] But what is "genuine theoretical form"? The answer to this question ultimately depends on how theories are understood. On this question, opinions widely diverge, as is shown in the extensive Section 2.4. Nevertheless, already at this point some things may be said concerning the topic of method. In Section 2.4, the concept of the structural-systematic theory is characterized as a triple, $\langle L, S, U \rangle$, consisting of the components L = language, S = structure, and U = universe of discourse. One can also provide a simplified formula by taking the first two components together (i.e., by integrating language into the component of structure). The theory then appears as the ordered pair $\langle S, U \rangle$. This theory form is nothing other than the totality of the relations between these two components, along with all of the individual elements of each of the two components. As is shown above, the first stage of the method provides the material for the theory, minimally ordered. "Minimally ordered" here means that it is simply assumed that all the elements constituting this material fit together in one way or another. The development of the genuine theoretical form then requires that the precise form of this "fitting together" be made explicit.

The currently most important theory forms, from the strictly theoretical perspective, are the *axiomatic* form and the *network* form; the latter can also be designated as *coherentist*. Thus, one can speak of the axiomatic and the coherentist (or network) method.

[2] The axiomatic form or method is logically the most stringent and exact. To formulate an axiomatic theory, one requires, according to the general characterization of the concept "axiomatic," three factors: a language (the terms of the theory), a logic (the rules of inference), and the axioms of the theory. Thus, T is an axiomatic (or axiomatizable) theory if T is a class of sentences or formulas and there is a subclass A of T consisting of the axioms of T, such that all of the sentences or formulas of T are provable (or derivable) from A. In the strict sense, an axiomatic theory is thus a deductively closed class of sentences or formulas.

This strict sense of axiomatic theory is used primarily, if not exclusively, in the formal sciences. In its application to empirical sciences and to philosophy, however, the expression "axiomatization of a theory" does not have exactly the same signification.   61
Wolfgang Stegmüller has distinguished among and analyzed five significations or five forms of axiomatization (see *Probleme* II-2/1973: 34ff.).

The first form of axiomatization is the Euclidean form. It corresponds to the general characterization formulated above, with the exception of an important specification of the concept of the axiom: according to the axiomatics traditionally termed "Euclidean," axioms are self-evident and therefore sentences that are universally true. This understanding of "axiom" is maintained up to the second half of the nineteenth century, although in philosophy the term "principle" is generally used instead. Axioms or principles are then principles evident or known in themselves (*principia per se nota*).

Stegmüller terms the second form of axiomatization *informal Hilbertian axiomatics*, also characterizing it as *abstract axiomatics*—"abstract" in the sense of not being based on intuition—and thereby opposing it to Euclidean *intuitive* axiomatics. To be sure, Hilbert, in his *Foundations of Geometry* (1899/1902), wherein he presents this axiomatics systematically, understands axioms as assumptions about three classes of things initially given in intuition (points, lines, and planes), and about three "intuitive" fundamental relations among them ("lies between," "coincides with," and "is congruent with"). These intuitive notions, however, play no roles with respect to the formulation of the axioms. Although the normal model for the formulated axioms is the domain that is characterized by the initial intuitions, additional models, not all of which are intuitive, are also possible. This form provides the noteworthy extension of the arena for interpretation that characterizes the transition from Euclidean to modern axiomatics. The fundamental concepts that are used (in the case of geometry, the six concepts introduced above) are *defined* solely by the requirement that the axioms hold for them. For this reason and in this sense, the well-known and often-used formulation "implicitly defined" is introduced for the characterization of the fundamental concepts.

Stegmüller designates the third form of axiomatization *formal Hilbertian axiomatics*. It has the form of an axiom system Σ and is based on the construction of a formal language that is (initially) constructed only as the syntactic system S. Presupposing a notation (a collection of signs), the well-formed formulas of S are determined in a metalanguage by the identification of a subclass A consisting of the axioms of Σ. Finally, a class of rules of derivation R is provided, the application of which makes possible the derivations of formulas from other formulas (initially, the axioms). The axiomatic system Σ is thus the ordered triple $\langle S, A, R \rangle$, also termed a *calculus*.

62   Such a system, in the genuine or strict sense, must satisfy several conditions. Above all, the class of well-formed formulas and that of the axioms must be decidable and the rules of derivation must be effective. Other requirements of such systems concern the dependence and independence of the axioms, freedom from contradiction, and the completeness of the systems.

It belongs to the essence of the formal axiomatic system that it leads beyond itself, in that a model (more precisely, models), in the strictly logical sense, is (are) assigned to it. Such assignment transforms the purely formal (i.e., here, purely syntactic) system into a semantic system. The purely formal system consists solely of meaningless signs, or, more precisely, of the relationships among such signs. Models are introduced when the signs are interpreted, that is, when they are given determinate semantic values. An interpretation is a function that maps the (not purely logical) signs (and/or the various categories of the not purely logical signs) of the formal system to appropriate semantic values. One can then define what it means for a formula of S, on the given interpretation, to be true. A model is then an interpretation of S within which all of the axioms of Σ are true.

The fourth form of axiomatization described by Stegmüller is significantly different from the first three. Stegmüller calls it *informal set-theoretical axiomatization* or *informal axiomatization by means of the definition of a set-theoretical predicate*. This method is familiar in mathematics. Thus, for example, an informal set-theoretical axiomatization of group theory is provided by means of the definition of the set-theoretical predicate "is a group." One of the possibilities introduced by Stegmüller is the following (39):

X is a group if and only if there are a B and a $\otimes$ such that:

1. $X = \langle B, \otimes \rangle$;

2. B is a nonempty set;

3. $\otimes$ is a function with $D_I(\otimes) = B \times B$ and $D_{II}(\otimes) \subseteq B$;[18]

4. for all $a, b, c \in B$, it holds that $a \otimes (b \otimes c) = (a \otimes b) \otimes c$;

5. for all $a, b \in B$, there is a $c \in B$ such that $a = b \otimes c$;

6. for all $a, b \in B$, there is a $c \in B$ such that $a = c \otimes b$.

An example from the natural sciences, the axiomatic set-theoretical definition of the predicate "is a classical particle mechanics," is introduced in Section 2.4.2.3.1, which provides additional explanations of axiomatization by means of the definition of a set-theoretical predicate.

Stegmüller emphasizes that the "axioms" of this fourth form of axiomatization are "strictly to be distinguished from what are called 'axioms' in the other forms of axiomatization" (40). In Euclidean axiomatics, axioms are sentences; in informal Hilbertian axiomatics they are sentence forms; in formal Hilbertian axiomatics they are formulas; in informal set-theoretical axiomatics, however, they are components of the definition of a newly introduced set-theoretical predicate.

The concept of the *model*, as understood and used in the context of the fourth form of axiomatization, must likewise be sharply distinguished from the concept of the model within the formal system in the sense of the third (and in part of the second) form of axiomatization. In the fourth form, a model is "simply *an entity that satisfies the set-theoretical predicate.*" Stegmüller completes his characterization of set-theoretical axiomatization by introducing the concept of mathematical structure: "All the axioms describe a *mathematical structure* that is expressed in the totality of the relations expressed in the axioms."

Only "for the sake of completeness" (41) does Stegmüller mention a fifth form of axiomatization. He refers to Carnap's procedure for the introduction of an *explicit predicate* or an *explicit concept* for an axiom system and confirms that it is the formal counterpart to informal set-theoretical axiomatization. The latter uses only the expressive means of nonformalized "naïve" set theory, whereas the fifth form of axiomatization articulates an explicit predicate within the framework of a *formal* system of set theory.

[3] As the preceding account makes clear, different things are understood by "axiomatic method." Before elaborating on the consequences of those different understandings for the second stage of the philosophical method, a different conception of theory-form is briefly to be introduced. This is the only alternative to the axiomatic method or theory-form that is presumably to be taken seriously: the coherentist or network method or theory-form. It does *not* differ from the axiomatic method with respect to several factors introduced above in the description of the various forms of axiomatization; most importantly, like axiomatized theories (broadly understood),

63

---

[18] 'D_I' is the domain of arguments, 'D_{II},' the range of values of the function.

theories with the network form can (but need not) introduce and rely on formal languages and can therefore develop models, etc.

64   The only fundamental difference is that axiomatic systems and/or the axiomatic theory-form (at least in the sense of the first three forms introduced above) have an exclusively *hierarchical-linear* structure, whereas the theory-form relying on the network method does not. Nevertheless, in a (comprehensive) theory with the basic form of a network, individual (local) hierarchical-linear and thus axiomatic interconnections are possible or can indeed be adequate. The deductive structure that characterizes strictly axiomatic theories is in one respect the consequence, result, or expression of the hierarchical-linear character of the theory, whereas in another respect it is the basis of this hierarchical-linear character. This should make it clear that the axiomatic method is characterized precisely by its moving from a (finite) series of basal theses (the axioms) to additional theses (the theorems). The coherentist or network method is based on a wholly different structure, i.e., the totality of the inferential interrelations of the theory; this method makes such a structure explicit. A diagram of Rescher's[19] makes these interrelations explicit (Figure 1.1 depicting the linear-hierarchical axiomatic method, Figure 1.2, the network-forming coherence method):

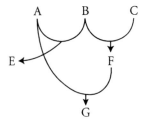

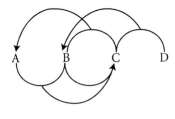

Figure 1.1   Figure 1.2

[19] For a thorough characterization of the coherentist or network method, see Rescher (1979), especially Chapters III and IV). To be sure, Rescher's conception is fundamentally *epistemically* rather than *ontologically* oriented. That is to say, Rescher characterizes the status of "systematicity" as "an epistemic desideratum for our knowledge regarding nature" *and not* as "an ontologically descriptive feature of nature itself" (115). But he elaborates on this thesis by saying that ontological systematicity is a "causal precondition of inquiry" (121) into nature and that cognitive systematicity is an "indicator of ontological systematicity" (124). This position is, however, not entirely intelligible, because it is not clearly coherent. How can ontological systematicity be a "causal precondition" for cognitive systematicity if ontological systematicity does not clearly exist and therefore cannot be assumed? Rescher's vacillating position is considered more closely in Section 5.1.2.2[5]. Here, it need be added only that it is a deficient solution to a problem that lies deeper: the problem of the relation between theory and the universe of discourse ("world," "actuality," "being"). This problem is articulated in the main title and the subtitle of this book and constitutes the central problematic of the structural-systematic philosophy here presented. As this book shows, the key to an adequate solution of this highly complex problem lies in the full acknowledgment of the necessity and the central importance, for every theory, of the concept of the theoretical framework, along with the full cognizance of the ontological consequences that arise from the plurality of actual and possible theoretical frameworks.

[4] What does all of this yield with respect to the second methodic stage of the   65
structural-systematic philosophy? To provide a well-founded answer to this question,
one must keep in mind the following two theses or insights: [i] the axiomatic method
(especially in the first three of the forms described above) is the most logically exact
of all; it is therefore the most demanding and appropriate theory-form, logically and
mathematically. The network method, in contrast, is far from having a comparably
unambiguous logical-mathematical status. [ii] It cannot be assumed that the relations
among the elements of a theory *qua* form of presentation *and* among the objective
(ontological) elements thereby articulated always simply have a linear-hierarchical
structure. Therefore, one must assume that there is *at least in the cases of many, if not
indeed in the cases of most subject matters*, a web (a network) of relations that therefore
cannot be understood and articulated in a linear-hierarchical manner. The more comprehensive a theory is, the less probable is it that it can be articulated in accordance with
the axiomatic theory-form. To be sure, one could consider expanding the number of
axioms, potentially infinitely, in order to articulate axiomatically the entire complexity
of the web of relations to be thematized by the theory. However, there are at least two
reasons for not doing this. First, articulating a number of axioms extending to infinity
would be impracticable with respect to the development of theories. Second, even an
arbitrarily large number of axioms could not comprehend the entire web of relations,
because, as Figure 1.2 presented above shows, many relations have a cyclical character:
they articulate cases where entities mutually support one another.

The two points just introduced support several important conclusions about the
philosophical method described here. In cases in which what is to be articulated by a specific theory have a linear-hierarchical structuration, the axiomatic method or theory-form
is the suitable one; in all other cases, the coherentist or network method or theory-form is
appropriate. For philosophy as a whole (i.e., as a comprehensive theory), only the coherentist method or theory-form can be used (this issue is treated extensively in Chapter 6).[20]   66
With those local theories for which, as is explained above, the axiomatic method or
theory-form is possible, the informal axiomatic theory-form is as a rule fully sufficient.

[5] A concluding remark on the concepts *formal language* and *formal system* is
appropriate at this point. A formal *system* is a totality of formulas presented in a formal
language that consists of pure (i.e., meaningless) signs. Because the signs as such are
meaningless, only the relations among them are relevant. In the terminology of this
book, the formulas of such formal systems express purely abstract structures, that is,
ordered pairs, each consisting of a set or universe whose elements are pure (meaningless) signs and a set of relations (and/or functions). Such structures, as is shown in
Section 1.2.3, are *pure* or *abstract* structures. If one says that they abstract from the
elements of the universe, then that is, if properly understood, correct. This means
that abstract structures are such that they can be applied to concrete contents. In the

---

[20] Rescher (1979: 45) quite accurately characterizes this state of affairs:

> [W]hile a network system gives up Euclideanism at the *global* level of its over-all structure, it may
> still exhibit a *locally* Euclidean aspect, having local neighbourhoods whose systematic structure is
> deductive/axiomatic. Some of its theses may rest on others, and even do so in a rigorously deductive
> sense. For a network system may well contain various deductive compartments based upon locally
> operative *premises* rather than globally operative *axioms*.

presentation of pure or abstract structures, one uses the artifice of articulating this feature of the structures by means of meaningless signs.

In the usual terminology, it is said that a formal system attains meaning in being provided with a model and that a model is an interpretation of the signs that, on the purely formal level, are meaningless. This is no doubt correct, but is formulated somewhat elliptically, because it explains only what happens to the initially meaningless signs: in that they are interpreted, they become meaningful. But it is not explicitly shown and said what the pure or abstract structures themselves are or what happens with those structures. The formal (in the sense of logical-mathematical) signs that are used to present the pure or abstract structures themselves of course have meanings, of a quite specific sort: they articulate networks of relations and/or functions (operations) and thus pure structures. These matters are presented more extensively and more precisely in Section 3.2.1, which describes and analyzes formal (logical-mathematical) structures.

67   The preceding considerations have several consequences that are quite important for the structural-systematic philosophy presented here. Two are now briefly to be introduced. The first is the significant fact that the structural-systematic philosophy *explicitly* implies and recognizes the possibility, in principle, that all its sentences (theses, theories) be subjected to precise and detailed analysis on the basis of all currently available theoretical mechanisms. Among such mechanisms belong the development and application of formal languages and/or formal systems. The second consequence is the insight that not everything that is possible in principle must or indeed even can be realized in a sensible manner. Philosophy is fundamentally a contentual discipline or science, not a formal one. The question, then, is that of the extent to which it lies within the concrete possibilities of philosophy to develop absolutely all of the aspects and details of its own sentences (theses, theories). The achievement of such an absolutely comprehensive development must be understood to be a kind of theoretical ideal and thus something whose realization is always a desideratum; even when not realized, however, that achievement is a theoretical ideal that has a significant regulative and clarificatory function with respect to the theoretical enterprise that is philosophy.

1.4.4 *Third Methodic Stage: Systematization of Theories*

At the first and second stages of the method (or with the first and second methods), individual theories are developed, each of which concerns a determinate theoretical domain. This question then presses, how do these individual theories relate to one another, or, differently put, how can the individual theories come to constitute a comprehensive theory, a network of theories or theoretical network, or—precisely—a system? The question first arises with respect to the informal theories initially formulated at the first stage, but the question presses the more urgently—and indeed becomes unavoidable—with respect to the properly formed theories that emerge from the second stage of the method (or from the application of the second method). Nevertheless, at least in philosophy it is neither requisite (nor realistically practicable) to accomplish explicitly what the second stage requires; such accomplishment would occur only in the ideal case.

For the most part, how theories relate to each other, introduced above as a pressing question, is addressed to informal theories. The method that responds to the question is the *system-constituting* method, which has a twofold character: first, it develops a structure of structures that it presents as a network of theories, and second, this

68

network of structures proves to be holistic—to be a whole (in a sense that must of course be clarified). The network model described in the preceding section also applies here—and in a certain respect particularly here, because the theories developed at this stage concern not individual states of affairs, but instead entire theories that themselves are interrelated within a far more comprehensive theoretical network. At this point, neither is it necessary nor would it be reasonable to explain this method in detail, because that would require reliance on many quite important factors that are treated only in the following chapters. This topic is treated extensively in Chapters 5 and 6.

1.4.5 Fourth Methodic Stage: Evaluation of the Comprehensive System or Network with Respect to Theoretical Adequacy and Truth-Status

The development of one or more theories, even in the theory-constitutive and system-constitutive sense that guides the third methodic stage, does not complete the theoretical undertaking. It cannot, because the project pursued here is not foundationalist, but instead coherentist, which means that there is no reliance on putatively fundamental truths that are presented at the outset as true, that hold throughout, and that, in conjunction with logical procedures of one sort or another, would establish as true, as they emerged, all of the additional theses or theories that were components of the ultimate system.

The coherentist project raises the genuine questions of theoretical adequacy and of truth only *at the end* of the theoretical undertaking (i.e., only after the development and presentation of a given theory). Methodologically, this is the point of application of the fourth stage of the comprehensive systematic-philosophical method. This procedure is best understood as an additional form of inference to the best systematization.

The question at the end concerns the confirmation of the theory that is presented as the comprehensive theoretical network. This question is extraordinarily complex. Even its adequate formulation requires more precise determination of what is meant by "theory." If "theory" were understood as an uninterpreted formal system (of sentences), the question would initially be formulated as follows: does the theory have one or more models? The next question would be whether or how one could develop the intended model(s). Decisive here is just how one understands "theory": for a theory in the sense of the "statement view," the questions of adequacy and truth are different from those for theories understood in the sense of the "non-statement view." This issue is considered in detail in Chapter 2.

The foundationalist project, in contrast, must confront the question of justification or grounding—must demonstrate its theoretical adequacy—at the outset, because this project depends on the truth of its initial sentences (be they axioms or empirical sentences) as somehow confirmed. That this project is untenable is clear from the recognition that the presentation even of the simplest sentence can be made only within some determinate theoretical framework, which is thus presupposed and therefore cannot be simultaneously justified or grounded. A foundationalist project could avoid the insuperable aporias that thereby arise only if it were understood as follows: the "truths" assumed at the outset would be viewed as absolutely self-evident, wholly independently of whether they were explicitly or implicitly articulated linguistically, logically, conceptually, etc.; they would be, so to speak, "truths" utterly unrelated to what this book terms any theoretical framework. These could be designated "bare truths."

Such a position would, however, be on the one hand wholly dogmatic and on the other deeply incoherent. It would be dogmatic in arbitrarily accepting its bare truths as, so to speak, floating in empty space, but nonetheless as definitive. The relevant "empty space" would be a putative locus for truths that would have to be devoid of any theoretical (linguistic, logical, epistemic, etc.) elements. Why should one acknowledge that there are or even could be any such "truths"? The position would be incoherent in that as soon as its bare truths were articulated, the problem would arise of whether the theoretical (linguistic, logical, semantic, etc.) *articulations* of these truths were *accurate to them*—a problem that would reintroduce the problem that was supposed to have been solved by the putative self-evidence of the putative truths.

In the structural-systematic philosophy, the fourth methodic stage is applied both on the *comprehensive theoretical level* and on individual *subsystematic levels*. In the former case, the question concerns the grounding of the *entire* systematic philosophy presented here (it can therefore be termed "comprehensive systematic grounding"). In the latter case, the question concerns the grounding of the individual theories that constitute the parts of the system ("subsystematic grounding(s)"). Neither of these groundings is absolute in the genuine sense, because each is relative to the relevant theoretical framework. Comprehensive systematic grounding is relative in only one way: it is relative to the comprehensive systematic framework. Subsystematic groundings have twofold relativities: they are relative both to their own subsystematic theoretical frameworks and to the comprehensive systematic framework. This cluster of issues, particularly comprehensive systematic grounding, is treated extensively in Chapter 6.

The general problem of the concept or process of grounding is treated in the following section, which does not extensively consider contemporary discussions of the problem. Section 1.5 and Chapter 6, in conjunction, provide the most comprehensive possible clarification of the fourth stage of the method.

1.5 (Self-)Grounding of Systematic Philosophy?

1.5.1 *On the Concept of Grounding in General*

This book uses the term "grounding" to designate a complex of issues that are usually presented as issues of *justification*. The book uses the less common term because of problems with the traditional terminology.[21] The problems begin to come into view when one notes that *justification* is a pragmatic notion (i.e., one that involves a relation to a subject and whose expression therefore requires an at least three-placed predicate; for example, "reason x justifies subject y in holding belief z"). What this book terms *grounding* in contrast, need not be pragmatic: thesis t may be well grounded whether or not a given subject recognizes it to be so. Unfortunately, *grounding* (like its German counterpart, *Begründung*) can connote *founding* or *establishing*, in the sense of providing

---

[21] The ambiguity of the term "justification" has led various authors to introduce wholly new terms to articulate their epistemic conceptions. Alvin Plantinga (1993a, b), for example, uses the term "warrant" in the epistemic theory he presents in two extensive volumes.

a firm or even unshakable foundation. This connotation is unfortunate in that, as is indicated already in the Introduction, this book rejects all forms of foundationalism (additional details are presented in the following subsections). *Grounding*, then, is chosen for lack of a better term and is used in this book in several senses that are clarified in this section. In the broadest sense, the one intended when the term appears without a qualifier, it is comprehensive in a specific respect: groundings include justifications, putative foundations, and various other procedures including those characterized in the following paragraphs, all of which can be taken to serve to clarify the theoretical statuses of truth-candidates (sentences, theories, or even comprehensive systems) with respect on the one hand to subjects as theoreticians, and/or on the other to the internal coherences of the theoretical configurations within which the truth-candidates emerge.

The first step toward the further clarification of groundings requires clearly and strictly distinguishing justifications as pragmatic from nonfoundational modes of grounding that are in no way pragmatic, but instead are appropriately termed *objective*. The strongest form of objective grounding is *proving* in the strict sense, according to which proofs are logically conclusive inferences. Other forms of objective grounding relied on in philosophy differ from proofs in *not* being logically conclusive; these other forms include inductive inference, material inference, and—of particular importance to the structural-systematic philosophy—inference to the best explanation or systematization.[22]

Objective groundings are explications of inner-theoretic or innersystematic interconnections that satisfy specific requirements established by the theories or systems—better, the theoretical frameworks—within which they develop. Thus, for example, the Pythagorean theorem is grounded, within the theoretical framework of Euclidean geometry, by its being *proved*; because the status of that theorem, within its framework, is fully determined, to ask in addition for justification—to ask what justifies the theorem, within the framework—would be senseless.

Objective grounding is not automatically justification (to or for a specific subject). A given objective grounding is a justification for a given subject only if, first, an additional premise is both introduced and at least implicitly accepted by that subject. The premise might appear to be the following: objective groundings provide the subject with compelling reasons for accepting their conclusions. This formulation of the premise is, however, misleading because of the way in which it mixes the pragmatic with the purely theoretical. As is noted above, the question of the status of the Pythagorean theorem within the theoretical framework of Euclidean geometry is in no way a *pragmatic* question; it is a purely *theoretical* one, whose answer is therefore determined by the framework itself (its definitions, axioms, rules, etc.), in conjunction with the relevant data (whose scope is also determined by the framework). The additional premise, then, involves the status of the framework. For the subject who—for whatever grounds or reasons of whatever kinds—does not accept the

71

---

[22] In most philosophical literature (including literature on logic and on the philosophy of science), the fact that proofs and other forms of inference (or objective grounding) are distinct from justifications is either not seen or is scarcely considered; often it is obscured by ways of talking that ascribe especially to proofs nearly mythical powers (such as the power of being incontrovertible justifications).

framework of Euclidean geometry, the fact that the Pythagorean theorem is proved within that framework is not a ground or reason to accept the theorem as true. Thus, more generally, the subject who rejects the theoretical framework within which (for example) a proof develops may well acknowledge that the series of sentences in question satisfies all the criteria required by this theoretical framework for proofs—that the series of sentences *is* a proof, or that the premises are axioms or theorems and that the conclusion follows from them—but can nevertheless reject the theorem.

The chief reason that objective grounding (including proof) and justification are so often confused is that the issue of theoretical frameworks is neglected or not even seen. Whenever this issue does not arise, and especially where a single theoretical framework is (usually tacitly) presupposed, proofs (especially) can appear simply to *be* justificatory. It must, however, be emphasized that even if one agreed in addition that *only* proofs were adequate providers of justification, a conceptual distinction between proof and justification would remain. To accept a proof as justificatory, the subject would have to accept an additional, extraordinarily important premise that is not a part of the proof: the premise ascribing to the proof's premises not the status merely of axiomatized hypotheses, but instead that of truths. For the subject who has made this ascription, all that can be derived from the premises must indeed be accepted as simply true. This subject has thereby accepted—at least implicitly—yet another premise: that every sentence that has the status of being a derived or deduced truth within the relevant theoretical framework must be accepted by this subject as true. Asked about the justification of this last premise, the subject—if, as is here assumed, he or she is fully rational—will rely on a final, highest premise: *veritas [demonstrata, manifesta…] index sui et falsi* (i.e., every sentence whose truth is manifest [here in the form of being a deductively attained logical conclusion] bears within it the ground for its acceptance). Having been proved or otherwise adequately grounded, objectively, *within an accepted theoretical framework*, it does not require *additional* justification; under these circumstances the objective grounding, within the framework, itself also functions as a justification.

1.5.2 *The Problem of Grounding in Philosophy*

The problem of grounding, often in the self-referential form of self-grounding, and then at times in the extreme form of absolute self-grounding (which could also be termed "absolute self-establishment"), is among those that have shaped the understanding and the task of systematic philosophy, particularly in modernity. Treatments of this problem, which have continued to appear since the time of Kant and German Idealism, have appeared in the quite recent past and into the present in increasing numbers and with great fanfare.

Within analytic philosophy, the problem is usually considered not in the extreme form of an absolute self-grounding, but in the restricted (and pragmatic) form of the justification of individual sentences and conceptions (theories). Even here, however, the problem of an absolute self-grounding is present, albeit generally only in an implicit manner. The following account does not consider the discussions in the current literature in anything approaching their full breadth. Instead, it presents the problem of grounding on the basis of the idea of the systematic philosophy envisaged here; this presentation requires the introduction and clarification of several fundamental distinctions.

1.5.2.1 On the Nonsystematic Concept of Grounding

The extreme degree to which the concept of (self-)grounding (or self-justification) is problematic stems from the fact that this concept is usually used in a purely intuitive, underdetermined, and, in a certain respect, absolute manner. It is such a concept—one used without reference to any adequately determined theoretical framework (whether implicitly presupposed or explicitly developed)—that is here termed the "nonsystematic concept of grounding." It comes into play when a philosopher has maintained something or other and is immediately challenged to *justify* (or ground) it. This challenge, in its utter generality, is senseless because it is insufficiently determined. What is the demand for "justification" (or "grounding") supposed to involve or require? If the concept is to have even a minimal content, it must be acknowledged to rely on a lengthy series of fundamental assumptions in the domains of logic, philosophy of language, epistemology, and even metaphysics. Formulated in the terminology used in this book: the concept of grounding (or justification) presupposes an adequately determined theoretical framework. This is clear from the fact that the user of the concept can and indeed must be asked, in every instance, what is to be understood by "justification/grounding"? Any adequate response will unavoidably appeal to a series of fundamental assumptions from the various disciplines just mentioned. The user must be prepared to identify the language being spoken; to specify the syntax, semantics, and pragmatics of this language; to describe the logic being relied upon, the conceptuality being invoked, etc. This makes it quite clear that the attempt to apply the concept of justification (or grounding) in this "absolute" manner—right from the beginning and with universal applicability—is doomed to failure. Justification (or grounding) can sensibly be demanded or provided only within the context of an adequately determined theoretical framework.

The considerations of the preceding paragraph yield, as the positive task for this book, the presentation of a *systematic* concept of grounding. Because, however, significant arguments have been introduced purporting to show that *no* concept of justification (or grounding) can be adequate, an initial *negative* task for this book is to show that these arguments fail to establish the futility of the project of adequately determining a systematic concept of grounding. The two arguments to be considered are [1] that of Hans Albert, according to which there are only three possible ways of providing ultimate grounding, each of which fails; and [2] the argument presented by Leonard Nelson and—in a completely different form—by Richard J. Ketchum, according to which developing or defining the concept of grounding is inevitably aporetic.

[1] In one of its more precise forms, the nonsystematic problem of grounding is understood to arise in conjunction with the procedure of deductive argument (i.e., deriving or deducing conclusions from premises). Given the demand that everything be grounded, if it is granted that the conclusion of a given argument is grounded by its premises, grounding is then required for the premises themselves. This leads to the problem Hans Albert (1968: 13) terms "the Münchhausen trilemma." The trilemma consists of three possibilities for ultimate grounding, none of which is satisfactory. The first perpetuates the demand for grounding: each premise must become the conclusion of an additional argument, but because this leads to an infinite regress, the process never provides genuine grounding for anything. The second possibility requires appealing to statements that are presented as self-grounding, but this is logically circular and

74

therefore also a failure. The third possibility involves simply stopping the procedure at some point; because the stopping point is arbitrarily determined, the resulting position is simply dogmatic.

Various attempts have been made to avoid or to move beyond Albert's trilemma; among the best known is that of Karl-Otto Apel, with his "transcendental pragmatics" (see Apel 1973/1980, 1990, Dorschel, Kettner, Kuhlmann, Wolfgang, and Niquet, 1993). Apel's basic strategy is the attempt to show that rejection of the possibility of an ultimate grounding (and indeed in *all* domains of knowing) involves a *performative self-contradiction*. This strategy is based on three theses. Apel first maintains that grounding does not necessarily consist in something (a true sentence) being derived from something else (from one or more true sentences serving as premises). Next, he contends that there is a determinate level beyond which it is impossible to go; this is the level of argumentation and of the necessary presuppositions thereof—on the whole, it is the level of the presupposed language, a level to be understood on the basis of the transcendental synthesis of the interpretation of signs. Third, he holds that these first two factors cannot be interpreted correctly within the framework of any philosophical solipsism, but only within some intersubjective framework. According to Apel, the maintaining of a sentence, a thesis, is a part of a *public* discourse; as such a part, it rests on immensely consequential presuppositions. From this he concludes that every rational participant in such a public discourse has *always already* made and accepted specific fundamental assumptions; any participant who rejected or denied them would commit a performative self-contradiction. This shows, according to Apel, that an ultimate grounding is not only possible, but absolutely unavoidable. The level of publicity of theoretical discourse is one beyond which no theoretician can possibly go; for this reason, the presuppositions that this level contains cannot be revised. Thus, an absolute foundation for truth is given that, according to Apel, completely avoids the Münchhausen trilemma.

Apel's argument does not, however, go far enough: he does not take into consideration the fundamental fact that *every* utterance, *every* assertion, etc., *always already* emerges within a determinate theoretical framework. If and insofar as *every* utterance is public and thus intersubjective, *and if and insofar as* the other participants in the relevant discourse presuppose and accept *exactly the same theoretical framework*, then the denial of any or all of the presuppositions of such a discourse indeed constitutes a performative self-contradiction. But Apel overlooks the fact that this is only a *relative*, not an *absolute* self-contradiction: the self-contradiction is *relative* to the presupposed theoretical framework. There is, however, a plurality of theoretical frameworks, and that this plurality exists introduces extremely difficult problems (problems thoroughly considered in Chapters 5 and 6). What functions as an ultimate grounding *within the scope of a determinate theoretical framework* so functions only *relative* to this theoretical framework, not absolutely. This fact totally undermines Apel's conception of an "ultimate transcendental-pragmatic grounding."

[2] That attempts to clarify concepts of grounding (as of justification, knowledge, epistemic criteria, etc.) are inevitably aporetic is argued in various ways throughout the history of epistemology.

[2.1] For example, in 1908 the German philosopher Leonard Nelson (1908: 444) presents his "proof of the impossibility of epistemology." Nelson understands *epistemology* as "the science whose task is investigating the objective validity of knowledge." This task arises from the assumption that "the presence of objectively valid knowledge presents a problem." Nelson maintains that "a scientific solution of this problem is *impossible*." His "proof" runs as follows:

> Assume . . . that there were a criterion that could serve to solve the problem. This criterion itself would either be something known, or not.—Let us assume that the criterion in question *is* known. It would then fall into the domain of problematic items the validity of which is supposed to be determined by epistemology. Thus, the criterion that was supposed to solve the problem cannot be known.—Let us therefore assume that the criterion is *not* known. In order to solve the problem, it would have to be recognized, i.e., it would itself have to be capable of becoming an *object* of knowledge. But whether the knowledge whose object is the criterion in question is itself valid would have to be determined in order for the criterion to be applicable. This determination would, however, require the application of the criterion.—*Establishing the objective validity of knowledge is therefore impossible.*

76

Is the problem Nelson deems insoluble a genuine problem? That depends on how the entire epistemic dimension is understood. The considerations that follow establish that the understandings of knowledge and therefore of epistemology that Nelson presupposes do not survive systematic scrutiny. Their failure to do so reveals that the task he takes to emerge from epistemology is based in a misunderstanding.[23] Nevertheless, if epistemology is understood as Nelson understands it (i.e., if it is independent of any and all presuppositions and theoretical frameworks), then his "proof" is conclusive. But, as made clear below, such an understanding of epistemology is fully incoherent and therefore untenable.

[2.2] Another wholly analogous problem—in a certain respect a variant of Nelson's problem—is more interesting and relevant with respect to the positive conception developed below (Section 1.5.2.2). This problem concerns the thesis that there simply can be no theory, analysis, or definition of "justification." The consequences of this thesis are considerable.

[2.2.1] Richard J. Ketchum (1991) presents a clear and astute defense of this thesis. Ketchum begins the "strict version" of his argument by clarifying the phrase "theory, analysis, or definition of justification."[24] He proceeds as follows (46):

> The task is to define this phrase broadly enough to include any answer that an epistemologist is likely to propose to the question, "What is it to be justified in believing something?" and yet precisely enough so that claims about any proposed analysis can be evaluated. By "theory, analysis or definition of justification" I will mean a clarifying, lawlike[25] sentence of the form:

> G: (S)(p) [S is justified in believing p true ——— if and only if —p—S—]

77

---

[23] See Franz von Kutschera's on-target critique (1982a: 47ff).

[24] The thesis speaks of *justification*, but the arguments, if conclusive, would extend to *grounding*.

[25] For purposes of the current discussion, it suffices to note that the qualifiers "clarifying, lawlike" rule out such utterly unilluminating "definitions" as "S is justified in believing p true if and only if S is justified in thinking p true"; see Ketchum (1991: 47–48).

Ketchum's thesis is that G cannot be the schema of any "theory, analysis, or definition" of justification, because G, when applied, leads to a peculiar but crippling circularity. He argues as follows:

> Let P be a clarifying sentence of the form, G, let L be a person who believes P true and let ΦLP be the *analysans* of P instantiated with names of L and P. L cannot become rationally convinced that he is justified in believing P true by means of the following argument:
>
> P
>
> ΦLP
>
> Therefore, L is justified in believing P true. (48)

Ketchum introduces two possible instantiations of P: P_1 and P_2.

> P_1: S is justified in believing p true if and only if S has a clear and distinct idea of what is expressed by p. (48)
>
> P_2: S is justified in believing p true if and only if S was caused by a reliable process to believe p true. (49)

The instantiation of G with P_1 and P_2, respectively, for P, and "I" for L, yields the following arguments:

> (I)　P_1
>
> I have a clear and distinct idea of what is expressed by P1.
>
> So, I am justified in believing P_1 true. . . .
>
> (II)　P_2
>
> I was caused to believe P_2 by a reliable process.
>
> So, I am justified in believing P_2 true. (49)

As Ketchum shows, the arguments beg the question in an unusual but quite determinate sense:

> If my interest in an argument is to use it either to become rationally convinced in believing its conclusion or perhaps to discover whether or not I am justified in believing its conclusion, I can legitimately assert as premises only statements which I believe myself to be justified in believing. A glance at the arguments reveals, however, that they each contain a premise (P_1 and P_2, respectively) that no rational person would assert *as premises in those arguments* unless s/he already believed their respective conclusions. (50)

On the basis of his proof of the impossibility of any "theory, analysis, or definition of justification," Ketchum concludes that philosophical epistemology must be abandoned. The alternative he offers is *naturalism* in Quine's sense:

> [N]aturalism: abandonment of the goal of a first philosophy. It sees natural science as an inquiry into reality, fallible and corrigible but not answerable to any supra-scientific tribunal, and not in need of any justification beyond observation and the hypothetico-deductive

method.... Naturalism does not repudiate epistemology, but assimilates it to empirical psychology. (Quine 1981: 72)[26]

Ketchum correctly remarks that this form of naturalism is a rejection of *epistemology* as first philosophy, but fails to note that it does not thereby rule out the possibility of a *metaphysical* first philosophy. Thus, Quinean naturalism is not the only alternative to epistemological first philosophy.

This book joins Quine and Ketchum in rejecting the thesis that epistemology could qualify as first philosophy, but diverges from them in that it rejects naturalism as well; its alternative is the introduction of a different understanding of grounding, one involving what it terms the *systematic* concept of grounding.

[2.2.2] Various philosophers have attempted to counter Ketchum's argument and with it his naturalistic position. Perhaps the most interesting criticism is that presented by Robert Almeder (1994). Almeder introduces two "acceptable" answers to Ketchum's demand. The first attempts to expose a contradiction within Ketchum's position, which Almeder identifies with the inconsistent position of "the general sceptic on the question of defining justification," that is, "the logically contradictory position of asserting as logically privileged or justified the claim that no beliefs are demonstrably and non-arbitrarily justified" (677). Almeder elaborates: "[I]f one's view is that what is wrong with generalized scepticism on the question of justification is that it is logically self-defeating because self-contradictory and incoherent, then that in itself would serve as a fine *reductio ad absurdum* of the need to justify one's definition of justification assuming that there is no non-question-begging definition to be offered" (678).

This first response offered by Almeder is far from compelling. It takes Ketchum's position to be self-contradictory in accepting, as logically privileged and thus as justified, the thesis that there are no beliefs that are demonstrably and nonarbitrarily justified. But this reading misses the main point. Ketchum does not present any such thesis, nor need he. Ketchum's argument demonstrates only a performative incoherence in the attempt to provide a comprehensive definition for the concept of justification (i.e., a definition that could be applied comprehensively). The argument (the thesis) is not presented as "logically privileged or justified"; the defender of the argument does not apply the concept of justification to the argument. Nor is it necessary that it be so applied, for the argument aims to *prove* the conclusion that a comprehensive definition of the concept of justification is not possible. The conclusion itself, as putatively *proved*, is not subsumed under a concept of justification that is in any way presupposed.

Almeder presents his second response as "more compelling," although not as more readily apparent. It reads as follows: "the question 'Are you justified in accepting your definition of justification' is actually self-defeating in a way not at all obvious." It becomes self-defeating if the one raising the question (henceforth, "the critic") is asked *what the critic* means by "justification." If the critic then fails to explain what he or she means by justification, then—given the plurality of candidate definitions of "justification"—the question is ambiguous and does not call for a response. If however the critic *does* offer an explanation, then the critic becomes open to his or her *own* initial question, that is, "Are you justified in accepting your definition of justification?"

---

[26] Ketchum (59) quotes the parts of the passage preceding the ellipsis.

80   Almeder concludes from all this that the person to whom the critical question is raised (henceforth, "the interlocutor") can and should simply ignore it as senseless.

Almeder does consider the possibility that the critic might respond to the counterdemand differently—with "Oh well, by 'justified' I mean just what you mean by 'justified.'" In this case, it might appear that the critic would be compelled to clarify neither what he or she understood by "justification" nor whether he or she deemed the concept meaningful, useful, or necessary. Almeder's response to this possibility is not convincing. He writes,

> [I]f his definition of justification is the same as mine, then his question amounts to an oblique way of asking "Are *we* justified in accepting *our* definition of justification?" To that question I can legitimately respond "What do *we* mean by 'justified' when we ask whether *we* are justified in accepting our definition of justification?" Presumably, answering this last question is simply a matter of repeating our original definition of justification. Otherwise our answer would be contradictory. But now does it make any sense to ask, "Are we justified in accepting this latter definition of justification?" if, as the original questioner asserts, it cannot be answered in a non-circular way? Thus the question "Are we justified in accepting our definition of justification" makes sense only if we have *already* justified our definition of justification, thereby rendering the question unnecessary. Either that or the question makes no sense whatever because the question demands an answer to a question which the questioner also claims cannot be given in a non-circular way. (680)

The problem with this response is that the critic, in specifying that the initial question concerns the interlocutor's concept of justification, need not make any commitment to that concept. The philosophically interesting response is, "Is your position—the position of a philosopher who defends a comprehensive definition of justification (i.e., *any* such definition whatsoever)—ultimately coherent? Can one defensibly take this position?" With this question—whose meaningful (and not inconsistent) character is incontestable—the critic relies on no concept of justification whatsoever.

What Almeder concludes from his two responses is the following: the question "Are you justified in accepting your definition of justification?" is senseless, because it is incoherent and contradicts itself. But he then adds some reflections that cast new light on the entire problematic of justification:

> Of course, none of this implies that we have no way of assessing the relative merits of mutually exclusive definitions of justification. What it does mean, however, is that when we are involved in the practice of making such assessments we remind ourselves that we are
81   not involved in the activity of justifying our definition of justification. Rather we are doing something else. We are determining which definition, if any, to accept as a more or less adequate generalization of our collective usage and practice in the relevant contexts. We do this because we believe that when it is done well it will produce a measure of understanding and enlightenment not otherwise available. Whether we are justified in this latter belief is an interesting question, but it is certainly not the question of justifying our definition of justification. For all the reasons mentioned above, we must not succumb to the temptation to characterize this activity as the activity of justifying our definition of justification. (681)

Interpreted literally, these remarks—like those in the preceding quotation—are not at all illuminating. Worse yet, Almeder's extremely condensed and indiscriminate formulations could easily lead readers to overlook what they involve and what they imply. Making these easily overlooked matters clear requires a careful, step-by-step analysis.

According to Almeder, there is something or other that qualifies as a "collective usage and practice" with respect to the term or concept "justification." Let us assume that we are confronted with a series of definitions (*definientia*) D_1, D_2, D_3 ... of this concept we putatively rely on in "our collective usage and practice"; let us then term *this* justification "justification$_1$." Let us further assume that we decide on D_1 as defining justification$_1$, and that D_1 (P_1, in the example from Ketchum cited above) reads:

> (D_1) S is justified in believing p true if and only if S has a clear and distinct idea of what is expressed by p.

Almeder does not make explicit which possible definition of justification he would accept as the one best fitting "our collective usage and practice"; D_1 is used here simply for purposes of illustration, but so using it conflicts in no way with Almeder's account. Although he has deemed the question demanding a justification for the definition of justification$_1$ to be senseless, he deems it possible and indeed requisite to engage in "assessing the relative merits of mutually exclusive definitions of justification." Such an "assessment," however, is nothing other than an answer to the question, which of the different and competing definitions of justification is best *justified*? In other words, it is the result of the consideration of the *justification* of a specific definition of justification.

Almeder characterizes what he terms "assessment"—which, as has just been shown, is a form of justification—as follows: "We are determining which definition, if any, to accept as a more or less adequate generalization of our collective usage and practice in the relevant contexts." This makes clear that this "assessment" or justification is nothing other than the application of a specific definition of justification; let us term *this* justification "justification$_2$," and the corresponding definition, "D_2":

> (D_2) S is justified in believing D_1 true if and only if D_1 provides a more adequate generalization of our collective usage and practice (with respect to "justification") in the relevant contexts. 82

Interestingly, Almeder does not stop here; instead, he next considers *this* concept of justification—thus, justification$_2$ or D_2—in that he recommends a *reason* for accepting it. Presenting such a reason is, however, implicitly responding to the question of D_2's *justification*. And indeed, Almeder introduces what he himself presents as an answer to a "why" question: "because we believe that when it is done well it will produce a measure of understanding and enlightenment not otherwise available." This introduces a new justification—"justification$_3$"—formulable in a new definition (D_3):

> (D_3) S is justified in believing D_2 true if and only if D_2 provides a degree of understanding and enlightenment not otherwise available.

Symptomatically, Almeder does not leave the matter even at this. He speaks explicitly of the "justification" of justification$_3$ (or D_3) in noting, "Whether we are justified in this latter belief [i.e., the acceptance of justification$_3$/D_3] is an interesting question." There is, then a "justification$_4$." Yet, although he has made justification$_2$ and justification$_3$ explicit, providing bases for their definitions, Almeder simply stops with justification$_4$, breaking off the inquiry: "Whether we are justified [justification$_4$] in this latter belief [i.e., justification$_3$/D$_3$] is an interesting question, but it is certainly not the question of justifying our definition of justification" (681).

The second contention in the sentence just cited is not only somehow off-target, but it is also clearly false. Almeder acknowledges that there is a question concerning the justification of justification$_3$/D$_3$, but he then maintains that this question is not a question concerning the justification of "our" (i.e., his) definition of justification. The preceding analysis reveals, however, that what Almeder terms "this latter belief" is precisely that "we believe that when it is done well it will produce a measure of understanding and enlightenment not otherwise available." But with respect to this (last!) conviction there arises once again, in Almeder's own words, the question of justification (i.e., the question of the justification$_4$ of justification$_3$/D$_3$, which Almeder presents as justifying justification$_2$/D$_2$, which in turn he presents as justifying justification$_1$/D$_1$). His abrupt termination of the procedure on the basis of the contention that the *new* question of justification—that of justification$_3$/D$_3$—is not a question of the justification of "our" definition of justification fully ignores the interconnection of his own theses.

In fact, Almeder's attempt clearly shows that the entire concept of justification must be newly determined. The decisive factor—as the following considerations clearly reveal—is the taking into account of the fact that every question, statement, contention, definition, etc., has a determinate, identifiable, and intelligible meaning only within a *theoretical framework*. What this means here, put quite globally, is that it is senseless to try to define an "absolute" concept of justification (or grounding), that is, one that would not be relative to a theoretical framework. Such a concept has within itself the contradiction of on the one hand *being relative to* a wholly determined theoretical framework, but on the other *purporting to be* understood and applied in such a way that its claim to validity and applicability is *free of any relativity whatsoever*.

[3] The justification concepts developed in the context of the consideration of Almeder's position show that justifications (or groundings) occur on different but interrelated levels. One could speak here of a hierarchy of levels of grounding, whereby the concept of grounding would always be relative to a level. The different concepts of grounding should then be correspondingly defined. Problems of the sort introduced by Ketchum would arise only if a criterion of grounding were applied to itself *in an inappropriate manner*. A self-application is inappropriate when the theoretical framework within which a specific definition is formulated is ignored. Taking the presupposed theoretical framework into consideration makes clear that inappropriate self-applications can and indeed must be avoided, how they can be avoided, and how appropriate self-applications can nonproblematically be accomplished. This is shown in detail in what follows.

One can introduce a corrected version of Ketchum's schema G. Instead of G,

G (S)(p) [S is justified in believing p true ——— if and only if —p—S—],

one can use G_S, a schematic systematic sentence taking into consideration theoretical 84
frameworks and the plurality of their levels:

G_S: (S)(p) [S is justified in believing p true on level——— within theoretical framework ——— if and only if —-p—-S—- on level ——— within theoretical
framework ———]

Then the problem that Ketchum treats arises: is G_S, for its part, "justified" ("grounded")? And if so, "justified" ("grounded") in what sense?

First, the idea of grounding and the resulting demand make sense, as has been emphasized more than once, only on the basis of a specific theoretical framework, because otherwise the idea and the demand cannot even be articulated. If then this idea/demand itself is to be grounded, two ways of proceeding come initially into view. Either grounding is again understood in the sense of G_S, which would have as a consequence that, if its application were understood as a deduction, G_S would have to appear both as premise and as conclusion—in which case there would be a clear circularity making the procedure untenable—or one says that one is dealing with a special or limiting case, with respect to which the application of the term/concept *grounding* in the sense of G_S to G_S itself is inadmissible or makes no sense. If one nonetheless wants, intuitively, to apply the word "grounding" to G_S, then the word must be used not in the sense of G_S, but instead in the following sense: G_S is grounded in that G_S presents an adequate characterization of a structuration of all the sentences that can emerge within the relevant theoretical framework. In other words, G_S is grounded in that it is a metasentence characterizing the relevant theoretical framework. However, to avoid conceptual and terminological misunderstandings, it would be best not to use the term "grounding" (and certainly not "justification") to characterize the status of G_S. More accurate and appropriate alternatives would include the likes of "metasentence conforming to the theoretical framework it concerns" and "metasentence presenting a component of the theoretical framework it concerns."

What this shows is significant for philosophy: it is naive to think that one can apply concepts (especially significantly consequential ones) comprehensively and unrestrictedly. The notion that one could arises from a fully unsystematic and indeed undisciplined manner of thinking—a manner of thinking that does not take into consideration the fact that all meaningful talk *is* meaningful *only* within the theoretical framework within which it emerges. The naive and unrestricted demand for "justification" can appear to 85 be demanded by thoroughgoing rationality, but in fact it is irrational to make demands without recognizing what those demands presuppose.

One consequence of these considerations can be briefly put as follows: one must distinguish strictly and strongly between groundings *within a given theoretical framework* and a or the grounding *of that theoretical framework itself*. This basic and central distinction is explained more fully in the next sections of this chapter.

The final point that must be noted here is that it has now been shown that the status of the theoretician concerned with justification or grounding is not neutral or *tabula rasa*, because it is based on an extensive series of central assumptions. The same holds for the kind of skeptics who demand all-around and immediate justifications. To be sure, these skeptics will vehemently deny making such assumptions, given that they take themselves to be free from all presuppositions. But this self-understanding is inaccurate. These skeptics are skeptics only in presenting themselves as such, for example, in explicitly articulating the demand for the justification of the contention that we actually know something or other. But this explicit articulation depends on a great many fundamental assumptions and presuppositions. These skeptics presuppose, for example, that they can express themselves precisely, clearly, and even argumentatively. But that depends on the assumption that the language they use, the logic on which they rely, and other such factors actually achieve what they must achieve in order to make the articulation possible. In other words: these skeptics presuppose that their skeptical questions can be formulated, and their negative sentences submitted, within a well-functioning theoretical framework, one that makes possible the expected achievements. These skeptics are comparable to hikers who follow their paths while maintaining that all of us, themselves included, are blind, and that there are no paths.

1.5.2.2 The Systematic Process of Grounding and Its Forms, Stages, or Levels

[1] The term "systematic process of grounding" designates a complex collection of tasks and procedures. The term "systematic" is here understood in the sense of "complex, comprehensive, and complete." The systematic process of grounding applies to every theoretical unit (i.e., to every part or section of philosophy to which the question of grounding may meaningfully be raised). According to the basic theses of the structural-systematic philosophy, it is in no way meaningful to demand the grounding of any single sentence *while completely ignoring* other sentences. Such a demand would be possible and defensible only within the framework of a foundationalist project. As, however, is shown in Section 1.4.5, such projects are untenable. In contrast to them, within the coherentist project undertaken here every *grounded* sentence (and every *true* sentence) *is* grounded (or true) only as embedded or embeddable within a constellation consisting of a plurality of strictly interconnected sentences. By means of this embedding, the sentence is *fully determined*: it attains, on the theoretical level, a wholly determined status and location.

Like the systematic philosophical method presented in Section 1.4, systematic grounding is a complex procedure; it involves three forms, stages, or levels that can be designated the *incipiently systematic,* the *innersystematic,* and the *metasystematic.* Although these forms/stages/levels are interconnected, in philosophical practice they are applied or concretized only rarely at best, and even then all more or less at once.

As has been indicated, the systematic concept of grounding can or more precisely must be applied to every theoretical unit, whereby every part or section of philosophy to which the question of grounding can and thus must meaningfully be raised is a "theoretical unit." This may be made more precise: both the structural-systematic philosophy *as a whole* and every theoretically relevant part or section *within this philosophy* are theoretical units of the relevant sort. The scope of application of the systematic task or process of grounding is thus comprehensive within this systematic framework. In the case of an individual, limited theoretical unit *within* the structural-systematic philosophy, the terms used to designate the three forms/levels/stages must be correspondingly reinterpreted: "systematic" with the forms of the incipient, the inner, and the meta then relate not to this systematic philosophy *as a whole*, but only to that smaller theoretical unit, as a unit that is embedded *within* this philosophy.

86

What follows are clarifications of each of the three forms/levels/stages of the systematic task or process of grounding. To simplify the presentation, this account treats the *highest case* of the accomplishment of the process, which is the one of greatest significance with respect to the purposes of this book. How the process applies to individual theoretical units is then in part obvious and is, in any case, visible at every step of the presentation.

[2] The three forms/levels/stages of the systematic process of grounding are nothing other than expressions of the fact that systematic philosophy is a whole whose articulation proceeds in three stages (or on three levels): in an incipiently systematic stage, an innersystematic stage, and a comprehensively systematic stage. First to be clarified is the incipiently systematic stage.

[i] A theory—and thus also a comprehensive philosophical theory—is not simply a Platonic entity that could be, so to speak, fetched out of a Platonic heaven. To say that a theory is an abstract entity is to make a purely abstract statement. Any actual theory, however, is a theory that is presented within the rational-theoretical space of a specific human-historical situation. How this space and this situation are to be understood is clarified below. Be that as it may, the presentation of a theory is a process whose realization is accomplished under the various conditions holding within such a human-historical situation. One consequence of this is that the (self-)presentation of a comprehensive philosophical theory cannot be accomplished all at once, so to speak, *in toto*. Instead, the process is one of increasing concretization or determination. Within this process, one can sensibly distinguish the three stages or levels introduced above.

The *incipiently systematic stage* is the stage at which the structural-systematic philosophy *initially* presents itself within the rational-theoretical space of a specific human-historical situation. A human-historical situation is the situation of human beings within a specific epoch of world history. This historical situatedness must be acknowledged to be a decisive factor; otherwise, the philosophical theory would have to be understood as an ahistorical abstraction. Within the historical-human situation, however, the philosophical theory presents itself in a wholly determinate respect; *as a theory*. This means that it is presented in a wholly determinate space within this situation (i.e., rational-theoretical space). And that means in addition that the presentation does and must proceed under rational-theoretical conditions.

In the initial stage there are three tasks whose accomplishment determines the status and form of the presentation of the structural-systematic philosophy: the first is its delimitation from nonphilosophical undertakings, and the second, the presentation and preliminary clarification of its global or basic concepts and an initial clarification of its basic position. These first two tasks are accomplished in the Introduction and in Sections 1.1–1.4. The third task is the one treated in this section. Clarification of the *grounding* of the basic insight determining the position taken here begins with the acknowledgment that *even at this initial stage* of the presentation the question of grounding is legitimate and indeed unavoidable. The reason is suggested above: the (self-)presentation of philosophy occurs in rational-theoretical space. To this space, however, the question of grounding essentially belongs. The problem thus arises of how to deal with the question of grounding at this initial stage; for immanent reasons, the grounding provided at this point can only be initial or *incipient*. Such a grounding is only a part of an adequate grounding; with respect to the comprehensive systematic-philosophical theory presented in this book, adequate grounding is possibly only *after* the presentation of the theory.

How such an incipiently systematic grounding is more precisely to be understood is basically a consequence of the fact that, as is noted above more than once, grounding is here understood as *coherentist* rather than as *foundationalist*. Of decisive importance for the more precise determination of this incipiently systematic grounding are two closely interconnected factors. The structural-systematic philosophy is to be viewed as grounded at its initial stage first if it attains at least sufficient clarity with respect to its external delimitation (i.e., vis-à-vis activities and undertakings that are not philosophical). The task of providing such grounding is undertaken in various ways in the Introduction and in Chapter 1. Second, a criterion for the incipiently systematic groundedness of the philosophy developed here is that the initially global (quasi-)definitions and theses express a comprehensive view that is presented already at the outset, at least in a global respect. In brief: with respect to the comprehensive structural-systematic philosophical theory, the explanations and global theories presented in the Introduction and in Chapter 1, taken together, exemplify the incipiently systematic process of grounding. To this process belongs in a wholly unique manner the all-determining concept of the *theoretical framework*. The groundedness of these various concepts (or quasi-definitions or procedures, etc.) results from the consideration of the two indicated factors—(1) critiques of other positions with respect to unclarities and incoherences and (2) the sketch of a positive alternative that is provided by the anticipatory view of the whole.

[ii] How this problematic is concretely visible in philosophical discussion can be seen from the well-known position that Rudolf Carnap (1950/1956) formulates in his essay "Empiricism, Semantics, and Ontology." Unfortunately, however, Carnap provides a response that is wholly insufficient, but—because the nature of its insufficiency has not generally been recognized—that has also significantly confused the issue. Carnap's thesis is that the introduction of a linguistic framework "does not need any theoretical justification" (214). If this is so, is such an introduction wholly arbitrary? Carnap has generally been held to have thought so. That cannot, however, have been his position, although his own misleading formulations are no doubt responsible for the misunderstanding. The phrase "theoretical justification" is the chief source of the misunderstanding. It is surprising that Carnap speaks so—or perhaps not, given that the "theoretical" and the "practical," which are distinguished in various ways throughout the history of philosophy, are fundamentally ambiguous. Beyond question, Carnap does recognize that the acceptance of a determinate linguistic framework requires *justification*:

> To be sure, we have to face at this point an important question; but it is a practical, not a theoretical question; it is the question of whether or not to accept the new linguistic forms. The acceptance cannot be judged as being either true or false because it is not an assertion. It can only be judged as being more or less expedient, fruitful, conducive to the aim for which the language is intended. Judgments of this kind supply the motivation for the decision of accepting or rejecting this kind of entities. (214)

Pace Carnap, judgments about the expedience or inexpedience of a language fully deserve the qualifier "theoretical," because their form is genuinely theoretical: "(It is the case that) linguistic framework L is more expedient (for the explanation of phenomenon P) than is linguistic framework M." One can expand this by speaking of the greater or lesser *intelligibility* of a linguistic framework. Questions of relative intelligibility are obviously and unambiguously theoretical questions.

Anyone who draws a *rational conclusion* concerning the acceptance of a determinate linguistic or theoretical framework thereby makes a *purely theoretical* judgment concerning the expedience or intelligibility of that framework. Such a judgment provides the rational basis for the theoretical conclusion. It is therefore clearly misleading, indeed false, to designate the question concerning the introduction or acceptance of a determinate linguistic (or theoretical) framework as a *purely practical* one. By so designating it, Carnap gives rise to a misunderstanding with serious consequences. For theoreticians—for scientists or philosophers—there is, strictly speaking, *no* practical question in Carnap's sense. Theoreticians would contradict themselves if they failed to accept the linguistic (or theoretical) framework allowing for a greater expediency or a greater intelligibility.

The *genuinely practical* question for philosophers and scientists arises at an earlier stage; it is the question whether one is to decide to *be* a philosopher or scientist and therefore to behave as one. If the response is positive, then along with it is made the *general* decision in favor of accepting—the commitment to accepting—whichever theoretical framework allows for or makes possible a (or the) greater intelligibility with respect to the subject matter upon which the philosopher or scientist focuses. One sees here that an apparently minor misunderstanding can have disastrous consequences.[27]

90

In sum: individuals who have *been pragmatically motivated*—no matter how or why—to be theoreticians will then, when they act consequently *as theoreticians*, as a matter of course accept (and use) the theoretical framework providing the maximal available intelligibility of the relevant domain of theorization. What this means, more precisely, is considered below in [4][ii].

[3] In contrast to the incipiently systematic process of grounding, which can have only a purely global character, the innersystematic process is far clearer and more precise. It involves the procedures, considerations, etc., on the basis of which the systematic status of a given truth-candidate (sentence, theory, etc.) is stabilized. To stabilize a truth-candidate innersystematically is to determine its systematic location. This stabilization or determination is a grounding for the speaker/knower who is already situated within the system. The innersystematic process of grounding thus relies only

---

[27] In another passage Carnap acknowledges to a degree the *theoretical* character of the deliberations that determine the decision for acceptance or rejection of a determinate linguistic framework:

> The decision of accepting the thing language, although itself not of a cognitive nature, will nevertheless usually be influenced by theoretical knowledge, just like any other deliberate decision concerning the acceptance of linguistic or other rules. . . . [Q]uestions concerning these qualities [i.e., efficiency, fruitfulness, etc.] are indeed of a theoretical nature. (239)

Carnap appears however fully to have missed the implications of this insight; he goes no further than this fleeting remark. His interest is exclusively in avoiding the misunderstanding according to which the question concerning the justification of a specific linguistic framework would be identified with the "question of realism." He does not arrive at the thought that such a question could and indeed must be understood and clarified as a question concerning the concept of reality itself. This is considered below in the main text.

on the inner structuration of a theoretical framework accepted and employed by the subject/speaker/knower. To be accurate, one would have to term the concept of the process of innersystematic grounding a blanket concept, in that one must ultimately speak of as many aspects of this process as there are modes of connection within the framework (a given framework might, for example, allow both deductive and inductive groundings). It is important to emphasize that these groundings are purely rational and theoretical matters, in no way pragmatic ones.

91

[4] The *metasystematic* process of grounding can here be characterized preliminarily from a *negative* and from a *positive* perspective.

[i] The former involves distinguishing the concept of grounding envisaged here from a foundationalist process, which could also be termed a process of establishment. The foundationalist process is, to put it one way, "archaeologically" oriented, whereas for the systematic process of grounding the designation "teleological" is appropriate. From the *archaeological* perspective, metasystematic justification or grounding—or, again, establishment—must be accomplished at the beginning of the theoretical enterprise. A sentence is formulated, perhaps in the form of a thesis or a principle, and *immediately thereafter* its justification is demanded and provided—it is somehow *established* as true. The beginning of Section 1.4.5 shows that the foundationalist position is untenable.

[ii] The *positive* perspective, one that can provide a first approximation of the metasystematic process of grounding, is that opened by the familiar and much-discussed question of the confirmation (or justification, etc.) of natural-scientific theories. This question can arise first and only after a theory has been developed and is thus available. Grounding as confirmation or support is thus the final step in the theoretical undertaking, and not—as in foundationalist grounding—an initial step.

These two perspectives yield a first, general determination of the metasystematic process of grounding: a systematic philosophy is metasystematically grounded not when it is foundationally established, that is, not when it is developed on the basis of foundational theses that are established as they are presented, but only when, after it has been completed, it is considered as a whole. Within the foundationalist enterprise the question of the justification or grounding of a systematic philosophy after the fact, so to speak, cannot arise, because the whole or developed systematic philosophy is, viewed foundationistically, established as true only if it develops from foundations accepted as true. The only question that the foundationalist can be faced with from the foundationalist perspective is whether derivative sentences have been correctly deduced from the foundational truths.

The more precise determination of the metasystematic process of grounding is an extensive and complex task that is considered in detail in Chapter 6. Requisite here are only some central considerations that must be kept in mind as the task is undertaken.

92

Significant problems arise already with the clarification of the task or process of confirmation of natural-scientific (empirical) sentences, theories etc. A radical or naïve empiricist conception, according to which a given theory is verifiable or falsifiable on the basis of the empirical data, is scarcely tenable because of the problematic status of such data. To take a datum as an X that is wholly independent of the relevant theory or indeed of any theory whatsoever is, as philosophy and the philosophy of science reveal, to proceed on the basis of a fiction. Both the empirical data and the observation language are always already theory-laden.

Because of the issue just described, the gap between philosophy and the natural sciences with respect to the problem of grounding is not as wide as has often been maintained. This is not to say, of course, that no differences remain. What follows sketches a solution to the problem as it applies to philosophy. As becomes clear, this solution must be developed along with the development of the system.

The dream that it would be possible or even sensible to develop the one true systematic philosophy and to exhibit or even establish it as such is one that must, finally, be abandoned. That this dream is not only unrealizable but nonsensical is demonstrated in the system developed here. But from this there follows neither (inconsistent) relativism (of the traditional sort) nor any sort of skepticism nor any other similar philosophical position or standpoint. This thesis, too, is grounded within the systematic philosophy developed here.

The metasystematic-philosophical problem of grounding arises from recognition that a developed philosophical theory is confronted not only in fact but also in principle with alternative theories both already developed and conceivable. This thesis is grounded at the outset because, if there were only a single theory, then the metasystematic problem of grounding could not arise because there would be no competitors in comparison with which the theory would need to be grounded.

At the same time, if there are or could be alternative conceptions, theories, or systems, then this question is unavoidable: how can a given theory or system be shown to be the superior one and thus the one to be accepted? The relevant criteria cannot of course be purely innersystematic (or not exclusively or even predominately so), because the very question concerns which system is superior. Proofs in the strict sense are here not to be expected nor indeed possible, because proofs presuppose precisely determined and established theoretical frameworks, only *within which* they can succeed in proving. But what then is left? 93

This book suggests that the central criterion, determined by the commitment to theoreticity, for the superiority of one framework in comparison with other available frameworks and even with additional frameworks envisioned as possible, is *greater* or *relatively maximal coherence*. This coherence is not meant to be synonymous with consistency (in the strictly logical sense); it has a broader signification. What follows is not a precise determination, but only a sketch of the basic idea (details and additional complications are introduced in Chapters 5 and 6). The notion of intelligibility is helpful in this respect: the greater the intelligibility (of a concept, a configuration, etc.), the greater is the coherence achieved by its means. For its part, intelligibility depends on a variety of factors that are relative in the sense that—in the extreme case—those that strike one theoretician as quite plausible might appear to another as utterly implausible (as is shown in Chapter 6, however, this initial "striking" does not make rational comparison impossible). One can distinguish [a] fundamental from [b] comparative criteria of intelligibility.

[a] *Fundamental* factors are ones that determine the boundary between intelligibility and unintelligibility. An illustrative example to which this book repeatedly returns: perhaps the oldest semantic-ontological pattern of thought, and one that suffuses not only the history of philosophy but also contemporary thought, relies on the assumption that two types of entities—substrata (substances, objects, things) and universals (properties and relations)—constitute the basic structuration of what is. This pattern takes the world to be the totality of objects (things, substrata, substances) that have

properties and stand in relations to one another. In the course of history and into the present, few philosophers have raised the question whether this semantic-ontological pattern of thought is ultimately intelligible. To many, the question might even appear senseless, for they will respond, "What could be more intelligible than what we are all already using, than what enables us to know what we are talking about when we speak of objects, properties, relations, and so forth?"

But do we indeed know what we are then talking about? What is a substratum/thing/object? Ultimately no more than this: an X to which are ascribed properties and relations and that is supposed to be involved in states of affairs. But the philosopher seeking intelligibility will continue to question, what is this X itself? It quickly becomes clear that this X is a quite peculiar entity: what it itself is simply cannot be said, because it serves only to provide the unspecifiable grammatical subject of which properties, relations (and facts) are grammatically predicated. Not even infinitely many properties/relations/facts introduced in the attempt to determine this X could succeed in actually determining it. To put it figuratively, the more properties, relations, and facts "determine" it, the further it withdraws. Is such an entity at all intelligible? Many, indeed presumably most philosophers would say that the X is of course intelligible, because only on the assumption of such an X do we have the concrete world of objects, things, etc. But this course of argumentation is completely circular. Given that the X withdraws in direct proportion to the number of properties and relations that are ascribed to it, the only option would appear to be to remove properties and relations from it. Once all properties and relations were removed, one would be left with the X, given that it is, *ex hypothesi*, not identical with its determinations. But what one is then left with is no longer intelligible as an entity.

[b] *Comparative* criteria are ones that exhibit greater and lesser degrees of intelligibility. It is probable that a consensus could be achieved on at least some of the chief such criteria. A leading candidate for the position of the most important criterion would be the following: a theory (a sentence, a system...) has a greater intelligibility than another (and thereby a greater coherence) if it develops and articulates a more detailed structuration of the subject matter (the issue, the domain, etc.) that is in question. One can also distinguish among accounts that are more or less appropriate to the subject matter, presupposing that a subject matter is not appropriately conceived and articulated if it is considered only, so to speak, by going ever deeper into it in some splendid isolation. The example introduced in the preceding paragraph makes clear that the substance-schema, taken strictly, is not compatible with the greater intelligibility required by the criterion under current consideration.

An additional comparative criterion is *completeness*, albeit relative rather than absolute completeness. This concerns both the set of data and the aspects of the data, in a comprehensive sense: the ways the data are available, their interconnections, the complexes or configurations in which they are involved, etc.

With this, the *entire* topic and problematic of the concept of grounding are not yet adequately treated. To this point, emphasis has been placed on the unavoidability of assuming one or another determinate theoretical framework. Because, however, it must be assumed that various theoretical frameworks in fact exist and yet more (indeed, in principle, infinitely many) are possible (i.e., are developable), the question arises, why accept one determinate theoretical framework and not another? This question is, in the best sense, a question of grounding.

As has repeatedly been noted, it is here assumed that one always thinks, speaks, presents sentences, develops theories, etc., within a determinate theoretical framework. It is, however, a fact that one of the theoretical capacities of our thinking is the ability to go beyond the borders of any determinate theoretical framework. This happens in particular when we compare a theoretical framework we accept to other actual or possible frameworks. Given that one can only ever think *within* a framework, the phenomenon of framework comparison reveals that when one steps beyond one's framework, one always steps into another one that is higher, more comprehensive—one that is, in a word, a *metaframework*.

It is within such a metaframework that one compares one's initial framework with some other (or others). What can the comparison yield? There are basically four possibilities. First, one's initial framework can emerge as superior. The superiority, globally viewed, can take two forms: it may be that one or more of the other frameworks can be incorporated or integrated into, or embedded within, the initial framework, or it may be that the initial framework radically excludes or undermines the other framework(s). Second, this can happen in reverse, as it does when some other framework emerges as superior to the initial one and to other alternatives. Third, it may be that one or more of the other frameworks proves to be equivalent to the initial framework. Fourth and finally, it may be that the comparison fails to establish any equivalency or superiority, that instead all the compared frameworks prove to be deficient and thus to require correction. In philosophical practice, this fourth case often arises, in forms that are complex in manifold ways. If the deficiencies are removed and the corrections made, the result is a case of one of the first three types. This issue is considered in additional detail in Section 6.3.2.1.

1.5.2.3 The Process of Systematic Grounding as an Idealized Form of the Practice of Systematic Grounding

As presented in the preceding sections, the systematic process of grounding, with its three stages (the incipiently systematic, the innersystematic, and the metasystematic), is an idealization. For reasons of technical presentation, this complex process is characterized above in such a way that it could be applied, hypothetically, to the entire field of systematic philosophy. If one were in a position to present this entire field with all its interconnections, the idealized form of the process of grounding would be identical to the real form. But it is extremely improbable that anyone will ever be in such a position. Considering philosophical practice—including that of the philosopher who thinks systematically in the sense articulated here—one must be more discriminating and concretize the systematic process of grounding in view of this practice. This must be done in the case of each of the three stages of the systematic process of grounding.

[1] The least problematic case in this regard is that of the *innersystematic* process of grounding. In this process, the systematic perspective and the practice of presentation for the most part coincide. The qualification "for the most part," however, must be made more precise. To ground something innersystematically indeed means to work out the place or location—to establish the exact status[28]—of an assumption, a thesis, an area-specific theory, etc., within the system's theoretical framework (for simplicity:the sys-

96

---

[28] These terms—"place," "location," and "status"—must be left vague, because precisely what they mean depends on the systems within which innersystematic grounding is in question.

tematic framework), and indeed exclusively with innersystematic means of explication and argumentation. From this it immediately follows that the presentation of a specific part of the system also involves—in a quite specific manner—the innersystematic grounding of every single sentence, area-specific theory, etc.

As is noted above, at this stage there is complete coincidence between justification (in the sense of a theoretical undertaking involving a subject or a speaker/knower) and proof (or, more cautiously, demonstration or exhibition—but still, as with proof, in the sense of a theoretical undertaking that does not make explicit reference to a subject or a speaker/knower, instead being purely immanent in the sense of being accomplished within, or being a consequence of the development or concretization of, the theoretical or systematic framework). That they coincide does not mean that they are mingled or confused. What is instead the case is that a given demonstration, in the strictly inner-systematic sense in which it provides a grounding, simply performs the function of justification without there being any need to make this explicit. It does so because the relevant theoretical subject is also within the system and thus has already accepted the theoretical or systematic framework.

97 　Given this position, no *subjective* factors—no factors of belief, conviction, prefer-ence, etc.—are in any way relevant; only logic and rationality are relevant, in that every-thing that stringently emerges within the accepted theoretical framework presents itself to the subject as simply true within the framework.[29] There is then, strictly speaking, no question of justification (as long as "justification" is understood in the pragmatic sense explained above).[30] The status of innersystematic truths thus cannot come into question pragmatically; innersystematic truths come into question only "from the outside," that is, from a perspective that brings the fundamental theoretical framework itself into question, thus, from the perspective of some *metaframework*. The notion of metaframework leads, again, to the question of *metasystematic* grounding.

[2] Philosophical practice diverges to the greatest extent from the above-sketched ide-alized process or accomplishment of systematic grounding with the incipiently systematic and metasystematic forms of the systematic process or accomplishment of grounding. In the typical concrete presentation of philosophical theories and conceptions, *both* forms of grounding in fact—and, one must say, sensibly—are always sought and indeed indispensable. Even when a basic theoretical or systematic framework is chosen in an appropriate manner, the task of concretizing this framework with respect to the various questions and domains of systematic philosophy remains far from accomplished. For this reason, at every new stage of systematic presentation the theoretician must develop *and*

---

[29] Chapter 3 (Section 3.3.4.3) and Chapter 5 more closely consider the precise sense and signifi-cance of this relativity of truth to theoretical framework(s).

[30] If this presupposition is not met, the innersystematic demonstration no longer functions as a justification. The most interesting case—and probably the most common—is the following: a theoretician investigates and reconstructs a philosophy innersystematically and discovers that the innersystematic connections are unobjectionably demonstrated ("proved"), thus, that the demonstrations are (innersystematically!) stable and thereby compelling; the theoretician, however, does not accept the theoretical or systematic framework within which this philoso-phy develops. In this case the demonstrations are, *for this theoretician*, not justifications: they do not provide *this theoretician* with grounds or reasons for accepting what they prove or demonstrate.

provide grounding for the introduction or acceptance of a *concrete or concretized* version of the general theoretical framework. The concept of grounding required here is essentially the incipiently systematic one, albeit not with respect to the general theoretical or systematic framework but with respect to the entirety of the systematic philosophy; this is so because in the (more) concrete case various assumptions will already have been   98 made, certainly with respect to the general framework but also, depending on where the theme to be treated fits within the framework of the systematic philosophy being presented, with respect to concrete and contentual assumptions. An additional, wholly specific concretization of the theoretical or systematic framework is required. Of course, the further developed the presentation of the systematic philosophy, the less latitude will there be with respect to possible candidates for the concrete theoretical framework for a specific domain.

Similar considerations hold in principle also in the case of the metasystematic process of grounding. In concrete philosophical practice it would not be defensible to provide the metasystematic grounding (thus confirmation) of a given sentence, theory, etc., only after the systematic philosophy had been presented in its entirety. For this reason, systematic philosophizing cannot require that one must already have presented a grand philosophical system as a whole before raising and responding to specific questions, above all questions of grounding. If that were possible, it would of course be preferable. But it must be added that philosophers who think systematically and who, in their philosophical practice, apply the "concretized" systematic concept of grounding described above, must not thereby lose sight of philosophy as a whole. Otherwise, their theories about particular subject matters would be problematic to a degree that the discipline of philosophy cannot tolerate. To be sure, in philosophical practice theses and distinctions may be introduced, etc., whose presuppositions and implications the philosopher cannot appraise completely or even at all, but such a procedure is sensible and permissible only when it is explicitly understood and described as provisional. Otherwise, the relevant philosophical thesis or theory can make no claim to even minimal rational and thereby genuinely grounded acceptability.

CHAPTER TWO

Systematics of Theoreticity: The Dimension of Philosophical Presentation

2.1 Theoreticity as a Dimension of Presentation

In the Introduction and the preceding chapter, the thesis that philosophy is a *theoretical* enterprise is preliminarily clarified and grounded, the grounding being, of necessity, incipiently systematic. The task of this chapter is to specify the precise sense in which the structural-systematic philosophy is theoretical.

Philosophy (more broadly, science) is an extremely complex phenomenon. It involves multiple factors, and its adequate understanding and presentation require explicit identification and clarification of those factors. The broadest of the relevant factors are the following: philosophy (science) is a specific type of activity; this activity presents (or aims to present) theories (in the ideal and ultimate case, a single, comprehensive theory); this activity and its products relate to the world as the unrestricted totality of subject matters and domains.

If one takes these three factors in conjunction, one has what can be termed a *dimension of presentation*; here, "presentation" is to be taken broadly, such that (for example) art, too, is a dimension of presentation in that it is an activity that presents works of art, and both the activity and the products relate in one way or another to the world. The same holds in principle for practical activity and its "products," i.e., most broadly, acts oriented with respect to goodness. Put briefly and simply, theoretical activity aims to present theories that are true; aesthetic activity, works that are (in a specific sense) beautiful; and practical activity, deeds that are good (or whose products or results are, or are foreseen as being, good).[1] What distinguishes the philosophical dimension of presentation from the aesthetic and practical dimensions, then, is precisely its theoretical character: it is a theoretical activity, with theoretical products (theories) that relate theoretically to the world, the latter thus thereby being theoretically articulated (i.e., presented).

100 With respect to the use of the terms "theory," "theoreticity," and their derivatives, it is important to note at this point that they are used here in accordance with the

---

[1] The aesthetic and practical dimensions are treated in Chapter 4.

significations that they have within contemporary philosophy—no matter how diverse the more precise determinations of these significations may be. It is of particular importance to emphasize that there is no concern here with anything like the "primordial signification" of the Greek word θεωρία; speculation or reflection on the etymology of the word and, above all, the extensive implications that some philosophers (above all Heidegger) seek to draw from such speculation, have no place in this investigation.

The use of the term "theoreticity" requires additional explanation and specification. In this book, *theoreticity* generally and fundamentally designates the status or character of theories. It thus serves to characterize anything and everything that is involved in theoretical presentation, be that (for example) as something presupposed by theorization (e.g., a specific language, a specific discursive mode, etc.) or as something that is a constitutive element of a theory (e.g., axioms, logical rules, etc.). On the whole, then, "theoreticity" as used here relates to "theory" as "rationality" relates to "reason."

This basic but quite general notion of theoreticity is, however, strikingly distinct from other less general conceptions that are sometimes associated with this term in contemporary philosophical literature. One such conception—a quite narrow and specific one—can serve as illustration. The so-called received view of theories, originally developed by Rudolf Carnap and Carl Hempel and later termed the "statement view" (see esp. Stegmüller 1980), is based on a language L (as a rule a first-order predicate language, in some cases expanded to include modal operators), which is divided into two sublanguages (or two distinct vocabularies): the observation language L_O (or the observation vocabulary V_O) and the theoretical language L_T (or the theoretical vocabulary V_T). On the basis of this distinction, the term "theoreticity" has been and continues often to be used to characterize the status of L_T (or V_T) and of everything that is directly connected to it (see Suppe 1977, esp. 50ff).

One of the alternatives to the received or statement view—the so-called non-statement view, developed by Joseph Sneed and Wolfgang Stegmüller as the "structuralist" conception of theory—also accords central significance to the term "theoreticity," but here the term again changes its signification. Within the traditional (received, statement) conception, the theoretical terms from L and the sentences containing them receive *partial interpretations* from postulates of the following types: the theoretical postulates T (i.e., the theory's axioms), which contain only terms from L_T, and the correspondence rules or postulates C, which are mixed sentences that must satisfy specific criteria. In opposition to this conception—which characterizes theoretical items purely negatively (as the domain of the *non*-observable, what is *not* fully intelligible, etc.)—the structuralist theory-conception introduces a so-called criterion for theoreticity that determines the role of the theoretical elements *positively* (i.e., by specifying the role that these elements play in the application of the theory). This role is that of being measured by *theory-dependent means*. For this reason, advocates of the structuralist conception often speak of "T-theoreticity." It is clear that the term "theoreticity" has for them a quite narrow signification (see Balzer 1986).

Ultimately, it makes little sense to argue about terminology. What is crucial is whether or not a specific terminology is clear and unambiguous. The basic context within which the term "theoreticity" is used in this book is not at all unclear or ambiguous: the conception is wholly basic and general, and thus clearly distinct from any narrower and more specific determination.

The task of determining or defining the dimension of theoreticity precisely and completely is an extensive one that reaches from the distinguishing of this dimension of presentation from other such dimensions to the precise definition of "theory," and to determining how theory or theories relate to the world. In a specific respect, this task coincides with the comprehensive presentation of a systematic philosophy; central parts of this task are undertaken in other chapters of this book.

The remainder of this chapter concerns theoreticity in an initially restricted sense: here, the term designates the specific character of the dimension within which theories are developed, thus the dimension of science (within which philosophy belongs). It begins by considering only two aspects of this dimension: that of language, as the medium of theoretical presentation, and that of knowledge, as the dimension of accomplishment of the presentation. These two aspects are not selected at random; instead, they are the two features that distinguish most decisively the dimension of theoreticity both externally (i.e., from aestheticity and practicity) and internally (i.e., with respect to how the dimension itself is articulated). These two factors, language and knowledge, are thus the two coordinates with respect to which what characterizes the basic and fundamental character of theoreticity comes into view in a particularly concise and unambiguous manner. In addition, important problems arise with respect to how language and knowledge relate to each other. How their relation is understood determines which of two basic starting points is accepted for philosophy. This chapter considers this issue in detail. The accounts of language and knowledge lead to two additional and also central aspects of the theoretical dimension: the concepts of theory itself (in a more developed sense) and of truth. In this chapter, the latter concept is treated only briefly, still from a global perspective.

102

An additional anticipatory description of the aims of Sections 2.2 and 2.3 may be helpful here. Given that philosophy presents theories that are ultimately universal in character, and that there are theories only if there are (one or more) theorists, it might appear—and indeed, it has appeared, to the majority of modern philosophers—that consideration of the theorist (or, more broadly, the subject who can be, among other things, a theorist) must be at least the first and foremost philosophical task, and per-haps indeed the only philosophical task. With respect to language and knowledge—the aspects of theoreticity considered in Sections 2.2 and 2.3 (and again in 5.1.4, especially 5.1.4.5)—this appearance might seem to be strengthened by the assumptions that there is knowledge only if there are (one or more) knowers, and that there is language only if there are language users. Despite the apparent strength and the undeniable popu-larity of the assumption or conclusion that the subject should or must be accorded philosophical primacy—whether as thinker/knower/cognizer, as speaker, as agent, or as anything else—this book denies the subject any such primacy. Sections 2.2 and 2.3 thus aim to establish the merely secondary and derivative status of the subject within, respectively, the linguistic and the epistemic domains.

2.2 Language as the Medium of Presentation for Theoreticity

Philosophy is concerned essentially with presentation; indeed, one can say that philosophy *is*, in a specific sense, presentation. The relevant presentation is theo-retical. Presentation as theory is the dimension of the expressibility of the world. The medium of this so-conceived dimension of presentation is language. Knowledge is to be understood as the dimension of the accomplishment of theoretical presentation. These matters are now considered in detail.

2.2.1 *Language, Communication, and Dimension of Presentation*

Although (as has just been shown) theory is an extremely complex phenomenon, language is yet more complex. Clarification of the phenomenon of language can begin with the distinction, introduced above more than once, between natural and artificial languages. Natural languages are the ordinary, everyday, spoken languages used within ordinary human communities; artificial languages are languages that are constructed, generally for theoretical purposes, and they may be either formal languages or adaptations of natural languages. The basic function of natural or ordinary languages is communication, which has at least the following four aspects: (1) descriptive or indicative presentation, typically involving descriptive or indicative sentences; (2) aesthetic presentation, which involves quite differently structured sentences; (3) pragmatic presentation, which consists of commands, instructions, etc.; and (4) what may be termed emotive (non-indicative) presentation, characterized by such things as expletives indicating attitudes or feelings of specific subjects. A complete philosophy of natural language would have to investigate and develop theories about at least these four aspects.

Artificial or technical languages are languages of theory that involve only the first of the four aspects listed above: indicative or theoretical presentation.[2] The general function of natural languages (i.e., communication) is reduced in theoretical languages to the dimension of purely theoretical presentation. Whether one can then speak of "communication" in a relevant or interesting sense depends upon how, precisely, one defines this concept. If communication is considered to be possible only if *all four* of the levels of natural language introduced above are involved, then it is clear that artificial and theoretical languages have no communicative function. If one in addition took the function of communication to be essential to language, then artificial "languages" would not even *be* languages. Alternatively, however, one can speak of a purely *theoretical* communication, and this appears reasonable in that artificial or theoretical languages enable language users to communicate on a specific level (i.e., the purely theoretical level).

More precise specification of the distinction between natural-language communication and theoretical communication is not necessary here, but one complication centrally important to the systematic philosophy presented in this book may be noted at this point. If the function of language is to serve communication, then one consequence appears to be that language must be learnable. Many, indeed most philosophers of language assume, however, that to be learnable, a given language must contain only finitely many expressions (or at most countably infinitely many expressions; see e.g. Davidson 1965/1984). This assumption is, however, problematic in that it is based

---

[2] Theoretical languages can of course—and in philosophical and scientific practice generally do—include pragmatic language (e.g., "assert," "argue," "maintain"), but such terms are used, in purely theoretical presentations, only in purely indicative sentences (e.g., "This book argues that substance ontology is unintelligible").

on an atomistic and thus naive understanding of learning. Moreover, acceptance of the assumption leads to the denial that semiotic systems with *uncountably* many expressions are learnable, because they can no longer serve what is generally understood as communication. Such systems would then be said not to qualify as languages. Chapter 5 considers this issue in detail.

This book rejects the assumption just introduced, contending instead that every formal or semiotic system that is interpretable can and must be acknowledged to be a language. This might appear to be a matter of superficial terminology, but it becomes clear below that it in fact has far-reaching consequences. At this point, it suffices to note that languages characterized by the uncountability of the expressions constituting them are abstract semiotic systems that can be constructed set-theoretically (see Hugly and Sayward 1983, 1986). To anticipate: one of the fundamental theses of the structural-systematic philosophy—that expressibility is a fundamental immanent determination of the world—has as a consequence that there must be not only one language (one semiotic system) containing uncountably many expressions but a plurality of such languages. This issue, like that of the learnability of language, is considered in detail in Chapter 5.

2.2.2 *Normal or Ordinary Language and Philosophical Language*

There is no question that at the beginnings of their accounts, as well as in various other places, philosophers must rely upon ordinary language—at the very least in those cases where they attempt to address readers who are unfamiliar with more theoretical languages. As concrete human beings, philosophers encounter others, philosophers and non-philosophers, within linguistic horizons that are not philosophical. Granted, however, that this is an indisputable fact, the important questions concern how it is to be interpreted and evaluated.

So-called ordinary or natural language is not an unambiguous, precisely determinate phenomenon clearly separate or separable from other phenomena. Fundamentally, it is a language of communication, in a non-theoretical sense (i.e., one not serving primarily to communicate theories). Yet because, as is noted above, it contains an indicative aspect, it also contains potentially or incipiently theoretical elements, including forms of argument and even semantic vocabulary. Included among the former are terms such as "if . . ., then . . .," "therefore," etc., and among the latter, "true" and "truth" and, in some uses, "mean" and "refer." Logic and semantics, central to theoretical languages, are thus not wholly absent from ordinary language.

The question thus arises: is philosophical language—or are philosophical languages—"ordinary" in the sense in which natural languages, the languages of everyday life, are ordinary? Or, put more radically, must philosophical language be understood to be a "normal" or "ordinary" language? This is a fundamental question, and different answers to it lead to widely diverging conceptions of philosophy. This book strongly distinguishes philosophical from ordinary language, and accords to ordinary language a role within philosophy, but not a fundamental one. This thesis is more closely explained and supported in Section 2.2.3; the remainder of the current section briefly presents several striking positions that reveal radical disagreements concerning the status and relevance of ordinary language with respect to philosophy.

[1] For the purposes of this book, important philosophical positions on this issue can be placed in three categories: 1. positions that *reduce* ordinary or natural language

to or transform it into a different, artificial (in some cases formal) language ([1.1]); 2. those that consider ordinary language to be either the sole measure for philosophy, or at least the decisive one ([1.2]); and 3. those that prefer some middle course between these two extremes ([1.3]).

The relationship between philosophical language and ordinary language has been a central topic in analytic philosophy since its beginning, but it is important to emphasize that this topic is not raised only in analytic philosophy, as the following account makes clear. Nevertheless, what is characteristic of analytic philosophy's treatment of the relationship—the distinction between the philosophy of ideal languages   106
and the philosophy of ordinary language—must be noted. In essence, this distinction dates from Frege, but it has explicitly been made a central topic only since the 1930s. Carnap is the grand theoretician of ideal languages, whereas the Oxford school and the later Wittgenstein are the best-known practitioners of the philosophy of ordinary language. At present, it would be wholly inadequate to situate analytic philosophy as a whole in either of these two camps; for the most part, analytic philosophers take specific variants of positions that mix these two extremes.

[1.1] Several examples illustrate positions that attempt to reduce ordinary language to or transform it into some artificial language.

[1.1a] Although Carnap is the greatest champion of the philosophy of ideal language, his position is not considered here. The reason for its omission is that Carnap's central concept, that of the linguistic framework, is transformed in this book into that of the theoretical framework, which is here equally central. "Theoretical framework" is, philosophically, a significant expansion and deepening of Carnap's "linguistic framework"; this book consequently considers Carnap in various contexts.

Instead of Carnap, then, this section briefly presents a different but equally striking example of a position reducing ordinary language to ideal language in the strictest formal-artificial sense. If strictly logical-mathematical methods are applied to the "reconstruction" of ordinary language, it can happen that the "language" thereby "reconstructed" is on the whole simply identified with the language that was to have been "reconstructed." This is done by Richard Montague, who takes a radical position concerning the relation between ordinary and theoretical languages. He writes, in an essay with the revealing title "English as a Formal Language" (1974/1979: 188): "I reject the contention that an important theoretical difference exists between formal and natural languages." Richmond Thomason (quoted in Montague 1974/1979: 2) comments as follows:

> According to Montague the syntax, semantics and pragmatics of natural languages are branches of metamathematics, not of psychology. The syntax of English, for example, is just as much a part of mathematics as number theory or geometry. This view is a corollary of Montague's strategy of studying natural languages by means of the same techniques used in metamathematics to study formal languages. Metamathematics is a branch of   107
> mathematics, and generalizing it to comprehend natural languages does not render it any less a mathematical discipline.

Or, as Montague elaborates in a different essay, "There is in my opinion no important theoretical difference between natural languages and the artificial languages of logicians; indeed, I consider it possible to comprehend the syntax and semantics of both kinds of languages within a single natural and mathematically precise theory" (1974/1979: 222).

Just what position Montague and Thomason want to take is not fully clear; to make it clearer, a number of distinctions must be brought into consideration. First, in a certain respect, Montague's central assertion, "I reject the contention that an important theoretical difference exists between formal and natural languages," is easily supported. That is, if one proceeds from the scarcely contestable assumption that both formal and natural languages are structured, then *in this respect* there is no difference between the two. One can (and must) further assume that the structures of these two kinds of languages can be identified and presented; this could be termed the *reconstruction* of the languages. There are, however, various ways in which such a reconstruction can be understood. The most precise would presumably be the one that relied upon formal means. In both cases (i.e., with natural and with formal/artificial languages), one would end up with formally reconstructed (presented) structures.

Does it follow from this that there is "no important theoretical difference" between formal and natural languages? That depends upon how one understands "important" and "theoretical." If "theoretical" concerns only the mode of presentation, particularly with respect to the *adequate* (in the sense both of *precise* and of *complete*) articulation of the language that is being theorized, it would be difficult to maintain that there is *no theoretical difference* between formal and natural languages. Formal languages can indeed be *fully adequately* (i.e., both precisely and completely) reconstructed, but the possibility of a theoretical reconstruction of a natural language that is adequate (in this sense) can safely be excluded. Only if one idealized a natural language would such a reconstruction be possible and indeed accomplishable, but *idealized* natural languages are not *real* natural languages. The reason for this is that real ordinary or natural languages differ from formal languages with respect to two factors that are of central relevance to the question of reconstructibility of the sort that is here in question. First, natural languages are comprehensive in every respect: they are not only and not even chiefly theoretical languages (in the sense of being languages that contain only indicative sentences); instead, they are languages of communication and thus contain everything that belongs to the articulation of this immense domain. Moreover, because of the way in which they are comprehensive, they are largely undetermined, vague, and imprecise. In this respect, natural languages are modes of articulation with absolutely open and undetermined borders. The second centrally relevant factor distinguishing natural languages from formal languages is that natural languages are living languages: they have histories, are constantly under development, and are conditioned and determined by the most disparate factors. That these ways in which they differ from formal languages are important is scarcely contestable.

As far as the philosophical language intended and postulated by this book is concerned, it is connected to natural language, but in such a way that it excludes as far as possible the incorrectnesses, indeterminacies, vagueness, etc., that characterize the natural language to which it is connected. Precisely these factors that radically distinguish natural from formal languages are excluded from this philosophical language. This holds particularly for contentual aspects, with respect to which the philosophical language envisaged here differs markedly from formal languages as well as from natural language. Formal languages, precisely because they are formal, do not at all thematize the contentual side, which essentially characterizes both natural and philosophical languages. On the contrary, formal languages introduce certain central contentual concepts as primitive concepts that they neither thematize nor bring into question as such (i.e., as contentual).

An example that plays a central role in this book can illustrate this point. In the formal language most widely used by scientists and philosophers—first-order predicate language—the concept "object" is used as the referent for names and/or singular terms and as the value of bound "individual variables" (which, for this reason, would better be termed "object variables"). In formalizations that are accomplished by means of the language of first-order predicate logic, this concept is simply understood and used as a kind of unanalyzed "primitive quantity." That this concept conceals radical onto-logical problems is thereby wholly ignored. This leads to a paradox that every serious philosopher should investigate: that a given informal presentation of a philosophical conception can be formalized and thus can be translated into formal language is insufficient for bringing the corresponding (contentual) conception into genuine clarity; quite to the contrary, formalizations can serve to conceal deep contentual problems and unclarities.

[1.1b] Overcoming ordinary language need not, however, take the form of a reduc- 109
tion to a formal language. Two examples from the nineteenth and twentieth centuries reveal that in philosophy there have been significant attempts, in aiming to develop languages appropriate to philosophy, to transform ordinary language into different languages—ones that are artificial, but *not* formal.

The first example is provided by Hegel, who writes (*Logic*: 708/6:406–07) the following concerning the relationship between ordinary and philosophical language:

> Philosophy has the right to choose from the language of ordinary life, which is made for the world of representations, the expressions that *seem to come close* to the determinations of the concept. It cannot be a matter of showing that a word chosen from the language of ordinary life is connected, in ordinary life, with the same concept for which philosophy uses it; for ordinary life has no concepts, but only representations, and it is [the task of] philosophy itself to recognize the concept where otherwise there is mere representation. It must therefore suffice if in the case of expressions that are used for philosophical determinations something approximating its own distinction hovers before representation, as may be the case with those expressions that one recognizes, in their representational adumbrations, as relating more closely to the corresponding concepts. (Translation altered)

Hegel's attempt to transform ordinary language is based on the central thesis that ordinary language is vague and underdetermined. This insight is also the starting point for the position, developed below, that is taken in this book. To be emphasized, however, is that the "philosophical language" Hegel develops, within which such terms as "representation" and "concept" are central, itself suffers from fundamental and irreparable unclarities (see Puntel 1996).

A second example from non-analytic philosophy is found in the work of Heidegger, who takes a completely different and quite idiosyncratic position. He writes ([1969/1972: 72/73),

> The difficulty [for philosophy] lies in language. Our Western languages are languages of metaphysical thinking, each in its own way. It must remain an open question whether the essence of Western languages is in itself marked with the exclusive brand of metaphysics

and thus marked permanently by onto-theo-logic, or whether these languages have other possibilities of utterance [*sagen*] and thus also the possibility of a telling silence [*sagenden Nichtsagens*].

A central feature of Heidegger's consideration of language is his attempt to get to what he often terms "the ground" of (ordinary) "Western" languages in order to rediscover the "primordial meanings" of their words, meanings that in the course of development of the languages have putatively slipped into ever deeper oblivion and have thus become deeply buried. One revealing example is his introduction and analysis of the word *Ereignis*, an ordinary-language term in German whose ordinary English counterparts are "event" and "occurrence." On the basis of obscure reflections involving the root of the term *eignen* (to be suitable, to possess), he arrives at one of the grounding theses of his later philosophy, the thesis of the "belonging together of man and Being [*Zusammengehören von Mensch und Sein*]." According to Heidegger (36/29),

> The *belonging* together of man and Being... drives home to us with startling force that and how man is delivered over to the ownership [*vereignet*] of Being and Being is appropriate to [*zugeeignet*] the essence of man. Within the combine [*das Gestell*] there prevails a strange ownership [*Vereignen*] and a strange appropriation [*Zueignen*]. We must experience this owning [*Eignen*] in which man and Being are delivered [*ge-eignet*] over to each other simply, that is, we must enter into what we call *the event of appropriation* [*Ereignis*]. The word "appropriation" [*Ereignis*] is taken from a mature language. "Event of appropriation" [*Ereignis*], thought of in terms of the matter indicated, should now speak as a key term in the service of thinking. As such a key term, it can no more be translated than can the Greek λόγος or the Chinese Tao. (Translation altered)

Heidegger is fully aware that for him, the word *Ereignis* no longer means "what we would otherwise call a happening, an occurrence." In part for this reason, Heidegger does the precise opposite of what Carnap calls "explication." By explication, Carnap means "rational reconstruction" (i.e., the development of a meaning related to natural language, but one for which natural language is not made the measure, and one that does not presuppose that the natural language of the time is somehow based on some "primordial" meaning). Instead, Carnap holds corrections, refinements, etc., not only to be possible but also to be unavoidable and indispensable (see, e.g., Carnap 1962, Hanna 1968, Puntel 1990: Ch. 2). The contrast to Heidegger could not be greater: Heidegger does not seek, in diverging from ordinary language, to increase rationality or intelligibility. Instead, as has been indicated, he aims at what he understands to be a historical and etymological reconstruction yielding the putatively primordial meaning that, again putatively, has been buried and forgotten in the course of history.

[1.2] An example from analytic philosophy is provided by the later Wittgenstein who, in his *Philosophical Investigations*, makes a particularly pointed and radical demand that ordinary language be recognized as the basic measure or standard for all philosophical questions:

> 116. When philosophers use a word—"knowledge," "being," "object," "I," "proposition," "name"—and try to grasp the *essence* of the thing, one must always ask oneself: is the word ever actually used in this way in the language that is its original home?—
> What *we* do is bring words back from their metaphysical to their everyday use.

Wittgenstein's demand, however cryptic it may be within the complicated context of the   111
Investigations, has been widely accepted. In addition to being cryptic, however, the demand
suppresses a premise and advocates a quite problematic reduction. Both are involved in
the move from the question, "[I]s the word [as used by some philosopher] ever actually
used in this way in the language that is its original home?" to the quasi-response that he,
as self-appointed representative of the philosophical community, provides for it: "What
we do is bring words back from their metaphysical use to their everyday use."

The suppressed premise is that all words have as their "original home" their
"actual or everyday use"; this premise seems clearly false, for a variety of reasons. Two
of the most important reasons are the following: first, if the premise is not to appear
as absurd, it must be based in the assumption that words like "knowledge," "being,"
"object," etc., that are originally used in ordinary language (their "home"), have clearly
delimitable and non-problematic ranges of uses in this language, so that their usage
therein does not raise questions, require clarifications, etc. Only given this assumption
would Wittgenstein's contention or demand be one that could be acted upon. Yet the
presupposition is in no way met, as is clearly revealed by the immense efforts philosophers have made to clarify just what such words mean. Second, "everyday use" is a
phrase that is hopelessly misleading. Whereas the question leaves open the possibility
of multiple uses—of equivocity—within "the language that is its original home," the
singular "everyday use" in the response suggests that there is a *single* such use back to
which any given word may be brought. In at least the overwhelming majority of cases,
ordinary-language words are used in multiple ways.

[1.3] As is indicated above, the vast majority of analytic philosophers take positions
lying somewhere between those considered in the two preceding sections. Generally,
ordinary language is considered to be the contentual foundation for philosophy in the
sense of providing, above all, information about how the world is constituted. With
respect to the logical structure of ordinary language, analytic philosophers generally rely upon first-order predicate logic. Quine is the definitive example in that he
subjects philosophical and scientific language to a regimentation by introducing a   112
canonical notation he considers first-order predicate logic to provide. Considered as a
whole, however, Quine's position is not fully clear, as is shown by examination of two
passages.

[i] On the one hand, he understands canonical notation as not depriving ordinary
language of its status as fundamental:

> [T]o paraphrase a sentence of ordinary language into logical symbols is virtually to
> paraphrase it into a special part still of ordinary or semi-ordinary language; for the shapes
> of the individual characters are unimportant. So we see that paraphrasing into logical
> symbols is after all not unlike what we do every day in paraphrasing sentences to avoid
> ambiguity. The main difference apart from quantity of change is that the motive in the one
> case is communication while in the other it is application of logical theory. (1960: 159)

On the other hand, he subjects the objects taken into the framework of the canonical
notation to a fourfold reductive interpretation that he summarizes (1981: 18) as follows:

> In the first example, numbers were identified with some of the classes in one way
> or another. In the second example, physical objects were identified with some of the

place-times, namely, the full ones. In the third example, place-times were identified with some of the classes, namely, classes of quadruples of numbers.

The fourth example is mind-body dualism, which Quine (19) presents as reducible to a physicalist monism: "[W]e can settle for the bodily states outright, bypassing the mental states in terms of which I specified them. We can reinterpret the mentalistic terms as denoting these correlated bodily states."

This procedure forces the question of what, from ordinary language, remains. The introduction above all of classes, classes of quadruples of numbers, etc., presupposes a wholly different language.

From consideration of the two passages in conjunction emerges the question of what more precisely it can mean to say that first-order predicate language, based as it is on the fundamental structures of natural language, is to be considered the language of scientific theories.

[ii] In various contexts, Quine maintains the "semantic primacy of sentences," whose introduction—which he credits to Jeremy Bentham—he compares with the Copernican Revolution in astronomy: "The primary vehicle of meaning is seen no longer as the word, but as the sentence" (69). If however this is accepted, it is no longer fully clear that first-order predicate language or logic can or indeed must still be viewed as *the* language of science (and, presumably, of philosophy). What Quine designates the "semantic primacy of the sentence" corresponds to what is usually termed the context principle. This issue is considered in Section 3.2.2.4.

The unclarity of Quine's positions concerning both natural language and the context principle is particularly visible from examination of his famous procedure for the elimination of singular terms. This too is examined in Chapter 3 (Section 3.2.2.3.2.3).

Within analytic philosophy, the relation between ordinary and philosophical language remains controversial. A quite interesting illustration of this is the utter disagreement between two of the best-known recent analytic philosophers, Quine and Davidson. Davidson (1985: 172) characterizes their difference as follows:

> Like Quine, I am interested in how English and languages like it (i.e., all languages) work, but, unlike Quine, I am not concerned to improve on it or change it. (I am the conservative and he is the Marxist here.) I see the language of science not as a substitute for our present language, but as a suburb of it. Science can add mightily to our linguistic and conceptual resources, but it can't subtract much. I don't believe in alternative conceptual schemes, and so I attach a good deal of importance to whatever we can learn about how we put the world together from how we talk about it. I think that by and large how we put the world together is how it is put together, there being no way, error aside, to distinguish between these constructions.

Davidson's position is the one generally taken by analytic philosophers who make little or no use that is worth mentioning of formal means and who rely exclusively upon ordinary language. As this book makes clear, the work of these philosophers, in all of their areas of inquiry, is significantly weakened precisely by this reliance.

To be noted in addition is that this book rejects Davidson's contention that there are no alternative conceptual schemes, at least if this contention is understood as the denial of the multiplicity of theoretical frameworks. This issue is considered in detail particularly in Chapter 5.

[2] Clarification of the relationship between ordinary and philosophical language requires consideration of a number of issues.

[i] The genuinely important and interesting philosophical question is not whether one uses a formal or a non-formal language *as a pure medium of presentation*. The use of formal language is completely compatible with a contentually naive and even scarcely defensible philosophy. An example that plays a central role in this book is the use of first-order predicate logic (with the semantic interpretation that is standard) to formally present a substantialistically oriented ontology. The genuinely decisive philosophical question is whether ordinary language is at all sufficient for the presentation of the contentual theses of philosophy. To this question, this book responds decisively in the negative. Attempts to develop a language appropriate to philosophy must therefore concentrate fundamentally on the examination of the contentual insights articulated one way or another in ordinary language. To be asked at the outset is whether the ontology and/or metaphysics implied by ordinary language can stand up at all to an analysis employing strict criteria of intelligibility. That it cannot is demonstrated in various ways in the following chapters.

114

The conception developed in this book, albeit thus far only intimated, does not hold that ordinary language has no significance or value for philosophy. On the contrary, it must be emphasized that ordinary language plays the extremely important (at least pragmatically scarcely avoidable) role of a starting point for the development of philosophical theories. The significance of such a starting point is not diminished by the fact that, within the framework of a developing theory, ordinary language is overcome. The following subsections consider several important aspects of the conception briefly described in this paragraph.

[ii] As is shown above, no ordinary language as a whole is a theoretical language, because each is a language of communication, with the many dimensions or aspects that characterize the phenomenon of communication. One of these dimensions, the indicative, is theoretical, but only incipiently so. This indicative dimension of ordinary language (including such sentences as, "Sunrise will be at 6:35 tomorrow morning," and "That car is speeding," etc.) enters as a rule as embedded in contexts of communication, not of theoretical discourse. A context of communication is a complex of various linguistic dimensions; it contains emotive, self-expressive, imperative, interrogative, indicative, and yet other elements. Because of this embeddedness within the communicative complex, the theoretical-indicative components of ordinary language cannot develop explicitly *as* theoretical-indicative.

The aspects of incipiently theoretical ordinary language most in need of refinement, with respect to the requirements of fully developed theoretical language, are its vagueness and underdeterminacy. To be sure, ordinary language meets the standards of pragmatic clarity and determinacy required for everyday communication. When we converse, we typically know, with pragmatic sufficiency, what we mean and what those with whom we speak mean; that we do is attested by the communicative success we typically achieve. This pragmatic clarity and determinacy cannot—at least not fundamentally—be doubted, because if they were, the phenomenon of mutual understanding would be fully inexplicable. But to confuse or equate this pragmatic sufficiency with adequacy for the purposes of theory is a fundamental error, albeit one that is present in many contemporary philosophical schools of thought and positions.

115

[iii] That the ordinary-language meanings of such terms as "object," "world," etc. are vague and underdetermined is notorious. The vagueness and un- or underdeterminacy of ordinary language arise with full clarity, however, only when ordinary-language meanings are considered from a strictly theoretical perspective. In the contexts of everyday life, within which ordinary-language expressions are usually used, their semantic values are fully determined, but only within the contexts of everyday usages. That the semantic values of ordinary-language terms are heavily dependent upon variable contexts that even when specific are generally unspecified results in the vagueness and indeterminacy of those terms when they are considered in abstraction from any such context. This explains why the countless attempts to "reconstruct" those meanings, as accurately as possible, yield radically diverging "semantics of ordinary language."

The vagueness and underdeterminacy of ordinary language visible from the theoretical perspective characterize its logical as well as its contentual aspect. This is clear from, for example, ordinary-language uses of such terms as "if . . . then . . .," "all," "only if," etc. Generalizing such observations, one can speak of an "informal" or "intuitive" logic that is immanent to ordinary language. Thus, for example, ordinary language obviously contains "natural" inferences. Analytically oriented philosophers and logicians tend first intuitively to ascertain the (semantic) correctness of such deductions, and then formally to justify them. The logician Ulrich Blau (1978: 13) raises the question of who or what is here in control, and concludes that the process involves a "certain circularity": "It appears to me that the controlling is reciprocal. Or, to be somewhat more precise: intuitive logic ultimately controls itself, at times revising itself; to accomplish the controlling or revising, it uses or can use formalizations of various sorts, but these are of course themselves revisable."

This view does not appear to do justice to formal logic as the latter is understood and employed at present. "Intuitive logic" is far too vague and indeterminate to "control itself," much less formal logic, because logical control requires a controller that has clear and determinate controlling factors, and "intuitive logic" has no such factors. It is no doubt the case that formal logic, strictly as such, can be linked to various intuitions articulable in ordinary language. Such linking is, however, exclusively a matter of discovery or of concrete use or application, not one that can or could determine the validity of logical structures. The finding of ordinary-language counterparts for logical structures can in some cases be interesting and/or instructive, but in no way controls or validates those structures.

Blau, in the work from which the above citation is taken, maintains that he wants to subject classical logic to a revision, but this formulation is misleading. He is concerned not with a revision of logic itself or as such, but with a different view of the relation between (classical) logic and ordinary language. Because Blau accords a central importance to ordinary language, he aims to take into account as many ordinary-language formations as possible; this leads him to introduce a three-valued logic. Yet he presents this, significantly, not as an alternative to classical (bivalent) logic, but as an expansion of it (see 15). The question, however, then arises whether this view can be brought into harmony with the just-described and -criticized position he takes concerning the "intuitive logic" he takes himself to find in ordinary language.

[iv] Many philosophers whose views diverge greatly on other matters agree that every formal, scientific, or philosophical language presupposes ordinary language as its

background language, and that for this reason there is no way to get beyond or beneath ordinary language. Differently stated: according to these philosophers, ordinary language is the ultimate and the only truly indispensable metalanguage. The claim is that all other "languages" build upon it, in one way or another; philosophical language, particularly, is understood by these philosophers as at most an extension of ordinary language, whereby, however, the precise meaning of "extension" remains unspecified.

This view, too, however, overlooks fundamental aspects of the relationship between ordinary language and philosophical languages. Ordinary language is, incontestably, self-referential; in Tarski's terminology, it is characterized by "universality" (1933/1956: 164) or "semantic closedness" (1944: 348–49), and therefore necessarily contains or   117 generates antinomies. In other words, it is at once, so to speak, foreground language and background language, or object language and metalanguage. But does this necessarily make it the ultimate metalanguage, or the indispensable background language for every theoretical language (every formal, scientific, or philosophical language)? Many analytic as well as hermeneutically oriented philosophers with otherwise widely diverging views hold—to be sure with quite divergent intentions and motives—that it does, yet in doing so, they ignore a number of central factors. That a theoretical language (e.g., a philosophical language) can adopt terms and grammatical constructions from ordinary language—that particularly in the course of its history, spanning more than two millennia, philosophy has regularly relied on such adoptions—cannot be denied. But two additional questions must be raised. First ([a]), what, precisely, does this "adoption" involve? Second ([b]), is such adoption, however understood, unavoidable? Is there no alternative?

[a] The "adoption" here in question cannot be understood as any sort of transposition of a piece or fragment of ordinary language into philosophical language such that the piece or fragment would remain intact. If it were anything of this sort, then all the vagueness and underdeterminacy of ordinary language would be retained. It would then, however, be impossible for philosophical language to overcome vagueness and underdeterminacy. The adoption must therefore be understood to involve at the very least some sort of purification, correction, or refinement of the pieces or fragments adopted into philosophical language. In part for this reason, the thesis that philosophical language is an extension of ordinary language is misleading in that it suggests that philosophical language retains all of ordinary language, but adds something to it. The reverse is true: precisely because it refines (modifies, corrects, excludes elements of, etc.) ordinary language, philosophical language ceases to be ordinary.

[b] It would be difficult to establish that adoptions from ordinary language of the sort just considered are indispensable components of any philosophical language. Why should it not be possible to develop a philosophical language that had no such components? If one first, with philosophical intentions, constructed a purely formal language and then developed a semantics for this language without relation to any pieces or fragments of any sort from any natural language, one would have such a philosophical   118 language. The question here is only whether this is possible in principle, not whether it is to be recommended, or even whether it actually could be done.

[v] Against the two positions just developed ([iv][a] and especially [iv][b]), many philosophers of various sorts would raise what they would take to be serious objections. This subsection considers two of these. The first is based on the contention that assuming and using an ordinary language is a condition of the possibility for

developing any philosophical language; the only alternative would be an impossible beginning with a *tabula rasa*. But if the slate (to continue the metaphor) is not blank, then the only "writing" that could be on it would be ordinary language.

To this objection there are two responses. First, although this objection appears initially to be illuminating, closer scrutiny reveals it to depend upon a confusion of two completely different linguistic aspects or levels: the pragmatic (or that having to do with the psychology of learning) and the level of genuine technical theory. To one considering the matter from the pragmatic perspective—which focuses on the language user, not on the subject matter—it will appear obvious that every language user must begin somewhere or other, and that the only possible point for beginning is one or another ordinary language. But this fact is, in principle, fully contingent. Reference to phenomena like language learning proves nothing here, because the question returns: if learning a second language always requires the prior mastery of a first language, then how is the first language itself learned (and how would it ever have developed)? No matter what explanation is given of the phenomenon of language learning, the problem of learning a theoretical language in the sense that has here been clarified is no greater than the problem posed by the learning of so-called ordinary or natural languages.

The second response to the contention that only an ordinary language can be a first language is that, if the objection were telling, then how it could be possible to learn formal logic would be inexplicable, because formal logic *as such* is absent from ordinary language. One can try to follow paths of various sorts to get to the level of formal logic, including examples, explanations, exercises, etc., but none can provide a continuous transition, precisely because the move is from one fully distinct level to another. One can only say, negatively, that formal-logical connections are not contentual connections—but no merely negative determination can make the formal-logical dimension intelligible; what is required is instead some sort of qualitative leap.[3]

A second objection to the positions presented above in [iv][a] and [iv][b] may be briefly formulated as follows: the position defended here has the simply unacceptable consequence of making mutual understanding impossible. Because the rejection of mutual understanding cannot be accepted, any position that rules it out in principle is *a fortiori* unacceptable. This objection, however, in all of its various forms, is based on a series of assumptions that in part are false and in part involve confusions and misunderstandings.

It is clear that mutual understanding presupposes a language common to the relevant participants. But why should this common language have to be an "ordinary" language (in the sense assumed here)? When philosophers or scientists come to mutual understandings, the question is whether they do this as users of ordinary language or instead as philosophers/scientists. If the first is the case, then clearly a non-theoretical language must be used. But if the second is the case, one must ask what, more precisely, "mutual understanding" can mean in this context.

---

[3] Hegel is fully clear about this matter. To be sure, he rejects formal logic (the formal logic of his time!), but he conceives of the logical as clearly separate from natural language and from other levels, such as the mental/psychological level—in a word, from everything "non-logical" (in his sense); this separation is similar to the one that holds between the formal-logical level (in today's sense) and every other level. See Hegel (*Logic: 70/5:68*).

To attain clarity with respect to this central question, Wittgenstein's idea of language games can be useful (although not all of Wittgenstein's conception need be incorporated). In brief, one can say that every language game involves its own possibilities for mutual understandings—that every speaker who participates in a given language game *eo ipso* "communicates" with other "speakers" within the game, that is, *eo ipso* attains mutual understandings with them (anyone who fails to understand is thereby established as a non-participant). The basic idea of the language game can even provide decisive clarification, indeed definition, of the concept of mutual understanding. The concept of mutual understanding emphasized by the second objection cannot be clarified by means of the conception of ordinary language as the ultimate, unsurpassable metalanguage, because ordinary language essentially and thus unavoidably encompasses a plurality of language games, such that no one of them simply grounds all of the others by serving as something like *the* metalanguage game. There is a relation of metalevel to object level, among language games, only in a relative respect. For example, from the perspective of everyday life, ordinary language unquestionably plays the decisive role, but this perspective is only one among many. In other respects, for example from the perspective of science (and thus of philosophy), ordinary language is not *the* metalanguage, but is instead merely an object language. When scientists or philosophers proceed as scientists or philosophers, they then proceed with respect to natural language from the perspective of a metalevel to an object level. But because they come to mutual understandings precisely *as* scientists/philosophers, it cannot be said that ordinary language is the condition of possibility of their mutual understanding *qua* scientists/philosophers. The second objection, despite its popularity, is thus the result of an evident confusion.

2.2.3 *Philosophical Language as Theoretical Language*

2.2.3.1 The Linguistic Criterion for Theoreticity

Linguistic form plays a simply decisive role in the characterization of a given discourse. Theoretical discourse, practical or pragmatic discourse, aesthetic discourse, literary discourse, and other types of discourse, superficially viewed, have their own linguistic forms, which appear to provide bases for distinguishing among them. These distinguishing features appear most clearly in the case of practical (pragmatic) discourse, because the sentences constituting this type of discourse (e.g., imperative sentences) are most clearly distinct from all other sentences and/or forms of sentences. More precise analysis, however, reveals that the situation is in some ways quite complicated. To develop an unambiguous linguistic criterion for theoreticity, it is necessary to investigate central structural moments that are *syntactic* as well as ones that are *semantic*.

[1] If there is a linguistic criterion for theoreticity, it must be decisively readable from the syntactic form of the sentence. In this respect, indicative sentences might appear to be clearly distinct from imperative sentences ("Always tell the truth!") and what may be termed "aesthetic" sentences ("How beautiful the Dolomites are!"), but examples reveal that in ordinary or natural languages, syntactic form *alone* is not sufficient. The sentence "I'll have two pounds of apples," uttered at a fruit stand, unambiguously has the syntactic form of the indicative, but the context makes clear that semantically, the sentence is (at least quite often) a polite form of a request or even indeed of an imperative: "Give me two pounds of apples." Moreover, sentences that contain evaluative terms can have the syntactic form of indicatives (e.g., "The good is

that toward which everything strives" (Aristotle) or "This is good"). Depending upon how such sentences are interpreted, they can be taken as indicative sentences or as practical sentences (e.g., the syntactically indicative sentence "It is good to tell the truth," if interpreted as equivalent to "Tell the truth," is practical). Sentences that have indicative forms and aesthetic components are also not simply unambiguous. The sentence "The Dolomites are beautiful" can be understood as a purely indicative-theoretical sentence ("It is the case that the Dolomites are beautiful"), but also as a purely aesthetic sentence ("How beautiful the Dolomites are!").

These ambiguities on the syntactic level can be explained only if one explicitly develops the factor that determines the precise status of the sentence. This factor is an operator that explicitly or implicitly precedes the sentence. The thesis that there are three equiprimordial modes of relation of the mind to the world (the universe, being as a whole)[4] provides the basis for the distinct thesis that there are three fundamental forms of sentences and correspondingly three fundamental sentence operators: the theoretical, the practical, and the aesthetic. At this point, these operators are introduced simply for methodological reasons, and are only briefly explained. Central emphasis is of course placed here on the theoretical operator; the other two operators are considered in detail below in the sections on ethics (4.3) and aesthetics (4.4).

To formulate a general linguistic criterion for indicativity/theoreticity, one can find inspiration in Wittgenstein's *Tractatus*: "That there is a general form for the [indicative] sentence is proved by the fact that there can be no sentence whose form one could not have foreseen (i.e., constructed). The general form of the [indicative] sentence is: It is the case that such and such"[a] (*Tractatus* 4.5; translation altered).[5]

A sentence φ is indicative (hence theoretical), then, if and only if a syntactically correct sentence results from the application to φ of the operator "It is the case that." At least three reasons qualify this as an appropriate articulation of a linguistic criterion for theoreticity on the syntactic level. First, syntactically, all sentences satisfying the

122

---

[4] This thesis is in accordance with the comprehensive architectonic that is the basis of the structural-systematic philosophy. It is considered in various places in this book.

[a] Wittgenstein's original phrase, of course quoted in *Struktur und Sein*, is "*Es verhält sich so und so.*" There is no English rendering that adequately captures and articulates the peculiar sense and force of the German wording, which is reflexive as well as impersonal, but the syntactic differences between "*Es verhält sich so und so*" and "It is the case that such and such" are irrelevant to their functioning, in their respective languages, as linguistic criteria for theoreticity. Syntactic and semantic differences between "*Es verhält sich so dass*" and "It is the case that" *do* become relevant at a later stage of the development of the structural-systematic philosophy; for details, see note a to Chapter 5 (p. 404).

[5] The introduction of this formulation at this point does not imply that Wittgenstein himself intends it as a criterion for theoreticity. His intention is presumably quite different. Here relevant is the remark that he later makes, in his *Philosophical Investigations*, to the cited passage from the *Tractatus*:

114. (*Tractatus Logico-Philosophicus, 4.5*): "The general form of the sentence is: It is the case that such-and-such."—That is the kind of sentence that one repeats to oneself countless times. One thinks that one is tracing the outline of the thing's nature over and over again, and one is merely tracing round the frame through which we look at it. (Translation altered)

criterion are unambiguously indicative-theoretical; the criterion thus forecloses the possibility of any such sentence being taken as a disguised form of any other sort of sentence. Second, that indicative-theoretical sentences can be preceded by the operator makes explicit precisely the decisive point with respect to the determination of theoretical discourse: this discourse, in distinction from all other discursive forms, literally brings to language just what is the case—and not anything like what ought to be the case, could be the case, must be the case, etc. Third, the criterion is completely non-restrictive with respect to the syntactic forms of sentences that can express states of affairs, and thus also to the forms of states of affairs they express; the sentences and states of affairs can have varying degrees of complexity. Of particular importance for this book is that the criterion is in no way bound to subject-predicate sentences, but allows also for the likes of "It's raining" and "It's redding," which, as is indicated above, have the form of primary sentences expressing primary propositions.

A possible misunderstanding must, however, be avoided from the outset. The operator "It is the case that" could be misread as applicable only to purely empirical sentences: "It is (implicitly understood: empirically) the case that. . . ." If this operator were used in this sense, however, one could not understand (for example) necessary, *a priori*, etc., theoretical sentences as having the form qualifiable by means of the operator. But this understanding or way of reading the operator is neither compelling nor here intended. "Empirically," "*a priori*," "*a posteriori*," "necessarily," etc., can indeed qualify the operator. That the "such-and-such's" of sentences of the form "(It is the case that) such-and-such," without further qualification, are often understood as empirical sentences is the result of a thoughtless habit. But this is a matter of a specific linguistic custom, not a factor given with the sentence form itself.

To attain maximal clarity concerning the status of theoretical sentences, it is advisable in certain important contexts to make the theoretical operator explicit by means of a special symbol. This book uses the symbol "Ⓣ." Thus, for example, the sentence "It   123 is the case that the earth revolves around the sun" is formalized as "Ⓣφ." Philosophy, as a theoretical science, presents only theoretical sentences. It is of the utmost importance that this fact be emphasized, because philosophy is often confused with one or another practical activity (e.g., advice, help with living, political engagement, etc.). This explains much of the contemporary confusion concerning the status of philosophy. But philosophy must also consider sentences that are not theoretical, but are instead practical and aesthetic. These sentences, however, are only *objects* of philosophical consideration. Likewise, to determine precisely the status of these sentences in specific contexts, this book uses the symbol "Ⓟ" for the practical[6] and "Ⓔ" for the aesthetic. "Ⓟψ" is to be read, somewhat awkwardly and artificially, as "It is ethically obligatory (or forbidden or permissible) that ψ" (e.g., "It is ethically forbidden that innocent human beings are executed"). The informal reading for the *aesthetic operator* is more difficult but equally artificial. "Ⓔ χ" can be informally paraphrased as "There is an aesthetic presentation such that χ" (e.g., "There is an aesthetic presentation such that Beethoven's Fifth Symphony is magnificent").

---

[6] Of the various forms and/or domains of the practical or pragmatic dimension, this book thematizes only the practical-*deontic*; it thus does not examine the technical, the artisanal, the political, etc. See Section 4.3.

To be noted with respect to syntax is that the basic operators can be combined: an operator can appear within the scope of another operator such that the latter is the determining factor for the former. For example, "Ⓣ(Ⓟ(ψ))" would be read as, "It is the case that it is ethically obligatory (or prohibited or permitted) that ψ," and "Ⓣ(Ⓐ (χ))" as "It is the case that there is an aesthetic presentation such that χ." Considered purely syntactically, every combination is in principle possible (i.e., thinkable), but it is not clear that every combination is sensible. Universal combinability appears to be philosophically excluded. The real reason for this is a semantic one that is presented in Subsection [2].

[2] In addition to the linguistic-syntactic form, an additional factor—a semantic one—must be explicitly provided if a secure linguistic criterion for genuinely indicative-theoretical sentences is to be formulated, because theoreticity involves language not only as a semiotic-syntactic system but also and decisively as a semiotic-semantic system. Theoreticity is essentially concerned with its relation to the world (to the universe), and in such a manner that this relationship itself is registered and articulated. This occurs essentially by means of language as a semantically structured system of articulation.

[i] The semantic factor is that every linguistically-syntactically indicative or theoretical sentence *expresses* something—more precisely, an informational content—in a *wholly determinate manner*. This *expressum* of the sentence is normally termed a "proposition" or a "state of affairs." The details of this centrally important issue are thoroughly considered in Section 3.2.2.4. In the present context, only the specification "expresses in a wholly determinate manner" need be clarified. The clarification is needed because practical and aesthetic sentence also "express" something in that such sentences, too, have what are ordinarily called "meanings"—they say or articulate something or other. To be sure, the opinion or—better—the habit has developed within the philosophical literature of saying that only indicative sentences that are clearly theoretical express anything, and most philosophers assume that what they express are propositions (or states of affairs). But such an opinion or habit is untenable, because it is wholly undifferentiated and thereby wholly ignores essential factors. Some of these factors are visible in the sentence "One should always tell the truth," when it is artificially and awkwardly paraphrased as, "It is ethically obligatory that the truth is always told," and formalized as $Ⓟ_{UD}(ψ)$, whereby "UD" symbolizes, "universally deontically." One would normally say that this sentence—which is of course syntactically indicative—articulates a general or universal ethical norm. This motivates the question of what such norms are if not propositions or states of affairs. In response to this question, this book characterizes these norms as expressa of (practical) sentences.

Concerning the issues introduced in the preceding paragraph, one finds in the philosophical literature immense confusion. Two considerations play decisive roles, and must therefore be clarified. [a] In general, it is accepted as some kind of fundamental dogma that between norms *qua* values on the one hand and states of affairs or facts on the other the strictest possible distinction is to be drawn. This distinction is generally understood as according to facts an ontological status, but to values only a twilight status that is not genuinely ontological. As this book shows, this is correct only in one respect and thus only in part (i.e., insofar as there is a *specific* distinction between "values" and "facts" (or "states of affairs")), but not a distinction

124

that conceives of them as two utterly different types of entities). It becomes clear below that within the transparent ontology and carefully developed terminology of this book's theoretical framework, "values" and "facts" are reinterpreted as distinct types of one and the same ontological category or structure: *primary fact*. The term "fact" thereby attains a transparently extended signification, fully in accordance with the contemporary philosophical tendency to use, without problems and without any danger of misunderstandings, such terms as "logical facts," "semantic facts," etc. To be sure, such a view is coherent only if an explicit and innovative ontology is developed and defended.

[b] The second factor is the correct insight that non-indicative sentences, and particularly practical sentences, can "mean" or "articulate," but what practical sentences articulate relates to *demands* or even involves or states demands. This insight is only partly correct because although it grasps an essential point, it does not explain that point correctly or, most importantly, sufficiently. Practical (deontic) sentences (and with them aesthetic sentences) "articulate" or "express" something or other and thus have their *expressa*, but their *modes* of articulation/expression are *wholly different* from the theoretical mode. The mode of articulation/expression of a *normative fact* by means of a practical (deontic) sentence is that of *demand*. The distinction is brought clearly to expression by means of the operators ⓉⓣⓉ and Ⓟ: "It is the case that ...," and "It is ethically obligatory (or prohibited or permitted) that...."

[ii] It is now possible to identify the fundamental *semantic* characteristic of the *theoretical* operator (and thus of the entire dimension of theoreticity): only sentences that are determinable by this operator *aim directly and completely at objectivity*. They bring exclusively the subject matter into view, not (also) anything else (e.g., the attitude of a subject, etc.). This fundamental characteristic is articulated absolutely clearly and explicitly in the *truth concept*. Only sentences that begin explicitly or implicitly with the theoretical operator can qualify as true. For this reason, the operator "It is true that ..." is an additional form—indeed, the fully determined form—of the theoretical operator "It is the case that...." The form of the theoretical operator "It is the case that" is thus not (yet) the fully determined form of the theoretical operator, because the form "It is the case that..." can also be applied when the sentence to which it applies expresses, at least initially, *only* a proposition or state of affairs, and *not* a fact.[b] In many contexts, such an "incomplete" application of the theoretical operator is appropriate because one wants to express something theoretical only generally (i.e., in a not (yet) fully determined manner). With the application of the truth operator, the theoretical presentation of the relation of the mind to the world (the universe, being) is brought to completion.

One significant consequence of the preceding considerations is the following: with respect to the use of such concepts as "articulate," "express," etc., in conjunction with sentences, one must strictly distinguish among three forms or modes: theoretical, practical, and aesthetic articulation/expression. Each of the three cases involves a relation

---

[b] In ordinary English, "It is the case that ..." and "It is true that ..." are often used synonymously. As is indicated more than once above, however, ordinary-language usage in no way governs the philosophical language used in this book; that language distinguishes between these operators in the manner described in the sentence to which this note is appended.

of the mind to the world (the universe, being), but the three remain fundamentally distinct. These are three equiprimordial ways of presenting the world (the universe, being), none of which is reducible to either of the others.

To be added is that at least within the framework of the structural-systematic philosophy, the theoretical operator Ⓣ is the absolutely universal and fundamental operator, precisely because both of the other two operators can meaningfully come within its scope. Examples would include such forms as "Ⓣ(Ⓟ(ψ))," to be read as, "It is the case that it is ethically obligatory (or forbidden or permitted) that ψ." Such sentences, taken as wholes, are *theoretical* sentences, but theoretical sentences with practical-deontic contents. Such sentences are primary constituents of philosophical theories of normative ethics; precise determination of the status of these sentences makes possible determination of the precise status of such theories (see Section 4.3). Also non-problematic are sentences of the form, "Ⓣ(Ⓔ(χ))"; these are sentences that would be primary constituents of philosophical theories of aesthetics. But sentence forms that have the practical operator as the main operator and, within its scope, the theoretical or the aesthetic operator (thus, "Ⓟ(Ⓣ(φ))" and "Ⓟ(Ⓔ(χ))") are philosophically senseless: how matters, including aesthetic matters, in fact stand within the world cannot be made dependent upon any sort of demand. On the other hand, the sentence form that includes the theoretical operator within the scope of the aesthetic operator seems to be wholly sensible. Many theoreticians (particularly mathematicians and natural scientists) often and fully intelligibly speak of such things as "elegant" formulas and proofs, and of mathematical and natural-scientific theories as magnificent works of art. If one translates this way of speaking into the terminology and notation used here, one gets the sentence form, "Ⓔ(Ⓣ(φ))": "There is an aesthetic presentation as a magnificent work of art that it is the case that φ."

2.2.3.2 Basic Features of a Program for the Development of a Systematic Philosophical Language

When a or the philosophical language comes into question, the notion of an *expansion* or *extension* of ordinary language plays a non-negligible role, but this notion is shown above to be ambiguous. In a quite broad sense, this expansion or extension can be understood to mean only that there is a certain dependence or reliance on ordinary language; such reliance is no cause for concern. Nevertheless, it appears preferable to avoid this potentially misleading term. What is decisive is that a philosophical language be (or strive to be) completely theoretically transparent. Here, a philosophical language that either possesses or strives for fundamentally complete theoretical transparency is termed a "systematic philosophical language." "(Completely) theoretically transparent language" here means the following: a language is completely theoretically transparent if *all of the factors* that characterize its theoretical status are precisely investigated and clarified; these are the factors that constitute its specific *theoretical framework*.

Three dimensions of language are commonly distinguished: the syntactic, the semantic, and the pragmatic. Complete philosophical consideration, however, requires the inclusion of two additional dimensions: the logical and the ontological. To be sure, the logical is intimately related to the syntactic and the ontological to the semantic, but in neither case is the relation one of identity. Be that as it may, the presumably most difficult problem concerns how the pragmatic dimension relates to the other dimensions. At this point in the account, only some general considerations concerning the interrelations among these five dimensions need be introduced.

With respect to the relation between the pragmatic dimension and the other dimensions, widely divergent positions have been and continue to be held. One extreme position ascribes to pragmatics an absolutely secondary and marginal role; the study of pragmatics then becomes a matter of purely empirical investigation of how a specific language is used by speakers and communities of speakers, such that this use is not accorded a role of any significance with respect to the other four dimensions of that language or of language in general. At the other extreme, pragmatics becomes the focal point, indeed the basis from which the other dimensions are conceived and explained.

The most important advocates of the former extreme include Tarski, Carnap, and Quine; this book, too, advocates a version of that position, but one that differs from any of its historical predecessors in ways to be specified. Its modifications involve two chief points. First, the pragmatic dimension is not here reduced to a purely empirical consideration of language; instead, everything that is connected with the pragmatic vocabulary of a language is recognized as belonging to the structure of the language itself. Second, the understanding of the syntactic, logical, semantic, and ontological dimensions of language presented here is markedly different from the understandings of those dimensions presented in the writings of Carnap, Quine, etc. Details are provided below.

The best-known advocate of the latter extreme, which ascribes centrality to pragmatics, is the later Wittgenstein (consider his slogan, "Meaning is use"), but he does not develop his conception systematically. Perhaps the most impressive systematic development of a radically pragmatically oriented conception of language and of philosophy as a whole is that of Robert Brandom, who, in his monumental book *Making It Explicit*, develops such a conception on the basis of the foundational contention, "Semantics must answer to pragmatics" (1994: 83, passim).

There are three particularly important reasons for rejecting conceptions according primacy—in whatever manner—to the pragmatic dimension. [i] The notion of "language use" or "linguistic practice," as it is presupposed by advocates of pragmatic primacy, is hopelessly underdetermined and vague. [ii] That *meaning*, however more precisely understood, cannot simply be identified with *use* is clear from the fact that one can use, in an intelligible and rational manner, only what one has already understood, that is, only that to whose meaning(s) one has prior access. If this were not so, then use would be some sort of brute fact, devoid of any content, intelligibility, or justification. [iii] The assumption of a fundamental primacy of the pragmatic dimension has significant and untenable consequences: such primacy would make impossible genuine apprehension of the world (reality, being), unrestricted recognition of the extent of anything like the mathematical dimension, etc. If pragmatics was indeed primary, then all of the dimensions of the actual and the formal (the logical/mathematical) would be "reduced," i.e., narrowed, to the narrow limits of our "lifeworld," i.e., the "world" that is subject to the scope of our "action," "action" itself having a highly questionable sense. The scopes of investigations and conclusions in mathematics and logic as well as in science and thus in philosophy are far broader than any such reduction would allow. Chapter 3 treats this topic both more extensively and in an explicitly systematic manner, particularly in conjunction with the theory of truth.

Of the five dimensions of language, the syntactic is clarified most easily, because the syntax even of ordinary language can be identified relatively precisely; a philosophical language can be inspired by this syntax, but not fully determined by it. The importance of the logical dimension can scarcely be overvalued, because presentations of connections—and such presentations are the central concern of science and philosophy—are revelations of logical relations. This also points toward the problem of how the relation of logic to semantics and ontology is to be understood. Shown below is that logic must ultimately be conceived in an ontological sense, although this to be sure requires an adequate and appropriate ontology that must of course be developed.

According to the central conception presented particularly in Chapter 3, semantics and ontology are intimately related or interconnected, indeed ultimately inseparable. Following Tarski, this book considers semantics to be the theory of the relation between language and world (see esp. Tarski 1944). To be shown, in opposition to various positions held in the past and at present, is that this determination of semantics involves a relation to ontology so fundamental and so tight that a quite determinate form of ontology emerges from it.

2.2.4 *The Centrality of Language to Philosophy*

As is emphasized above, the structural conception of systematic philosophy accords language a simply central position. This section formulates the most important arguments in favor of this thesis, and introduces and briefly explains several points that are of central significance with respect to developing a system on its basis.

Philosophy is essentially concerned with presentation. A philosophy putatively only in thought, in intuition, or in one's head is an absurdity. To be sure, it is possible and even natural to develop or possess a conception before it is (and even without its being) presented in the sense of being expressed aloud or appearing in written form; in this case, "presentation" basically means articulation, and that can occur, in principle, without any use of or reliance on visible or audible signs or symbols. The question then arises of how a so-conceived articulation can be explained (e.g., whether such explanation would require a language of thought or something of the sort). Important however as this question is in various respects, in the current context it is not of decisive significance. Of importance here is only the thesis that presentation, at least in the current sense of articulation, is an essential element of any philosophical conception. Additional explanation of and grounding for this thesis are provided below.

Language is often conceived of as a pure *medium* of presentation. This view is problematic, however, because it is generally understood as based on the assumption that there is a fundamental difference or separation between what is presented and the medium that presents it. If "language" is reduced to a so-understood medium of presentation, however, then there can be talk of an unbreakable bond between philosophy (science) and language, if at all, only in an extremely weak—and thus largely uninteresting—sense. Even, however, if one conceives of language as a medium of presentation in the sense of a semiotic system structured syntactically, semantically, and pragmatically, it can be shown that the assumption just introduced is questionable, or indeed—seen fundamentally—false. Between the "medium" and what is presented by means of the medium there cannot be a relation as loose as that presupposed by the assumption, because if there were it would be inexplicable how what was presented could *be* presented. If the medium were not suited to present the "thing itself," then it would not *be* a medium of presentation (with respect to the relevant "thing"). Between the content that is or is to be presented and the medium of presentation there must be an internal relation. This insight yields among other things the consequence that from the structuration of language can be read the structuration of the content presented by means of language. This latter insight can be designated the most fundamental assumption underlying the entire analytic concentration on language—which does not mean that analytic philosophy has made this assumption clear, or that the procedures of analytic philosophers do or have always or even generally done justice to this assumption.

One might think that one could speak in a certain sense of a separability between a medium of presentation and what the medium presents, if by "medium of presentation" were understood not language overall or in general, but instead a specific language, and if one also presupposed that for one specific linguistic formulation a different one could be substituted without there being any effect on the presentation of the content. Despite the difficulties in this domain to which, among other things, Quine has pointed by considering the "indeterminacy of translation," most contemporary philosophers make some such presuppositions. Yet doing so is questionable, precisely because it depends upon the presupposition that one linguistic formulation can be translated *literally* into another, as if language were like some kind of clothing, such that one piece of clothing could be replaced by a different one while the "clothed" body remained absolutely identical. If for example one used the English sentence "Snow is white" and the German sentence "*Schnee ist weiß*," and assumed that sentences of this form express propositions, one would usually also assume that the two sentences express precisely the *same* proposition. The questionability of this assumption is considered in Section 3.2.2.4.2 and also Puntel (1990).

131

2.2.5 *From the Dimension of Language to the Dimension of Knowledge: The Roles of Speaker and Subject*

The dimension of language or presentation is one that is in many respects (although not absolutely) self-sufficient. To be sure, such a dimension involves a relation to subjects—speakers and users of language—in some respect. The fundamental question, however, is whether there can be a pure dimension of presentation or of language (i.e., a dimension either without or independent of subjects/speakers). It might appear that such a dimension would be possible and intelligible only in or as a so-called Platonic (i.e., subjectless) world. Yet the mere fact that there are speakers or subjects does not make the question of the possible independence of language from them superfluous, hair-splitting, or unanswerable, because the question ultimately aims to clarify the status of those speakers/subjects.

The thought of the *expressibility* of the world (being, the universe, the unrestricted universe of discourse) is absolutely central to the structural conception of systematic philosophy. Expressibility, however, can be understood only as a relation that has an inverse, the relation of expressing. The latter, for its part, appears to be intelligible only if one assumes a factor that accomplishes the expressing. In other words, the world can intelligibly be expressible only if there is some means by which it can be expressed ("It is expressible but there is no way in which it can be expressed" is contradictory), and any such means of expression would qualify as a language. This insight is here viewed as the *locus originarius* of the understanding of language (better: of linguisticality). As is demonstrated thoroughly below, language in this originary sense can be conceived of as an abstract semiotic system (abstract in the sense of involving no essential relation to any user(s) of the system).

This view decisively diverges from all notions that determine language as, in essence, a semiotic system of *communication*. Whether "language" is used exclusively for semiotic systems of communication, or also for abstract semiotic systems, is initially a question of terminology. Henceforth, "language" is often used in this book to mean language in the sense here identified as originary, thus, as a semiotic system attained from the idea of the universal expressibility of being (the world, the universe).[7] As Chapter 5 demonstrates in detail, language in this sense is in principle or primarily an abstract semiotic system that contains uncountably many expressions. This broad conception of language does not of course exclude speakers/subjects, but it does give them a status fundamentally different from the one accorded them by the other (generally accepted) conception.

Before this fundamentally different status of speakers/subjects is considered more closely, there is a question to be raised: given the existence of originary language, does it follow that there must be subjects/speakers at all? The basic assumption of the expressibility of the world analytically yields the thesis that originary language articulates the structuration of the world, but does the existence of this language also entail the existence of one or more speakers of it? At this point, only the following need be noted: because originary language is not primarily conceived of as a system of communication—although it does not preclude communication—there appears to be no such direct or analytic entailment. If, however, such an entailment were to emerge, or were the necessity of speakers to be established on some other basis, that would have fundamental consequences of various sorts. One of these consequences concerns the theme of comprehensive systematicity, which is treated in Chapter 5. The current consideration begins with the fact that there are speakers and subjects; the task here is to clarify their status.

At this point, two of the most important differences between the two conceptions of language introduced above—between language in the more familiar sense and originary language—must be considered briefly. [i] The first is fundamental: speakers/subjects are determined and understood on the basis of originary language, not vice versa. Speakers/subjects are thereby totally disempowered in the sense that they play no constitutive role with respect to originary language; on the contrary, they adequately understand themselves as speakers/subjects only when they conceive of themselves as embedded or enmeshed within this language (or within the linguisticality corresponding to this language). To be sure, this relation of being embedded or enmeshed must be clarified; it is clarified below. At this point, a slogan may prove helpful: it is not the case that the world is expressible because speakers express it; on the contrary, speakers can express the world only because the world is expressible. [ii] The second difference

---

[7] When the context does not make clear the sense in which "language" is used, the qualifier "originary" is used. As should be evident but bears emphasizing, this use of "originary" is utterly different from Heidegger's.

consists in the fact that originary language has an incomparably greater expressive power than does ordinary language. As is noted above, one consequence of the fact that ordinary language is essentially a semiotic system of communication is that it can have only a finite number (or, in a determinate respect, at most a countably infinite number) of expressions, whereas originary language must—and can, non-problematically—be   133
conceived of as a semiotic system that contains uncountably many expressions (this issue is considered in detail in Chapter 5).

The thesis of the primacy of originary language brings with it the subordination, within the linguistic dimension, of the subject *qua* speaker. As is noted above, however, some philosophers have accorded and some continue to accord the subject philosophical primacy not (only) *qua* speaker, but also or instead *qua* knower (i.e., within the epistemic rather than the linguistic dimension). That this view is not tenable is shown in Section 2.3.

2.3 The Epistemic Dimension as the Domain of the Accomplishment of Theoreticity

Because it is a fact that there are subjects and because only subjects develop theories, subjects play an indisputable role in theoreticity. Quite globally, subjects accomplish theoreticity by (subjectively) producing, considering, or taking epistemic stances with respect to (objectively extant) theories. The highest subjective degree of accomplishment of theoreticity is knowledge. But how is the relationship between theoreticity, on the one hand, and the attitude or accomplishment of knowledge, on the other, to be conceived?

Within modern philosophy globally, and within a significant sector of contemporary philosophy, the dimension of theoreticity is viewed, conceived, and situated—in a word, determined—with primary reference to knowledge. The structural-systematic philosophy is fundamentally opposed to this form of determination; its emphasis is instead upon the *theoretical framework* within which a given theory develops and the *content* that is articulated within that framework. Of far less importance is that the theory or its content is known, along with whoever knows it. The structural-systematic thesis is that the accomplishment of theoreticity, particularly in its highest (subjective) form, that of knowledge, is a factor not to be neglected, but one that ultimately plays only a derivative role within the total process or the total dimension of theoreticity. This thesis is now to be clarified and supported.

2.3.1 *On the Problem of the Epistemic Subject (or of Epistemic Subjectivity)*

However one conceives of knowledge, it always involves subjects or subjectivity (the following account primarily uses "subjectivity," but not in any sense restricted by anything other than the emphasis on the subject as epistemic rather than, primarily, as linguistic). The "involvement" can be one of activity, of accomplishment, of being   134
a condition, etc. How this is to be understood more precisely depends upon how subjectivity itself is understood.

The concept of subjectivity pervades modern philosophy, and cannot be ignored with respect to contemporary philosophy. Because of the status it has been accorded within philosophy, its consideration must precede that of knowledge. No exhaustive examination—particularly, no exhaustive historical examination—is here undertaken. Instead, the three most important ways of considering and presenting subjectivity are briefly characterized: the subject as intentional, the subject as determining a decisive standpoint, and the subject as constituting objectivity transcendentally. From these arise three different types of conceptions or theories, such that subjectivity is then determined more precisely with one of the relevant designations.

[1] Subjectivity conceived as *intentional* or as *taking positions* corresponds most closely to what may be termed the normal or usual concept of subjectivity. In analytic philosophy, which considers the problem primarily in light of the phenomenon of language, one speaks of the subject's "propositional attitudes," which are articulated linguistically in such formulations as, "(Subject) S knows (believes, doubts, assumes, . . .) that such-and-such." As is shown below, the definition of knowledge as "justified true belief," formulated by Gettier in 1963 and now considered to be classic, is based on a variant of this notion of subjectivity. This conception has of course had a great variety of distinct forms.

[2] A second approach understands subjectivity as taking a or the *standpoint* that is decisive in the domains both of theory and of practice. Here too there are many variants and interpretations. For present purposes it suffices to introduce two: the *foundationalist* and the *pragmatist*.

135 [i] The *foundationalist* position is also termed the Cartesian, thanks to its classical formulation by Descartes; for the foundationalist, whatever is adequately *evident* to the subject is simply *true*. *Cogito ergo sum*, as expression of the subject's absolute self-certainty, is taken by Descartes to provide the unshakeable foundation (*fundamentum inconcussum*) for the entire edifice of philosophy and the sciences. Simultaneously with the self-certainty of the I, the first and most fundamental truth is given or attained, the truth from which all other truths can be won or attained. The self-certainty of the ego thus provides the standpoint that is absolutely irrefutable and that is decisive for philosophy and the sciences.

[ii] A great many in part quite diverse, even disparate positions are currently termed "pragmatic." Here, three are briefly characterized: classical pragmatism, Habermas's pragmatically oriented discursive-rational theory, and Brandom's semantics based in pragmatics.

[a] For the classical pragmatist, the subject who determines what is *useful* or *good* for it, as subject, has thereby determined what the subject will accept (as true). Two central passages by the founders of classical pragmatism, Charles Peirce and William James, reveal the sense in which subjectivity is thereby understood as providing the decisive standpoint. In "The Fixation of Belief" (1877/1965: 232), Peirce writes,

> [T]he sole object of inquiry is the settlement of opinion. We may fancy that this is not enough for us, and that we seek, not merely an opinion, but a true opinion. But put this fancy to the test, and it proves groundless; for as soon as a firm belief is reached we are entirely satisfied, whether the belief be true or false. . . . The most that can be maintained is, that we seek for a belief that we shall *think* to be true. But we think each one of our beliefs to be true, and, indeed, it is mere tautology to say so.

In 1903 (232n2), Peirce appends to this passage the following note:

> For truth is neither more nor less than that character of a proposition which consists in this, that belief in the proposition would, with sufficient experience and reflection, lead us

to such conduct as would tend to satisfy the desires we should then have. To say that truth means more than this is to say that it has no meaning at all.

It is here explicit that the subject, as having desires that may be satisfied, is the sole and absolute measure. To be noted, however, is that Peirce undergoes a remarkable development that leads him, among other things, to distance himself from the "pragmatism" of William James; he emphasizes the distance by coining the term "pragmaticism" to designate his own position.   136

According to William James (1907: 46–47), the "pragmatic principle" is the following:

> To attain perfect clearness in our thoughts of an object . . ., we need only consider what conceivable effects of a practical kind the object may involve—what sensations we are to expect from it, and what reactions we must prepare. Our conception of these effects, whether immediate or remote, is then for us the whole of our conception of the object, so far as that conception has positive significance at all.

James emphasizes "clearness" rather than Peirce's "fixation," but both are pragmatic— hence, center on the subject—in that what is at stake is clearness *to* or fixation *for* a subject or subjects, as desiring (Peirce) or as practical (James).

[b] Jürgen Habermas advocates a different form of pragmatism, wherein the decisive standpoint is that of subjects who communicate rationally with one another—the standpoint of communicative rationality. Because this book considers Habermas's position in various places, it is here merely mentioned.

[c] As is indicated above, Robert Brandom has developed a radically pragmatic position guided by the slogan, "Semantics must answer to pragmatics" (1994: 83). For Brandom, active subjects within the contexts of social practices play the absolutely central role and determine the perspective from which not only a semantics but also an ontology and a philosophy of mind develop.

[3] In many respects, the most important conception of subjectivity is the one developed by Kant, who understands subjectivity as transcendentally constitutive. Kant's position with respect to the constitutive role of the subject is sufficiently well known that not even a brief characterization of it is needed here. Moreover, much of what is said in the following subsections relates directly to the Kantian position.

2.3.2 *On the Systematic Status of the Epistemic Dimension—The Dimension of Knowledge*[8]

2.3.2.1 The Ambiguity of "Knowledge"   137

The term and concept "knowledge" are beset by a noteworthy ambiguity and obscurity. As is well known, terms like "assertion" have double meanings: on the one hand,

---

[8] Under the heading "2.3.2," *Struktur und Sein* includes a paragraph about the two German terms *Wissen* and *Erkenntnis*, both often translated into English as "knowledge"; that paragraph is not relevant to *Structure and Being*. Footnote 8 in *Struktur und Sein* introduces passages containing *Wissen* and *Erkenntnis* from contemporary German philosophical literature; those examples are likewise not relevant to *Structure and Being*.

the active or verbal meaning of the performing of an act (e.g., "Assertion without justification is his favorite mode of argument"); and on the other, the resultative or nominal meaning of the product of the act, the "object" that it yields (what is asserted). Something similar holds with respect to "knowledge." When for example there is talk of "the encyclopedia of human knowledge," or of philosophy as "the system of philosophical knowledge or of rational knowledge," the objective aspect is intended: what is known. But when "know" is understood in the active sense, as a substantivized verb, then the act or condition of an agent (i.e., a subject or speaker) enters the picture. It is shown below in conjunction with the treatment of the definition of knowledge that both aspects, the strictly subjective and the resultative-objective, affect the definition.

138

The generally undifferentiated use of the term "knowledge" is symptomatic in that precisely this lack of differentiation marks the central characteristic of the epistemic dimension. This characteristic consists precisely in the relation between a *content* (what is known), on the one hand, and a (mental) performance or condition on the other. Various positions have been taken with respect to this relation, linking the two moments more or less closely (more so with *conviction*, for example, less so with *doubt*) or, in extreme skepticisms, denying any such link. But the highest or most determinate form of this relation is knowledge. In the current context, only this form of the relation is considered.

Neither of the two meanings of knowledge introduced so far—thus, neither the active nor the objective—can alone be identified as the genuine or complete meaning. The adequate determination must encompass both of the two. If the systematic status of the epistemic dimension is to be determined adequately, all three meanings (the third being the encompassing one) must be taken into consideration. Lack of distinction or imprecise distinction among the three meanings is largely responsible for the frequently noted widespread confusion within the domain of epistemic inquiry and theory in contemporary philosophical literature. This is shown in what follows.

2.3.2.2 Knowledge as a Philosophical Problem

[1] The question that must first be addressed is, what is knowledge? Usually, the question is formulated this way: how is the concept of knowledge to be explicated or defined? But this formulation, in all its variations, is far from unambiguous. Frequently or indeed usually philosophers appear to proceed on the basis of the illusory assumption that "knowledge"— along with "truth" and other central terms and concepts—denotes an "entity" of some kind that is somehow fixed and fully determinate, such that the task is to apprehend and then to articulate it. What the theoretician actually encounters in each such case is, however, in the first place a word, and indeed a word that has been and is used in various ways and associated with different conceptual contents in both theoretical and ordinary languages. A better starting point is therefore this question: what conceptual content do speakers of a given ordinary or philosophical/scientific language attach to these words? But this epistemic-semantic question quickly leads to its own replacement, in philosophy and the sciences, by this deeper critical question: what meaning or conceptual content *should* be connected, in a philosophical/scientific language, with the term "knowledge"?

139

Explicating terms that are used in natural language presents philosophers with a quite specific problem. Assuming the explication to have been accomplished successfully for a given term, the philosopher must raise the question whether the conceptual content that has been revealed is at all interesting or appropriate with respect to philosophical aims. If it is not, then the philosopher must either correct it by means of improving it (refining it, making it more precise, etc.), or cease to use the term.

In this respect, the case of "knowledge" is central and exemplary. If one attends carefully to the current state of affairs, one sees that the intensive debates on the question, "What is knowledge," are to a great extent based in fundamental misunderstandings. This is shown in what follows.

A basic ambiguity of "know" in ordinary English is indicated by the use of parentheses in the following passage, which begins the *Oxford English Dictionary*'s entry for "know, *v.*": "*Know*, in its most general sense, has been defined by some as 'To hold for true or real with assurance and on (what is held to be) an adequate objective foundation.'"[c] If the phrase enclosed within the parentheses is removed, then knowing is said to require that the objective foundation of the holding-to-be-true *actually* be adequate; if the phrase is retained but the parentheses are removed, then the foundation must only be *held* to be adequate. To be fully explicit:

(K₁)   To know is to

    (a) hold for true or real

    (b) with assurance and

    (c) on an adequate objective foundation.

(K₂)   To know is to

    (a) hold for true or real

    (b) with assurance and

    (c) on what is held to be an adequate objective foundation.

[2] The definition of knowledge generally accepted among philosophers today, in one form or another, is similar to (K₁) rather than (K₂) with respect to (c), but differs from both (K₁) and (K₂) by (in effect) combining (a) and (b), and adding a third condition: that what is held to be true actually be so.[d] Stemming explicitly from Edmund Gettier (1963: 121), who presents it *not* as a stipulation, but instead as an analysis of what he takes to be the most widespread philosophical understanding of "knowledge," this definition reads as follows:

(K_G)   S knows that *p* iff

    (i) *p* is true

    (ii) S believes that *p*, and

    (iii) S is justified in believing that *p*.

---

[c] *Struktur und Sein* does not of course introduce the passage from the *OED*, which concerns English usage, nor does it include counterparts to (K₁) and (K₂), introduced below. *Structure and Being* cites the *OED* passage as clear evidence of the ambiguity (common to German and English) considered in this section; (K₁) and (K₂) make the ambiguity, as presented in the *OED* passage, explicit.

[d] Depending upon how "on an adequate objective foundation" is determined, K₁ might or might not entail that this third condition be satisfied.

To the discussion generated by Gettier's article Alvin Plantinga (1993b: 6–7) remarks, wholly correctly, as follows:

> After 1963 the justified true belief [JTB] account of knowledge was seen to be defective and lost its exalted status; but even those convinced by Gettier that justification (along with truth) is not *sufficient* for knowledge still mostly think it *necessary* and *nearly* sufficient for knowledge: the basic shape or contour of the concept of knowledge is given by justified true belief, if a quasi-technical fillip or addendum ("the fourth condition") is needed to appease Gettier. Of course there is an interesting historical irony here: it isn't easy to find many really explicit statements of a JTB analysis of knowledge prior to Gettier. It is almost as if a distinguished critic created a tradition in the very act of destroying it. Still, there are *some* fairly clear statements of a JTB analysis of knowledge prior to Gettier.

As Plantinga notes, Gettier's article purports to establish that the JTB definition is defective by demonstrating that the conditions that it formulates are insufficient. Gettier's argument relies upon counterexamples focusing on problems relating to condition (iii); the counterexamples are cases that appear to satisfy the conditions listed in the definition, but that are not instances of "knowledge," in the intuitive sense that the definition attempts to make precise. Over the years, the number of putative counterexamples has increased considerably. Various authors have attempted to reformulate the clearly problematic condition (iii), but as yet nothing approaching consensus has been attained.

Over the past twenty or so years, the discussion has taken a new turn that must be considered. First, however, something must be noted that, given the background provided by the lengthy debate about Gettier's definition and so-called Gettier problems, reveals a remarkable aspect of the situation. In applying the definition, Gettier—along with others who have engaged in the debate—appears to assume that the satisfaction of condition (i) is somehow straightforward or is in any case non-problematic. This condition says that for there to be knowledge, p must be true. That this assumption should not be made is shown in what follows.

Because this is to be a definition of *knowledge*, all of the factors constituting the definition must be carefully considered. Gettier's definition includes three such factors— three conditions that must be satisfied if knowledge is to be defined. What, then, about condition (i)? The condition that p be true must count as satisfied, because otherwise the definition makes no sense. But how can this be? Or, to question it more radically: how is this to be understood? There can be talk of satisfaction of the condition that p be true only if the truth of p can be *presupposed as given* or indeed is in fact presupposed. But what does that mean, and what does it entail? Here the point is reached where it becomes clear that Gettier's definition contains a basic error. Gettier, along with those who follow his interpretation of condition (i), appears to understand this condition as follows: condition (i) is to be considered to be satisfied if the truth of p is a fact or a condition in the world (or "corresponds" with a fact/condition in the world, etc).[9] Of this, only the following could be said: this truth (this fact, this condition) in or of the world either *subsists* or *does not subsist*. Whether or not it does would be fully independent

---

[9] In discussions of the definition, as a rule nothing is said about the concept of truth it presupposes. That too is remarkable.

of any attitude of anything like a subject. Given this understanding of the satisfaction   141
of the condition, no more than this could be said of the matter, but this is taken to be
completely sufficient, completely adequate.

Popular and widespread though such an explanation of the satisfaction of condition (i) may be, it fully misses the point that is decisive with respect to the problem
that arises here. This point is the following: it in no way suffices simply to assume
or to presuppose that p is true, for example in the sense that p *actually* articulates a
given fact or condition in the world (or is, in any other manner, interpreted as true).
Such an assumption or presupposition would be wholly arbitrary and empty. Instead,
indeed, the definition requires that condition (i), which requires that p be true, must
be *satisfied*. This satisfaction requires that the truth of p be *expressed* or *articulated*, as it
indeed is, precisely in the explicit *naming* of it as condition (i). But this makes it clear
that for there to be any talk of the satisfaction of condition (i), the truth of p must be
presupposed already to have been *grasped*. But what could this *grasping* of the truth
of p mean other than that the truth of p must be presupposed already to be *known*, if
condition (i) is to count as satisfied? The argument is thus circular.

The defender of Gettier's definition could respond by denying that the truth of p
being grasped is the same as that truth being known. But this defender would then have
to explain what this being grasped would involve. Other terms could of course be used,
but all raise the same problem once again: how are they to be explained? Come what
may, one cannot here avoid the fact that Gettier's definition contains a fundamental
petitio principii, a manifest circularity: it presupposes what it intends to define.

An additional point must be added to this critique. Even if one could succeed in
explaining the truth of p being grasped as somehow different from the truth of p being
known, an additional and comparably serious problem would arise: if we are already
"in possession" of the truth of p (no matter how this "possession" might be understood
or designated), what could "knowledge" then add? If one said that "knowledge" would
add the factor "justified belief," that would be an utter redundancy, because if the truth
of p is already taken to be grasped, then it is fully irrelevant whether or not one also
speaks of "justified belief." The sole and exclusive aim or goal would already have been
attained. If one nevertheless wanted to add the factor of justified belief, at no matter
what cost, one would be caught up again in the circularity introduced by all undertakings like Gettier's attempt to define knowledge.

[3] This book relies on a completely different definition of *knowledge*, a definition that is in perfect harmony with this book's conception as a whole, particularly   142
with respect to the dimension of the subject or of subjectivity; in another respect, this
definition arises directly from this comprehensive conception. The definition is based
on the following central thesis: knowledge is the accomplishment of theoreticity by a
subject; knowledge is thus an intentional or—more precisely—a propositional attitude
or a propositional mental state that is essentially directed toward something objectively
subsistent (i.e., something true). From this directedness of the accomplishment of
theoreticity, one cannot subtract that which belongs essentially to the subject—that
which characterizes the attitude as an attitude (i.e., belief or conviction, etc.). From this
it follows that all of the components of the propositional directedness of the subject to
what subsists objectively, the true, *including* what subsists objectively, the true itself, are
to be situated *within the scope of the belief*. If "knowledge" designates this directedness,

then it is a relation. It is only consequent to say that the "objective relatum" of the relation still belongs within the scope of the relation; conversely, it is inconsequent to remove from knowledge, as such a relation, any component—specifically, here, what objectively subsists, the true—from the scope of the relation. If one uses the concept of the operator to articulate this matter, then one can say, using terms from Gettier's definition, that both the component "true" and the component "justified" are to be situated within the scope of the operator, "Subject S believes that . . . ." What results is a definition of knowledge involving the satisfaction of only two conditions. A welcome side effect of this definition is that it completely avoids the problems raised by Gettier's definition. The definition is as follows:

(K)   S knows that p iff

   (a) S believes that p is true, and

   (b) S believes that S's belief that p is true is justified.[e]

It is important to attend to the precise status of the two conditions. In opposition to Gettier's definition—and indeed to all the definitions provided within analytic philosophy—the truth relevant to condition (a) enters the definition not as an independent or self-sufficient factor—not, so to speak, as *de re* truth—but instead within the scope of the operator "S believes that," thus as dependent upon it, and thus as only *de dicto* truth. Equally important is the distinction of (b) from its counterpart in Gettier's definition: the justification of S's belief that p is true is not a factor independent of S, thus again not a *de re* factor; instead, the justification is itself the object of a higher-order belief of S, in that it lies within the scope of a higher-order belief. This notion is further clarified below.

[4] Throughout the discussions engendered by Gettier's article, (K_G)'s condition (i) is never questioned. Instead, in the course of the discussions first (iii) and then (ii) are abandoned. Crispin Sartwell (1991, 1992), for example, determines knowledge as "merely true belief"; this position continues to be controversial.[10] Ultimately, even condition (ii) is abandoned, so that only the factor "truth," still in the determinately independent (i.e., *de re*) perspective, remains.

This latter position is taken by Timothy Williamson, who, in the imposing book *Knowledge and Its Limits*, develops a fully original conception of knowledge at whose center is only the factor *truth*, albeit in connection with mind. His basic thesis, supported with powerful arguments, is that "knowledge" can be analyzed syntactically but not semantically (such semantic "analysis," in Williamson's sense, would involve the division or separation of the concept into multiple other concepts). This thesis leads Williamson to reject, as wholly wrong-headed, accounts like Gettier's "definition," which is a characteristic example of the type of analysis Williamson deems impossible.

---

[e] (K) is obviously an analogue to (K_2), not to (K_1).

[10] A possible ambiguity of the term "true belief" should be noted. It could be understood as the conjunction of two conditions (1. p is true, 2. S believes that p (is true)) or as *one* condition (S believes that p is true). The same holds for the phrase "to believe truly," which is often used in the analytic literature. Authors who want to define knowledge as "merely true belief" understand "true belief" as conjoining the two conditions.

Yet Williamson, having rejected analysis, does consider a "non-analytic account" of knowledge to be possible (2000: 34):

> The main idea is simple. A propositional attitude is factive if and only if, necessarily, one has it only to truths. Examples include the attitudes of seeing, knowing, and remembering. Not all factive attitudes constitute states; forgetting is a process. Call those attitudes which do constitute states *stative*. The proposal is that knowing is the most general factive stative attitude, that which one has to a proposition if one has any factive attitude at all. . . . The point of the conjecture is to illuminate the central role of the concept of knowing in our thought.

To clarify further, Williamson introduces the concept of the *factive mental state operator* (FMSO), which can be represented as "Φs that." "S Φs that A" can thus represent "S knows that A." Williamson denies that FMSOs can be defined, but provides some of their characteristics: (i) a given FMSO has as a deductive consequence its own argument (thus, from "S Φs that A," one can deduce A); (ii) FMSOs are states, not processes; (iii) a given FMSO ascribes to a subject an attitude toward a proposition; and (iv) FMSOs are semantically non-analyzable (35ff). Knowing is an "example of an FMSO," but it is unique in that it is the sole occupant of a particular place within the class of FMSOs: "Knowing is the most general stative propositional attitude such that, for all propositions *p*, necessarily if one has it to *p* then *p* is true" (39). Thus, S cannot *remember* that *p* without *knowing* that *p*, cannot *see* that *p* without *knowing* that *p*, etc., but S's knowing that *p* does not entail either S's remembering that *p* or S's seeing that *p*.

Williamson emphasizes (41) the following basic feature of his conception: "The present account of knowing makes no use of such concepts as *justified*, *caused*, and *reliable*. . . . [I]t makes no essential mention of believing. Formally, it is consistent with many different accounts of the relation between the two concepts [i.e., knowing and believing]."

From this brief characterization of the current discussions concerning the concept of knowledge may be seen what is asserted above: *in a specific respect*, made precise in the following paragraph, it is senseless to argue about which conception articulates the "true concept" of knowledge and which conceptions do not. All the available accounts may be associated, in one way or another, with the word "knowledge." If a given philosopher wants to use the word "knowledge" to denote, say, the concept *true belief* rather than, as does Williamson, *the most general factive stative attitude*, that cannot be prohibited. But from this it does not follow that, for that philosopher, Williamson's conception would have to be senseless; Williamson's concepts would merely have to be linked to different words.

According to the preceding paragraph, it is senseless *in a specific respect* to argue about the "true concept" of knowledge; the qualification is based on one matter of principle and one empirical fact. [i] The matter of principle is that one can associate any given word with any concept whatsoever, if one does this clearly and unambiguously: a given account stipulates specifically how it uses a given term. Of course, if one chooses to use an ordinary-language term in an idiosyncratic manner, difficulties are likely to arise; such idiosyncratic uses will likely appear arbitrary, and easily give rise to misunderstandings and confusions. [ii] The empirical fact is this: none of the participants in the contemporary discussions proceeds in this stipulative fashion;

this is made clear by the fact that all these participants support their positions—all claiming to get at something like the heart or the truth of the matter, with respect to knowledge—by providing ordinary-language examples (this point cannot here be supported in detail).

Even if all of this is acknowledged and taken into consideration, it appears to remain possible to develop a conception that, *empirically*, attains the best fit (not a perfect fit) with the intuitive understanding of the concept associated most broadly with the word "knowledge," and to establish this breadth by presenting the most extensive array of specific examples. This is the conception that connects the word "knowledge" with the concepts *belief*, *true*, and *justified*. Philosophers seeking philosophical terminology containing terms stemming from ordinary language should not simply ignore this fact. Among its consequences is that the definition of knowledge to be preferred would be one that explicitly included all three of these concepts in the *definiens*. A *decisive modification* should, however, be made to the Gettier definition (and similar definitions); it is briefly introduced above in (K). The concept *truth* should not appear as a self-sufficient factor, thus not outside but instead within the scope of the operator "believes that" (such that truth has a *de dicto* status). One would then have only two conditions, because "belief" ("believes that") and "true" would enter together within a *single* condition.

2.3.2.3 Knowledge and Cognition in Kant

Although Kant's uses of "knowledge" (*Wissen*) and "cognition" (*Erkenntnis*) are rarely even considered in contemporary philosophical discussions of the definition of knowledge, examination of them reveals that they clarify relevant issues. Kant relies heavily on the word "cognition," both in the singular and in the plural, whereas his significant uses of "knowledge" are quite rare and importantly distinct. That the distinction is important to the interpretation of Kant's work makes it the more significant and unfortunate that the most widely used English translation of the *Critique of Pure Reason*, that of Norman Kemp Smith, uses "knowledge" to translate both German terms. Because "knowledge" has no plural form, Smith renders *Erkenntnisse* as "diverse modes of knowledge."[11]

Kant's "cognize" is similar to Williamson's "know" (introduced in Section 2.3.2.2), above all in that Kant's "cognition" and "cognitions" have nothing to do with the likes of belief and justification in Gettier's senses of the terms. As is shown in what follows, the concept Kant associates with the word "cognition" is much closer to Kant's concept of truth, and indeed in a sense coincides completely with it. In this respect Kant differs significantly from Williamson, who does not explain his truth-concept in detail, but who clearly does not understand truth in Kant's transcendental sense.

First, however, Kant's concept *knowledge* requires more careful investigation. As is indicated above, orthodox analytic philosophy understands knowledge as *justified true belief*. The two passages most commonly identified as historical predecessors of Gettier's account of knowledge as justified true belief are one from Plato (*Theaetetus* 201c-d) and the following one from Kant:

---

[11] Fortunately for Anglophone readers of Kant, the recent translation of Paul Guyer and Allen Wood (Kant 1998) does distinguish between "cognition(s)" and "knowledge."

Taking something to be true, or the subjective validity of judgment, has the following three stages in relation to conviction (which at the same time is valid objectively): **having an opinion, believing,** and **knowing.** **Having an opinion** is taking something to be true with the consciousness that it is subjectively **as well as** objectively insufficient. If taking something to be true is only subjectively sufficient, and is at the same time held to be objectively insufficient, then it is called **believing.** Finally, when taking something to be true is both subjectively and objectively sufficient it is called **knowing.** Subjective sufficiency is called conviction (for myself), objective sufficiency, **certainty** (for everyone). I will not pause for the exposition of such readily grasped concepts. (A822/B850)

It should first be noted that Kant is mistaken in speaking of "such readily grasped concepts," because it is obvious that his characterization of *conviction* is incoherent. He begins by asserting, "Taking something to be true, or the subjective validity of judgment, has the following three stages in relation to *conviction* (*which at the same time is valid objectively*)" (emphasis added); he then reverses himself, without noting that he has done so: "Subjective sufficiency is called **conviction** (for myself)."

R. K. Shope (1983: 18–19) maintains that the cited passage *explicitly* formulates the three "conditions" for "knowing"; he then contends,

This definition incorporates a belief condition when it requires that one's holding the judgment to be true involves subjective sufficiency. It includes a justification condition because it requires objective sufficiency, which is defined shortly before this passage: the grounds of a judgment are said to be objectively sufficient when the judgment is "valid for everyone, provided only he is in possession of reason" (A820, B848), and the validity is said to concern whether the judgment has the 'same effect on the reason of others' (A821, B849)....

Aaron maintains that Kant omits a truth condition from the above explanation of knowledge (Aaron 1971, p. 8).... However he has misconstrued Kant because he has overlooked the stipulation that one's propositional attitude be "at the same time objectively valid." Kant generally uses "objectively valid" to indicate that something is related to an object. He accepts the "nominal definition of truth, that it is the agreement of knowledge with its object," i.e., the knowledge must "agree with the object to which it is related" (B82–83). Since Kant recurs to such a definition just before the above explanation of knowledge, he intends the requirement of objective validity in that passage to involve the actual truth of one's judgment.

147

Whether this interpretation is correct is questionable. Kant unambiguously distinguishes among the three mental conditions he terms "taking something to be true"—opining, believing, knowing—but he does so in such a way that "knowledge" ("knowing") cannot be "belief" ("believing") of any sort, not even if "belief" is qualified as "justified" and "true." The reason is that Kant determines or defines "belief/believing" as the *conjunction* of two components (i.e., subjectively sufficient holding to be true and objectively insufficient holding to be true). If one takes *one* of the components of this conjunction (in this case, subjectively sufficient holding to be true) and considers it as an element or component of *another* conjunction, then the new conjunction has nothing to do with "belief/believing" in Kant's sense, because one of the determinations of the former (objective sufficiency) contradicts one of the latter (objective insufficiency).

One might say that "belief/believing," in analytic terminology, corresponds to Kant's "taking something to be true," and not to his "belief/believing." Kant could then be said to define "knowledge" as "justified true belief," although only with the qualification that the justification would have to be sufficient both subjectively and objectively. But this apparent agreement would be merely terminological; one would still have to determine (1) just what Kant means by "taking something to be true" (which he himself appears never to make clear) and (2) precisely how analytic philosophers understand "belief/believing" (and that is also far from clear).

Kant's concept *cognition*, unlike his concept *knowing*, relates neither directly nor indirectly to his *taking something to be true*. Granted, Kant provides no strict definition of cognition, but his many assertions about it, thus his theory of cognition, have quite different concerns: in the terms of this book, for Kant cognition is *the complete (theoretical) structuration of subjectivity*. What Kant's theory seeks to explain thus involves nothing like "attitudes," including "taking something to be true," which are one and all empirical. What he wants to answer is instead the *transcendental* question that can here be formulated as follows: what are the conditions of possibility of the empirical subject's having such attitudes as doubting, believing, and taking to be true? With respect to the complete theoretical structuration of subjectivity, the "attitudes" of empirical subjects are thus utterly irrelevant. Kant appears to understand and to determine *knowledge* as an attitude of the *empirical* subject; *cognition*, on the other hand, he understands as an absolutely central *transcendental* matter (or structuration).

A closing remark on knowledge/cognition and truth in Kant. Shope's remarks on the "truth" component of Kant's "definition" of "knowledge" may be accepted as basically accurate. But if—unlike Shope—one distinguishes strongly between Kant's "knowledge" and "cognition" in the way that this is done above, then there is a surprising, indeed astonishing consequence: the "cognition" of the *Critique of Pure Reason* proves to be *equivalent* to that work's transcendentally (re)interpreted notion of truth, which is reflected in its "nominal definition" of truth. "The complete theoretical structuration of subjectivity," introduced above as appropriately characterizing Kant's concept *cognition*, is equally appropriate as a characterization of the transcendentally (re)interpreted concept *truth*.

2.3.2.4 Subjectivity and Knowledge with Respect to Systematicity

[1] The aims of this book require that, at this point, the systematic status of the epistemic dimension be determined. This is the dimension of the theoretical link or connection between subjectivity and anything and everything within the dimension of things, states of affairs, language, and structures of all sorts. One can get clear on what, more precisely, this means by asking what is involved in being a subject within the world or the universe. To answer the question somewhat technically: to be an epistemic subject is to be a point in the world/universe from which reference can be made to anything and everything in the universe.

The references can, of course, be of various sorts, as becomes clear from the description of the ways the subject within the world or universe can relate to other items within the world or universe. Such a subject is in one basic respect situated at or in a particular spatiotemporal point. From this point it has the option of relating only to itself; it can indeed accomplish a self-relation from the perspective of its own self, ignoring all other perspectives. But it can also expand its perspective or horizon *at will*, in that it can think about anything and everything in the universe. And this change of horizon can bring with it a change of perspective: the subject can consider whatever it does consider from the perspective or standpoint of its neighborhood, its city, its country, its continent, the various institutions to which it belongs, etc. Ultimately, its perspective can be that of the universe itself; at this ultimate stage of expansion of its perspective or horizon, subjectivity is *intentionally coextensive with the world* (*with the universe, with being as a whole*). For this reason, Aristotle is fundamentally correct in describing the subject as being "in a certain way all beings" (*De Anima* 431b21).

[2] How subjectivity is understood determines the significance and systematic status of knowledge. If subjectivity is understood *particularistically*, then all the knowledge it produces or acquires is likewise particularistic. If however subjectivity is understood as *universal*, then its knowledge has a correspondingly universal character.

The status of the epistemic subject can be clarified by means of a subject-operator Ⓢ taking as its immediate argument another operator Φ ("knows that"), whereby this second operator takes as its arguments sentences (or propositions expressed by sentences). These operators yield linguistic configurations of the form Ⓢ$\Phi_S p$, read, "It is the case from the perspective of subject S that S knows that *p*." The distinction introduced above reveals, however, that in this configuration, the operator Ⓢ remains vague and underdetermined. It attains a precise meaning only when further determined with respect to the levels of subjectivity located on the scale introduced above. The goals of this part of this book require that only two such stages be distinguished: the *particularistic* and the *universal*, such that the particularistic is to be understood as encompassing an extensive array of substages (i.e., all stages lower than the universal stage), whereas the universal stage excludes any particularity of any sort. The relevant operators are then Ⓢ$_P$ and Ⓢ$_U$ for, respectively, the particularistic and the universal.

With only one possible exception, all the forms of subjectivity introduced above and all their variants—each of which corresponds to a form of knowledge—are *particularistic*. The possible exception is Williamson's conception of subjectivity via the notion of the factive mental state. To be sure, it is not fully clear that this determination characterizes subjectivity as *universal* in the sense introduced above, but the possibility that it does remains open. All of the other treatments take subjectivity and with it the epistemic dimension as limited in one way or another, that is, with respect to what is centrally relevant here: as *non-coextensive with the universe* (*being as a whole*). Because they share this feature, they are collectively vulnerable to criticisms of it. The focus of the critique to follow is related directly to the third of the forms of subjectivity introduced above: subjective as transcendentally constituting (the position first developed by Kant).

[3] To focus on subjectivity as particularistic is to privilege the active/subjective moment of knowledge—the fact that every item of knowledge *is known by someone or other*—and thereby to subordinate the resultative-objective moment—the fact that every item of knowledge *is knowledge of something or other*. What is the basis for this privileging? One could criticize it by noting that it indefensibly limits the entire dimension of contentuality, thus of the world (the universe, being as a whole), to what is accessible from what

can be a quite narrow point of view. Instead of developing this line of thought, however, Section [4] introduces two other arguments; both are comprehensive and fundamental, and both undermine all possible privilegings of particularistic forms of subjectivity.

Before the privileging of particularistic forms of subjectivity and thus of knowledge can be criticized, it is necessary to clarify precisely what such privileging involves. The particularistic operator aids this clarification. Particularistic positions are ones that, whether explicitly or implicitly, place all sentences, all assertions, all theories, indeed anything whatsoever that is theoretical, within the scope of the particularistic epistemic operator $\textcircled{S}_P$. Concrete examples could include, "From the particularistic epistemic perspective of the Greeks, it is the case that the Trojans are in the wrong," and "From the particularistic epistemic perspective of the Trojans, it is the case that the Trojans are not in the wrong." In a general characterization,

> From the (intentional or standpoint-determining or transcendentally constituting) particularistic epistemic perspective of S, it is the case that [the sentence] p [is true].

Or, more simply:

> From the particularistic epistemic perspective of S, the sentence p is true.

In the notation introduced above, this yields $\textcircled{S}_P \Phi_S p$, to be read as,

> From the particularistic epistemic perspective of S, it is the case that S knows that p [is true]

whereby "S knows that p is true" is to be analyzed as satisfying the two conditions identified in (K) above (Section 2.3.2.2): (a) S believes that p is true, and (b) S believes that S's belief that p is true is justified. A complete formalization would thus be

(1) $\textcircled{S}_P\{B_{S,P}(Tp) \wedge B_{S,P}[J_S(B_{S,P}(Tp))]\}$,[12]

read (somewhat awkwardly, but with maximal precision) as:

$\textcircled{S}_P$	It is the case from the particularistic perspective of S that
$B_{S,P}$	S particularistically believes that
Tp	it is true that p
$\wedge$	and
$B_{S,P}$	(that) S particularistically believes that
J_S	S is justified
$B_{S,P}$	in particularistically believing that
Tp	it is true that p.

---

[12] "True" (T) is here understood as a *sentential operator* (in the sense of "It is true that. . ."—see Chapter 3). If one were to take "true" as a predicate, the formalization would be $(T \ulcorner p \urcorner)$.

Just how true/truth is to be understood depends upon the truth-theory that is relied upon (see 2.5 and 3.3, below). To be noted, however, is that although "Tp" appears in both conjuncts without further qualification, it is within the scope of a particularistic (and thus restricted) belief that itself lies within the scope of the operator $\widehat{S}_p$. Also to be noted is that the factor "belief" ($=$ "B") appears in both conjuncts with two indices ("$B_{S,P}$"), the second of which restricts it; it thereby differs from the factor "justification" ($=$"J"), which is indexed to subject S but *not* to S as particularistic, and is therefore unrestricted. This is to be understood as follows, omitting for the sake of clarity the first of the conjuncts:

(1) The particularistic (hence restricted) subject S believes that

   S's particularistic belief that

   it is true that p

   is unrestrictedly justified.

Unrestricted justification is universally conclusive grounding, such that every rational subject can and must have access to it. Yet this makes sentence (1), and the proposition it expresses, appear paradoxical: how can a particularistic subject believe that its restricted belief that a determinate sentence (p) is true is unrestrictedly or universally grounded? Yet that is a thesis that is ultimately defended or presupposed by every argumentatively formulated position that is "finitistically" oriented, i.e., here, that takes the cognitive capacities of the subject to be radically restricted or limited (e.g., Kant's transcendental philosophy[13]). As a rule, such positions argue that they are universally rationally justified, and that they therefore can and must be accessible to all rational subjects. Formula (1) above articulates just such a position.

It is clear that no position determined by the operator $\widehat{S}_p$ can coherently understand any single utterance as standing outside of this operator. In other words, such positions can include *no* utterances of the *unrestricted* form "(It is the case that) such-and-such," because these positions *always* explicitly or implicitly precede their utterances with the operator $\widehat{S}_p$. This does not prevent a given sentence of the form, "(It is the case that) such and such"—which, considered in isolation, has an unrestricted status—from appearing within the scope of the particularistic operator. In this case, however, its status is *de dicto* rather than *de re*: the sentence is *believed to be* unrestricted. This state of affairs is here articulated with the following formulation: no position determined by the operator $\widehat{S}_p$ can present sentences of the *utterly* unrestricted form, "(It is the case that) such and such."

[4] Particularist positions bring with them at least two weighty problems.

[i]The first problem is that such positions appear to be fundamentally relativistic in an unambiguous and emphatic sense. To be sure, many philosophers would presumably see this not as a weakness, but as a great advantage. But one must consider

152

---

[13] It is in no way coincidental that one of the most central methodological or systematically theoretical sentences in Kant's *Critique of Pure Reason*—a sentence cited above—reads, "The **I think** must **be able** to accompany all my representations" (B131). A precise and comprehensive interpretation of this sentence would reveal that it articulates an operator $\widehat{S}$, to be understood in a transcendentally epistemic manner, in a characteristically traditional mode of expression; this operator would have to be understood as preceding every sentence in Kant's transcendental philosophy, if one wanted to make this philosophy clear.

what the relativism that accompanies the operator $\textcircled{S}_P$ actually involves and entails: a self-imposed restrictedness of perspective that, in its very imposition, surpasses the restraint it purports to expose.[14] One can accept without dispute that there *is* a perspective of the subject that can be represented by the operator $\textcircled{S}_P$. But the pressing question then becomes: what is the status of this perspective at the level of theoreticity? This perspective is anything but clear, because it has not been determined how the subject actually understands itself if it takes this perspective, how it can understand itself in other cases, and—if it is to develop all of its capacities—how it not only can but must understand its status as theoretical. The scale of determination for how the subject can and indeed must be understood reaches from that of an entity that understands itself as isolated—as excluding all other entities and excluded by them—to that of an entity that understands itself as universal in the sense of being intentionally coextensive with the whole (i.e., the whole of being, of the actual, of the universe). Nevertheless, throughout the subdiscipline of epistemology, the subject is assumed to be somehow restricted and particularistic. The subject as universal, as coextensive with the whole of actuality, is of course fully incompatible with any restriction to any particularistic perspective and thus to any restrictedly particularistic understanding of the subject.

153   These considerations yield a significant consequence. As is indicated above, depending upon how "subject" is understood, the significance and the systematic status of the subject-operator $\textcircled{S}$ change, and indeed fundamentally. Only the particularistically understood operator $\textcircled{S}_P$ yields one-sided, untenable positions, because it considers the objective or contentual aspect either not at all or only insufficiently. If on the other hand the subject-operator is understood *universalistically*—as $\textcircled{S}_U$—then it introduces no theoretically untenable one-sidedness, but instead constitutes an indispensable component of an adequate concept of knowledge, as is shown shortly below.

The operator $\textcircled{S}_U$ is the logical-theoretical expression of a "subject-perspective" in accordance with which the subject is understood and determined with respect to a universal framework or dimension. One can term this the "framework of rationality" if one thereby avoids understanding "rationality" as an empty cliché or as a canon of decisionistically proclaimed rules, demands, standards, etc., and understands it instead as a methodic expression of an anticipated comprehensive conception of objectivity. "Subject" then loses the limitedness that it appears to have when it is taken particularistically.[15]

It is clear that the perspective *of the subject*, *when* the subject understands itself as *universalistic*, makes itself superfluous (i.e., such a perspective is redundant in the sense that it need no longer be named as such or introduced as relevant, *because* it simply coincides with the completely *objective* "perspective").[16] It must also be noted,

154   however, that the problematic of the concept of "perspective" must be reconsidered in

---

[14] This only superficially paradoxical—Hegel would say "dialectical"—relation is the epistemic counterpart to the linguistic maneuver of identifying a limit for a given language; see Section 1.2.4.

[15] To revert again to the example, the Trojans and the Greeks, if they seek to determine *rationally* whether or not it is unrestrictedly (i.e., absolutely universally) the case that the Trojans are in the wrong, do so from a perspective that is universal rather than particularistic in that what they seek to determine is precisely what is the case with respect to the relevant actions and to the criteria that determine rightness and wrongness.

[16] This state of affairs can be well illustrated by considering the position of Jürgen Habermas. Scarcely any other philosopher has accorded the "subject" factor so central a role in rational

conjunction with the concept of the theoretical framework. Theoretical frameworks, as this book understands them, are *not* determined by any subject-operator, but they *are* related to the dimension of the subject. *No* theoretical framework that is *actually presupposed and used* can stand within the scope of any particularistic subject-operator, because if it did, then the theoretical framework putatively presupposed and used would actually be one that was only *meant or believed* to be used—thus, a merely *de dicto* theoretical framework—and thus not identical with the theoretical framework that *realiter*—*de re*—was presupposed and used. From this it follows that the sentences of every theoretician who can be taken seriously are presented as being *objectively* and thus *unrestrictedly*—rather than in any way particularistically—true. This becomes clear from the analysis of any theoretical framework *actually* presupposed and used by any (empirical or transcendental) subjectivist, relativist, skeptic, etc.

   The result of these considerations can be summarized as follows: any and every *actually presupposed and used* theoretical framework is *incompatible* with any all-determining particularistic subject-operator; every genuinely presupposed and employed theoretical

---

discourse as has Habermas (see esp. Habermas 1973, McCarthy 1978, esp. 347ff). He contends that the two factors "subject" and "rationality" condition and determine each other reciprocally (one could almost speak of an interdefinability of the two factors). Yet it is highly revealing that this view ultimately makes the subject a redundant factor. This is easily shown. For Habermas, the concept of communicative rationality is simply central. It is of the essence of rationality so conceived that it is attained when a consensus is reached among speakers or among those taking part in a given discourse. Habermas evaluates the rationality of speakers/subjects by their ability to reach consensus in a discursive-argumentative manner. To make this clearer, Habermas relies on the "transcendental fiction" of an "ideal speech-situation," by which he understands the totality of conditions under which a rational consensus is possible (and thus to be viewed as attained). Differently stated: the ideal speech-situation is one in which the arguments and proofs are so constructed that any competent and rational speaker would have to come to the same conclusion (i.e., for example, that a speaker who did not initially agree would be brought to agreement by the other rational speakers in that he would simply allow himself to be swayed by the strength of the better argument).

   Strangely, the extensive literature about Habermas fails to note that he thereby makes the speaker or subject superfluous. To present and characterize the thoughts or issues Habermas envisages, one needs only criteria for what he calls the "better" argument. The reason is that for Habermas what makes speakers rational is their allowing themselves to be guided by the force of the better argument. What is then however decisive—and ultimately, alone decisive—is how the concept of the better argument is characterized, because the clarification of this concept is presupposed by any talk of the rationality of speakers/subjects. For just this reason, the concept of the better argument is not characterizable in terms of any modes of behavior of speakers/subjects. Habermas has never acknowledged the significance of this decisive point. He continues to focus on the idea of an *ideal* consensus, attained rationally and discursively. *As what and how* this (ideal) rationality and the universal consensus it aims for are *ultimately* to be understood is not an issue on which he focuses. Habermas's concept of ideal rationality and of the consensus it makes possible thus floats in empty space. He characterizes actually *universal rationality* only procedurally (with respect to the type of discourse that must allow itself to be guided by such rationality). Habermas has not even seen, much less thematized the fact that genuinely universal rationality is grounded in the structuration of being, of what is. That it is so grounded is immensely significant for the clarification of the fundamental interconnections among logic, semantics, pragmatics, epistemology, etc.

155   framework *entails* a *universalistic* subject-operator and is entailed by such an operator. As this formulation indicates but as must also be emphasized, this universalistic feature of every theoretical framework that is in fact presupposed and employed does *not* prevent there being a possible and indeed actual *plurality* of theoretical frameworks. Chapters 5 and 6 treat this issue in detail.

[ii] The second problem with particularistic positions is a serious inner incoherence in any conception that accords a central or determinative role to the operator S_P. S_P limits S's knowledge to what can be known from the viewpoint that S has as a subject understood not to be universal. If however S were indeed such a subject, then it would be unable to know or even think of itself as such a subject. Therefore, even its entertaining of the claim that it is such a subject establishes that it is *not* such a subject. The reason for this is that the claim is about what is the case, simply as such, and not about what-is-the-case-for-a-(particularistically-understood-)subject.

[5] To these arguments, advocates of the determinative role of the operator S_P have at least two possible objections. First, they can point out that, for the subject, having a perspective is simply unavoidable: knowledge is accomplished by concrete subjects, and concrete subjects are inevitably at specific standpoints that involve specific perspectives.

This objection is easily countered. From the fact that every subject has a wholly determinate standpoint it in no way follows that any given subject is incapable of attaining a universal perspective. A concrete standpoint need not necessarily be a particularistic standpoint (i.e., a restriction on or denial of a universal perspective); on the contrary, it can be a richer determination—indeed, a concretization—of a universal perspective. From the fact that every European has a wholly determinate, wholly concrete standpoint in a specific country and indeed in a specific region of that country, it in no way follows that his or her perspective is restricted or particularistic. A Bavarian, a Portuguese, a Viennese, etc., can of course have more comprehensive perspectives (e.g., a European perspective) without thereby ceasing to be Bavarian or Portuguese or Viennese.

The particularist's second objection would be the claim that any conception articulated in universalist formulations of the sort "It is the case that ϕ" must presuppose a dogmatic or absolutist standpoint, and is for this reason unacceptable. The counter to this objection brings to light fundamental and decisive considerations. As the introduction of
156   the subject-operator S makes clear, the issue is that of a *determinate* qualification of all the sentences that are formulated under the governance of the corresponding conception (in the case of the operator S, a subject-related qualification). But the sentence putatively articulated from a dogmatic/absolutist standpoint, "It is the case that so-and-so," contains no qualification at all. It also cannot be said that the sentence implicitly involves a dogmatic/absolutist perspective, as though the sentence were to be read, "It is *absolutely* or *irrefutably* the case that so-and-so," or something of the sort. To be sure, the sentence can be so read, but it need not be. What is remarkable about this sentence as an expression of theoreticity is, in part, that it remains open to qualification or determination. The sentence involves—at least implicitly—the negation of a purely restricted or particularistic perspective as the determinative perspective; in that it does, it has a universal status. But this universal status is not fixed once and for all. Instead, it remains open, in the following sense and for the following reason: every sentence with the operator "It is the case that" presupposes some specific theoretical framework, and, as is shown above, every theoretical

framework has a universalistic status. The account above also, however, indicates that a *plurality* of theoretical frameworks must be acknowledged. The operator "It is the case that" would have an absolute and/or dogmatic sense only if it were connected with something meriting designation as the super or absolute theoretical framework, and thus with a super or absolute theory—a theory that would be simply ultimate and definitive. But such an understanding of the operator "It is the case that" contradicts the fundamental theses of the structural-systematic philosophy developed here.

[6] That a super theory is not possible even in principle can be shown by reference to Gödel's famous incompleteness theorem. The theorem says that any formal system that has sufficient expressive power includes truths that cannot be proved to be true within the system (i.e., with the means of the system). Elaboration is provided by Dummett:

> Gödel's theorem [. . .] shows that provability in a single formal system cannot do duty as a complete substitute for the intuitive idea of arithmetical truth. (Dummett 1975/1978:172)
>
> By Gödel's theorem there exists, for any intuitively correct formal system for elementary arithmetic, a statement U expressible in the system but not provable in it, which not only is true but can be recognized by us to be true: the statement being of the form $\forall x A x$ with Ax a decidable predicate. (186)

In Quine's words:

> Gödel proved that a complete deductive system was impossible for even so modest a fragment of mathematics as elementary number theory. (Quine 1987: 82)
>
> What Gödel proved, then, is that no axiom system or other deductive apparatus can cover all the truths expressible even in that modest notation; any valid proof procedure will let some true statements, indeed infinitely many, slip through its net. (83)

157

One can apply this result—the impossibility of the absolutely comprehensive system, thus of the absolutely complete provability of all of the truths of a given formal system with the means of the system—to the case of an absolutely comprehensive (i.e., complete) qualification of the standpoint articulated in the universalizability of the sentence "(It is the case that) ϕ." This standpoint is objectively universal, but its universalizability cannot be fixated and is in this sense incomplete. Its incompleteness does not undermine its objectivity; it only rules out the naive qualification of its objectivity and the acceptance of it as implicitly qualified. It seems strange that many philosophers (but also scientists and other learned and unlearned people) speak as though philosophy could be taken seriously only under the condition that it proved anything and everything, thus including its own bases, to be *absolutely true*. This view exhibits a basic lack of recognition of Gödel's demonstration that such a demand cannot be met even in the case of a system as simple as that of elementary arithmetic.

2.3.2.5 A Reversal of Perspective: The Indispensable but Secondary Theoretical Status of the Epistemic Dimension

The thesis announced in this section's title emerges from the preceding section; this section explains it in additional detail.

[1] The epistemic dimension is the dimension of the perspective of the subject. Depending upon how the subject understands itself or is understood, there are different

determinations of this perspective. The two determinations "particularistic" and "universalistic" designate the alternatives on the scale of determinations possible in principle; their status is clarified above with the aid of the two operators $⑤_U$ and $⑤_P$. The universalistic operator $⑤_U$ introduces no fundamental problems because, given its universality, it encompasses within its scope the entire dimension of objectivity. The arbitrariness or randomness of the particularistically engaged "epistemic subject," the subject who understands itself and acts as particularistic, has no effect on it, because the universalistic subject understands itself as fully included within the dimension of objectivity, and acts on the basis of that understanding—it does not introduce or consider any theses or other theoretical items presented as true only from some particularistic perspective. This has as a consequence that mentioning the subject is, strictly speaking, superfluous, because the subject is taken simply as a rational subject who therefore acts in accordance with the objective guidelines of rationality. The subject who understands itself and acts as particularistic, on the other hand, can best be characterized as the subject that is guided either not at all or only insufficiently by objective guidelines. There are therefore various forms of the particularistic epistemic subject.

The term "dimension of objectivity" has been introduced but not precisely explained. As long as, as in this chapter, the mode of consideration is global, only global remarks may be made concerning the dimension of objectivity. One of the tasks of the following sections and chapters is to provide added precision. What is here designated quite generally as the "dimension of objectivity" is constituted by two subdimensions: one that is (more) formal or structural and one that is contentual. The formal/structural subdimension encompasses the so-called formal sciences in the narrower sense (logic and mathematics) and all disciplines that are concerned with conceptual-theoretical, methodological, presentational, etc., factors. The contentual subdimension is the dimension of the subject matter(s) that are understood and articulated by means of the formal/structural dimension. This distinction remains at this point quite vague and is still largely underdetermined, because there are various items that, without closer specification, cannot be placed in to either of the two subdimensions. An example is the item that, in the history of philosophy, is called "category." Chapter 3 considers this item in detail. For these reasons, the distinction between the two subdimensions has, at this point, a purely programmatic and heuristic character.

As the various theories of rationality show, there is no consensus concerning the concept to be associated with this term. At this point, the following may be noted: if rationality is the dimension that is opposed to the dimension of subjective whim or arbitrariness, then rationality has to do with the dimension of objectivity. To put this more precisely, rationality is a concept that designates, in a global manner, all that the dimension of objectivity, in its entirety, involves. "Rationality" that is not grounded in objectivity, and not an expression of objectivity, is an empty concept.

[2] At this point, some concluding remarks must be added.

As has been shown, if a definition like Gettier's is accepted, then the following implication is valid: "S knows that $p \to p$ (in the sense of: p is true)." According to this book's definition (K), however, such an implication must be reckoned invalid as long as the status of the subject remains unqualified. Moreover, the implication is always invalid whenever the subject S is understood as particularistic, thus as $⑤_P$. The following is thus always valid: $⑤_P \Phi_K Tp \nrightarrow Tp$. If however the subject is taken as universalistic, thus as $⑤_U$, an additional distinction must be introduced.

With respect to precisely determining this distinction, reference to Williamson's position is helpful. The concept of knowledge he develops on the basis of a factive mental state operator (FMSO), and indeed the most general FMSO (thus, $Ⓢ\,\Phi_{WI}$ ("WI" for "Williamson")) does *not* contain the partial definientia "belief" and "justification." There is therefore no reason not to accept the implication $Ⓢ\,\Phi_{WI}Tp \to$ Tp as unrestrictedly valid. If however one presupposes or accepts the definition of knowledge (K), the situation changes completely, because "belief" and "justification" *are* partial definitia in (K); "believes" is indeed the main operator. The distinction between Williamson's definition and the one suggested here is immediately visible if the question is posed concerning the status of the subject S. According to Williamson, the subject S is presumably to be understood as $Ⓢ_U$—thus, as the subject that understands itself as universal. But Williamson accords to the universality of this subject a quite peculiar meaning that must be clearly noted: the subject is understood universalistically insofar as it is released, so to speak, from all attitudes such as belief, etc. This status of the subject may be interpreted as follows: the so-understood subject is the subject that is, in an absolutely idealized manner, in possession of the truth in the sense that it finds itself *immediately* related to truth. To be sure, Williamson characterizes knowledge as "the most general factive stative *attitude*," but in his interpretation this attitude reduces to an immediate relatedness to truth.

In opposition thereto, according to definition (K), knowledge is an attitude of the subject S that is of a wholly different sort: it is determined as "belief that p is true" in conjunction with "belief that the belief that p (is true) is justified." In the current context, noting this difference raises the following question: is the *knowing* subject S, according to definition (K), particularistic or universalistic, or could it be either? That it could there be understood as particularistic is clear beyond question, but can it also be understood as universalistic?

To answer the question just introduced, it is necessary to introduce an important distinction between two types of belief, which can be termed *normal* belief and *idealized* (in the sense of *fully realized*) belief. Normal (or concrete) belief involves an *ineradicable distance* between the subject and whatever it is that the subject believes. Reasons are the means introduced to narrow this distance, but this narrowing 160 does not eradicate the distance; instead, it merely introduces a stronger or weaker connection between the two poles of the relation called "belief." The stronger the grounding, the tighter the connection, but whatever is believed is always articulated from outside itself, from the perspective of the subject, which remains irreducible. The attitude of the subject thus always remains as a determinative factor that cannot be reduced to whatever is believed and from which whatever is believed does not compellingly follow. Thus, in this case the implication $Ⓢ\,\Phi_KTp \to Tp$ is *not* valid; instead, the following holds: $Ⓢ\,\Phi_KTp \nrightarrow Tp$. This state of affairs can also—and better—be articulated as follows: the subject that has the attitude of normal or concrete belief does not simply identify itself with the dimension of unrestricted rationality or objectivity. *This subject is not a universalistic subject* (not an $Ⓢ_U$) in the sense in which this concept is explained in Section 2.3.2.4[4][i]. For the universalistic subject, the following is unrestrictedly valid: $Ⓢ_UTp \to Tp$. "Belief" in no way appears in this formula.

In opposition to normal (or concrete) belief, idealized belief involves a distinction but not a distance, in the sense described above, between the subject and whatever the subject believes. The attitude of the subject to whatever it believes is here simply the identification of the subject with the universal perspective of unrestricted rationality and objectivity. What is believed is *not* articulated *from the perspective of the subject*, insofar as the latter is *not* understood as identical to the dimension of rationality and objectivity; instead, what is believed is grasped and presented from the universal perspective. The adequate articulation is thus, "The subject matter presents itself such that so-and-so" (or "It is the case that so-and-so"). The formulation "Subject S believes that it is the case that so-and-so" is then inappropriate, indeed incorrect, *if* "belief" is understood as normal or concrete in the sense introduced above. If, however, one understands "belief" as *idealized* belief, in the sense likewise introduced above, then the formulation is correct. To avoid misunderstandings, one should attach to the term "belief" or, in the formula, to the symbol for "knowledge" (i.e., "Φ"), an index. One would then have $\text{ⓈΦ}_{I\_K}Tp \rightarrow Tp$ ("I" for "idealized"). In this case, however, the formula is completely equivalent to one given above (i.e., $\text{Ⓢ}_UTp \rightarrow Tp$)—thus, to a formula that contains *neither* the factor "knowledge" *nor* the factor "belief."

It is clear that the formulation "idealized belief" is quite problematic, because the concept exclusively associated in English with the term "belief" is what is termed above normal or concrete belief. Nevertheless, there are two reasons for introducing the concept of idealized belief (in the sense explained above). First, this concept can be understood as a limiting case of the general concept *belief*, just as *identity* can be interpreted as a limiting case of correspondence (see Section 3.3.3.2.2[2]). Second, introducing the concept *idealized belief* illuminates an important aspect of the concept *knowledge*. It is often assumed both in the history of philosophy and at present that, to put it one way, with the application of the concept *knowledge* the highest theoretical level of the rational activity of a subject is attained. The intuition behind this assumption must not be either ignored or misunderstood, despite the difficulty involved in grasping it accurately. This difficulty arises from the fact that widespread indeterminacy, unclarity, and indeed misunderstandings rule with respect to the use of the term and concept "knowledge." The concept *idealized belief*, clarified above, serves to articulate the intuition alluded to above. It makes clear that putatively purely objective formulations like "It is the case that φ" do not imply that no subject is in any way involved. Any such formulation certainly expresses an attitude of a subject (i.e., the attitude that can be characterized as the *idealized* form of knowledge, in accordance with the definition (K), but with the specification that the attitude of the subject is an *idealized belief* in the sense that has been explained).

To be noted is that the view just presented has a purely global character and that it relates to the purely theoretical domain. It of course requires further development and concretization. It should be clear that in the practical domain, the related issues are considerably more complicated. What this section aims to show is only that, although the epistemic dimension is indispensable, it can ultimately claim only a secondary significance.

If one compares the dimension of language with the epistemic dimension, the considerations introduced in this section show that the linguistic dimension must be accorded central theoretical significance. What this means, in detail, is a fundamental theme of the sections that follow.

2.4 The Dimension of Theory in the Narrower Sense

2.4.1 *On the Proper Concept of Theory in General*

Chapter 1 distinguishes between two fundamentally different categories of characterizing factors important to the concept of theory: external and internal (immanent) factors. The first category, that of purely external factors, characterizes the relation of theory to domains that are themselves not theoretical. Section 2.2.3.1 examines a factor that does not belong exclusively in either category; this is the linguistic criterion for theoreticity, which is in one respect an external factor and in another an internal factor.   162

   This section treats the purely *internal* factors determining the concept of theory. As is noted in Section 2.1, this does not require or involve consideration of the history of the term or concept "theory." When philosophers—most notoriously, Heidegger—look to the etymology and the original meaning of the term θεωρία ("contemplation," "consideration," then "knowledge") in the attempt thereby to gain something significant and deep with respect to the "essence" of "theoretical" thinking, they mix etymology with the philosophical clarification of concepts. The meanings and uses of the term "theory" have indeed had long and complicated histories (see König and Pulte 1998). Yet from at the latest the beginning of the nineteenth century until relatively recently, the term has been used in philosophy and science in a basically unambiguous manner, at least as far as a general, intuitive understanding is concerned; from this it of course does not follow that a consistent explanation or definition has been provided or is at present available.

   In contemporary philosophy, unfortunately, the term "theory" is used in an inflationary manner: any account presenting and/or defending a position or conception appears to qualify as a "theory." Fortunately, the term is used more carefully at present in the sciences, both formal (logic and mathematics) and empirical. Here there are several significant—even if not wholly consistent—characterizations and definitions of the theory-concept. The following account first considers the theory-concept within the frameworks of metalogic/metamathematics and of the philosophy of science. On that basis, it then considers the problem of the application of the theory-concept in philosophy.

2.4.2 *The Theory-Concept in Metalogic/Metamathematics and in the Philosophy of Science*

2.4.2.1 The "Logical" Theory-Concept

The logical or in a specific respect metalogical theory-concept is the one that is defined most precisely; it can also be termed the (meta)mathematical theory-concept. For the sake of simplicity, the following account relies primarily on the designation "logical theory-concept." According to this concept, a theory is a set of sentences or formulas[17] of a formal language that is deductively closed, or closed under logical consequence.   163
The characteristic form for theories of this sort is axiomatic form. To construct such a theory T, one must explicitly identify three factors:

(1) the language of T, $L(T)$,

(2) the logical and nonlogical axioms of T, and

(3) the rules of derivation for T's logic.

---

[17] There is also often talk of "statements."

Essential to this theory-concept is the concept of a *model* for T. The theory T is initially an uninterpreted formal system. A *model* for T is an interpretation of $L(T)$ or a structure for $L(T)$ (often called a "model-structure") within which all of T's non-logical axioms are true. A formula is true in T just when it is true in every model or structure of T, or—equivalently—when it is a logical consequence of T's non-logical axioms.

A classical example can illustrate this logical theory-concept (see Shoenfield 1983: 22f.). Consider a first-order number theory N that formalizes a classical axiom system for the natural numbers. The non-logical symbols are the constants 0, the unary function-symbol S, the binary function-symbols '+' and '·', and the two-placed predicate '<.' The non-logical axioms of N are as follows:

N1. $Sx \neq 0$ N6. $x \cdot Sy = (x \cdot y) + x$

N2. $Sx = Sy \rightarrow x = y$ N7. $\neg (x < 0)$

N3. $x + 0 = x$ N8. $x < Sy \leftrightarrow x < y \vee x = y$

N4. $x + Sy = S(x + y)$ N9. $x < y \vee x = y \vee y < x$

N5. $x \cdot 0 = 0$

An interpretation or a model for—or a structure of—N can be constructed if one interprets that of which the theory speaks as the set of natural numbers (i.e., all values of x are natural numbers), and if one interprets S as the successor function ("is the successor of"), + as the addition function, · as the multiplication function, and < as the relation "is less than." The result is a first-order number theory that formalizes a classical axiom system for the natural numbers.[18]

2.4.2.2 The "Scientific" Theory-Concept I: The "Received View"

In the philosophy of science, the logical empiricism of the Vienna Circle develops a conception of scientific theory conceived basically as an application of the logical theory-concept. After numerous discussions that cannot be examined here, a theory-concept arises that, since 1962 (Putnam 1962), has been known as the standard concept or the "received view." According to this view, theories are classes of sentences; this view is therefore a version—indeed, the most important version—of what Joseph Sneed and Wolfgang Stegmüller later term, in opposing it, the "statement view of theories" (e.g., Stegmüller 1980: 2).

Frederick Suppe (1977: 16–17) presents the original formulation of the received view as follows: a scientific theory is an axiomatic theory, formulated in the mathematical logic (or language) L, that satisfies the following conditions:

(i)   The theory is formulated in a first-order mathematical logic with equality, L.

(ii) The non-logical[19] terms or constants of L are divided into three disjoint classes called *vocabularies*:

---

[18] N is weaker than the usual Peano-arithmetic, because it does not include the axiom of induction.

[19] Suppe's use here of "non-logical" is incorrect, given that the first of the classes of such "terms or constants" is the "logical vocabulary." Presumably, the error is typographical.

(a) the *logical* vocabulary consisting of the logical constants of L (including mathematical terms).

(b) the *observation vocabulary*, V_O, containing observation terms.

(c) the *theoretical vocabulary*, V_T, containing theoretical terms.

(iii) The terms in V_O are interpreted as referring to directly observable physical objects or directly observable attributes of physical objects.

(iv) There is a set of theoretical postulates T whose only nonlogical terms are from V_T.

(v) The terms in V_T are given an *explicit definition* in terms of V_O by *correspondence rules C*—that is, for every term 'F' in V_T, there must be given a definition for it of the following form:

$$(x)(Fx \equiv Ox),$$

where 'Ox' is an expression of L containing symbols only from V_O and possibly the logical vocabulary.

To put this theory-concept in the right light, one must attend to the concept of *explanation*. "Theory" and "explanation" are by no means synonymous or equivalent concepts, yet they are related closely. Within the framework of the just described standard concept of scientific theory, a specific concept of explanation has become famous, and must here be described briefly; this is the model of explanation developed by Hempel and Oppenheim, often termed the *H-O-schema of scientific explanation*. According to this schema, explanation has a *deductive-nomological* (DN) character; hence, one also speaks of *DN-explanations*. The H-O schema can be presented summarily as follows:

Explanans $\begin{cases} I_1, I_2, \ldots, I_n & \text{(Initial Conditions)} \\ L_1, L_2, \ldots, L_r & \text{(Laws)} \end{cases}$

Explanandum E (Explained Event)

It is apparent that this explanation-schema and the received view of the theory-schema are a perfect match: the standard theory-concept is concretized in a consequent fashion by means of application of the DN-schema of explanation.

The received view has been much criticized, with the following points being particularly questioned: the distinction between "analytic" and "synthetic," by which this theory-concept is decisively supported; the distinction between "theoretical" and "observable"; the analysis of the correspondence rules; and the axiomatization of theories. This critical questioning of the received view has led to the development of two positive countermovements, one designated by Frederick Suppe as yielding "*Weltanschauungen* analyses," the other as consisting of "semantic approaches."

The first movement has been developed by authors including Stephen Toulmin, Thomas Kuhn, N. R. Hanson, Karl Popper, and Paul Feyerabend. One of the central ideas developed within the framework of this movement is that of the theory-ladenness of observation languages—in general, of all (talk about) data. Another central idea is the inclusion of every scientific theory within the history of science, whereby all the problems enter that arise from the changes in or evolution of scientific theories (see Suppe 1977: 125–221, 633–49, and 1989).

The second movement, that of the "semantic approaches," is of incomparably greater importance with respect to this section and indeed to this book. Its best-known advocates are Frederick Suppe, Bas van Fraassen, and Ronald Giere. But the structuralistic theory-conception developed by Joseph Sneed and Wolfgang Stegmüller may also be considered a "semantic approach" in the sense characteristic of this movement, although Suppe explicitly considers this structuralism *not* to be semantically oriented.[20] The positions of van Fraassen and of scientific-theoretical structuralism are now to be considered in somewhat greater detail.

2.4.2.3 The "Scientific" Theory-Concept II: "Semantic Approaches"

2.4.2.3.1 *Bas van Fraassen's Constructive-Empiricist Position*

Van Fraassen presents his conception in *The Scientific Image* (1980), written for a general audience, and especially in the two imposing works *Laws and Symmetry* (1989) and *Quantum Mechanics: An Empiricist View* (1991). Like all who take semantic approaches, van Fraassen begins (1991: 6) with a thesis that Patrick Suppes (1957) has been formulating and propagating since the 1950s: "[T]he correct tool for philosophy of science is mathematics, not metamathematics." Van Fraassen elaborates, "Suppes's idea was simple: *to present a theory, we define the class of its models directly*, without paying any attention to questions of axiomatizability, in any special language, however relevant or simple or logically interesting that might be."

Van Fraassen often writes as though the step from the received view to a semantic approach consists in rejecting the contention that theories should be axiomatic. This is incorrect (see Section 1.4.3). Semantic approaches, insofar as they emerge from Suppes's insight or slogan, thoroughly appreciate axiomatization. In this respect, the semantic approaches differ from the received view in that the former do not undertake their axiomatizations within the framework of a formalized (uninterpreted) language (usually, the language of first-order predicate logic), but instead within the framework of set theory. The central thought with respect to the axiomatization of theories is formulated by Suppes (1957: 249) as follows: "[T]o axiomatize a theory is to define a predicate in terms of notions of set theory." This is illustrated by an example that Suppes himself provides and that also plays an important role in the structuralist conception. The theory known as "classical particle mechanics" is characterized as

---

[20] The reason Suppe gives (1989: 10) is that "Sneed's solution appeals to correspondence rules" or that Sneed "explicitly retains some correspondence rules in the form of Ramsey-sentences" (19; see 20). One of the advocates of the Sneedian/Stegmüllerian structuralism, Werner Diederich (1996: 17), rightly criticizes Suppe's contention.

$\mathfrak{P} = \langle P, T, s, m, f, g \rangle$, where P and T are sets (the sets, respectively, of particles p and of real numbers representing times t), s is a binary function indicating the position of p at t, m a unary function indicating the mass of p, f a ternary function yielding the force a particle p exerts on a particle q at time t, and g a binary function indicating an external force acting on p at t (293–94).[21] The system's definition (in part) is the following (294):

> A system $\mathfrak{P} = \langle P, T, s, m, f, g \rangle$ is a system of (classical) particle mechanics just when the following seven axioms are satisfied:
>
> Axiom P1. The set P is finite and non-empty.
>
> Axiom P2. The set T is an interval of real numbers.

The list continues through Axiom P7.[22]

The point that is decisive with respect to the semantic approaches is that they deny that theories are linguistic entities—more precisely, that they are systems or classes of linguistic entities in the sense of sentences (or formulas); the counterthesis is that theories are classes of models. Unfortunately, however, the term "model" is used in widely diverging ways within the literature of the philosophy of science. Van Fraassen (1991: 7) understands models to be "structures in their own right" (i.e., entities utterly unrelated to language): "A model consists, formally speaking, of entities and relations among those entities" (1989: 365n2). He thus identifies models with mathematical structures. This distinguishes his model-concept from the one that is standard in formal logic and that van Fraassen accurately characterizes when he notes that models, in the formal-logical sense, are "partially linguistic entities, each yoked to a particular syntax" (366n4). Van Fraassen (1989: 221) speaks in this connection of a "tragedy" that philosophy of science has undergone: "In any tragedy, we suspect that some crucial mistake was made at the very beginning. The mistake, I think, was to confuse a theory with the formulation of a theory in a particular language."[23]

Given this position, it is surprising that van Fraassen accepts an approach that is designated as "semantic." But van Fraassen adopts the designation introduced by Suppe,

---

[21] The summary to which this note is appended follows Suppes quite closely, and some phrases are taken verbatim from his text. This is noted here, rather than indicated by means of quotation marks, to make the text more easily readable.

[22] Suppes sometimes uses a terminology that is easily misunderstood. As Frederick Suppe notes (Suppe 1977: 100f), Suppes simply identifies "axiomatization" and "formalization" (see also the related discussion between Hempel and Suppes, Suppe 1977: 244–307). When Suppes approves of "formalization" for theories and declares it to be requisite, he does *not* thereby mean the presentation of a given theory within the framework of any formalized language or, as van Fraassen says, "in any special language" (1991:6).

[23] In a note (365n3) to this passage, van Fraassen writes,

> I use the word ["tragedy"] deliberately: it was a tragedy for philosophers of science to go off on these logico-linguistic tangles, which contributed nothing to the understanding of either science or logic or language. It is still unfortunately necessary to speak polemically about this, because so much philosophy of science is still couched in terminology based on a mistake.

and seems to be the only one who wonders about its appropriateness. Van Fraassen (1991: 5) notes that this approach, "despite its name," "de-emphasizes language,"[24] and indeed, this designation is unsuitable and misleading.

As van Fraassen notes, the choice of a "semantic approach" over the various alternative conceptions is wholly independent of all disputes about scientific realism and anti-realism.[25] But van Fraassen himself—in opposition to Suppe, Giere, and other advocates of semantic approaches—champions a decidedly anti-realistic position that he calls *constructive empiricism*. The central thesis of this position can be formulated as follows: the goal of science is not truth as such, but only *empirical adequacy* (i.e., truth relative to observable phenomena). Van Fraassen reaches this thesis by pursuing the question, what is a scientific theory? His answer: a scientific theory is an object that we can accept or reject, and that can be an object of our belief or disbelief.

How is this answer precisely to be understood? According to van Fraasen, it is one of the two possible answers to the question, what is a scientific theory? His answer is empiricism (anti-realism); the other answer is scientific realism. According to realism, a theory is an entity that is true or false, whereby the criterion for success is truth itself. This means, according to van Fraassen, that acceptance of a theory as successful requires the conviction that the theory is true—and it means in addition that the goal of science is to develop literally true theories with respect to what the world actually is.

The other answer, which van Fraassen defends, is empiricism. It is based on several assumptions. Among the most important of these is the distinction between acceptance and belief. For van Fraassen, acceptance rests on grounds that do not directly involve truth. Concretely: when we develop and evaluate theories, we follow both our own needs for information and our striving for truth. We therefore want, on the one hand, to develop theories that have maximal powers of empirical prediction and, on the other, to develop theories that are true. Van Fraassen contends that because there are grounds for acceptance that are not grounds for belief, acceptance is not the same as belief.

From this distinction van Fraassen draws the additional conclusion that the goal of science consists in developing *acceptable* theories, not (literally) *true* ones. An acceptable theory is one that is empirically adequate in the sense that it coherently articulates observable phenomena. Van Fraassen (1991: 4) puts it thus: "Acceptance of a theory involves a belief only that the theory is empirically acceptable." Differently put: acceptance of a theory involves truth-relativized-to-the-observable-phenomena.

Here, a critical note with respect to constructive empiricism, one that concerns the determination of the theory-concept, is sufficient: constructive empiricism includes a radical distinction that must be designated, more precisely, as a dichotomy or gap; the gap is between (observable) phenomena and "the world." If a theory articulates models (in van Fraassen's sense—thus structures), the question must be raised, what sort

---

[24] The reason for the introduction of this wholly unsuitable and misleading designation appears to be the desire to have a "non-syntactic" approach (a hint in this direction provided by van Fraassen (1989:217): "[I]t is 'semantic' rather than 'syntactic'"). But an approach that one can sensibly term "non-syntactic" is one that remains determined within the framework of language—otherwise, one would have to speak of a "non-linguistically oriented approach." The designation "semantic" for an approach that accords to language no place worth mentioning is a misnomer.

[25] To the following, compare especially van Fraassen (1991 Chapter 1 "What Is Science?").

of entities are such models or structures? The question is whether or not one is here confronted with an inflation of entities that reside somewhere and somehow *between* language and the world (see esp. Rosen 1994).

2.4.2.3.2 *The Structuralistic Conception of Theory*

[1] The second subform of semantic approaches is the so-called structuralistic conception of theories.[26] This subform appears to be the most extensively developed alternative to the received view; it also emphasizes most strongly the "non-statement view" of theories. Van Fraassen—who, as has been indicated, likewise rejects the received view—explicitly opposes the structuralist thesis that theories have a "non-statement character" (e.g., 1989: 191).

To be sure, in the most recent developments of the structuralist conception the view is maintained that it is not adequate to characterize theories as classes of statements, but it is emphasized that this does not constitute a rejection of the claim that one extremely important task for science is to produce statements as "things that can be true or false" (Moulines 1996:9). In the more recent words of C. U. Moulines (2002: 6):

> This characterization ["non-statement view"] has led some critics (who apparently only know of this approach that it has been called a "non-statement view") to dismiss it immediately as a grotesque view: science would not make any statements about the world, it would not consist of things that are true or false—which is obviously absurd. Of course, this interpretation of structuralism's view comes from a gross misunderstanding.

Understanding this self-interpretation of the structuralists requires that two states of affairs be taken into consideration. (i) One of the founders of the structuralist theory-concept regularly clarifies the concept *non-statement view* with optimal clarity (to cite but one instance of many): "*A* [*structuralist*] *theory is not the sort of entity of which one can sensibly say that it is falsified (or verified)*" (Stegmüller, *Probleme* II-2/1973: 23). The removal of the concept of truth from the structuralist concept of theory is, according to Stegmüller, of absolutely central significance with respect to rejecting Thomas Kuhn's contention that the history of science is an irrational process. Stegmüller rejects this contention on the basis of the claim that theories are mathematical structures that are either *successful* or not. A theory that is unsuccessful has in no way been falsified.

(ii) When structuralists speak about theories that make empirical claims (many use the term "propositions"; Moulines (2002:10) refers to "the so-called central empirical claim of the theory") what is meant is that *with the aid* of the relevant theories, empirical claims or statements are (or can be) made (by scientists). Theories *as such* are defined by the structuralists in such a way that the components "statement," "assertion," etc. are not explicitly involved. These components are added subsequently and remain external; they are not integrated into the genuine concept of theory. About this more must be said.

The following presentation is limited to what is absolutely essential and relates to the most recent form of the structuralist theory-conception, particularly the concise accounts of C. U. Moulines (1996, 2002[27]).

---

[26] The most important works are Sneed (1979), Stegmüller (*Probleme* I/1983, *Probleme* II-2/1973, *Probleme* II-3/1986, 1980, 1986), Balzer et al. (1987), and Balzer and Moulines (1996).
[27] Moulines (2002) is the introduction to an issue of *Synthese* devoted in its entirety to structuralism.

171   [2] The structuralists prefer no longer to use the term "theory," on the one hand because they take this term/concept to be irremediably polysemic in philosophy and the sciences (and indeed in ordinary language), and on the other because they maintain that the complexity of the theoretical enterprise cannot adequately be grasped even with the aid of a corrected explication of this term/concept. They prefer instead to speak of "theory elements," "theory nets," etc. In practice, they do use the term "theory," explaining, however, that in structuralist accounts the term should be taken to mean "theory elements" (the latter understood structuralistically).

The structuralists stringently follow Peter Suppes's recommendation that theories be defined by means of the introduction of set-theoretical predicates. From the outset, they emphasize that they use an informal (naïve) set theory and not a formal language. In not using formal languages they are being consistent, because structuralist theories are not collections of sentences/statements, but instead collections of models (in the sense of mathematical structures). This is repeated with increasing vehemence by Sneed and especially by Stegmüller. Contemporary structuralists make the conception more precise by noting that their reasons for not using formal language or for using informal set theory are practical and methodological: "[I]n most developed scientific theories this procedure [i.e., the use of formal languages] would be very clumsy if not fully impracticable. As soon as you employ a bit of higher mathematics in your theory, a complete formalization of the axioms becomes an extremely tedious task" (Moulines 1996: 5–6).

From this clarification emerges a question: what significant difference remains between the structuralist position and the "logical" theory-conception presented above, or even the received view concerning scientific theory? If it is possible in principle to use a formal language, then—even if it may not be practically or methodologically optimal to use it—with what is "the theory" then identified? Is it then a class of sentences? But then no fundamental difference could be detected from the logical concept or from the received view. In any case, it appears that extensive internal modifications would be required if a formal language were to be added as an internal component of the structuralist theory-concept.

172   In the structuralist concept of theory, the set-theoretical predicate is explicated by means of a series of internal components that are specified by axioms. By this process, a mathematical structure—the structural core of the theory ("K")—is defined. "Models" are simply identified with structures, and structures are, as is usual in mathematics, understood as sequences of the form, $\langle D_1, \ldots, D_m, R_1, \ldots, R_n \rangle$, where D_i designates the basal sets (i.e., the basal ontology) and R_i the relations among these sets.

[3] According to the most recent form of the structuralist conception, the core structure K of any theory consists of six components. First, a distinction is introduced between two different types of axioms: those that determine the relevant conceptual framework and those that establish the substantial laws of the theory. These then yield two types of models/structures: those that satisfy axioms of the first type are the *potential models* ("M_p"—concepts and axioms), and those satisfying the second are the *actual models* ("M"), models that say something about the world. The identity of a structuralistically understood theory is thus given by an ordered pair $\langle M_p, M \rangle$. This ordered pair is termed a "model-element"; it is the most fundamental basic unit that must be given or grasped if a theory is to have a specific "essence." A full-fledged theory (or again, theory element) must have four additional components.

The third component is also a determinate type of model or structure. To characterize this component, one must first distinguish two conceptual and methodological levels within one and the same theory. The first is the level of the concepts that are specific to the relevant theory (the theory's conceptual framework)—concepts that can be determined only if the theory is actually developed; the set of these is the set of *T-theoretical* concepts. The second level involves concepts that are initially found outside the theory; as a rule, these are determined by other (more basic) theories. These constitute the class of *T-non-theoretical* concepts. The class of submodels or substructures that satisfy the axioms for the *T-non-theoretical* concepts are *partial-potential models* ("M_{pp}"—the set of all the empirical factors, elements, etc.). These provide the relevant *data basis* for the theory.

The status of the three other constitutive components is different from that of the first three. Here, they are explained only briefly. The elements of the fourth type are "constraints" ("C"). They are introduced on the basis of the assumption that the models of one and the same theory exist not in isolation, but instead as connected by means of certain second-order conditions. C is the class of these *intratheoretical* conditions or connections. Because, however, it is also the case that the theories themselves do not exist in isolation, the models attached to different theories are connected with one another. *Intertheoretical* conditions or connections of this sort are termed "links" ("L"). 173

Finally, the introduction of an additional component—the last—is motivated by the following consideration. It can justifiably be assumed of any empirical theory that it can be *applied* to our world. Because such a theory (structuralistically understood) is (or in principle can be) applied to various kinds of phenomena or to various sectors of the world, no such application can be determined without explicit acknowledgment of a certain degree of *approximation* (to the corresponding sector of the world). From this the structuralists conclude that any theory's models must be to some extent "blurred," in the sense that certain limits must be set; otherwise, the theory would be wholly unusable. The most general way in which such "blurs" can be characterized is by making them elements of a so-called uniformity ("U"), understood topologically. Not all of the elements of such a uniformity can, however, be accepted in the case of an empirical theory; the acceptable elements must be subjected to certain restrictions. These acceptable elements are termed "admissible blurs" ("A").

The structural core of any theory is thus a sextuple:

$$K: = \langle M_p, M, M_{pp}, C, L, A \rangle$$

[4] The structuralists add to the structural core the domains (or classes) of intended application ("I"): the totality of phenomena to which the theory is supposed to apply. This reveals that with the aid of (structuralistically understood) theories, empirical claims are (or can be) made—i.e., that a given domain of intended applications can be subsumed under the concepts and principles (laws, constraints, and links) of the relevant theory. Let "$Cn(K)$" be the symbol for the (theoretical) content (Cn) of the theory having structural core K. The central claim of the theory is thus to be articulated as $I \in Cn(K)$. Moulines (2002: 10) comments: "We may write the so-called central empirical claim of the theory as follows: $I \in Cn(K)$. This formula expresses a statement 'about the world,' and this statement may be checked by means 174

independent of K." The elementary definition of a structuralistically understood theory (more precisely, of a "theory-element") consists of a theory core (i.e., a mathematical structure K) and the domains of its intended applications I. More precisely: a theory (a theory-element) T is an ordered pair, consisting of the two elements K and I. Thus, $T = \langle K, I \rangle$.

One takes an additional step when one asks about the relations among the components of the structural core and the relations between these components and the element I. The structuralists answer these questions by applying naïve (i.e., non-axiomatic) set theory. They set-theoretically relate the other components to the component M_p, with the following results:

a) $M \in M_p$

b) $C \subseteq \wp(M_p)$

c) For every $\lambda \in L$ there is an $M_p' \neq M_p$ such that $\lambda \subseteq M_p \times M_p'$

d) There is a many-one (non-injective) function r such that $r : M_p \to M_{pp}$

e) $A \subseteq U \in \wp(M_p \times M_{pp})$

The critically important point is the following: one can understand that M, M_p, and M_{pp} are internal components of the theory-concept. But the other components, L and A (and in certain respects C) and especially the set I of intended applications, seem to have a status that *presupposes* the theory *as already defined or constituted*—but defined or constituted *without* these three components or this other member. Their task is, after all, to locate the theory in relation to other theories or to the domain of applications. It appears that the structuralists consider *every* factor that is associated with a given theory in any manner whatsoever to be a *component* of the theory-concept. This appears to lead to a definition of "theory" that becomes ever more complex.[28]

2.4.3 *A Structural Theory-Concept for (Systematic) Philosophy*

This section develops a *structural* theory-concept that is appropriate for systematic philosophy. It is designated the *structural* theory-concept to distinguish it from—but also to reflect its proximity to—the just presented and critically examined *structuralistic* theory-concept.

2.4.3.1 The Problematic

What do the preceding considerations of the theory-concepts in logic and in the sciences and the philosophy of science yield with respect to the theory-concept in philosophy?

[1] It must first be noted that the term "theory" is used in contemporary philosophy—unlike in the earlier epochs of the history of philosophy—with extraordinary frequency. Every conception that is presented appears to want to glorify itself

---

[28] John Forge (2002: 119n4) notes correctly, "We might have some reservations about including A as an element of K ... [T]he object A is 'historical' in a sense in which other elements of K are not ..."

by being a "theory of" something or other. But what is thereby said? In virtually all cases when this designation is used in philosophy, either nothing specific is meant, or at most something only vaguely reminiscent of the concepts presented above. With the unspecific use of the term "theory" in philosophy, usually no more than the following is meant: an opinion or conception concerning a topic considered to be philosophical is formulated in such a way (1) that one or more concepts are somehow thematized, and perhaps explained, in that "theses" are presented; (2) that these concepts and/or theses are supported—most often, "justified"—with reasons and/or arguments of one kind or another; and (3) that—finally—reasons and/or arguments in favor of one or more alternative conceptions are criticized and/or refuted. In the vast majority of cases, there can be no talk of a "theory" in the sense of a configuration that is even minimally *explicitly* determined with respect to its constituent elements (linguistic, logical, conceptual, argumentative, ontological, etc.).

Even a glance at the history of philosophy reveals that many philosophers have taken the various constitutive elements of theoreticity explicitly into consideration; examples span from Aristotle through Thomas Aquinas, Descartes, Spinoza, Kant, etc., into analytic philosophy. The form of philosophical presentation visible in this history that shows some proximity to the more contemporary theory-concept is the "axiomatic method": the method—and its in part quite distinct forms—that operates in a specific logical-conceptual-argumentative respect with principles (i.e., axioms). A famous example is Spinoza, with his determination and application of the philosophical method *more geometrico*. Useful and illuminating as this "theoretical" procedure is, however, it cannot be said that it suffices or even approaches sufficing to meet, in a way worth mentioning, the demands of contemporary theoreticity.

176

The question presses: at present, how should philosophy—and particularly systematically oriented philosophy—proceed? Should it rely upon one of the stringent theory-concepts introduced above? If so, upon which should it rely? If not, how else should it proceed?

[2] The following account provides answers to these and other questions that arise in conjunction with the determination of a theory-concept for philosophy. Because the systematic philosophy envisaged here understands itself as *theoretically* oriented in an eminently and uncompromisingly strict sense, it should be clear that it must develop on the basis or foundation or within the framework of a clear concept of *theory*. That this philosophy should be theoretical in this sense is argued above in various ways; that course of argumentation is not here repeated or expanded. The task here is instead that of determining how a clear theory-concept for philosophy can be developed. No attempt is made to accomplish this task in such a manner that at the end a perfect or even sufficient *definition* would be provided for a concept of theory that was itself sufficiently clear; whether that goal could be attained at present is not considered. But such a concept is at least basically *characterized* here, although it is essential to emphasize that the characterization of a concept is not to be equated with its *definition* (taken in the strict sense). One could speak at most of an approximative definition.

To be noted at the outset is that a problem arises with the adoption or application, in philosophy, of the theory-concepts presented above because those concepts were developed for the formal sciences (logic and mathematics) and/or the empirical sciences. An immediate consequence of the fact that philosophy is neither a formal nor an empirical science is that it cannot simply appropriate the theory-concept of any such science. This insight retains its force even if one—as philosopher of science—concludes that one of the concepts presented above is correct or appropriate for the science with which it is associated.

[3] The position taken here is that none of the presented theory-concepts, in isolation, can count as correct or appropriate *in every respect*, even for the science(s) to which it is intended to apply. Given this, the philosopher clearly cannot simply appropriate any one of these concepts. The philosopher can, however, nevertheless develop at least the basic features of a theory-concept appropriate for philosophy *under the explicit inspiration of each* of the theory-concepts presented above. Here, "under the explicit inspiration" means the following: each of the theory-concepts must be subjected to corrections on the basis of comparisons made among them; the process of comparison reveals and makes possible avoidance of the one-sidednesses and weaknesses of each.

As far as the domains of application of the presented theory-concepts are concerned (i.e., for the one the formal sciences, for the others the empirical sciences), it must be kept in mind that, as is indicated above, philosophy is neither a formal nor an empirical science; this is made more precise by the qualification that, although philosophy is neither *purely* nor *predominantly* either a formal or an empirical science, it certainly does contain both formal and empirical elements or characteristics. This itself provides a reason for philosophers to consider carefully the theory-concepts introduced above.

This leads to an additional relevant point: although it is the case that the presented theory-concepts were *in fact* developed for formal and for empirical sciences, it does not follow that they would have to be appropriate *exclusively* for applications in the domains of those sciences; it is completely possible that they have theoretical potential that extends far beyond the purely formal and purely empirical sciences in the narrower sense. Hence, it is more accurate to say that these theory-concepts were conceived *at first* or *primarily*, albeit not *exclusively*, as appropriate for formal or empirical sciences. How broadly their theoretical potential might extend—here, with respect to philosophy—must be determined case by case.

[4] If one compares the presented theory-concepts with one another, several limitations and defects of each quickly become visible—particularly from a determinate philosophical standpoint that can be termed the *semantically realistic* standpoint, the standpoint adopted and defended in this book.

The "logical" theory-concept is fundamentally syntactically oriented; this raises the question of how one could move from it to semantics and to ontology. The classical or standard adaptation of this concept to the empirical sciences (i.e., the received view) tackles this question and attempts to answer it. That it introduces the distinction between a theoretical vocabulary and an empirical or observation vocabulary makes clear that the purely syntactic level has been surpassed and the semantic-ontological level broached. But it is doubtful whether this move has genuinely succeeded.

The thesis that the move has not succeeded is supported, adequately for this book's purposes, by Bas van Fraassen's (1989: 221ff) concise critique of the received view. Van Fraassen radically questions the thesis (based on the received view) that scientific theories are interpreted theories. According to van Fraassen, the problems arising from this thesis are insoluble. As illustrations, he briefly introduces the two variants of the positions based on this thesis that he considers the most important. The first insists on the formal character of theories as such, and connects theories to the world by means of *partial interpretations*; van Fraassen deems the most interesting explication of this concept that of Hans Reichenbach, according to whom theoretical relations have *physical correlates*. The partial character of this correlation becomes clear, according to van Fraassen, if we consider the paradigmatic example of this relation: light rays are the physical correlates of straight lines. Because it is immediately evident that not every line is the path of a extant light ray, it is clear that the language-world connection is only partial.   178

The second variant van Fraassen considers is that developed in the later writings of C. G. Hempel. Hempel holds that axioms are stated already in natural languages. Principles of interpretation lead to the division of the class of axioms into two subclasses: (1) axioms that are purely theoretical—those in which all the non-logical terms are introduced specifically and exclusively for the sake of formulating the theory—and (2) axioms with a mixed status; these also include non-theoretical terms. Concerning both variants, van Fraassen (221) remarks,

> It will be readily appreciated that in both these developments, despite lip-service to the contrary, the so-called problem of interpretation was left behind. We do not have the option of interpreting theoretical terms—we only have the choice of regarding them as either (*a*) terms we do not fully understand but know how to use in our reasoning, without detriment to the success of science, or (*b*) terms which are now part of natural language, and no less well understood than its other parts. The choice, the correct view about the meaning and understanding of newly introduced terms, makes no practical difference to philosophy of science, as far as one can tell. It is a good problem to pose to philosophers of language, and to leave them to it.

The alternative developed as the "semantic approaches" seeks to avoid any relation to language that is essential or even worth mentioning by instead *from the outset* considering models, understood as structures. This alternative, however, has at least *two* significant problems: first, the models/structures must themselves be *linguistically articulated*, and in fact they are linguistically articulated. This articulation, however, has deep implications for this alternative, because the role the immense linguistic dimension plays with respect to this theory-concept must be explained. Second, there is the question of how this alternative conceives of the models/structures. In every model/structure there is a basic set of entities. But what (kind(s) of) entities? Things? Objects? Substances? Processes? And how are such entities precisely to be understood? These questions reveal enormous deficiencies in all the presented theory-concepts, but the deficiencies are particularly noticeable in the semantic approaches: how can this fundamental semantic-ontological question be clarified if language is accorded no significant status? There is in addition the question of which language is to be chosen. That reliance upon natural language is untenable is shown in various places in this book.   179

[5] The strength of the various so-called semantic approaches is their reliance on models (or, synonymously, on structures). This is a progressive step that should be

retained. But many questions must be addressed. The following account considers the two that appear to be the most important. The first concerns the exclusive reliance on informal set theory. This book ultimately concludes that set theory—when it is used either as the exclusive or even as the fundamental formal instrument—is inadequate for philosophical purposes, because it is not rich enough to articulate the complex relations of the models/structures either among themselves or in their relations to the world. This point has enormous implications.[29]

The second question is of decisive importance with respect to the basic thought guiding the development of this book; it is the question, raised above, of what basic entities are recognized. Or, more generally formulated, what status is accorded to ontological issues? Three factors in the structuralist theory-concept can be seen as gestures toward the ontological level: two components of the structure-core K—the potential-partial model (M_{pp}) and the admissible blurs (A)—and the second member in the ordered pair that serves to define the theory-concept, the set of intended applications (I).

M_{pp} represents the class of substructures that satisfy only the axioms for the T-non-theoretical concepts; M_{pp} is thus the (relative) data basis for T. This data basis is always to be understood as T-relative, thus as relative to the theory in question. Current structuralists (Moulines 2002: 7) emphasize that M_{pp} is the class of "concepts coming from 'outside,'" thus the empirical concepts. "Typically, the latter [the concepts from "outside"] are determined by actual models of other, 'underlying' theories." But this seems to entail an infinite regress, because the "underlying" theories themselves would have to rely on underlying theories—and so *ad infinitum*. Such a regress appears avoidable only on the assumption of something like a base-M_{pp} that would have to be interpreted as a "data base" in a generally classical sense—thus as immediately available empirical data, data that would be immediately related to phenomena or sectors of the world.

The second factor in the structuralist theory-concept that is ontologically relevant is A, the class of admissible blurs—the degrees of approximation allowed between different models. Here, too, the concern is with the application of the theory to the real world, but the structuralists contend that no such application is or can be complete, that any must involve the leeway of a certain degree of approximation. Talk of the "real world" and of "approximation" to it, and indeed of (the relation to) the ontological dimension, is insufficient as long as the truth-concept that is presupposed remains unclarified. The structuralists appear to presuppose a truth-concept drawn from formal (Tarskian) model-theory. That no such concept is adequate for philosophical theorization is, however, thoroughly demonstrated in this book (see especially 2.5 and 3.3).

The third factor that points to the ontological level is the domain of intended applications I: the domain of phenomena or things or sectors of the world to which the theory is or can be applied. Moulines explains that structuralism makes the following basic *epistemological* assumption with respect to the domain of intended

---

[29] It is significant that even within structuralism (see Moulines 1996: 8n3), the question is raised whether the components of M_p, M_{pp}, and A, all belonging to the structure-core, can be characterized fully or even adequately in set-theoretical terms.

applications: the domain is not to be identified with anything like "pure reality," "pure experience," preconceptually given "things in themselves," or sense data; "[r]ather, the assumption is that the domain of intended applications of a theory is conceptually determined through concepts already available" (Moulines 1996: 8).

These three factors fail to provide a minimally and much less a fully adequate answer to the question, posed above, concerning ontological issues. To be sure, the structuralist philosopher of science can say that clarification of these issues is a task for a specific philosophical discipline: ontology. But one additional and indeed decisive point still requires clarification: the relation between a given empirical theory and the empirical claims made by scientists *with the aid* of the theory. How is this relation to be understood, and how is it to be justified? Here, there appears to be an unsolved problem. If a theory T, defined on the elementary level, is an ordered pair $T = \langle K, I \rangle$, this implies—at least for scientific and philosophical purposes—that there is a relation between K and I *and* that this relation can be thematized explicitly: that it can be articulated by means of sentences. In the case of a concrete theory—an empirical theory relating to a determinate sector of the world—it is said that this sector is an element in I and thus a subclass of M_{pp}. Moreover, to repeat the passage cited just above, "the assumption is that the domain of intended applications of a theory is conceptually determined through concepts already available." This last assumption, however, does not relate (only) to a pure "conceptual sphere" in abstraction from what is thereby conceptually articulated, it relates to the conceptual dimension of the data that constitute the sector. This is confirmed by a second passage, also cited above: "[W]e may write the so-called central empirical claim of the theory as follows: $I \in Cn(K)$. This formula expresses a statement 'about the world,' and this statement may be checked by means independent of K" (Moulines 2002: 10).

This appears to mean that the "empirical claim of the theory" is a constitutive factor of the theory itself. To be sure, however, the term "claim" is quite problematic in this context because, strictly speaking, it designates not a semantic concept but a pragmatic one. A claim is not the same thing as a truth; on the contrary, the claim in question presupposes truth: the truth of the theory (see Section 2.5.2.2). It would therefore be more appropriate to speak of empirical sentences, and still more appropriate to speak of the empirical propositions expressed by empirical sentences. Pragmatic factors like (empirical) claims are not constitutive structural moments of theories; instead, they articulate the attitude (of belief, of conviction) that a subject (the theoretician) takes concerning the truth-status of a given theory. Empirical sentences and propositions, in contrast, *are* constitutive factors within theories.

Whenever "theory" is used to designate only the abstract structural-conceptual level, thus the level that excludes semantics, that is a terminological stipulation, and therefore fundamentally unobjectionable. This stipulated sense is not, however, the one that is generally intended when, in the sciences and in philosophy, there is talk of theories. The appropriate term for what the structuralists term "theories" would be, in the vocabulary of this book, "theoretical frameworks."

The reason that advocates of the structuralist theory-conception do not address these issues is presumably that that conception lacks an explicit semantics; the reason for that lack, in turn, is that language is not included within this conception.

2.4.3.2 The Essential Components of a Structural Theory-Concept for Systematic Philosophy

The structural theory-concept now to be developed is intended to be appropriate for systematic philosophy. As is indicated above, the following account does not attempt to present an absolutely unobjectionable, perfect, and complete *definition* of a theory-concept that would be fully appropriate to and adequate for philosophy. The goal instead is to provide an explanation or characterization of such a concept.

[1] At the heart of the structural theory-concept is, as is emphasized above more than once, the concept of *structure*. Although this conception thus adopts the insight guiding the semantic approaches—and especially the structuralist variants of this approach—it diverges from all of these approaches in three essential ways. First, it conceives of *structure* differently, not with respect to the purely abstract definition of the concept, but with respect to the more precise understanding of the *definiens* and of the types of structures that are essential or relevant to the determination of the concept of theory. Second, it does not simply acknowledge the relation of structure(s) to language, but instead presents it as essential. Third, it takes explicating the connection of structure(s) to the ontological domain (to the universe of discourse, to the world) to be indispensable and simply central.

The elementary philosophical concept of theory involves three basic components defined as an ordered triple, $\langle L,S,U \rangle$ where L = language, S = structure, and U = universe of discourse. Of course, these three components must be determined (i.e. explicated) accurately and in detail. Alternatively, one can combine the components L (language) and S (structure), because the two are intimately connected or, more precisely, because complete articulation of the concept of structure requires the inclusion of the dimension of language. One can thus speak simply of "structure," because in abstraction from its connection to language, structure cannot adequately be defined. What results is a *dyadic* concept of the philosophical theory: the ordered pair $\langle S,U \rangle$. This reflects—and articulates, as the elementary concept of philosophical theory—the thought that provides this book with its title (which, for stylistic reasons, uses "being" instead of "universe of discourse"), and that underlies the conception of systematic philosophy presented in this book.

[2] As is shown in the following sections and in the remainder of this work, there are three fundamental types of structures: *formal* (*logical-mathematical*), *semantic*, and *ontological*. Chapter 3 is devoted to these structures.[30]

With respect to the connection between structure(s) and language, three important points must be made at the outset. (i) If one begins with a purely syntactically determined language—a purely formal language—then there will be in principle innumerable possibilities for interpreting the language, that is, for introducing models for the language. Philosophers of language have devoted a great deal of energy to this issue and the problems that come with it, and this expenditure has led to various

---

[30] The structuralist distinction among M_p, M_{pp}, and M proves useful and indeed important in many respects, but it is oriented more toward surface structures than to deep structures.

theses that have become famous, including those of the indeterminacy of language, of ontological relativity, etc. The famous Löwenheim-Skolem paradox must also be considered in this context (see esp. Putnam 1980/1983). With any given expression in a purely syntactically determined language, one can connect or associate any possible meaning or any possible semantic value. This leads to problems of various sorts. As is indicated above, Bas van Fraassen has forcefully characterized this gigantic complex of problems, speaking of a "misunderstanding" and even of a "tragedy." But everything changes if one begins not with a naked (i.e., uninterpreted) language, but instead with a determinately interpreted language. One then speaks of the "intended model" of the language. And with this, one easily avoids the endless problems that otherwise arise in this domain.

(ii) The integration of language into the dimension of structure brings with it all of the advantages that are provided by the purely logical theory-concept presented above.

(iii) The integration of language into the dimension of structure makes possible the introduction *and clarification* of the central concept of truth. Once this is done, it makes sense to speak of theories being true or false.

In the proposed dyadic theory-concept, the component U (the universe of discourse) represents the *ontological level*. This topic is considered later in this chapter.

[3] The central task for the theory envisaged here is to determine precisely the relation between the dimension of structure (S) and that of the universe of discourse (U). This is the chief or genuine task of systematic philosophy, and it is undertaken in the rest of this work. The concept of truth, as becomes clear, plays the concluding—and in this sense the decisive—role.

At this point, only the following question need be treated: if the relation between the dimension of structure and the dimension of being is as central as has just been maintained, should not this factor—the relation—be recognized as *itself* an *essential* component of the concept of philosophical theory? Should it not be *explicitly* introduced and identified by name? One would then have three essential components: the dimension of structure (S), the ontological dimension (U), and finally the dimension of the relation between S and U, for which the symbol "$\leftrightharpoons$" could be introduced. The resulting theory would be a triple: $\langle S, \leftrightharpoons, U \rangle$.[31]

What is essential is that this relation—and thereby the ontological dimension itself—be explicitly thematized. Whether the relation between S and U should be

184

---

[31] The German physicist G. Ludwig, in his important book (1978), presents a conception of theory whose orientation is decidedly structuralistic and that, although developed in complete independence of that of Sneed and Stegmüller (introduced above), nevertheless exhibits a fundamental similarity to their concept. Unlike Sneed and Stegmüller, however, Ludwig explicitly takes up the task of clarifying the relationship between the dimension of structure and the ontological dimension. Ludwig conceives physical theories essentially as applications of mathematics to the real world. Any such theory, he contends, consists of three parts: a mathematical theory ("MT"), a domain of the real world ("W"), and a rule of application, for which he uses the symbol "$\subset\!\!-\!\!\supset$." Ludwig determines the component $\subset\!\!-\!\!\supset$ more precisely by means of detailed *principles of mapping*. This is a significant advantage of this theory-concept over structuralist concepts of the Sneed-Stegmüller sort.

named as a third component of the theory-concept is, however, a separate question. No doubt it *could* be so named, but this *need* not be done. One reason *not* to do it is the following: just as language belongs essentially to the dimension of structure, the relation of the dimension of structure to the ontological dimension belongs essentially to the dimension of structure, and vice versa: the relation of the universe of discourse to the dimension of structure belongs essentially to the universe of discourse. That this must be so, at least within the theoretical framework of this book, should already be clear. The domain within this framework that is perhaps least obviously structured is that of the world (actuality, the universe), but it is argued above that that domain is *expressible*, and the complex designated by the concept of *universal expressibility* is unthinkable save as precisely structured internally. As far as the dimension of structures is concerned, it is to be noted that it includes, in addition to purely formal structures, also contentual structures (semantic and ontological structures), whereby structures of all these types belong within the unrestricted universe of discourse. One can therefore *adequately* determine *neither S nor U without* explicitly thematizing the relation between the two, in both directions. For this reason, the additional component "$\leftrightharpoons$" is not necessary; adding it would not be incorrect, but would be redundant.

₁₈₅ ### 2.4.3.3 The Structural Theory-Concept as Regulative, and Its Approximative-Partial Realization

The suggested theory-concept for a systematically oriented philosophy is just that: a concept. Moreover, it has not been defined precisely here, but only characterized generally. As with every such concept, the question arises concerning the extent to which it is or can be realized or instantiated. Only a very few concepts can be said to be perfectly realized, and even those, as a rule, only in few cases. Presumably, achievement of such perfection is possible (and perhaps even in some cases actual), if at all, exclusively within the domain of the formal sciences (e.g., with the realization of the concept of connectives in propositional logic, with that of the concept of logical consequence, etc). Similar examples can of course be found in mathematics. But in the *contentually* oriented sciences and *thus in all likelihood in philosophy as well*, one can scarcely count on anything like the perfect realization of concepts.

Such considerations hold in a quite particular manner for the concept of philosophy itself and within this concept for the concept of philosophical theory. The reason for this is that philosophy both in its demands and in its subject matter is an extremely complex undertaking. It would be utter presumption—combined with a total ignorance of the task and of the requirements for its realization—to attempt to proceed in the manner of the German Idealists (i.e., to attempt to develop something like "the absolute system of philosophy"). Two relevant issues are briefly considered.

[1] Even when a concept cannot be viewed as perfectly realized (or indeed realizable), that does not mean that it must lack decisive and indispensable theoretical benefits. Of the grandest concepts, we know thanks to Kant that they are indispensable in their function as *regulative* (for Kant: *regulative ideas*). But the regulative character of the structural concept of theory differs from the Kantian notion in that it does not absolutely exclude, in every respect, the possibility of realization, including perfect realization; instead, it indicates that *as a rule* no full realization is to be counted on.

What does this mean for the systematic philosophy envisaged here? This question can be clarified in terms of the four-stage method presented in Chapter 1 (Section 1.4). A first point is easily extracted: the *second* method or methodical stage is termed the *theory-constitutive*, and the theory-forms appropriate to it are identified as the *axiomatic* and the *coherentist* ("network") theory-forms (see Section 1.4.3). On the basis of the preceding explanations and examinations concerning the theory-concept, it is clear that the *fundamental* crux of the matter is the status of these theory-forms within the framework of a structural conception of theory. This conception is developed above for philosophy and, as distinguished from the *structuralistic* theory-concept, is termed the *structural theory-conception* or *structural theory-concept*. *Axiomatic or coherentist methods or theory-forms* in the purely syntactic sense are not thereby fundamentally excluded, but if those methods or theory-forms are used, they have in every case only a purely derivative and thus secondary status.

In the ideal case a given philosophical conception should ·be presented as an axiomatic and/or as a coherentist theory as this is determined by the structural theory-concept. But in philosophical practice—at least in the foreseeable future—this ideal case will likely not be approachable save in a few relatively small theoretical domains (e.g., a *formal* theory of truth). But the structural theory-concept should always be understood and presupposed as the *regulative* concept that establishes the benchmark for its various interrelated theories.

In this sense, it can be said that the structural theory-concept should be realized at least in a *partial-approximative* form. This concept of partial-approximative form is a quite serviceable one, particularly given that many different degrees of its success can be conceived of and realized. Use of this form appears both to be realistic for philosophy and to provide a disciplinary procedure requisite for this enterprise.

[2] Also to be emphasized is that a structural theory in the strict sense cannot simply appear as "finished" without having depended upon an extensive milieu; such a theory must be so constituted or construed that initially all the constituents belonging to it (i.e., its concepts, its instruments, axioms, structures, data, etc.) must be provided *before* and *so that* it can be formulated *as a theory*. This is the task of the first methodical stage, that of assembling data and of collecting it into informal theories. If a thus-developed philosophical conception were formulated *as* (*axiomatic or coherentist*) *theory* in the sense explained above, the task would arise of integrating that theory into the totality of individual philosophical theories. This is the task of the *third* methodical stage, the *system-constitutive* stage that can also be designated a *method of structuring a holistic net*. Thereby, in essence, a *complete philosophical theory* here termed *comprehensive systematics* is envisaged.

The final major question would be that of the truth-status of the theory thus developed and formulated, which would then appear as a comprehensive philosophical system. The fourth and last methodic stage, the *truth-testing method*, has determining this truth-status as its task.

The final part of the structural-systematic philosophy here under development has the designation *metasystematics*; this part or dimension has a wholly unique theoretical status that is characterized in part in the Introduction and in Chapter 1, and more fully in Chapter 6.

The milieu of a philosophical conception *qua* theory, mentioned in passing above, can also be fully articulated only in the ideal case.[32]

As is noted above, but may be worth repeating in light of the background provided by the structural theory-concept and the four-stage method, philosophy, even in its best representatives and products, tends (tacitly) to rely on conceptions that in the vast majority of cases have a theoretical status that corresponds to a (generally only partial) application of the *first* and the *fourth* methodic stages. The best among these conceptions introduce relevant concepts, clarify them, formulate theses, provide support for the resulting informal theories, introduce responses to and criticisms of opponents—and then generally term the result "theory of...." And they may well continue to do so; it appears scarcely possible or rational to prescribe to other philosophers what terminology they may or should employ and what they must or should reject. Given the fact that the word "theory" is used throughout philosophy, the following attitude appears the most prudent: when the designation "(philosophical) theory" is used, one should understand it—in virtually all cases—in the sense of *partial-approximative theory*. Finally, it may be granted that from the common usage of the term "theory" in contemporary philosophy one can also gain something beneficial. Given the irrationality rampant in many domains of contemporary "philosophy" (even in universities), an irrationality that generally presents itself as radically *anti-theoretical*, the use of the term "theory" to characterize a given philosophical conception has at least a valuable *suggestive* meaning.

188

---

[32] Between the structural theory-conception developed here and the structuralist theory-conception there is, with respect to what is here termed "the milieu of the theory," both a remarkable similarity and a no less remarkable difference. The structuralist conception recognizes, in a certain respect, the factor that is here termed "system-constitutive," insofar as this conception thematizes intertheoretical relations (called "links" and symbolized as 'L'). But the deep difference is that the individual theories are related by L only *to one another*, and are *not* integrated into a *total theory*:

> [T]he intended applications of any given theory don't cover the "whole universe." . . . They represent "small pieces" of human experience. There is no such thing as a theory of everything . . . (Moulines 2002: 9).

What this doubtless means is that there is no natural-scientific theory of everything; this is not the place to discuss that issue. Systematic philosophy, as it is understood in this book, would (if complete) be a philosophically understood theory of everything, but that theory would not provide detailed explanations of all of Moulines's "small pieces" of human experience or of all the small pieces of anything else. What the theory would have to thematize and could succeed in thematizing are the universal or most general structures of the unrestricted universe of discourse.

The factor corresponding to the fourth methodic stage—test of truth—also has a counterpart of sorts in the structuralist theory-concept, to the extent that the thought of *gradual approximation* is thematized there. But there is no counterpart of any sort, in the structural theory-concept, to structuralism's "admissible blurs." In the application of the structural concept, the thought of approximation or relativity does—as shown below—play an important role, but not in anything like the sense of "admissible blurs." The "milieu-factor" of "construction/constitution of a theory" in the sense of making available the materials for the theory has no counterpart in the structuralist theory-concept.

2.5 Fully Determined Theoreticity: First Approach to a Theory of Truth

In the preceding sections of this chapter, the dimension of theoreticity is given an initial, general clarification. First, the dimension of language is analyzed as the dimension of presentation of theoreticity. Second, the dimension of knowledge is presented as the dimension of accomplishment of theoreticity, but in such a manner that the subject (speaker/knower) is shown to have a subordinate rather than preeminent status with respect to theoreticity. Third, the concept *theory* is closely examined, and a theory-concept appropriate to systematic philosophy, the structural theory-concept, is developed. It remains to be said what results when the interconnections among these three domains or aspects or dimensions of theoreticity are determined.

According to the conception developed in this book, what then arises is the full determination of the dimension of theoreticity; this determination, which has the concrete form of theories and ultimately (ideally) of a comprehensive theory, results only from the interconnecting of language, knowledge, and the structural theory-concept. The theory in this global perspective is, namely, a determinate linguistic account that, for the subject, is an accomplishment of knowledge. In this general sense, the theory unites within itself all the elements that constitute the structuration of (indicative) language and the accomplishment of knowledge. To work out in detail this structuration and this accomplishment is the task of systematic philosophy. The highest stage of the determination of the dimension of theoreticity, the stage of full determination, is not, however, attained simply with the presentation of a theory or of theories, but only with the determination of the truth-status of the theory or theories. Truth is thus the capstone-concept of the dimension of theoreticity.

2.5.1 *Preliminary Questions*   189

2.5.1.1 The Word "Truth" and the Problem of the Concept of Truth

[1] As a rule, one poses the question concerning the clarification of a concept by beginning with a word (which can, of course, have various forms, such as "true/truth"). Word and concept are not the same; on the contrary, as a rule multiple concepts are associated with a given word. This should be beyond philosophical question, but it is nevertheless often ignored in the philosophical literature; its being ignored leads to pseudo-questions and questions that are misleadingly formed. The question, "What is truth?," for example, is deeply ambiguous. For the most part, the question is understood as asking for a "thing" or "entity" or, at times, for a "concept." Thereby scarcely noticed is the presupposition that this "thing/entity" or this "concept" actually is associated with the word "truth." It is overlooked that—in no matter what manner—the concept in question can in principle be associated with various other expressions. That is, it is not recognized that a wholly unsupported and indeed false presupposition has been made: that the word "truth" is bound to precisely one fully determined concept, and conversely that to the intended concept is bound precisely one fully determined word.

In asking about *truth*, one starts with this word. If one wants to understand the word as associated with a single, *unitary* concept, the question becomes, which is the right concept? No matter what concept is selected or presented, it becomes immediately clear that it cannot do justice to all uses of the word "truth" and its variants: "It is true that snow is white" and "A true democracy rules in this state" exhibit no relevant conceptual unity. This problem is particularly clear and acute when one considers certain domains of human history, particularly the religious dimension and especially that of language and tradition of Christianity. Here, "true/truth" has an unrestricted use that surpasses the boundaries of any conceptual unity.

[2] Three characteristic attitudes or phenomena illustrate how problematic it is to deal with the word "truth" in conjunction with developing a theory of truth.

[i] The extensive domain of analytically oriented truth-theories exhibits three major approaches. The first simply does not consider uses of "true" that appear to be philosophically anomalous or unruly, such as the attributive use in formulations like "true democracy." The immense realm of Christian-theological uses of "true" is also generally ignored. A second approach seeks to recognize and develop a plurality of truth-predicates *and* truth-*concepts* that putatively exists because "true" and "truth" are used in so many different areas or disciplines (in the social sciences, the natural sciences, the formal sciences, etc.).[33] The third approach, finally, attempts to extend a use of the truth term into all the domains of the theoretical dimension in the narrower sense, even if the word in fact never or only rarely occurs in some of those. The efforts of analytic philosophers who attempt to clarify or defend the concept of "moral truth" showcase this approach.[34]

[ii] Specific problems arise from the use of "truth" in religious language and Christian theology (see Puntel 1995). Many twentieth-century Christian theologians speak of two distinct truth-concepts, both of which they see as legitimate: a Greek-philosophical concept and a biblical or Judeo-Christian one (see the references in Puntel 1995: 24ff). This is an extremely important development particularly with respect to the theorization of truth, because it shows just how far off track one can go when one ignores quite basic issues. The theologians who embrace the two distinct concepts think primarily in terms of biblical language and the biblical tradition, within which the *word* "truth" does indeed have meanings that are peculiar and that are wholly unrelated to the philosophical concept stemming from Greek philosophy. But how can these theologians conclude from this fact that there are two utterly different truth-*concepts*? Only because they confuse word and concept.

This is the more astounding in that this confusion can be precisely documented historically and philosophically. When the Old Testament books of the Bible written in Hebrew were translated into Greek (the Septuagint, LXX, 3rd-1st century BCE), the Hebrew word *'ämät* was usually replaced with the Greek ἀλήθεια. But the original or basic meaning of this Hebrew word (and of other expressions related to it, such as *'ämunā*) involves fixedness, the enduring, the reliable, and that by which one must take one's bearings. To be sure, there is controversy among exegetes concerning the precise meaning of *'ämät* (and *'ämunā*), in their various occurrences[35]; the fact, however, is that the history of Septuagint translation has led to a totally contradictory tradition. That

---

[33] Characteristic of this undertaking are various works of Crispin Wright and Gila Sher (see esp. Wright 1992 and Sher 1999). For discussion see Tappolet (1997, 2000) and Beall (2000).

[34] See esp. Wright (1996) and Puntel (2004).

[35] Michel (1968: 56) concludes, "Nowhere is it necessary to understand *'ämät* as a personal property (*'ämunā* is different!). *'ämät* appears always to relate to a verbal element (assertion, promise, command)." See also Landmesser (1999: 209ff).

the Septuagint renders 'ämät with ἀλήθεια in no way justifies the contention that there is a *Hebrew (biblical) truth-concept* that is fully different from the Greek philosophical truth-concept.

[iii] A third case worth mentioning of a catastrophic confusion of word and concept with respect to the truth-problem is that of Martin Heidegger, who, as is well known, devoted a considerable amount of his philosophical attention to this issue. Many of the works he himself published and of his posthumously published works have this problem as their central theme. His initial and long-held thesis is that the truth-concept dominant in the history of philosophy and still today is only a "derived" one, not the primordial truth-concept; the primordial truth-concept is supposed to be articulated in the writings of the earliest (pre-Socratic) Greek philosophers. According to Heidegger, "ἀλήθεια" originally means *unconcealment* ("in the sense of the opening," 1969/1972: 70/76).

After championing this thesis for most of his life, Heidegger gives it up in 1964 (1969/1972: 70/76), making the following self-correction:

> To raise the question of ἀλήθεια, of unconcealment as such, is not same as raising the question of truth. For this reason, it was inadequate and misleading to call ἀλήθεια, in the sense of opening, truth.

Here, Heidegger could and should have said clearly that he had confused a word and a concept.[36]

[3] The preceding considerations reveal the pressing importance of asking whether the question of how to understand the truth-concept is even a sensible one—that is, whether its status is sufficiently clear. If so, how best to proceed? To be emphasized in advance is that here an exclusively *unified* concept of truth is developed. 192

The account proceeds from the following central insight or thesis: although it is of course possible in principle to stipulate a semantic value for the term "true"—it is possible to connect the term with any arbitrary meaning whatsoever or any conceptual content whatsoever—it remains the case that in ordinary and technical languages from the history of philosophy and throughout Western intellectual history, the term has a determinate and demonstrable core signification. One can—following Tarski (1933/1956: 153), among others—speak of an *intuitive* understanding of truth that is associated with the word "true." Tarski provides what is likely the briefest and most accurate formulation of this intuitive understanding in what could be termed his pre-definition of the term: "*a true sentence is one which says that the state of affairs is so and so, and the state of affairs indeed is so and so*" (155). This pre-definition articulates what could be termed the most primary and fundamental trait of the concept intuitively associated

---

[36] Heidegger was led to his self-correction by Ernst Tugendhat's (1967) critique of his truth-conception. In the introduction to his collection (1992: 14f), Tugendhat justifies as follows his decision to omit his 1969 essay:

> I have not here included [the essay "*Heideggers Idee von der Wahrheit*"], because that essay is simply a central section from [Tugendhat 1967]. Nevertheless, this essay was important because Heidegger himself read it in manuscript and it led him to make one of his few self-corrections, although in making it he made no mention of me. [Tugendhat here cites the passage from Heidegger (1969/1972) reproduced above.] After sending Heidegger the manuscript, I visited him in 1966 and we had a conversation that was, by his standards, quite open.

with "true": to be true, whatever is true (sentence, proposition, etc.) must relate to actuality, to the world, to the ontological dimension. In *Grundlagen einer Theorie der Wahrheit* (Puntel 1990: 302ff), the author develops three other moments of the intuitive understanding of truth—three in addition to the relation to the ontological dimension just articulated: (2) a distinction between two domains or dimensions (i.e., language [thought, mind...] and world); (3) a discursively redeemable claim of validity; and (4) maximal determinacy. These three additional moments follow from the first, the basic feature of the concept. To be sure, this is at first only the articulation of a merely *intuitive* understanding. The real task is to make this understanding explicit.

The task of a truth-theory has now been sketched in its outlines: a determinate concept is presupposed to be given *intuitively*; it is then maintained that throughout the entire history of philosophy and science this concept is coupled with the word "true"; finally, it is noted that this intuitive understanding is in no way sufficient, that it must be explained—and possibly in part corrected and improved.

193 ### 2.5.1.2 Substantialism and Deflationism

The most broadly accepted classification of current truth-theories is based on the distinction between substantialism and deflationism. Normally designated "substantial" are the traditional conceptions of truth, particularly—and sometimes exclusively—the correspondence theory, according to which a given item (sentence, proposition, etc.) is true if and only if it agrees with or corresponds to whatever aspect of reality it purports to be about. More generally, all truth-conceptions that are not "deflationist" are termed "substantial." It is, however, highly questionable whether the division of truth-theories into these two types is adequate. The structural conception developed here certainly does not instantiate either. As becomes clear below, in one specific respect this conception is radically deflationistic; in another, it is radically substantialistic. Sufficient at this point are the following introductory remarks on truth-theoretical deflationism.

Deflationism is not a unitary position. Nevertheless, all the stances that can be designated "deflationistic" have two basic factors in common. The first is the rejection of the correspondence theory of truth. Deflationistic stances broadly criticize this theory, which intuitively seems so illuminating, but the ultimate basis of the criticism is not clear. One would presumably have to say that the critiques develop from a vague notion of an assumption attributed to the correspondence theory: that language and world are two utterly heterogeneous domains that are fully independent of each other. The correspondence theory then appears to be a conception that intends or attempts to achieve a comparison between these two domains or an interrelating of the two. And that is seen—relatively vaguely—as an impossibility.

The rational core of this vague anti-correspondence notion appears to be the thesis that the attempt to achieve the comparison or relating has no chance of success, because the attempt undermines its own presupposition: to compare language and world, to confirm or articulate a correspondence between them, language must be used, and this shows that there is no separation between language and world of the sort presupposed by the correspondence theory, as understood by the deflationists. We find ourselves always already, so to speak, within the two domains to be put into relation: language and world. Put briefly: world is always already related to language, and language is always already related to world.

From this legitimate critique deflationists overhastily draw a conclusion rich in   194
consequences: that "true" is not a "proper (i.e., substantial)" predicate designating
a positive or contentual property. But this conclusion follows only from the rejection
of the correspondence theory as understood by the deflationists. The development of
the structural theory makes clear that the central idea of the correspondence theory
need not be so understood.

The second factor from which deflationism has grown and by which it continues to
be nourished is a putative insight that can appear, at first glance, astonishingly illuminating, indeed obvious and quite simple; it is one that for nearly one hundred years (at
least since Frege) deflationistically oriented authors have continued to repeat and that
has become the core of the deflationist conception. The contention is this: "To say 'It
is true that Caesar was murdered' is *the same* as to say 'Caesar was murdered'" (semi-
formalized: "To say 'It is true that p' is the same as to say: [simply] 'p'"). This shows,
according to the deflationists, that truth is not a *positive* property. This central thesis is
made more precise as follows: the truth term is in general (i.e., in all cases of its simple
ascription) fully redundant; only in relation to specific sorts of sentences, namely gen-
eralizations ("Every sentence of the form 'p or non-p' is true"), is the word "true" to be
seen as *useful*, in that it may there be considered to be a logical device with whose help
sentences of this sort can be formulated.

In anticipation, it may be noted that this putative insight concerning the sameness of the two sentences is correct and acceptable only on the *semantic surface*, where
"semantic surface" means the level of speaking and understanding that provides only
quite unspecific and elliptical formulations. The deflationists have always made the
calamitous error of nevertheless accepting this "insight" as foundational—as correct
and not subject to further analysis—even with respect to deep semantic structure.

2.5.1.3 "Truth" as Predicate and as Operator

A third preliminary question concerns the kind of language for which a or the truth-
concept is to be explained (or defined). Tarski maintains that a truth-concept for
ordinary language (natural, spoken language) cannot be defined, because this language
is "semantically closed": it can talk about itself, for example in "This sentence is false,"   195
which cannot, without paradox, be assigned a truth-value. Many attempts have been
made to solve this problem, but this issue is not treated in this book. Instead, a different
aspect of ordinary language is accorded more importance: that ordinary language is
only quite restrictedly appropriate for theoretical aims.[37] Ordinary languages are essentially languages that are useful for communication in the life-world and that are
correspondingly structured. For theoretical purposes they are therefore only minimally
useful. A theory of truth for ordinary language would be, if not impossible, extraordinarily complex and almost certainly incomplete, given that it would have to take into
account the well-nigh limitlessly many contingencies and indeed extravagances of ordinary language that emerge from all the possible situations in the life-world. The truth-
concept sought here cannot and does not attempt to do justice to this chaos of ordinary
language; it is appropriate for use in ordinary language only to a quite restricted

---

[37] This point is introduced above and developed and defended further in Chapter 3.

extent. What is sought is instead a truth-concept for languages that are fundamentally theoretically oriented—in traditional German terminology, for scientific languages, thus for philosophical, empirical-scientific, and formal-scientific languages.

Given present purposes, the truth-concept developed here need not (and does not) provide an adequate or positive account of all the phrases in ordinary language in which "true," "truth," etc., appear. In addition—and above all—it must involve a *reduction* of certain ordinary-language phrases; what follows explains and supports one of the most important of these reductions.

Most truth-theoretical works are concerned with the "truth-predicate," which has two forms. In the first, "true" is a predicate in a sentence: "[The sentence] 'Snow is white' is true," formalizable with Quine-corners as $T\ulcorner p\urcorner$. In the second, "true" is the predicate of a phrase constructed with the particle "that"—more precisely, a nominal-ized sentence that denotes a proposition: "[The proposition] that snow is white is true"; semi-formalized (following Horwich 1998: 10) as "$\langle p \rangle$ is true" (or: "$T\langle p \rangle$," whereby "$\langle p \rangle$" is to be read, "That p" or, more explicitly, "The proposition that p").

196   There is however a use of "true" that has the syntactic-grammatical form not of a predicate, but of an operator; the operator is indicated by the phrase, "It is true that . . . ." There is a fundamental syntactic distinction between "true" as predicate and "true" as operator. In the former case the arguments of "true" (that to which "true" is applied) are names or single terms—more precisely, nominalized sentences either of the simple form that results from the placing of sentences between quota-tion marks, or of the form that emerges when sentences are preceded by the word "that." In the second case, the arguments of "true" are not nominalized sentences, but genuine (complete) sentences. At first glance, this distinction might appear to be of little significance. Actually, however, it has wide-reaching consequences, as becomes clear in what follows.

To provide a brief indication: from the two different syntactic forms arise, coherently, two fully different semantic and ontological conceptions: in the case of the predicate version compositional ones (based on the principle of compositionality) and in the case of the operator version ones that are contextual (based on a strong version of the context principle). Compositional semantics determines the semantic value of the sentence as a function of the semantic values of the subsentential components of the sentence. Contextual semantics, as understood by and defended in this book, rec-ognizes only sentences that do not have the subject-predicate form (they are termed "primary sentences"); according to this semantics, the semantic value of the sentence is the direct *expressum* of the sentence lacking the subject-predicate form.

To be sure, it is in principle possible, with respect to syntax, to attach either the truth-predicate or the truth-operator to sentences of any form, thus, to ones with the subject-predicate form, like "The USA is a rich country" and to ones without the subject-predicate form, like "It's raining." But if the predicate version were used in conjunction with sentences without the subject-predicate form, the semantic and ontological interpretation would encounter grave difficulties that could be avoided only by means of the introduction of quite implausible *ad hoc* assumptions. The more natural (i.e., more coherent and more intelligible) interpretation associates the truth predicate with compositional semantics and ontology, and the truth operator with contextual semantics and ontology. Be that as it may, however, the interpretation of the truth operator articulated in this book is in no way, even in principle, compatible with compositional semantics and ontology.

The ontology corresponding to the contextual semantics defended here recognizes only primary facts (including complex configurations of simple primary facts). This does not, however, mean that subject-predicate sentences may no longer be used; it means only that the book's semantics cannot take its bearings by sentences of this form. One can continue to use sentences with this syntactic structure, but only if one interprets and understands them as convenient abbreviations of sentences lacking subjects and predicates. At this point, these programmatic remarks suffice. Chapter 3 thoroughly considers compositional and contextual semantics and ontology.[38]

This is the place for a remark on the problem of so-called truth-bearers. As the word "bearer" shows, the use of the expression "truth-bearer" presupposes, strictly speaking, a language of predicate logic: the truth term is understood as a *predicate*, and thus as predicated of an "*x*," its bearer. But if the truth term is understood—as in this work—not as a predicate but as an operator, then this presupposition is unfounded. Nevertheless, one can continue to speak of "truth-bearers" if one uses this expression in a quite broad and unspecific sense.

This work recognizes three "truth-bearers"—or, again (now more precisely), three possible arguments for the truth operator: the proposition, the sentence, and the utterance (of a sentence). The proposition is the fundamental or primary argument, the sentence the first derivative argument, and the utterance the second derivative argument. Chapter 3 shows that the grounding and more precise explanation of this richly consequential assumption are possible within the framework of the structural semantics and ontology whose outlines are formulated in that chapter. Better: such a grounding coincides with the development of structural semantics and ontology.

To avoid initial misunderstanding as far as possible, it must be kept in mind that the terms "proposition" and "state of affairs" are used synonymously and that they receive a stipulated, theory-specific meaning within the framework of structural semantics and ontology. To make this sense adequately clear, this book relies upon the terms "primary proposition," "primary state of affairs," "primary fact," and "primary structure."

2.5.1.4 Comprehensive Theory of Truth and Subtheories of Truth

A final point that is important for understanding the approach taken here must briefly be considered. Although it might appear at first glance that the currently common expression "theory of truth" has a precise meaning, this is by no means the case—indeed, quite the contrary. Richard Kirkham (1992) convincingly shows that this designation has generated considerable confusion because it is used without differentiation to name a great number of in part quite heterogeneous projects andsubject matters. Of particular significance here is the distinction between comprehensive theories of truth and various individual subtheories of truth. What are currently presented as "theories of truth" are as a rule individual subtheories of truth. The two best known arise from the traditional distinction between the concept and the criterion of truth. Theories thematizing the criterion question are conveniently termed *criteriological* theories of truth,

---

[38] See Sections 3.2.2, 3.2.3, 3.3, and Puntel (2001).

and theories having the concept of truth as their subject matter, *definitional* theories. This work is concerned exclusively with a definitional theory, in this sense.[39]

2.5.2 *The Basic Idea of Truth*

2.5.2.1 The Fundamental Fact About Language: Linguistic Items Require Determination

No matter how one may conceive of truth, one cannot thereby avoid considering *language*. The truth term is indeed the central term of semantics, thus of the discipline that has language as its subject matter. The approach taken here begins with a phenomenon that can be termed the *fundamental fact about language*: the fact that linguistic items require determination.

In its most general or abstract form, any language is a system of mere signs or symbols. Symbols can be inscriptions or sounds, although for the sake of simplicity, languages as systems of sounds are not considered here. Any language determined only as a system of inscriptions is determined, trivially, as such a system, but remains to a great extent underdetermined—even what the inscriptions are has not been determined. Language as we use it and are familiar with it is, of course, a highly determined and indeed in various respects fully determined system of signs. Here, that a language is determined or indeed fully determined means the following: language users know how to deal with the system of signs, and they understand one another as they use it. For a given linguistic item (a string of symbols), (full) linguistic determination involves syntactic correctness and semantic significance (whereby the term "significance" expresses what is usually expressed by terms such as "meaning," "reference," and above all "truth"). The following account focuses on the semantic determination of linguistic items, hence semantic value, and treats the syntactic determination of linguistic items only occasionally.

The following decisive thesis grounds the approach taken here: clarifying the truth concept requires clarifying how to explain, achieve, and best understand the semantic determination of linguistic items. The account that follows shows that the so-called truth-concept, functioning as an operator, is the means by which *a language semantically determines its own indicative or descriptive items*. It remains of course to be said exactly how this is to be understood. The rest of this chapter and the final sections of Chapter 3 develop this thesis and what follows from it.

How are linguistic items determined? The simplest answer is, by means of determiners. This answer, however, immediately motivates the questions of what the determiners are and of how they can be characterized more precisely.

A brief initial account of how semantic determination (in the sense relevant here) is attained for formal languages in logic and in formal semantics may be helpful. A given language as a purely formal system can be determined by means of the introduction of a valuation semantics or of an interpretation semantics. The former assigns truth-values to the sentences (or formulas) of the language. The latter (in the Tarskian

---

[39] Its definitional theory of truth is a component of this book's theoretical framework. Within the framework, the theory emerges, or is innersystematically grounded and situated (see Section 1.5 and Chapter 6), as the one that maximizes the framework's intelligibility and coherence by explicitly interrelating its semantics and its ontology.

tradition) introduces a model: an ordered pair consisting of a domain and an interpretation function such that the individual non-logical expressions of the language are mapped to semantic-ontological values. The singular terms are thereby assigned denotations, and the predicates, extensions. Thus, for example, the constant 's' might be assigned to *Socrates*, and Socrates might be included (along with Plato) in the extensions of the predicates P ("is a philosopher") and G ("is Greek"), but not (unlike Plato) in that of A ("is an author"). A given sentence is determined as true just in case the semantic values of the two components of the sentence are combined in the following manner: the denotatum of the singular term (or of the constant) is contained within the extension of the predicate. Because determination by means of valuation semantics presupposes truth values, it presupposes the concept of truth. Also in model-theoretical determination, it is not clear exactly how the truth concept is determined.[40]

2.5.2.2 The Three Levels of Semantic Determination

This account proceeds differently in that it approaches the question in a philosophically more primordial manner. Oversimplifying somewhat, the determination of language occurs fundamentally on three levels or in three forms. To get to the root of the task of presenting a comprehensive conception concerning the semantic determination of language *as a whole*, and to get that task correctly in view, it is advisable to begin from an important critical insight that emerges from the preceding considerations. All deflationistically oriented truth theories unavoidably introduce the self-standing sentence p ("Snow is white") on the right-hand side of the famous equivalence, "'p' is true if and only if p" ("'p' is true $\leftrightarrow p$" or "T$\ulcorner p \urcorner \leftrightarrow p$"). But how is this putatively "self-standing" sentence, implicitly presented as semantically autonomous, to be understood? What is its genuine semantic status? It is said to be a sentence in a metalanguage, into which the sentence p found on the left-hand side of the equivalence sign, between the quotation marks or the Quine corners (and thus nominalized), is said to have been "translated." But how is one to understand this sentence that is putatively self-sufficient in the sense

---

[40] This is clear from the way that the notation for the model-relation is given or "read." For a given sentence, the model-relation is indicated as a rule by the notation $\mathfrak{A} \models \varphi$, which is read in three ways: (i) as "$\mathfrak{A}$ is a model (or a model structure or simply a structure) for φ"; (ii) as "φ is true in (the structure) $\mathfrak{A}$"; (iii) as "φ is satisfied in the structure $\mathfrak{A}$" (see, e.g., Hodges 1993: 12ff) This shows that the "truth-concept" really has no unambiguous status here. It would be most illuminating if one were to say that the truth of φ is defined by means of the concept of satisfaction of φ in a structure $\mathfrak{A}$, but that is not universally accepted. Thus, for example, Gabbay and Guenther (1983: 14) presents the following definition:

"(3.5) For each sentence letter φ, $\mathfrak{A} \models \varphi$ iff $I_{\mathfrak{A}} (\varphi) = T$"

where T indicates the truth value "true" and $I_{\mathfrak{A}}$ designates the function that assigns the sentence the truth value "true" within the structure $\mathfrak{A}$. But what is the precise status of this "truth value"? Wilfrid Hodges, the author of the essay that contains this definition, explicitly and correctly remarks, "Nobody claims that (3.5) explains what is meant by the word 'true.'" But then the question presses once again: what is meant by the truth term, in formal logic and/or in model theory? To this it may be said—and this book indeed demonstrates—that ultimately both forms of determination are based in thoughts that are wholly correct, but that must be presented differently, above all without circularity.

of determining itself to be fully intelligible and determined? Its semantic status cannot be determined until the semantic status of the metalanguage has been determined. Differently put, what is the result of the translation? Translated, does the sentence become a sentence to be used in an advertisement, or in a fairy tale, or as an example in a grammar book, or something quite different? Sentences that—like sentence p, as presented above—are formulated or uttered or presented no matter how, if they are presented without explicit or clearly implicit qualification, remain underdetermined until a determining factor is somehow provided. When the deflationists simply rely on the isolated sentence p as though it were a solid foundation, when they in no way characterize or qualify its status, their procedure is literally senseless.

[1] The first of the three abovementioned levels of semantic determination is here termed the *lifeworldly-contextual* level. When language is used exclusively for uttering sentences in everyday life, it cannot be contested that the uttered sentences are semantically determined, because as a rule we proceed successfully by their means: we communicate, understanding what we mean and being understood by other language users. The language used on this level—used as it most normally is, as is here presupposed—contains no pragmatic and particularly no semantic vocabulary (thus, no terms like "assert" or "true"). Here, this language is named language L_0.

L_0 is determined; to deny that it is would be simply absurd. The question is not whether it is determined, but instead what precisely its determinacy means or how its determinacy is precisely to be understood. Sentences of language used purely *contextually* succeed in hitting their targets, which means that they say, in a determinate manner, what they are meant to say. These linguistic items thus have the status of being determined, and indeed fully determined. And this status is brought about for each such sentence by a factor purely external to language but one of which the conversants are fully aware: the life-worldly context. That is to say, there takes place no linguistically articulated reflection concerning what happens when language is used, how the use or the language itself is to be understood, etc. Language is simply used; its status is determined by a dimension external to language: the lifeworldly context.

Consequently, this determination is in no way theoretical, scientific, philosophical, etc. It is simply a lifeworldly-contextual determination. Much more could of course be said about it, in many respects. Decisive here, among various other factors, is that this determination is neither evaluated nor explicated. This state of affairs may also be characterized as follows: the uttered sentences of the purely contextually used language are to be thought as always already preceded by a context operator that can be expressed as follows: "It-is-contextually-presented-that" (e.g., "snow is white"). This implicitly presupposed operator is what determines the language of contextual communication.

[2] Everything changes at once when, for example, questions of whatever sort arise concerning lifeworldly linguistic utterances. These interrupt the flow of the natural, communicative lifeworld—there is a break between language in its naturalness or its immediate use and another level or other levels.[41] That the lifeworldly context is

---

[41] Jürgen Habermas (1963/1974: 19/26) accurately characterizes the occurrence here designated as a "break" within the perspective of communicative action that he thematizes; it is "the critical threshold between communication (which remains embedded to the context of action) and discourses (which transcend the compulsions of action)."

interrupted means that a new level emerges or arises. Thereby, the lifeworldly-contextual determination of language at least becomes questionable—in any case it is no longer definitive, and in the strict sense it has, as such, disappeared. What the relevant new level *is* cannot be said or derived *a priori*, because there are various possibilities. But at least frequently it is a common and indeed utterly familiar level that is characterized by the use of pragmatic vocabulary. This linguistically pragmatic level is the second level at which linguistic items can be determined. It can be designated as internal/external to language, thus as a mixed level. It is mixed in that, at this new level, the determination of linguistic items is accomplished or indicated in the following manner: an act is accomplished by one or more speakers, and this act is simultaneously linguistically articulated as the act that it is. This linguistically articulated act determines the uttered sentence.

This second level is, then, the level of application of pragmatic vocabulary. A classical example: if on the lifeworldly-contextual level the simple sentence "p" (e.g., "Snow is white") is given (i.e., uttered), a transformation arises that can be articulated as follows: "I (or person S) assert(ed) that p (that snow is white)." That is, the uttered sentence is explicitly preceded by a linguistically pragmatic operator that can be articulated or read as follows: "It-is-linguistically-pragmatically-posited-that . . . (e.g., 'snow is white')." The individual sentence p attains its linguistic-semantic determination by means of a factor external to language (i.e., in the example, the act of asserting) and a factor internal to language (i.e., the linguistic articulation of this act). Because in this example sentence p represents language as a whole, on this level language is determined language-internally and language-externally: by means of the application of pragmatic vocabulary. The linguistic expressions are given "meaning" (i.e., semantic determination) by an act linguistically articulated as a sentence-operator. From this "material" that is language-external and language-internal essentially arises the currently widespread pragmatically oriented semantics.[42]

[3] The contextual and the linguistically pragmatic levels are not the only levels at which the determination of language takes place; there is an additional level, and it is by far the most important, because it is the ultimately fundamental level. This is the level on which semantic vocabulary, quite particularly the central semantic term "true," is located. It should be immediately clear that the operator "It-is-true-that" also has a determinative or determining character with respect to language—one analogous to the determinative or determining character of the linguistically pragmatic operator "It-is-asserted-that." There is, however, a radical distinction between these two operators that, as becomes clear shortly below, plays a decisive role in the approach developed here. Unlike the linguistically pragmatic operator, the truth-operator has no relation to any factors external to language, such as lifeworldly context, subjects (speakers), agents, acts, accomplishments, etc. The truth-operator—and semantic vocabulary in general—is a linguistic determiner that is purely internal to language; indeed, it is the determiner of language that is genuinely internal to language. It is by means of the

203

---

[42] As indicated above, the best example for this pragmatic position is the ambitious semantics developed by Robert Brandom (1994: see 83), which follows the slogan, "Semantics must answer to pragmatics."

occurrence of semantic vocabulary, correctly understood, that language determines itself. Differently stated: semantic vocabulary is the linguistically determinative dimension of language itself. In the course of the development and dissemination of epistemic and generally pragmatic movements of all varieties—including, among many others, classical pragmatism, Brandom's expressive pragmatism, and Habermas's consensus-oriented pragmatism—this fundamental factor has been decreasingly noted and the more stubbornly ignored to the point that by now it has utterly vanished from the horizons of these movements. The obvious fact that language determines or interprets itself, indeed by means of its own semantic vocabulary, is thus wholly overlooked.

Ironically, however, language's self-determination is acknowledged in a striking fashion by the founder of strict (formal) semantics, Alfred Tarski, who views it as a truism. That he does so is clearly visible from striking formulations in his works, including the following (which has already been cited): "A true sentence is one which says that the state of affairs is so and so, and the state of affairs indeed is so and so" (1933/1956: 155). The sentence—or, generalizing—language itself, *says that* . . . This is language's semantic self-determination, its semantic determination of its own items.

2.5.2.3 The Interconnection of the Three Levels: The Explicitly Semantic Dimension as Fundamental

The preceding section considers the three levels of semantic determination largely separately. There is, however, the question whether they are interrelated and, if so, how. The position taken here is that they are indeed interconnected. Linguistic determination at the lifeworldly-contextual level presupposes or is grounded in linguistic determination at the pragmatic level, and the latter presupposes or is grounded in linguistic determination at the purely linguistic or explicitly semantic level, whereas semantic determination at the purely linguistic or explicitly semantic level neither presupposes nor is grounded in either of the other two, and linguistic determination at the pragmatic level neither presupposes nor is grounded in linguistic determination at the lifeworldly-contextual level.

To be at all intelligible, situational semantic determination must be such that the spontaneously uttered sentence p has a semantic status that is articulable linguistically-pragmatically. How else could the determination of language on the lifeworldly-contextual level be explained? If, for sake of simplicity, one considers only the mode or status of assertions, then a given spontaneously uttered sentence can be understood only if it is (implicitly) identified as an assertion rather than, say, as a quotation or a pronunciation exercise. This shows that the lifeworldly-contextual level of semantic determination presupposes the linguistically pragmatic level. Briefly, to say that an uttered sentence p has lifeworldly-contextual determination is (in the strongest case) to say that p has the status of having been asserted.

The next question to arise is analogous to the one considered in the preceding paragraph concerning lifeworldly-contextual linguistic determination: what does it mean to say that a sentence has the semantic status of an assertion? What indeed *is* an assertion? Crispin Wright (1992: 34) champions the view that clarification of the truth-predicate requires recognition of a series of "platitudes," of which one of the most important is "[T]o assert is to present as true." This formulation stems from Frege, who writes, for example (1979: 233), "In order to put something forward as true, we do not need a special predicate: we only need the assertoric force with which the sentence is uttered." To be sure, one cannot say that such a formulation contains a direct characterization—much less a definition—of "assertion," but it indirectly articulates how "assertion" is to be understood.[43] But that shows that to assert something is to do something with respect to truth (i.e., to put something forth as true). "Asserted" is thus not equivalent to "true," or differently stated, the operators "It is asserted that . . ." and "It is true that . . ." are not even equivalent, much less synonymous.

205

From this it follows that linguistically pragmatic determination of language is possible (i.e., here, intelligible) only under the presupposition of the language of truth. Thus, the linguistically pragmatic level of linguistic determination presupposes or is grounded in the semantic level. And this reveals the explicitly semantic level to be fundamental. The uniquely specific character of explicitly semantic vocabulary is again to be emphasized: this vocabulary is absolute in the sense that it has no relation, no relativity (no matter how conceived) to any factors of any sort that are external to language. By means of semantic vocabulary, language speaks about itself: it qualifies or determines its own items.

2.5.2.4 Informal-Intuitive Formulation of the Fundamental Idea of Truth

It is now possible to present a first—still general and informal—formulation of the fundamental idea of truth. Presupposing that language is understood as the totality of primary sentences, each of which expresses a primary proposition, then the word "true," understood by way of its appearance in the operator "It is true that . . .," accomplishes both the transition of language from an un- or underdetermined status to a fully determined status and the result of that transition: fully determined language.

---

[43] Note, however, that Frege's position is extremely complicated and indeed confused. In light of the passage to which this note is appended, others of Frege's formulations appear scarcely intelligible or indeed incoherent, for example,

> If I assert that the sum of 2 and 3 is 5, then I thereby assert that it is true that 2 and 3 make 5. So I assert that it is true that my idea of Cologne Cathedral agrees with reality, if I assert that it agrees with reality. Therefore it is really by using the form of an assertoric sentence that we assert truth, and to do this we do not need the word "true." Indeed we can say that even when we use the form of expression "it is true that . . ." the essential thing is really the assertoric form of the sentence. (1897: 228–29/39–40)

Here, the acute logician and philosopher Frege confuses a number of things, as can easily be shown by careful analysis of this passage. Here is only the main point: the first sentence, "If I assert that the sum of 2 and 3 is 5, then I thereby assert that it is true that 2 and 3 make 5," contains an error fraught with consequences. If one accepts that "assert" means "put forward as true," then it is not the case that when it is asserted that the sum of 2 and 3 is 5, it is thereby asserted that it is true that 2 and 3 make 5. This becomes clear if one substitutes "put forward as true" for "assert": "When I put forward as true that the sum of 2 and 3 is 5, I thereby put forward as true that it is true that 2 and 3 make 5." This sentence is not nonsensical, but it is clearly not equivalent to, "When I assert that the sum of 2 and 3 is 5, I thereby assert that it is true that 2 and 3 make 5."

This formulation speaks of "language" quite generally; this is not wrong, because language is ultimately to be understood as a system of sentences, however this may be understood more precisely (see Section 3.3[3]). Nevertheless, it is advisable to opt for a more specific formulation and, instead of speaking so generally of "language," to speak of the so-called truth-bearers (proposition, sentence, utterance). One would then say, for example, that "The proposition expressed by a sentence from the chosen language qualifies as true" means that the proposition has accomplished a transition from an un- or underdetermined status to a fully determined status; the truth term designates both this process of transition and its result. Chapter 3 completes this book's consideration of this matter by treating it in greater detail.

206

Systematics of Structure:
The Fundamental Structures

3.1 What Is the Systematics of Structure?

3.1.1 *The Basic Idea*

The Introduction provides an initial clarification of the title of this book—*Structure and Being*—and Chapter 1 presents a more fully developed determination both of the concepts used in the title and of various other concepts that they implicitly presuppose. This chapter aims to clarify the central concepts as fully as possible. Because Chapter 1 works out the general concept of structure, this chapter can focus on the concrete forms and types of structures that play determinative systematic roles and that are designated, in this sense and for this reason, as the *fundamental* structures; they are the foci of the systematics of structure. These prove to be of three types: fundamental *formal*, *semantic*, and *ontological* structures.

The first task is to explain how and why the concept of structure is accorded an all-determinative systematic status within the conception of systematic philosophy developed in this book. As Chapter 1 explains, ascribing to structure the decisive systematic role is grounded in one basic idea that this section aims to clarify further. That clarification requires the taking of a broader view.

The task of any theoretical enterprise is the conceptualizing of a specific X; here, "conceptualizing" is used broadly to include explaining, articulating, and the various other activities that are or can be involved in the development of theories. "X" stands for whatever it is that is to be conceptualized; in the course of the history of philosophy, the relevant X has been characterized in various ways, on the one hand as "being," "actuality," "reality," "world," "nature," "universe," "the absolute," etc., and on the other as some "question," "problem," etc.

As what, and how, can the conceptualizing of X, the subject matter, be accomplished? This question pervades the entire history of philosophy, and different philosophical positions can be grouped according to how they answer it. Most globally, the question is whether the conceptualizing is to be determined on the   208 basis of the subject matter to be conceptualized, or vice versa. One extreme position would characterize the dimension of conceptualization as a *tabula rasa* such that everything ever to be found within it would have come to it from the subject matter conceptualized (thus, from the dimension of being). The opposed extreme would make the subject matter a *tabula rasa* in the sense that it would be nothing other than what was constituted by the activity or process of conceptualizing. It is doubtful that the history of philosophy contains pure examples of either of these extreme positions. Even so-called absolute idealism, which one might be inclined to view as exemplifying the second extreme, proves upon more careful historical and philosophical analysis not to do so. The vast majority of positions are intermediate forms falling at varying distances from the two extremes.

Kant's transcendental idealism provides a famous and characteristic example of an intermediate position. Kant presents the relation between conceptualization and subject-matter-to-be-conceptualized as follows:

> Up to now it has been assumed that all our cognition must conform to the objects; but all attempts to find out something about them *a priori* through concepts that would extend our cognition have, on this presupposition, come to nothing. Hence let us once try whether we do not get farther with the problems of metaphysics by assuming that the objects must conform to our cognition, which would agree better with the requested possibility of an *a priori* cognition of them, which is to establish something about objects before they are given to us. (*CPR*: Bxvi)

Kant's position is thus that for humans, conceptualization provides all the "formal" elements, whereas the dimension of subject matter presents only purely sensory material. Cognition is accomplished by the application of pure, *a priori* concepts to the matter provided by the senses and initially formed by space and time (the pure, *a priori* forms of intuition). Kant sees quite clearly that one consequence of this conception is that it makes the "thing in itself," thus the subject matter itself, inaccessible to humans. He formulates this consequence in various ways, emphasizing, for example, the following:

209

> [F]rom [the] deduction of our faculty of cognizing *a priori* in the first part of metaphysics, there emerges a very strange result, and one that appears very disadvantageous to the whole purpose with which the second part of metaphysics concerns itself, namely that with this faculty we can never get beyond the boundaries of possible experience, which is nevertheless precisely the most essential occupation of this science. (*CPR*: Bxix)

This consequence of course has additional consequences:

> But herein lies just the experiment providing a checkup on the truth of the result of that first assessment of our rational cognition *a priori*, namely that such cognition reaches appearances only, *leaving the thing in itself as something actual for itself but unrecognized by us.* (Bxix–xx; emphasis added)
> [A] light dawned on all those who study nature. They comprehended that reason has insight only into what it itself produces according to its own design; that it must take the lead with principles for its judgments according to constant laws and compel nature to answer its questions, rather than letting nature guide its movements by keeping reason, as it were, in leading strings . . . (Bxiii)

Kant's position is not considered further here, but it is worth noting that all post-Kantian philosophy is influenced in one way or another by his "transcendental turn," although the focus has ceased to be the transcendental structure of the subject, becoming instead language and formal instruments.

The structural position is the result of a more precise and subtly differentiated view of the relationship between conceptualizing and the subject matter to be conceptualized. It begins with the thesis that the dimension of the subject matter to be conceptualized is *initially*, at the outset of the theoretical undertaking, still *empty* in the sense that the thematization of its features is the task that conceptualization is to tackle. The dimension of conceptualizing is what should be considered first because the entire theoretical enterprise unavoidably develops within the framework of this dimension. Investigation of this dimension thus appears to be the first great task of philosophy. At first glance this contention might appear to have a consequence even more radical than Kant's position, according to which the dimension of the subject matter to be conceptualized remains wholly inaccessible. Yet precisely the opposite turns out to be the case: detailed investigation and explication of the dimension of conceptualization are ultimately nothing other than the presentation of the framework within which the structures of the subject matters of theories are made manifest. The central task of this book is to show, in detail, how this is to be understood and how it is to be achieved.

The basic idea of this book may be further clarified by an additional historical comparison. In his *Science of Logic*, Hegel aims to present the system of all of the "determinations of thought." By "determinations of thought" Hegel means what the philosophical tradition generally terms "categories," although he expands this notion considerably in that he fully integrates into his comprehensive concept the development of doctrines of categories from Aristotle through Kant. Whereas Kant recognizes only twelve categories and takes them to be functions of the thinking determined by the I or the subject, Hegel aims to get at the determinations of thought "in themselves":

> [B]ecause the interest of Kantian philosophy was directed to the so-called *transcendental* determinations of thought, the treatment of them remains empty; what they are in themselves, without the relation to the I that is the same for them all—their determinacy in the face of one another and their relations to one another—is not made an object of investigation; the knowledge of their nature has thus in no way been furthered by this philosophy. (*Logic:* 63/60–61; translation altered)

With respect to Hegel's endeavor, only two points need be made here. The first is that it is a failure: his *Science of Logic* is ultimately incoherent in a number of respects and simply does not accomplish its task; that it fails in these ways is not argued here, but is argued in various of the author's other works (see esp. Puntel 1996). Be that as it may, the second point to be made here is that the idea of the task itself—of the systematic development of all the determinations of thought—is a valuable one, particularly in light of the identification of the determinations of thought with the fundamental determinations of *things*. Hegel remarks quite correctly: metaphysics "is superior to the later critical [i.e., Kantian] philosophy in that it presupposes that what *is* is, in being *thought, known as it is*" (*Enc. Logic:* §28). To be sure, Hegel does not show how the determinations of thought he purports to have displayed systematically can be coherently thought of as the fundamental determinations of things, but this does not detract from the importance

210

of his recognition that the fundamental structures he calls "determinations of thought" indeed have an ontological status.

Hegel's position, like Kant's, results from the attempt to grasp and to make explicit the dimension of conceptualization. What is distinctive about Hegel's attempt is that he appears at first glance to allow the subject matter to be completely eliminated by the dimension of conceptualization—appears to reduce the former dimension to nothing by maximizing the latter dimension. In this respect, this book's structural-systematic enterprise is similar to Hegel's. But it aims to demonstrate—clearly, coherently, and consequently—that the explication of the dimension of conceptualization (of structure) itself develops the theoretical framework within which the subject matter (being) is made manifest. The technical terminology of this book thus uses "dimension of structure" to designate what is more commonly termed the dimension of conceptualization; the explication of the dimension of structure is the task of the *systematics of structure*. The following sections accomplish this task step by step.

3.1.2 *Preliminary Clarifications of Terms and Concepts*

English, like the other languages used by Western philosophers, offers, in both its ordinary and its philosophical uses, a host of terms (often cognates, among these languages) that might appear to be suitable on the one hand for the domain of conceptualization (of structure, as opposed to being), and on the other, for that of the unrestricted universe of discourse. Those apparently suitable for the former would include "concept," "meaning," "thought," and "proposition," those for the latter, "object (thing)," "property," "relation," and "fact." "Category" has a place of its own. None of these terms, however, may be simply appropriated into an adequately clear philosophical account: not only are all too vague and have all been used in too great a variety of ways, but—and this is a far more radical fault—their general employment implicitly presupposes a systematic conception that is at best inadequate. As is shown below in this chapter, this conception involves the semantic principle of compositionality and the substantialist ontology that comes with it—both of which this book rejects on the basis of reasons presented in Section 3.2.2.3.

3.1.2.1 "Concept," "Meaning," "Sense," "*Bedeutung*," "Semantic Value," "Thought," "Proposition," and "State of Affairs"

[1] The term "concept" belongs indisputably among the most commonly used expressions in all of philosophy. Yet scarcely any philosopher would be able to provide a clear, adequate, and convincing answer to the question of what is meant by it, or how precisely it is or should be understood. To see that there is not even minimal agreement among philosophers, one need only consider the explications of "concept" that have been provided in the history of philosophy. For example, what Hegel means by "concept" has little or indeed nothing to do with what "concept" means for philosophers like Kant, Frege, and Carnap. Despite the confusion, however, the term continues to abound in the philosophical literature.

This practice need not necessarily be viewed only negatively: the use of this term often attains, in an informal and intuitive manner, the positive result of giving something like an indication of the intended meaning. The term thus has a programmatic status, as is clear from the fact that one can think or maintain that one has a "concept" of anything and everything, from, for example, a (or the) "concept 'God'" to a (or the) "concept of truth." But as soon as this general and programmatic level is abandoned, further use of the term proves to be not only not a help, but instead a hindrance to the understanding of what is meant or intended by it. For this reason, this term should be used in philosophy only with great care, at least whenever one attempts precisely to articulate what one means or intends. It is symptomatic of the problem that what has been here termed the "meant or intended," and can also be termed "what is (or is to be) understood" by the term "concept" is as a rule designated with the term "concept": it seems quite natural to speak of the "concept of the concept."[1]

The task of developing a theory of concepts has been recognized and tackled only quite recently.[2] The task involves not only philosophy but also other disciplines, most important among them psychology and cognitive science. For the purposes of this chapter, however, only one question is crucial, and it is the question that constitutes the core of the philosophical theory of concepts; this is the question concerning the *ontological status* of concepts. This question for its part involves two specific and essentially inter-related questions: what kind of entity are concepts themselves, and to what kind(s) of entities do concepts relate?

To the first question there are, chiefly, two completely different answers. The first, found throughout the history of philosophy, is that concepts are mental representations: mental entities that represent contents of some sort or other. The second answer is at best implicit in the history of philosophy and is first explicitly formulated by Frege; it is that concepts are abstract objects or entities. Frege also presents the classical arguments against the first answer (see esp. Frege 1918); those arguments are not repeated here. The structural-systematic philosophy accepts a version of the second answer, albeit a fundamentally revised one. "Abstract" objects or entities are generally taken to have two characteristics: they are not located in time and space, and they "exist" independently of their being thought, grasped, or known. These two characteristics make this conception wholly inadequate. It could be accepted, if at all, only after precise specification, partial correction, and thorough explanation. This chapter treats this issue within the framework of the gradual development of a structural semantics and ontology.

For present purposes, the second question concerning the ontological status of concepts, briefly introduced above, is of far greater and indeed of decisive significance. This is the question concerning what it is to which concepts relate. The concept of "relating,"

---

[1] An additional problem with this term arises in the German language. English and the Romance languages inherit from Latin at least two different terms that are normally rendered in German with the one term "*Begriff*": "*notio*" (French/English "notion," Italian "*nozione*," Spanish "*noción*," Portuguese "*noção*," etc.) and "*conceptus*" (respectively, "concept," "*concetto*," "*concepto*," "*conceito*," etc.). This fact sheds further light on the peculiar history of semantic equivocations on the German "*Begriff*."

[2] See, e.g., Margolis and Laurence (1999). In its extensive opening chapter the authors distinguish five theories of concepts: the "classical theory," the "prototype theory," the "theory-theory," the "neoclassical theory," and "conceptual atomism." See also Peacocke (1995).

for its part, is ambiguous in this context. It can be understood in two wholly different ways. On the one hand, it can be understood as "applying to." So understood, it suggests as an answer to the question, at least with most concepts, that concepts relate to objects by applying to them. But "relating" can also be understood in the sense of "referring to," "designating," etc.—and this understanding has been the more common. Given this understanding, concepts "relate to" contents in that they *have* "contents." The whole question then becomes that of how to interpret this conceptual content. This chapter focuses on that question.

Distinguishing between the intension and the extension of a given concept is often done. A purely extensional understanding of concepts reduces their content to their scopes, such that a given concept is equated with the members of the set of objects to which the concept applies. According to understandings of concepts that are not exclusively extensional, a given concept "denotes" or "designates" an entity that can be of various sorts, corresponding to the linguistic expression with which the concept is associated: an attribute (property or relation), a function, etc.

214

[2] Thus far the concept of the concept has been considered without explicit consideration of language; this is the mode of consideration that characterizes the traditional (preanalytic) perspective, according to which the concept is basically a mental entity, a representation (Hegel is an exception). Introducing language wholly transforms the issue of the concept. Concepts are then associated (although not equated) with linguistic expressions. According to the most common view, the concept associated with a given expression is nothing other than the meaning of this expression. But what is "meaning"? Differently put, how is the "concept of meaning" to be determined? Simply to equate concept and meaning is problematic, because such an equation presupposes the thesis that every concept is associated with a linguistic expression, and this thesis is, at the very least, not obviously true; might or even must there be concepts that are not linguistically expressible? This is a difficult question. If one understands "language" as so-called ordinary or natural language, taken as it is in fact used as a semiotic means of communication, it appears at least plausible to say that there are concepts that are not linguistically expressible. But, as arguments presented above reveal, that is a superficial understanding of language. As is thoroughly demonstrated later in this chapter and particularly in Chapter 5, at least for philosophical and scientific purposes, language must be understood and determined otherwise. With respect to language as it is determined within this work, it is demonstrably the case that every concept is or can be associated with a real or possible linguistic expression. This decisively shifts the problem of the determination of the concept of the concept into the arena of language.

As is well known, the philosophy of language, and particularly philosophical semantics, distinguishes theories of meaning from theories of reference. Many philosophers, for example Quine, either reject theories of meaning or are extremely skeptical of them. Yet what one can or should call "meaning" cannot simply be presupposed, but must itself be the object of an explicit semantic theory. Because of the extreme ambiguity of the term "meaning," evidenced by the multiplicity of theories of "meaning," this book avoids, as far as possible, relying upon it. Its adopted alternative is "semantic value."

Clarity concerning the concept of the concept is increased, to a certain point, by

215   consideration of Frege's famous distinction between sense and *Bedeutung*, whereby

Frege uses the word *Bedeutung* in a completely unusual manner.[3] It is to be noted that, on the whole, Frege's acclaimed treatments of concept, object, sense, *Bedeutung*, truth value, etc., quite clearly manifest a conflict of terminologies; one is based in mathematics, the other in language. This conflict holds particularly for the term "concept." According to Frege, the fundamental conceptual-ontological distinction is that between function and object. Concepts are functions of a specific type—those that have as possible values only "the True" and "the False": "A concept is a function whose value is always a truth value" (1891a: 15/139). For Frege, truth values are simply *objects*.

The conflict of terminologies introduced above emerges when the level of language is explicitly taken into consideration. It then becomes clear that the terminology clarifying "concept" in terms of function and object comes from mathematics. Frege remarks elsewhere that when "predicate" and "subject" are understood linguistically, "a concept is the *Bedeutung* of a predicate; an object is something that can never be the whole *Bedeutung* of a predicate, but can be the *Bedeutung* of a subject" (1892c: 198/187). But *Bedeutung* does not have the same meaning here that it has when Frege uses it in the conjunction (or opposition) "sense [*Sinn*] and *Bedeutung*." In that usage, according to Frege, every expression has both a sense and a *Bedeutung*: "It is natural . . . to think of there being connected with a sign (name, combination of words, written mark), besides that which the sign designates, which may be called the *Bedeutung* of the sign, also what I should like to call the *sense* of the sign, wherein the mode of presentation is contained" (1892a: 26/152). "By employing a sign we express its sense and designate its *Bedeutung*" (31/156). Frege later analyzes "conceptual content" in terms of sense and *Bedeutung*.

Frege's case shows in exemplary fashion that concepts like "concept" undergo fundamental transformations when the linguistic dimension is explicitly taken into consideration. It does not suffice to acknowledge that the distinction between concept and object is not consistent with that between sense and *Bedeutung* (in Frege's sense; see 1892b: 128/172–73). Instead, there is a lesson to be learned: the traditional terminology is all too unclear, and a terminology drawn from a philosophical approach focusing on the linguistic dimension is preferable in every respect. This is not to say that terms like "concept" cannot be used as intuitive indications or programmatic anticipations of clear semantic-ontological structures and interconnections; indeed, such usages can scarcely be avoided and therefore continue to occur in the following pages.

[3] As is indicated above, to avoid the scarcely reparable unclarities and ambiguities of the traditional terms "concept" and "meaning," the term "semantic value" is here used in a manner that makes explicit the content most usually associated, intuitively, with the former terms. Those terms are henceforth used only as convenient abbreviations for "semantic value."

[4] Talk of "thinking" and thus also of "thoughts" is without question found throughout philosophical discourse. The two most common starting points for determining how "thoughts" are to be understood are the mentalistic and the linguistic-semantic.

---

[3] Frege's *Bedeutung* corresponds to what is generally termed "reference" in analytic philosophy (for a discussion of the problems involved in translating Frege's term into English, see Beaney 1997: 36–46). Frege holds that to every non-syncategorematic term is to be ascribed both a sense and a reference. This is a quite technical and controversial semantic thesis.

The first proceeds from the assumption that thoughts are products, contents, or, simply, the objective aspect of thinking. The thought is then understood as a specific kind of mentalistically understood representation; such a conception is found in Kant, and in classical German philosophy in general. How "thoughts," so understood, can be genuinely ontologically oriented—how they can be in any way "about" what is real—remains ultimately unexplained and indeed unexplainable. It is no accident that the philosophers who adopt this perspective most stringently defend global conceptions that are idealistic. This option is not considered here either critically or systematically.

The linguistic-semantic approach also uses mentalistic terms, but explains them not on the basis of thinking, but on that of language and/or semantics (as is made clear below, "thoughts" apprehended in this manner are simply identical with "propositions" and "states of affairs"). The best-known advocate of this approach is Frege, who uses the term "thought" extensively, characterizing it as follows:

> Without offering this as a definition, I call a "thought" something for which the question of truth can arise at all. So I count what is false among thoughts no less than what is true. So I can say: thoughts are senses of sentences, without wishing to assert that the sense of every sentence is a thought. The thought, in itself imperceptible by the senses, gets clothed in the perceptible garb of a sentence, and thereby we are enabled to grasp it. We say a sentence *expresses* a thought. (1918: 60–61/327–28)

217

According to Frege, then, the most general characterization of the thought is—because Frege also interprets "sense" linguistically/semantically—that it is the *expressum* of a sentence. His notion of "sense" can be clarified only in conjunction with *Bedeutung*, which, for Frege, is (so to speak) "complete objectivity"—what he terms "the object" in the most objective sense, thus the genuinely ontological dimension. "Sense," on the other hand, is the "mode of presentation" of the object, such that one and the same object can be presented in various ways. But the sense, too—the thought of the sentence—because understood ontologically/semantically, is ontologically oriented, as Frege makes clear in, for example, the following passage:

> We can never be concerned only with the *Bedeutung* of a sentence; but again the mere thought [hence, sense] yields no knowledge, but *only the thought together with its Bedeutung*, i.e., its truth value. Judgments can be regarded as advances from a thought to a truth value. (1892a: 35/159, emphasis added)

In his "Comments on *Sinn* and *Bedeutung*" (composed between 1892 and 1895), Frege notes the following:

> The intensional logicians are only too happy not to go beyond the sense; for what they call the intension, if it is not an idea, is nothing other than the sense. They forget that logic is not concerned with how mere thoughts, regardless of truth-values, follow from thoughts, that *the step from thought to truth value—more generally, the step from sense* [Sinn] *to* Bedeutung—*has to be taken.* (1892b: 133/178; emphasis added)

Finally, a passage from Frege's letter to Husserl of 25 May 1891: "Judgment in the narrower sense could be characterized as a *transition from a thought to a truth-value*" (1891b: 150; emphasis added).

Frege's terminology—"advance," "step," "transition"—is clearly ontologically oriented. Of course, Frege advocates a somewhat idiosyncratic semantics and ontology. The *Bedeutung* of a sentence is taken to be its truth-value, and the truth-value an object, such that there are two truth-values and thus two objects, "the True" and "the False" (1892a: 96/158). Moreover, Frege accepts a compositional semantics and ontology—a conception that apprehends the sentence fundamentally as combining subject and predicate such that the thought expressed by the sentence is the entity that arises from the combining of the semantic values of those components of the sentence. This is considered more closely below. The aim at this point is only to provide initial clarification of the "thought." In this book, "thought" and "proposition" are understood and used as synonymous.

[5] "Proposition" is contained within the general terminological repertoire of analytic philosophy, but it is not consistently used. Sometimes "proposition" is simply identified with "sentence," but usually these terms/concepts are strongly distinguished. In general, "thought" and "proposition" are—as in this book—used interchangeably.

Currently, there are two main approaches to determining "proposition." The first follows Frege in taking the proposition to be the *expressum* of a sentence, whereas the second determines the proposition as the object of a propositional attitude (see Puntel 1990: 358ff). Because this object must also be articulated linguistically, the more fundamental determination of "proposition" appears clearly to be the one understanding the proposition as the *expressum* of a sentence. But those holding this general view vary considerably with respect to how they understand this entity.

The term/concept "state of affairs" is not usually simply identified with "proposition." Sometimes distinctions emerging from the philosophy of language (semantics) being relied upon are made or presupposed. There is, however, no reason, within the semantic conception developed in this chapter, not to view "proposition" and "state of affairs" as synonymous.

3.1.2.2 "Object," "Property," "Relation," "Fact," and Other Entities

The terms listed in this section's title are from another group often used in the history of philosophy in conjunction with "concept"; although these terms, too, are used in various ways, on the whole they differ in a specific respect from the terms considered in the preceding section. This specific respect is their directedness toward the dimension of being (of content, objectivity, ontology); this directedness is explicit, but is of various sorts. It encompasses the entire murky and extremely complex network of relations that interconnect the two poles of (a) subjective content as pure representations or thoughts and (b) objective content as "pure" reality. To show the variety of ways these terms/concepts have been used, one would have to present the history of philosophy virtually in its entirety. For present purposes, exhaustive consideration of how these and other related terms have been used in the history of philosophy is unnecessary; instead, a few remarks suffice.

[1] The terms/concepts "object (thing)," "property," and "relation" are basic to the substance ontology that dates back to Aristotle. It is symptomatic that whereas throughout most of the history of analytic philosophy the term "substance" is used rarely if at all, "object" being the preferred alternative, of late "substance" is being used with increasing frequency. This is presumably a consequence of increased attention

to ontological issues. Be that as it may, "object" is a word that is convenient but that conceals and thus conjures away an enormous range of philosophical problems. "Object" can be determined more precisely only by means of the concepts "property" and "relation": an object is the X of which something must be said to determine it, whereby that "something" is a property it possesses or a relation within which it stands. It thus becomes clear that "object" is ultimately understood as "substance." "Predication" is the logical-semantic term for this semantic-ontological connection or determination, and first-order predicate logic—according to its standard interpretation—is the logical instrument used to articulate it formally. The purposes of this book do not require that the property-relation distinction be made more precise, but it is worth noting. For the sake of completeness, it is also to be noted that there are both a more traditional understanding of properties and relations and one that is determined by first-order predicate logic. The former is intensionally or conceptually oriented; according to it, a predicate designates an attribute. According to the latter, extensionally oriented conception, a predicate does not in the strict sense designate anything like an attribute; the predicate does not designate anything at all. Instead, it has an *extension*: the set of the objects to which the one-place predicate or the set of tuples of objects to which the many-placed predicate applies (in the former case, the property is defined, in the latter, the relation; see Section 3.2.2.3.1). Section 3.2.2.3.1 introduces fundamental objections to this semantics and ontology.

[2] "Fact" is an everyday term that also plays a vital role in philosophy, particularly in analytic philosophy. It may be that no philosopher has provided a better brief characterization of the intuitive understanding of *fact* than Wittgenstein, in his *Tractatus*: "What is the case—the fact—is the existence of states of affairs" (*Tractatus* 2). This characterization is not the last word philosophically, but it does identify just what philosophy must clarify to make sense of *facts*. On the whole, the *fact* may be said to be the ontological moment of or counterpart to the thought and/or proposition. Different semantics and ontologies yield different notions of "fact." Within the framework of a substance ontology, "facts" can be understood only as composites of at least one substance and one or more properties and/or relations. In this book, "fact" is understood in a wholly different sense (as *primary* fact, introduced above and explained more fully in this chapter).

[3] To be noted in preliminary conclusion is that in addition to "substances/objects," other "fundamental" entities are often introduced: above all, "processes," "events," etc. Traditionally and preponderantly, and with few exceptions (see, e.g., Seibt 2004) however, these entities enter into and remain within the general framework of an ontology that is fundamentally substantialist (see, e.g., Davidson 1985).

3.1.2.3 "Category"

In the course of the history of philosophy, distinctions have been drawn among various types of concepts. As noted above, the term "concept" is so universally used that it can always provide a metadesignation for whatever subject matter or domain is being considered. Thus, there are distinctions between empirical and *a priori*, material and formal, logical and non-logical, particular and universal, etc., concepts. The term "concept" thus serves as a convenient abbreviation for whatever is or is to be grasped theoretically.

Among the various kinds of concepts discussed in the history of philosophy, *categories* are accorded a particular and systematic status. Every philosophy uses specific categories, although generally without recognizing them explicitly. Many philosophers, however, do explicitly identify categories as central to their teachings, above all Aristotle—who presents the first genuine doctrine of categories—and then Kant, and Hegel. Currently, "category" is generally used in a quite global and therefore quite vague sense to designate fundamental concepts or domains that are not provided with further specification. Nevertheless, the history of doctrines of categories yields important insights.

With Aristotle and then in the metaphysical tradition there is a vacillation between the following two determinations or understandings of categories: categories as modes of predication and categories as modes of being, indeed the highest modes of being, because the highest genera of being. Taken as modes of predication, categories correspond to the various significations of the copula "is" in ordinary atomic, subject-predicate sentences (see Weingartner 1976: 268ff.). According to Aristotle, who lists ten categories, one of these significations is fundamental—that of *being in the sense of substance*; the other nine signify *being in the sense of attributes* (see especially *Categ* 4–5 and *Metaph* Δ7, Z1–4). It is immediately clear that if categories are so understood, the assumption behind their identification is that the basic form of the sentence is the subject-predicate form. A semantics that does not accept this assumption cannot conceive of categories as modes of predication of this sort. At the same time, it is symptomatic that within this tradition, categories come to be located with increasing frequency in the ontological dimension: categories are modes of being or of beings.

The fundamental tension between categories as modes of predication and of being is retained even after Kant's "Copernican turn," although under fundamentally changed conditions. According to Kant, on the one hand categories are functions of the understanding or modes of judgment, but on the other they are constituents of objects of cognition. Kant characterizes them as follows: the categories "are concepts of an object in general, by means of which its intuition is regarded as **determined** with regard to one of the **logical functions** for judgment" (*CPR*: B128). To be found here are transcendental analogues of both of the above-described understandings or determinations of categories: categories as concepts of an object in general (in analogy to modes of being) and categories as logical functions of judgings (in analogy to modes of predication).

As the history of doctrines of categories shows, such doctrines involve not only clarifying the concept of the category but also determining the number of categories. Kant is the philosopher who has claimed most clearly and decisively to have presented and indeed to have "deduced," strictly and systematically, the complete table of categories. By means of a "metaphysical deduction" (which he distinguishes from "transcendental deduction"), Kant derives the categories from (he himself says that he "discovers" them "in") the twelve basic forms of judgment identified by traditional logic. The "metaphysical deduction" is said to establish "the origin of the categories a priori . . . through their complete coincidence with the universal logical functions of thinking" (B159).

221

Kant's conception is to be rejected for two primary reasons. [i] Kant bases his categories on then-traditional logic and thus on a table of judgments that, given the post-Kantian discovery and development of modern logic, is no longer tenable. [ii] The coincidence Kant purports to find between the forms of judgment and the categories is in no way compelling.

Here, a wholly different conception is developed, proceeding from a wholly different starting point. For reasons presented in Chapter 2, the entire mental dimension, wholly determinative for Kant, is not ignored, but is accorded no more than secondary status. In the foreground is not the logical-mental dimension, but the logical-linguistic dimension.

When language becomes philosophically central, the concept of category is watered down. In part for this reason, to speak today of anything like a complete table of categories is scarcely possible save in discussions of the history of philosophy.[4] The ultimate reason for this near impossibility is that there is no current consensus concerning the concept of category. Thanks to the history of philosophy, the term and thus the concept retain great intuitive potential, but realization of the potential requires complete clarity.

A particularly important domain within which a transformed version of the traditional concept of category should find broad application is that of scientific research. Here, the concept of category is widely used to characterize both the whole structure of the process of theorization and the results of this process. Granted, however, opinions on those results are quite divergent (see, e.g., Rescher 1982).

A final contemporary use of "category" is one that diverges strongly from the traditional philosophical usage: the term at times characterizes what is specific to a particular domain of actuality. Thus, for example, one speaks of the "categories" of the mental and the material or physical, of the semantic and the pragmatic, etc. Within the framework of the traditional understanding of categories, this makes no sense.

---

[4] An example of a conception of categories that interrelates, in a manner that remains unexplained, quite divergent perspectives is Chisholm (1996). To the question of what a category is, Chisholm introduces the following table (3):

ENTIA
1. Contingent
   a. States
      i. Events
   b. Individuals
      i. Substances
      ii. Boundaries
2. Necessary
   a. States
   b. Nonstates
      i. Substance
      ii. Attributes

Hoffman and Rosenkranz (1994: 18, 21) present similar tables of categories. See also Rescher (1982: 61–79).

Substance, the central category of traditional metaphysics, contains both the domain of the mental or spiritual and that of the material; according to this metaphysics, there are both spiritual substances and material substances.

The often-used contemporary notion that appears to be connected most closely to philosophical categories, as traditionally understood, is that of the *conceptual scheme*.[5] For reasons given above, this book relies instead upon *theoretical framework*.

3.1.3 *The Systematic-Architectural Status in Philosophy of the Expanded Concept of Structure*

All of the terms considered in the three preceding subsections, despite the variety of ways in which philosophers have used them, nevertheless have, within the *lingua franca* of philosophy, familiar intuitive senses that make them helpful and easily intelligible. For this reason, this book relies upon all of them, but it reinterprets them—and uses them exclusively as—*brief or elliptical abbreviations* of the theoretical content (so to speak) fundamental to it; this content is designated by means of the technical term "structure." That the other terms, when used in this book, do not designate anything other than structures is particularly to be emphasized with respect to the terms "concept" and "category." The basic semantic structures for which they serve as such abbreviations are termed *primary propositions*, and the basic ontological structures, *primary facts*.

Chapter 1 introduces *structure* in a provisional and global manner, presenting and briefly explaining its usual mathematical definition. This chapter shows that and how structure attains a comprehensive systematic status within the global conception presented in this book. Its task is to explain and unfold a *systematics of structure*. This section contains basic introductory comments.

Philosophy cannot simply appropriate the mathematical definition of structure, because mathematics is a purely formal science, whereas philosophy is both a metascience and a contentual science. But philosophy's differences from mathematics should not be exaggerated, because in mathematics, too, as it is concretely practiced, there is no thoroughly explicit and explicated use of "structure" that philosophy could, even in principle, simply appropriate. This is true even for the mathematical approach that is associated with the name Nicolas Bourbaki, the approach that purports to develop the methodic and systematic foundations of this science in the most radical and complete manner that is possible.[6]

A reminder of what Kleene (1974: 24–25) says about the concept of system or structure, first presented in Chapter 1, may be helpful, particularly if Kleene's "system" is replaced (as appropriate) by "structure" or "concrete structure," and his "structure" by "abstract or pure structure." Kleene's "structure of the system" is thus "abstract or pure

224

---

[5] See, e.g., Davidson (1984: 183–98). See also the critique of Davidson in Rescher (1982: 27–60).

[6] As is well known, "Nicolas Bourbaki" is not the name of a person; it is a pseudonym for a group of predominantly French mathematicians. The status of *structure* for Bourbaki is examined in the outstanding article Corry (1992). See also the treatment in Section 3.2.1.2.

pure structure." In the following passage, Kleene's terms are replaced by those enclosed within square brackets:

> By a [structure] S of objects we mean a (non-empty) set or class or domain D (or possibly several such sets) of objects among which are established certain relationships. . . .
>
> When the objects of the [structure] are known only through the relationships of the [structure], the [structure] is *abstract*. What is established in this case is the [abstract or pure structure], and what the objects are, in any respects other than how they fit into the [abstract or pure] structure, is left unspecified.
>
> Then any further specification of what the objects are gives a *representation* (or *model*) of the abstract [or pure structure], i.e., a [concrete structure] of objects which satisfy the relationships of the abstract [or pure structure] and have some further status as well. These objects are not necessarily more concrete, as they may be chosen from some other abstract [or pure structure] (or even from the same one under a reinterpretation of the relationships).

One central concern of the conception here being developed is how to understand these "contents" ("objects," "entities"). As is explained in detail below, the primordial contents that are structured by the pure structure(s) are here to be conceived of semantically as 225 *primary propositions* or *primary states of affairs*, and ontologically as *primary facts*. Among these primordial contents are not only configurations of primary propositions and primary facts—not only *complex* primary propositions and primary facts—but also *simple* primary propositions and primary facts (see esp. Sections 3.2.2.4.1.2, 3.2.2.4.1.3, and 3.2.3.1). Simple primary propositions and primary facts are themselves *structures* of a specific sort: structures on the zero level, or "null-structures," in some ways analogous to the empty or null set. Because no structure structures anything other than structures, and because there is nothing that is not a structure, *structure* has an unrestrictedly comprehensive systematic status.

3.1.4 *The Program of a Philosophical Systematics of Structure*

[1] An excellent introduction to the following considerations is provided by a passage from Quine's *Word and Object*, a passage that, strangely and symptomatically, has been almost wholly ignored by analytic philosophers (including Quine, in other works), and whose extensive consequences thus remain to be developed. In §33, titled "Aims and Claims of Regimentation," Quine writes,

> [T]he simplification and clarification of logical theory to which a canonical logical notation contributes is not only algorithmic; it is also conceptual. Each reduction that we make in the variety of constituent constructions needed in building the sentences of science is a simplification in the structure of the inclusive conceptual scheme of science. Each elimination of obscure constructions or notions that we manage to achieve, by paraphrase into more lucid elements, is a clarification of the conceptual scheme of science. The same motives that impel scientists to seek ever simpler and clearer theories adequate to the subject matter of their special sciences are motives for simplification and clarification of the broader framework shared by all the sciences. Here the objective is called philosophical, because of the breadth of the framework concerned; but the motivation is the same. The quest of a simplest, clearest overall pattern of canonical notation is not to be distinguished from a quest of ultimate categories, a limning of the most general traits of reality. Nor let it be retorted that such constructions are conventional affairs not dictated by reality; for may not the same be said of a physical theory? True, such is the nature of

reality that one physical theory will get us around better than another; but similarly for canonical notations. (1960: 161; emphasis added)

This passage is remarkable in many respects. [i] To be sure, Quine is primarily concerned with logical theory, but his concern is explicitly situated within "the inclusive conceptual scheme of science." What Quine here calls "conceptual scheme" corresponds to the basic idea of what this work calls the dimension or framework of structure, or the systematics of structure. Canonical notation is for Quine nothing other than the articulation of this "inclusive framework," this framework made explicit.

[ii] The two characteristics of traditional doctrines of categories, often mentioned above, that in the history of philosophy are usually either maintained in juxtaposition or strictly separated, are grasped by Quine, in one of the strongest formulations to be found in the history of philosophy, as inseparably united: "*The quest of a simplest, clearest overall pattern of canonical notation is not to be distinguished from a quest of ultimate categories, a limning of the most general traits of reality.*" It is to be noted, however, that Quine's "is not to be distinguished" is to be understood as "is not to be held or considered to be separate from," such that the two quests are not to be taken not to form a fundamental unity. Also to be noted is that Quine's quite determinate understanding of the "broader framework" is not retained in this book. 226

[iii] Particularly remarkable is that Quine explicitly includes (formal) logic within ontology; he does so by according it an ontological status.

[2] The structural-systematic philosophy also recognizes logical structures as ontological. The inclusive or comprehensive *structural framework* thus consists of two types of structures: formal structures and contentual structures. The formal structures are the logical and the mathematical ones, the contentual, the semantic and ontological ones. But the distinction between the formal and the contentual is here in no way foundational, despite its putative status as such throughout much of the history of philosophy. Also rejected here is the assumption on which the tradition has based this distinction, i.e., the often vague notion of an unquestioned, comprehensive "hylomorphism" holding at all levels, including the ontological; this hylomorphism understands "matter" as fundamental, and "form" as the moment that "shapes" the otherwise formless matter. Talk of formal logic within this perspective presupposes that this logic is *purely* formal, separate from matter in the sense that it simply has no "content" of its own. Logical structures then appear to be "laws of thinking," "laws of argumentation," or the like. The radical distinction between the logical and ontological dimensions—the utter separation of the two—then becomes a fundamental dogma that governs much of the history of philosophy, and that continues to govern even now. As is shown by the Quine passage cited above, there have been other conceptions, but as a rule they have been and continue to be ignored. 227

Although the terminological distinction between formal and contentual structures is here retained, it is understood in a wholly non-traditional sense—as a distinction between the most abstract and the most concrete, such that there are various intermediate levels between these extremes. To encapsulate the view in a slogan: Whatever is, is a structure, but not all structures are on the same level. How this is precisely to be understood is shown in what follows.

[3] The fundamental structures at the *formal* end of the spectrum are *logical* and *mathematical* structures; those at the *contentual* end are *semantic* and *ontological* ones.

The following sections show how and why this distinction is accepted here as a starting point. The rest of this chapter shows how these different structures are precisely to be conceived, and how they interrelate.

It would be futile and indeed senseless to attempt to find some absolutely privileged standpoint from which to comprehend logic and mathematics, or to discover something from which they could be derived. Instead, they are at first simply to be accepted as facts. To be sure, this does not suffice for philosophy; one must attempt to understand them—to determine what types of entities or structures they present, how they fit into the theoretical enterprise, etc. Here the attempt is made to understand logic and mathematics on the basis of *structure*. Logical and mathematical structures prove to constitute the fundamental formal subdimension of the dimension of the most universal fundamental structures.

The sciences, and thus philosophy, are not however concerned only with this fundamental formal subdimension; instead, their concern is usually with what is often termed *reality* or *actuality*, and thus with matters that are quite concrete and determinate. How are the comprehension and articulation of the dimension of the concrete to be accomplished? The answer provided here: by means of the comprehensive thematization of language, which involves (among other things) the identification and presentation of *semantic* and ultimately of *ontological* structures.

3.1.5 *The Status of Language and Semantics Within the Systematics of Structure*

The task of this section is to explain why and in what sense the contentual dimension of the comprehensive structural framework also includes *semantic* structures. This is a central problem with respect both to method and to content, and one that must be clarified if the conception developed here is to qualify as thought through and well grounded. The clarification involves a distinction between language/semantics in a comprehensive sense and language/semantics in a narrow sense.

[1] The inclusion of semantic structures as fundamental is based on the thesis that anything that appears or counts as conceptual content (in any broadly traditional or intuitive sense) must appear or be articulated within the medium of language or in connection with language. This thesis presupposes a quite determinate—and radical—conception of the status of language, according to which language is not a medium for the presentation or expression of conceptual contents that are in any manner whatsoever already present or possessed independently of language. Language is of course the *indispensable medium* for expression/presentation of conceptual contents, but these contents, *adequately* or *strictly* understood, cannot be present or possessed without their linguistic articulation (or articulability). In other words, the conceptual contents cannot be isolated from their linguistic articulation, which is an essential ingredient of the conceptual contents themselves. To be sure, conceptual contents are not themselves linguistic entities, but that must be correctly understood: conceptual contents are not linguistic entities in the sense that they are not purely syntactic constellations of uninterpreted (linguistic) signs; they are non-linguistic entities, but nonetheless entities that are *dependent upon language* (in the sense of being linguistically articulated or articulable). Accepting Quine's famous demand, "No entity without identity," one can say the following: linguistic articulation belongs essentially to the conditions of identity for conceptual contents.

To be sure, the relevant linguistic articulation need not be given *explicitly*; it need not be uttered or expressed. But if one says that a conceptual content can be had without its being linguistically articulated, the claim appears plausible only if what is meant is *explicit* articulation. But if, as often, the having is supposed to be *purely mental*—thus, such that there is and can be no relation to linguistic articulation—then the claim becomes indefensible. If it were true, what explanation could there be for the fact that in many cases—indeed, in the normal case—there is, incontestably, linguistic articulation of conceptual contents? The only "explanation" available to one accepting purely mental possession of conceptual contents would present the linguistic articulation as wholly external or extrinsic to those contents, as something somehow added on and thus purely accidental—something whose presence or absence would be utterly irrelevant to the contents themselves. But this would be no explanation, because it would make of linguistic articulation a kind of miracle: how can a conceptual content be linguistically articulated *as conceptual content* if linguistic articulation is purely accidental to it? To say that an X is purely accidental in relation to a Y is to say that there is no internal relation between the X and the Y. But if linguistic articulation were such an X with respect to conceptual content, then the linguistic articulation of conceptual contents would be a miracle.

No matter how one determines more precisely the relation between the mental and the linguistic dimensions, it is incontestable that it is ultimately the linguistic dimension within or by means of which conceptual content is manifest. But from this it follows that there is a close, indeed an unbreakable relation between the two. It follows that what conceptual content is, how it is structured, etc., can be gleaned from the examination of linguistic articulation.

These considerations reveal that what can be called the *concretization* of the fundamental structures can result only from the thematization of the semantic dimension.

[2] Here arises an important and interesting question: if language or semantics must be accorded the vital and comprehensive status outlined above, what becomes of the relation between language and the fundamental *formal* structures? Should not language or semantics play a comparably fundamental role with respect to formal structures? But if so, wouldn't one have to *superordinate* the *semantic* structures not only to the ontological structures but also to the formal ones, in that the conceiving and defining of these other structures would be dependent upon semantic structures? The sense in which semantic structures are superordinate to ontological structures is indicated above; only the case of formal structures appears to remain problematic.

Because of its comprehensive centrality, language/semantics is indeed superordinate, in a sense requiring precise determination, to logical and mathematical structures; the basic reason is that, as is shown below, logical and mathematical structures are what those linguistic forms called logical or mathematical sentences express. The conceptualization of the formal structures thus presupposes the semantic structures that determine such sentences. In this sense or respect, semantics is comprehensive. As becomes clear below, however, semantics is thus superordinate only in a relative respect (a pragmatic-methodological one), because the relation of the semantic and ontological dimensions—whereby formal structures are within the latter—can adequately be grasped only if the semantic and ontological dimensions are conceived of as ultimately inseparable (in a metaphor: as two sides of the same coin).

Most commonly, however, the term "semantics" is used in a *narrower* or *restricted* sense: it then designates, at least generally or primarily, the relation of language to the "real world," in abstraction from the question whether logical and mathematical structures belong to that world. The usual reason for this is either that the question is avoided or that a negative answer for it is simply presupposed. In this work, "semantic structure" is sometimes used in the narrower sense, although not for either of these reasons. [i] To explain and ground its use here with respect to the globally systematic perspective taken in this work: the fundamental formal structures constitute a central and largely autonomous dimension that is of simply decisive importance with respect to method and structure; it is therefore desirable to present this dimension, in accordance with its fundamental status, within the comprehensive structural framework. That logical and mathematical structures are also expressed linguistically is not ignored, but also not accorded priority within the systematic conception as a whole. [ii] The semantic-ontological status of logical and mathematical structures is explicitly treated below in the appropriate place. What is demonstrated about semantic structures is shown to have immediate and unrestricted application to logical and mathematical structures.

3.2 The Three Levels of Fundamental Structures

3.2.1 *Formal Structures*

3.2.1.1 Logic, Mathematics, and Philosophy

[1] If in a philosophical work there is talk of logic and mathematics, then it must be kept in mind that these are disciplines in their own right, independent of philosophy, no matter how the relationship between the two may be determined (a question considered in [2], below). Regardless of the history of the relationship between philosophy on the one hand and logic and mathematics on the other, the thesis of the complete independence from philosophy of contemporary logic and contemporary mathematics can no longer be questioned seriously. To be sure, philosophy has had, both throughout its history and at present, a far closer relationship to logic than to mathematics; for this reason, the former discipline's independence from philosophy must be articulated more subtly. But the fact that contemporary logic, taken as a whole, is no longer thinkable without mathematical forms of presentation speaks decisively in favor of that discipline's independence from philosophy. This situation has as a consequence that philosophy cannot claim that it must or can either define or develop logic or mathematics. The autonomy of logic and of mathematics is a factor that philosophy can in no way affect.

This does not mean, however, that philosophy has nothing to say about logic and mathematics. On the contrary, logic and mathematics, like every other science and every other domain, are *topics* for philosophy. In the special cases of logic and mathematics, philosophy is confronted chiefly with two tasks. The first is that of interpreting or conceptualizing the logical/mathematical dimension. An essential aspect of this task is determining what logical and mathematical "entities" are. This aspect is treated particularly by contemporary philosophy of mathematics. The second task, which largely presupposes the satisfactory accomplishment of the first, is clarifying the extent to and manner in which philosophical theories are dependent upon logic and mathematics, or can or indeed must rely upon logic and mathematics. In this book, the first task can be treated only marginally; this treatment is found in this section, and in Section 3.3.4.2. But the second task is a central topic for the conception of the structural-systematic philosophy. This topic is treated in Sections 3.2.1.2 and 3.2.1.3. The general thesis of this book concerning this topic is the following: logical and mathematical structures are the fundamental formal structures that play, within the structural-systematic philosophy, a central and therefore indispensable role. The remainder of this section clarifies and supports this thesis.

[2] To begin, a few remarks must be made concerning the question whether logic and mathematics are two distinct disciplines. This question is extremely complex and   232
cannot for that reason be treated here in a way that is at all adequate.[7] Instead, what is presented here, in a basically thetic form, is the starting point from which a fundamental answer can be developed. From the outset, one must attend to a fundamental distinction: that between the concrete (or pragmatic or institutional) situation of logic and mathematics, on the one hand, and, on the other, the fundamental question whether the two disciplines, despite being currently pursued as distinct, nonetheless constitute a unity. If they do constitute a unity, that unity would have to be explained. As far as the concrete situation is concerned, it appears clear that as a rule logic and mathematics are considered to be two distinct disciplines. This is particularly clear with respect to their institutionalization: the two disciplines tend to be pursued in different departments, they play quite different roles in curricula and with respect to exams, etc. In this respect it is also clear that logic, but not mathematics, has always been and continues to be, in one way or another, a constituent of the curriculum of philosophy.

This concrete or pragmatic situation is not, however, the exclusive or decisive criterion for answering the *fundamental* question. If one considers all the relevant aspects of the problem, then the most plausible conception appears to be the one that understands logic and mathematics *fundamentally as a single discipline*; this is the discipline that can be designated the *fundamental formal discipline*. As is made clear below, this discipline can be designated more adequately as the discipline of the *fundamental formal structures*. *Within* this fundamentally unified discipline it is meaningful and indeed unavoidable to distinguish between the two dimensions or subdisciplines of logic and mathematics.

If one accepts this conception, then one must pay close attention to the use of the terms "logic" and "mathematics." If either of these two terms is used to designate the postulated *fundamental formal discipline* or the discipline of *fundamental formal structures*, then one can no longer speak of a distinction between "logic" and "mathematics." In the current context, and in this book as a whole, the terms "logic" and "mathematics" are used as designations for the two disciplines that, in the concrete or pragmatic or institutional sense introduced above, are considered to be two distinct disciplines.

The assumption of a comprehensive, unitary formal dimension or discipline with the two subdisciplines logic and mathematics excludes the possibility that either of the subdisciplines be reduced to the other. Thus mathematics (as it currently exists) cannot be   233
reduced to logic in the sense of being derived from logic, as the famous position known

---

[7] For a detailed and extraordinarily well-documented discussion of this question, see Weingartner (1976: 24–86).

as *logicism* maintains that it can. But logic also cannot be reduced to mathematics. That these reductions are impossible cannot be shown here in detail; only two considerations are introduced briefly.

The first consideration is the recognition that logic is concerned with *more comprehensive*, unrestrictedly universal concepts and/or structures, ones that current mathematical discourse for the most part presupposes and relies upon. These include the central logical concepts of derivation and of logical consequence, with all the factors that they involve.

The second consideration is the difference in the status of *language* within the two disciplines. As do all other disciplines, logic and mathematics use language or, better, their own languages. It is therefore completely natural and non-problematic to speak of mathematical and logical languages. These languages are languages that are used, but languages can be used without being thematized. As a rule, mathematics does not thematize its language; instead, it proceeds on the basis of the assumption that the sentences of the mathematical language express that with which mathematics is concerned: mathematical structures. For this reason, one can say that mathematics is objectively oriented in the sense that it focuses on the *expressa* of the sentences that it uses: the domain of mathematical structures.

In opposition, logic *thematizes* the *formal structuration* of logical language. It does so by making a distinction between the so-called syntactic and the semantic construction of logic or, more generally formulated, between the syntactic and the semantic levels of the language of logic. The syntactic level is thematized in that the concept of the *formal system* is developed: a formal system is an uninterpreted system that involves only the introduction of a language and the fixation of specific connections between sequences of linguistic signs; such sequences are called *formulas* (or sentences). In this manner, basic logical concepts, such as axiom, theorem, rule of inference, derivation, derivability, proof, etc., are explained and defined. This mode of procedure is always initially understood and characterized *negatively*, in that it is said that a so-developed formal system has no "meaning," and is thus *uninterpreted*. *Positively*, the formal system is often characterized as being the result of the manipulation of (sequences of) (linguistic) signs.

On the *semantic* level, the logical (and non-logical) symbols are endowed with meanings. There are two types of logical semantics: semantics of valuation and semantics of interpretation. The first establishes the meanings of the logical symbols by ascribing truth-values to their occurrences. The second develops a model such that, by means of an interpretation function, all of the symbols are provided with meanings. The details of these two semantic procedures are not relevant in the current context. To be noted is that one can move back and forth between the two types of semantics.[8]

Section 3.2.1.3 considers more closely some of the aspects of the syntactic and semantic levels that are important for this book. This section's task is only to show that the status of language in logic and in mathematics is particularly important for making clear the distinction between the two disciplines. This distinction also provides the basis for the introduction and clarification of the concept of *formal* (i.e., *logical and mathematical*) *fundamental structure(s)*. Because the concept of mathematical structure is far more familiar and also far less problematic than the concept of logical structure, it is considered first.

234

---

[8] See Stegmüller and Varga (1984: Chapter 9, 295ff, Chapter 14, 403ff).

3.2.1.2 Mathematical Structures

[1] That the concept of structure is a central *mathematical* concept is a widely known fact. More controversial is its precise status within mathematics as a whole. That status depends upon the specific conception of mathematics that is held, where "the conception of mathematics" is taken in a twofold sense: as the conception that mathematicians themselves have of mathematics and as the philosophy of mathematics. In light of the central theses of this book concerning semantics and ontology, the most adequate position in the philosophy of mathematics is the one named "(mathematical) structuralism," whose central thesis is that "mathematics is the science of structure" (Shapiro 1997: 5). This thesis must, however, be made more precise: mathematics is the science of *mathematical* structures—or, to avoid the merely verbal circularity—mathematics is the science of one of the two types of fundamental formal structures. As made clear above, the term "mathematics" in such formulations is taken in the concrete or pragmatic or institutional sense, such that it names a discipline distinct from logic. With this characterization, all is said that is required for basic understanding of the comprehensive theory presented and defended in this book. Nevertheless, a few additional clarifications are added in the following two subsections.

[2] The term "(mathematical) structuralism" is usually associated with the name   235
Nicolas Bourbaki.[9] In the 1930s, Bourbaki undertook a complete restructuration and unification of mathematics as a whole on the basis of the axiomatic method, at the center of which stands the concept of structure. There is at present a broad consensus that mathematics has a "structural character," but there is no consensus concerning how "structure" is precisely to be understood and how its status within mathematics as a whole is to be determined systematically. This holds even within the work of Bourbaki. Leo Corry maintains the remarkable thesis that the often-emphasized structural character of contemporary mathematics has a clearly identifiable meaning only if it is understood as a way that mathematics is done, such that it can be described only in a *non-formal* manner (Corry 1992: 316). According to Corry, since the 1930s there have been various attempts to develop formal theories within theoretical frameworks within which the non-formal idea of "mathematical structure" can be explained *mathematically*. He adds that the failure to consider the distinction between the non-formal and the formal significations of "structure" brings with it a great deal of confusion. His surprising central thesis is the following: "Bourbaki's real influence on contemporary mathematics has nothing to do with the concept of *structure*" (316). If this thesis is not to be misunderstood, it must be noted that Corry explicitly connects the emphasized term "structure" (printed in italics) with the *formal* concept of structure.

To understand the actual status of the concept of structure within Bourbaki's conception, one must—as Corry has demonstrated persuasively—distinguish among *three kinds* of writings in Bourbaki's work. The first kind consists of general, non-formal writings in which the (non-formal) concept of structure plays a comprehensive role. The volume *Theory of Sets*—the first volume of the collection *Elements of Mathematics*—belongs to the second kind; its Chapter 4 contains an explicit formal

---

[9] As is noted above, this name is a pseudonym for a group of predominantly French mathematicians.

definition of "structure." The third type consists of the individual volumes that treat specific domains of mathematics.

The most important of the texts belonging to the first category is the often-cited essay, "The Architecture of Mathematics." In this essay, there is a remarkable non-formal definition of the concept of structure:

236   It can now be made clear what is to be understood, in general, by a mathematical structure. The common character of the different concepts designated by this generic name, is that they can be applied to sets of elements whose nature[10] has not been specified; to define a structure, one takes as given one or several relations, into which these elements enter[11]; ... then one postulates that the given relation, or relations, satisfy certain conditions (which are explicitly stated and which are the axioms of the structure under consideration.) To set up the axiomatic theory of a given structure, amounts to the deduction of the logical consequences of the axioms of the structure, excluding every other hypothesis on the elements under consideration (in particular, every hypotheses as to their own nature. (Bourbaki 1950: 225–26)

Astonishingly, Corry does not cite this non-formal definition. One must, however, say that it articulates, in a non-formal manner, precisely the same understanding of "structure" that is also the basis of the *formal* definition of structure common elsewhere in mathematics, the one that is presented in this book in Chapter 1 (Section 1.2.3). One can term this the "simple" or "basal" definition. As the passage from Bourbaki cited in footnote 11 makes clear, this definition is "not general enough for the needs of mathematics." One must note that the distinction between the non-formal and the

---

[10] Here Bourbaki includes the following footnote, extremely important with respect to philosophy:

> We take here a naïve point of view and do not deal with the thorny questions, half philosophical, half mathematical, raised by the problem of the "nature" of the mathematical "beings" or "objects." Suffice it to say that the axiomatic studies of the nineteenth and twentieth centuries have gradually replaced the initial pluralism of the mental representation of these "beings"—thought of at first as ideal "abstractions" of sense experiences and retaining all their heterogeneity—by a unitary concept, gradually reducing all the mathematical notions, first to the concept of the natural number and then, in a second stage, to the notion of set. This latter concept, considered for a long time as "primitive" and "undefinable," has been the object of endless polemics, as a result of its extremely general character and on account of the very vague type of mental representation which it calls forth; the difficulties did not disappear until the notion of set itself disappeared (and with it all the metaphysical pseudo-problems concerning mathematical "beings") in the light of the recent work on logical formalism. From this new point of view, mathematical structures become, properly speaking, the only "objects" of mathematics . . . .(1950: 225–26n†)

This cluster of issues is considered critically below.

[11] At this point Bourbaki includes the following footnote:

> In effect, this definition of structures is not sufficiently general for the needs of mathematics; it is also necessary to consider the case in which the relations which define a structure hold not between elements of the set under consideration, but also between parts of this set and even, more generally, between elements of sets of still higher "degree" in the terminology of the "hierarchy of types. . . ." (1950: 226n*)

formal definitions is significant and must therefore be taken into consideration; but it   237
is nevertheless only a relative distinction, because one can in principle formalize *any* non-formal definition, and present every formal definition non-formally. To be sure, Corry does not understand the distinction in this manner. According to him, "formal structure" is not simply the "formalized definition of the concept of structure"; instead, a "formal structure" is to be sure a "formalized structure," but beyond that—and utterly essentially—it is an element of an axiomatic, deductive mathematical system.

To the second kind of texts of Bourbaki treating the concept of structure belongs Chapter 4 of the volume *Set Theory*. The book defines structure in its final chapter. The definition is complicated because it presupposes various other concepts that Bourbaki first introduces and explains. Among these, particularly important is "species of structure." Corry reconstructs Bourbaki's definition quite clearly (323f.). This reconstruction shows that it in no way contradicts the simple or basal definition; instead, it either makes it more detailed or—if one will—generalizes it, but in any case it presupposes it.

Corry's thesis that in the individual volumes of Bourbaki's collection *Elements of Mathematics* the concept (*formal*) *structure* plays no significant role may well be correct, presupposing that one attends to the precise and narrow sense of "formal structure" that Corry presupposes.

Two additional comments on aspects of Bourbaki's conception are appropriate here. First, it should not be difficult to see that the non-formal concept of structure in the cited essay and the formal concept in Chapter 4 of *Set Theory* articulate what Chapter 1 (Section 1.2.3) of this book, following Kleene, terms "pure or abstract structure." Second, as the text cited in footnote 10 shows, Bourbaki turns away from (philosophical) questions concerning the "nature" of mathematical "objects" or "entities (beings)" by characterizing them as "metaphysical pseudoproblems concerning mathematical 'beings'" ("entities," in contemporary terminology). Interestingly, he then maintains that within the framework of the axiomatic investigations of the nineteenth and twentieth centuries, mathematical concepts are reduced first to the concept of natural number and then to the concept of set, and also that later, even the concept of the set itself disappears—and with it all putative "metaphysical pseudoproblems" concerning mathematical objects/entities. With his talk of the disappearance of the concept of the set, Bourbaki may have in mind the emergence of the mathematical theory of categories (see Corry: 332ff.). From this he concludes: "According to this new standpoint, mathematical structures are really the only 'objects' of mathematics." Here   238
it is clear what Bourbaki's conception of ontology is: for him, ontology's domain is that of physical or material things or entities. Only on the basis of such a conception is it at all intelligible that mathematical structures are not understood as entities, and thus that *in this sense*, they are denied ontological status. But on what could such a conception be grounded rationally? The ontological status of formal (logical/mathematical) structures is considered in Section 3.3.4.2.[12]

[3] Bourbaki's brief presentation of the problem of mathematical structure clarifies the status of the project of developing a systematic philosophy with a structural character. At the end of his essay, Corry writes,

---

[12] On this topic, see also the excellent Shapiro (1997).

What I have shown here is only that Bourbaki's *structures* could have no influence at all. But I have also shown why, if we want to describe Bourbaki's influence properly, we must then concentrate on what I have called the "images of mathematics."[13] This in no way belittles the extent or importance of Bourbaki's influence. On the contrary, the images of mathematics play a decisive role in shaping the path of development of this discipline . . . . [I]t should be clear now that the rise of the structural approach to mathematics should not be conceived in terms of this or that formal concept of structure. Rather, in order to account for this development, the evolution of the nonformal aspects of the structural image of mathematics must be described and explained. (341–42)

If even so grand a project as Bourbaki's *Elements of Mathematics* does not succeed in developing an exact and comprehensive coherence (not even in terms of presentation), one should not simply expect from philosophy a better, much less a perfect, result.

3.2.1.3 Logical Structures

The concept *logical structure* is often used, but does not have a status within logic comparable to that of the concept of mathematical structure within mathematics. This book is not, however, concerned with the frequency of usage of the concept. The only important factor is substantive: because logic, like mathematics, is a formal discipline, to speak of logical structures is not only justified but also highly appropriate. This must, however, be explained more precisely.

[1] A good beginning may be made by avoiding a rarely seen unclarity, indeed ambiguity, in the distinction between the syntax and the semantics of a language. One considers such notions as "express a proposition (by means of a sentence)" to belong to the semantics of a language. If, however, as is common practice, one distinguishes between syntax and semantics without differentiating them quite subtly, a question that appears puzzling can arise: what is the status of sentences that express syntactic contents? Do such sentences not express propositions? With respect to the syntactic level of logic, one such question would be, does the sentence by means of which the syntactic structure *derivation* is articulated have no *expressum*? Does it not express a proposition? But what then would be the difference between what is articulated or formulated by means of this sentence and the proposition expressed by the sentence, that is, its *expressum*? What is articulated or formulated by means of such a sentence is something syntactic, a determinate syntactic combination of (linguistic) signs. But one must then say that such a sentence has a syntactic *expressum* and thus expresses a syntactic proposition. It is easily seen that the distinction usually made or presupposed between syntax and semantics requires more detailed differentiation.

The distinction that is likely most important in the current context is that between *two aspects* of the syntactic and semantic levels: the *objective* aspect and the *formulation* aspect. If the focus is exclusively on the objective aspect, the syntactic level is understood

---

[13] Corry explains this concept as follows:

> "Images of mathematics" denote any system of beliefs *about* the body of mathematical knowledge. This includes conceptions about the aims, scope, correct methodology, rigor, history and philosophy of mathematics. (342n6)

as the dimension of the syntactic (briefly: as the syntactic) and the semantic level as the dimension of the semantic (as the semantic)—the two dimensions are then considered *in themselves*, without consideration of such factors as how the dimensions themselves can be formulated. If, however, the focus is on the formulation aspect, the two dimensions are considered *along with* the factor of themselves being formulated or articulated. This yields the following: the syntactic and the semantic are two well-distinguished dimensions. The syntactic is the dimension of the logical interconnection of pure (linguistic) signs, the semantic, the dimension of the values (the semantic values) ascribed to the logical signs. If the formulation aspect is taken into consideration, it must be said that insofar as the syntactic is itself formulated or articulated, it proves to be a determinate mode of the semantic, because in this case the syntactic has the status of a (precisely semantic!) value that is ascribed to every sentence that articulates the syntactic. This "determinate mode of the semantic" is a *syntactic expressum* and   240
thereby a *syntactic proposition*. According to the semantics that is developed in Section 3.2.2.4, propositions (more precisely: primary propositions) are understood as (primary) *structures*. It can therefore here be said that, if the focus is on the second aspect, a (logical-)syntactic sentence expresses a *(logical-)syntactic* structure. Conversely, from the preceding considerations it is clear that, with respect to the relation of the semantic to the syntactic, insofar as the semantic is formulated or articulated, it proves to be (also!) a determinate mode of the syntactic. That is because in this case the semantic (the expressed proposition or structure) is (so to speak) drawn into the dimension of the syntactic, because the formulation *also* has a syntactic form.

If sufficient attention is paid to the just-explained interrelationality of the syntactic and the semantic levels, then the terms/concepts "(syntactic-)logical proposition or structure" and "(semantic-)logical proposition or structure" are disambiguated, and thus clear and free of misunderstanding.

That logic contains both types of structures should also be clear. A problem could possibly arise from that fact that what qualifies as normal logic explicitly recognizes and applies logical rules (rules of inference). But can a rule be understood as a proposition or structure? This is not a genuine problem for the position taken here concerning logical structures. A logical rule is, as could be said, the flip side of a structure. Every structure can, so to speak, be pragmaticized in the sense that it can be pragmatically read *and* used. This is easily shown in the case of the rule of inference named the *rule of detachment* or *modus ponens*. Often, this rule is written:

$$p \rightarrow q$$

$$p$$

$$\therefore q$$

Pragmatically, this formula is read as follows: given the premises $p \rightarrow q$ and p, it is allowable (or one may, etc.) conclude that q. But one can also use another, non-pragmatic way of writing this rule—$(p \, \& \, (p \rightarrow q)) \rightarrow q$—or, more exactly, for all formulas α und β: $\{\alpha, \alpha \rightarrow \beta\} \vDash_0 \beta$ (see Bell and Machover 1977: 34). The symbol "$\vDash_0$" here designates logical consequence. One can also understand the formula as a law (one speaks then of the "law of detachment," etc.; see Suppes 1957: 32). One thus sees that the concept of the rule of inference can in no way be the basis of any objection to the assumption that all logical rules are structures.[14]

[2] From the preceding considerations in this section can be derived the legitimacy *and* the necessity of an extremely important distinction concerning logic: one must distinguish between logic as it exists in its factual or concrete form and is used, and logic as it can be seen and comprehended from a more comprehensive perspective. Within that perspective, the concept *logical structure* attains something like a *particular programmatic status*; the concept does not simply fit into the familiar and trusty framework of logic. As the preceding considerations show, it can be explained only with care. That it has such a status is best understood as an interpretation of a program yet to be carried out; this is the program leading to a deeper and more comprehensive theory of logic.

---

[14] Whereas at present the symbol "⊢" always designates logical derivability, the symbol "⊨" is not used in a uniform fashion (on the history of the two symbols, see Hodges 1993: 83f.) Many authors use "⊨" exclusively in the model-theoretical sense as a relation between sentences and structures, i.e., as $\mathfrak{A} \models \varphi$, that can be read in three ways (see 2.5.2.1): first, as "$\mathfrak{A}$ is a model (in the sense of a structure) for (the sentence) φ," second, as "φ is true in (the structure) $\mathfrak{A}$," and third, as "φ is satisfied in (the structure) $\mathfrak{A}$." Some authors use "⊨" to designate both logical consequence and a model-theoretical relation (for logical consequence, Bell and Machover write "$\models_0$"; see 1977: 24). These notational differences result from an interestingly problematic situation in logic. The concept of logical consequence defines a specific relation between semantically interpreted sentences. If in a formula (sets of) sentences appear to the left of the symbol "⊨," the symbol designates logical consequence. An example: let Δ be a metavariable for theories and let theories be sets of sentences. The formula $\Delta \models \varphi$ is then to be read, "(the sentence) φ is a logical consequence of Δ (or: Δ logically implies φ)." But this formula is explained by reference to the concept of *structure*, which itself is articulated by means of the symbol "⊨" used in the model-theoretical sense. If Δ is a theory and $\mathfrak{A}$ is a structure, then $\mathfrak{A} \models \Delta$, i.e., (in one reading), the structure $\mathfrak{A}$ is a model for every sentence in Δ. The formula $\Delta \models \varphi$, articulating logical consequence, then means (or is explained as), every structure that is a model for Δ is also a model for φ (see Hodges 1983: 56).

If however one asks how this relation between the formula that articulates logical consequence and the formula that articulates a model-theoretical relation (between sentence and structure) is more precisely to be understood, one finds no satisfactory answer. How this relation is explained within the framework of the semantics and ontology presented below (see Section 3.2.2.4) can be indicated briefly. Within that framework, every sentence of theory Δ expresses a primary proposition and thus a semantic primary structure (as is noted above, primary propositions are primary structures); the totality of these primary propositions/structures constitutes a highly complex semantic primary proposition/structure that, if true, is identical to a highly complex ontological primary structure. This explains what it means for sentence φ to be *true* with respect to the highly complex ontological structure $\mathfrak{A}$: the ontological structure expressed by the true sentence φ is a component structure of the highly complex ontological structure $\mathfrak{A}$. Logical consequence is then a matter of a specific semantically interpreted sentence being included within (and therefore logically following from) a specific configuration of semantically interpreted sentences. The "interpretation" of sentences thereby presupposed is precisely the structure.

Notational clarity would be improved by the adoption of the practice of various authors (e.g., Stegmüller *Probleme* I/1983: 76, Stegmüller and Varga 1984: 600, 84, etc.) of using the symbol "⊩" for logical consequence or validity.

Such a theory can neither be demanded nor developed within the framework of this book. Here, it appears appropriate to introduce an example of such an undertaking: Arnold Koslow's book *A Structuralist Theory of Logic* (1992). Koslow's theory is based on the concept *implication structure*. This structure (*I*) consists of a non-empty set *S* and a finite relation on the set that Koslow names "implication relation" and that he designates by means of the symbol "$\Rightarrow$." The definition is $I = \langle S, \Rightarrow \rangle$. The implication structure is fundamentally characterized initially by two negative-positive factors. First, the implication relation is not restricted to the familiar relations of logical derivability and logical consequence, because the set *S* does not contain only syntactic and semantic elements, thus linguistic signs and their interpretations; instead, this relation is *ubiquitous* in the sense that every non-empty set can be structured by means of an implication relation. Second, the logical operators are not explained by recourse to truth-values, logical form, conditions of assertibility, conditions of *a priori* knowledge, etc.:

> [W]e shall focus on a question that appears to be much simpler: What are the conditions under which a certain item is to count as a hypothetical, a conjunction, a negation, or a disjunction, or as existentially or universally quantified? Our question concerns what it is like to be a hypothetical or a conjunction, for example, not what it is for a hypothetical, conjunction, or other logical type of item to be true, false, assertible, or refutable. (3–4)

On the basis of this central concept of the implication relation, the logical operators are determined or (newly) explained. This occurs in strict relativity to an implication relation. A theory of each logical operator is developed that describes

> how each of them acts on implication structures (sets provided with implication relations). The logical operators are then to be thought of as certain types of functions that are defined on each structure and whose values (if they exist) are the hypotheticals, conjunctions, negations, and other logical types of that structure. (4)

It is thus clear that logical operators can be understood as logical structures (or substructures), or, as one can say, as concrete or determinate structures of logical implication. An implication relation on set *S* is determined by Koslow as any relation $\Rightarrow$ that satisfies the following conditions:

1. Reflexivity: $A \Rightarrow A$, for all *A* in *S*.
2. Projection: $A_1, \ldots, A_n \Rightarrow A_k$, for any $k = 1, \ldots, n$.
3. Simplification: If $A_1, A_1, A_2, \ldots, A_n \Rightarrow B$, then $A_1, \ldots, A_n \Rightarrow B$, for all A_i and *B* in *S*.
4. Permutation: If $A_1, A_2, \ldots, A_n \Rightarrow B$, then $A_{f(1)}, A_{f(2)}, \ldots, A_{f(n)} \Rightarrow B$, for any permutation *f* of $\{1, 2, \ldots, n\}$.
5. Dilution: If $A_1, \ldots, A_n \Rightarrow B$, then $A_1, \ldots, A_n, C \Rightarrow B$, for all A_i, *B*, and *C* in *S*.
6. Cut: If $A_1, \ldots, A_n \Rightarrow B$, and $B, B_1, \ldots B_m \Rightarrow C$, then $A_1, \ldots, A_n, B_1, \ldots B_m \Rightarrow C$, for all A_i, B_j, *B*, and *C*. (5)

Koslow introduces "a simplified account of the conjunction operator" (6) to serve as an example for the procedure of forming the logical operators on the basis of the implication structure or relation and the six named conditions. Because conjunction is thereby determined as *logical-structural*, it is, as Koslow emphasizes, a case of "an abstract view of conjunction."

> The conjunction operator C acts on the implication structure I in the following way. To any elements A and B that belong to the set S, C assigns a subset $C_\Rightarrow (A, B)$ of S whose members (if they exist) satisfy the following conditions [we shall use "$C_\Rightarrow (A, B)$" to indicate conjunctions . . .]:[a]
>
> C_1. $C_\Rightarrow (A, B) \Rightarrow A$, and $C_\Rightarrow (A, B) \Rightarrow B$, and
> C_2. $C_\Rightarrow (A, B)$ is the weakest member of S that satisfies the condition C_1. That is, for any T contained in S, if $T \Rightarrow A$, and $T \Rightarrow B$, then $T \Rightarrow C_\Rightarrow (A, B)$.

All the other logical-structural operators are defined in similar fashion.

Compared with the usual or standard conception of logic, Koslow's theory is a remarkable attempt to understand logic at a higher level, from the dimension of pure, formal structurality. It clearly shows that, structurally understood, logic and mathematics are subdisciplines not of mathematics, but of a fundamental, unitary formal discipline, as characterized in Section 3.2.1.1[2]. Like mathematics, logic thereby attains an objective character in the sense that it is fundamentally concerned with what is expressed by means of logical language (i.e., with logical structures). There thereby occurs a *certain* uncoupling of logic from language, indeed in the sense that sets of syntactic and semantic items are only among the subsets of the comprehensive set to which the implication relation and the logical operators derived from it are applied. Koslow thus forms, for example, a logical operator that is applied not to sentences, but to individuals (see Koslow's Section 22).

244   This theory yields the result that logical structures and/or logical-structural operators can have a directly *ontological* status: they can also structure "real things." The precise understanding of this result depends upon the ontology that is accepted. The ontology presented in this book (see Section 3.2.2.4) recognizes only primary facts, which are understood as ontological structures. Here, then, ontological structural-logical operators would be operators that (further) structure primary facts and thus ontological primary structures.

Above, there is mention of a "certain uncoupling" of logic from language; the qualifier is meant to indicate that no total uncoupling occurs or can occur. As Section 3.2.1.3[1] shows, neither mathematics nor logic can wholly ignore language. Koslow's complete ignoring or non-thematization of the linguistic dimension is therefore a central error that has many consequences. Among these is his thesis that the logical structures (the logical operators) are to be defined without any relation to semantic (and ontological) factors like truth-values. As shown above, in the case of mathematics and logic one must distinguish between two factors, the *objective* aspect and the *formulation* aspect. These aspects are correlative: they are only relatively distinct, in that they are also related to each other. The implication structure and the logical-structural operators in Koslow's sense are to be sure objective entities, but they are also formulated or articulated *as* such entities—i.e., they are the *logical expressa* of logical sentences.

---

[a] The passage in square brackets is in Koslow (1992); the ellipsis indicates that the passage is not quoted in its entirety.

These *logical expressa* can also be termed *logical propositions* and, more precisely, *logical structures*. But then the question arises: how are such sentences to be qualified or determined? The truth-theory presented in this book (see Sections 2.5.2 and 3.3) makes clear that the truth of sentences must be understood as the attainment by the sentences of fully determined status; because every indicative or theoretical sentence expresses a proposition, it follows that every true sentence expresses a true proposition; the proposition is true—fully determined—when it is identical to a *fact*.

A logical sentence thus attains its highest logical qualification when it is determined as true. The logical sentence is, however, true only when the logical proposition it expresses is also true. It is true when it is fully determined. In addition, a logical proposition is fully determined only when it is identical with a logical fact. This logical fact is therefore the genuine logical structure; it is fully determined when it is definitively situated within the logical system of which it is a component. It is definitively situated by means of the provision of conditions.

Surprisingly, this interpretation of the truth of logical sentences is not essentially   245
different from Koslow's determination of logical-structural operators. Koslow proceeds on the basis of a *comprehensive* logical structure (the implication structure $\Rightarrow$) and derives various forms that concretize this structure. This concretization of the implication structure, which is accomplished by the provision of determinate conditions to be satisfied, means, in the terminology used above, the fixation of a determinate, definitive place that the concretized form in question takes within the implication structure of the presupposed logical system and by means of which it is made fully distinct. This in essence corresponds to what is presented above as the truth of logical sentences or logical propositions. Koslow's rejection of reliance on truth (values) in defining the logical operators is therefore based on a misunderstanding, which results from his failure to consider the role of language, described above. Even the just-presented interpretation of the truth of logical sentences in no way rules out the structuration of factors or entities other than syntactic and semantic ones by means of logical operators; precisely therein consists the *ontological* status of logical operators. But such logical structurations are formulated/articulated, and that presupposes language in the manner described above.

The conception and the critical evaluation presented here of Koslow's structuralistic theory cannot make the claim of being fully intelligible and grounded. For that, more extensive treatment would be required. The presentation above can be understood and situated as an interpretation sufficient for the purposes of this book.

3.2.2 *Semantic Primary Structures*

3.2.2.1 General Characterization

[1] The designations "syntactic category" and "semantic category" are widely used. The former designation concerns the qualification of linguistic expressions from purely grammatical viewpoints (for this reason, reference is often made simply to "grammatical categories"); the latter concerns the interpretation of linguistic expressions. In this book, the syntactic level is not treated explicitly and directly, but only brought into view within the framework of the treatment of semantics. In addition, for reasons presented above, the concern is not with semantic *categories*, but with semantic *structures*, whereby   246
"structure" is taken in accordance with its strictly determined semantic value.

To be emphasized at the outset is that neither in the following sections nor elsewhere in this book is there developed a detailed critique of other views of semantics. The presentation is systematic.

[2] The concept of structure introduced in Chapter 1 must now be employed within the domain of semantics: the task is to develop *semantic structures*. First to be shown is that the concept of structure can be explicated or presented, within the domain of semantics, in a rigorous fashion. A wholly normal predicate language serves as an illustration. For such a language, Stegmüller and Varga (1984: 410) define the concept of (normal) *semantic structure* (as opposed to *full semantic structure*) as follows:

> A semantic structure for the symbols S—a semantic S-structure—is an ordered pair $\mathfrak{A} =$ $\langle A, \mathfrak{a} \rangle$, such that:
>
> 1. A is a non-empty set, termed the *object-domain* or *universe* of $\mathfrak{A}$.
> 2. $\mathfrak{a}$ is a function that assigns to S-symbols as arguments denotations in A that are appropriate in the following sense:
>   (a) For each n-ary predicate symbol P^n from S, $\mathfrak{a}(P^n)$ is a set of ordered n-tuples of elements of A.
>   (b) For each m-ary function-symbol f^m from S, $\mathfrak{a}(f^m)$ is an m-ary function on A, i.e., $\mathfrak{a}(f^m): A^m \to A$.
>   (c) For each constant c from S, $\mathfrak{a}(c)$ is an element of A.

$\mathfrak{a}$ is termed the *designation-function* of $\mathfrak{A}$ (or the *denotation-function* or the *interpretation-function* of $\mathfrak{A}$). Applying set-theoretical symbols more strictly, these authors define the designation-function (with the symbols to be read as follows: $D_1(f) = $ the domain of arguments for the function f, $\hat{S} = $ the set of S-symbols (i.e., the set Pr_s of S-predicates, the set Fu_s of S-function-symbols, and the set Co_S of S-constants; for sake of simplicity, $\hat{S}$ is identified with S). $\wp = $ power set; $\mathfrak{a}|_{Pr_s} = $ the restriction of the function $\mathfrak{a}$ to Pr_s; $\cup = $ symbol for union; $\mathbb{Z}^+ = $ the set of positive integers):

$(\mathfrak{a}_1) \quad D_1(\mathfrak{a}) = \hat{S}$

$(\mathfrak{b}_1) \quad \mathfrak{a}|_{Pr_s}: Pr_s \to \cup \{\wp(A^n) \mid n \in \mathbb{Z}^+\}$

$\qquad\qquad P^n \mapsto \mathfrak{a}(P^n) \in \wp(A^n)$

$(\mathfrak{b}_2) \quad \mathfrak{a}|_{Fu_s}: Fu_s \to \cup \{A^{(A^m)} \mid m \in \mathbb{Z}^+\}$

$\qquad\qquad f^m \mapsto \mathfrak{a}(f^m) \in A^{(A^m)}$ (i.e., $\mathfrak{a}(f^m): A^m \to A$)

$(\mathfrak{b}_3) \quad \mathfrak{a}|_{Co_s}: Co_s \to A.$

247   If one adds *variables*, the result is, in the terms of Stegmüller and Varga, the *full semantic structure*. For that one requires also the concept of *the assignment of values to variables*, which can be defined as follows: an assignment for (a set) A is a function $\eta: Var \to A$ that assigns to variables elements of A. The concept of *full semantic structure* is then defined as follows:

> A *full semantic structure* is an ordered pair $\mathfrak{A} = \langle A, \mathfrak{a}^* \rangle$ such that $\mathfrak{a}^*$ is the union of the two functions $\mathfrak{a}$ and η, thus $\mathfrak{a}^* = \mathfrak{a} \cup \eta$, and η is a function whose values are from A.

A semantic structure is thus a functional structure consisting of a function and a set of arguments for the function. The elements of the set of arguments are linguistic signs. It is immediately clear that the semantic structure depends essentially on the linguistic signs that are assumed. In the language of first-order predicate logic, presupposed above, the central and indispensable signs are, as the definitions reveal, singular terms (names), predicates, and function-symbols. Sentences are not recognized at all. The deeper reason for this is that this language is based on the *principle of composition*, which the following sections consider in detail. The purpose of this section is to show that and how the strict concept of structure can be applied to the domain of language or of semantics.

3.2.2.2   The Decisive Option: Ontologically Oriented Semantics for Philosophical Language

The decisive question, in this context, is this: what language, or what semantics, is to be accepted?

[1] The answer to this question that most philosophers in fact rely upon, although it is rarely explicitly present, leads to the acceptance and use of so-called natural or ordinary language, and of the semantics of that language. This acceptance of natural or ordinary language is, however, highly complex, as is revealed by its consideration in Chapter 2. As indicated in Section 2.2.2, ordinary-language indicative sentences as a rule have the subject-predicate form. For this reason, first-order predicate logic is, for virtually all analytic philosophers, the logic that is simply fundamental. This, however, has enormous semantic and ontological consequences, as is shown in more detail below. Nevertheless, scarcely any analytic philosopher has questioned this subject-predicate structure of the sentence, with the consequence that the semantics and ontology accompanying that structure are simply assumed to be adequate and without contradiction. When analytic philosophers do suggest corrections for ordinary-language structures, the suggestions as a rule remain restricted to superficial corrections without real semantic-ontological consequences; Section 3.2.2.2.1 provides a revealing example.

The structural-systematic philosophy is committed to a philosophical language that is wholly different from the ground up in its semantics and ontology. Its radical difference from ordinary language consists in its radical and consequential abandonment of the subject-predicate structure for sentences. In this precise sense, it diverges fundamentally from ordinary language.

An additional characteristic of the structural philosophical language is the strongly ontological orientation of its semantics. To use familiar terminology, one could say that this semantics is *realistic*. But the word "realistic" is used in contemporary philosophy in so many different senses that it is inadvisable to rely upon it, without extensive clarifications, as a technical term. This ontologically oriented semantics is fundamentally connected to the Tarskian tradition, but with essential alterations necessitated by the dependence of Tarski's semantics on the subject-predicate structure. Structural semantics is sharply opposed to all forms of semantics that fail to recognize or thematize the ontological import of language.

A final thesis centrally relevant to structural-systematic semantics, one mentioned above, requires further attention: this is the interconnection between semantics and ontology, an interconnection so strong that it is appropriate to describe them as two

248

sides of the same coin. The basic thesis that they are thus interconnected arises from the acceptance of specific premises concerning the central status of language for philosophy and for every theoretical enterprise. This book is primarily concerned not with grounding this thesis, but with showing how it is to be understood and what are its consequences. This showing itself, however, provides grounding for it.

The task now is to show how the basic features of the envisaged structural semantics (and ontology) are to be understood. The relevant considerations are decisively based on the just-formulated thesis that semantics and ontology are so closely related as to be, in a specific sense, as inseparably interconnected as are the two sides of a single coin. If a specific ontology proves unacceptable, it immediately follows that its semantics must be rejected; conversely, if a given semantics proves untenable, its ontology is untenable as well.

[2] The following sections take a lengthy detour in working toward structural semantics. The detour begins with the presentation and critical examination of the semantics that results from a normal interpretation of ordinary language; this semantics is based on the *principle of composition*, which provides the key to the standard interpretation of subject-predicate sentences. The ontology emerging from this semantics is here termed *compositional* ontology; it conforms to the long-standing tradition of substance ontology. The course of argumentation in Section 3.2.2.3 develops within the framework of a large-scale *modus tollens* syllogism: acceptance of a compositional semantics requires acceptance of a compositional or substance ontology; this ontology is untenable; therefore compositional semantics is also untenable. Differently stated: compositional-semantic structures entail compositional-ontological structures; compositional-ontological structures are untenable; therefore compositional-semantic structures are untenable. The first and particularly the second premises of the argument require extensive support.

3.2.2.3 Critique of the Semantics and Ontology That Are Based on the Principle of Compositionality

3.2.2.3.1 *Basic Features of Compositional Semantics: Compositional Semantic Structures*

[1] If one begins with languages containing subject-predicate sentences, the semantics that seems natural is compositional; the reason it seems natural is that it appears to account for the *composition* and complexity of this linguistic form. This seemingly natural view has led, in contemporary semantics, to the quite widely accepted *principle of compositionality* (CPP). This principle articulates the semantic structure of every complex or compound expression, and reads, in its simplest form, as follows:

(CPP)   The meaning (or semantic value) of a complex or compound expression is a function of the meanings (or semantic values) of its parts or components.

For sake of simplicity, the following account considers only by far the most important form of complex expression: the indicative sentence.[15] The CPP for such sentences (CPP$_S$) can be formulated as follows:

---

[15] In the account that follows, the qualifier "indicative" is usually omitted, for sake of brevity; it nevertheless remains in force. This chapter does not consider non-indicative sentences.

(CPP$_S$) The meaning (or semantic value) of a sentence is a function of the   250
meanings (or semantic values) of its subsentential components.

The relevant subsentential components are the subject (the singular term) and the predicate. The composition of the semantic values of these two components yields the semantic value of the sentence as a whole. The semantic value of the subject or singular term is generally designated the *denotation* of this term, and considered to be a (real) object. With respect to the semantic value of the predicate, two opposed views are defended, one purely extensional, the other sometimes (misleadingly) termed"intensional." The extensional interpretation presents as the semantic value of any unary predicate the set of objects to which the predicate applies (in the case of any *n*-ary predicate the semantic value is identified with the set of tuples of objects to which the predicate applies). The intensional interpretation identifies the semantic value of each term with some specific entity, a property (in the case of any one-place or monadic predicate) or a relation (in the case of any *n*-ary predicate, where $n > 1$); the general term covering both properties and relations is "attribute." There are widely diverging opinions concerning just what attributes are. The pertinent debates, and also those between extensionalists and "intensionalists," are not centrally relevant to the current endeavor and so are not further considered here.

[2] The basic idea behind the CPP can be formulated as follows ("[[ ]]" indicates the semantic value of the enclosed expressions):

Consider, for example, a syntactic formation rule of the form "if α is an A, and β is a B, . . ., and μ is an M, then $f(\alpha, \beta, \ldots, \mu)$ is an N" (where A, B, . . ., M, N are syntactic categories). The function f specifies how the inputs are to be mapped into the output, i.e., it specifies the modes of combination of the arguments. Corresponding to this syntactic rule there will be a semantic rule of the form "if α is an A, and β is a B, . . ., and μ is an M, then $[[f(\alpha, \beta, \ldots, \mu)]]$ is $g([[\alpha]], [[\beta]], [[\mu]])$." Here, g is a function which, so to speak, specifies the "semantic mode of combination" of the semantic values which are its arguments. (Dowty et al. 1981: 42)

The compositionality so described is represented by the following diagram (see Dowty et al.: 1981: 43):

Syntactic formation rule	Semantic rule	
$\alpha, \beta, \ldots, \mu$	$[[\alpha]], [[\beta]], \ldots, [[\mu]]$	251
$\downarrow\downarrow \quad \downarrow$	$\downarrow \quad \downarrow \quad \downarrow$	
f	g	
$\downarrow$	$\downarrow$	
$v = f(\alpha, \beta, \ldots, \mu)$	$[[v]] = [[f(\alpha, \beta, \ldots, \mu)]]$	
	$= g([[\alpha]], [[\beta]], \ldots, [[\mu]])$	

A homomorphism is usually assumed to hold from the syntax to the semantics; when it is, the same semantic value may be assigned to different syntactic structures.

This basic idea is made concrete in an example involving the one-place predicate *F* and the individual constant *a*, by means of which the predicate of the sentence *Fa* is saturated. The semantics based on the CPP introduces an interpretation function *I* that assigns to every expression a semantic value. It then holds that if the semantic value of $Fa = s_3$ (i.e., if $I(Fa) = s_3$), then there is a semantic value s_1 (i.e., $s_1 = I(a)$), a semantic value s_2 (i.e., $s_2 = I(F)$), and a composition function *C*, such that the semantic value s_3 (of the sentence *Fa*) is the value of *C* with the arguments s_1 and s_2 (i.e., $s_3 = C(s_1, s_2)$).

[3] The question arises: *as what* is the semantic value of the sentence (i.e., s_3) to be determined? On this question, current thinkers hold two basic positions. The first relies on the extraordinarily vague notion of truth-conditions and can therefore be termed the *truth-conditional* position. Its central thesis can be formulated as follows:

(CPP$_{TC}$)   The semantic value of the sentence *Fa* is the value of a function that maps the semantic values of the subsentential components *a* and *F* either to the truth-value *true* or to the truth value *false*. Thus: either $s_3 =$ true or $s_3 =$ false.

The second position can be termed *propositional-conditional* and can be characterized as follows:

(CPP$_{PC}$)   The semantic value of the sentence *Fa* is the value of a function that maps the semantic values of the subsentential components *a* and *F* to the proposition P expressed by Fa. Thus, $s_3 =$ P.

252   The entity "proposition" is considered below.

How then is the composition of these two semantic values to be understood more precisely? The spectrum of views on this issue is highly complex and involves various quite subtle distinctions. Not all need be considered here.

The high degree of complexity of the different positions currently held results essentially from the fact that Frege in certain respects holds both of the two positions just described when he introduces the terminology—much of which is considered above—that has by now become classic. As indicated above, Frege contends that not only every non-syncategorematic term but also every (indicative) sentence has not only a sense [*Sinn*] but also a *Bedeutung*. The *sense* of a sentence is the *thought* that it expresses, whereas its *Bedeutung* is "the object" to which it relates, and that "object" must be either the True or the False. What Frege terms "thought" is a purely intensional entity (an entity restricted to the dimension of pure meaning—whatever that may be), devoid of any "reality"; the thought is therefore generally identified with a particular conception of a "proposition."

What is decisive lies in the question, "Why is the thought not enough for us?" (1892a: 33/157). Frege responds to this question with a course of argumentation that at first appears somewhat confused in that it appears to provide two different answers. On the one hand, he writes the following:

The thought loses value for us as soon as we recognize that the *Bedeutung* of one of its parts is missing. We are therefore justified in not being satisfied with the sense of a sentence, and in inquiring also as to its *Bedeutung*.

Frege's second answer is this: "Because, and to the extent that, we are concerned with its truth-value."

Despite this apparent confusion, careful consideration reveals Frege's two answers to be different formulations of a single complex thought that has three ingredients: value, *Bedeutung*, and truth value. Frege's course of argumentation, examined carefully, analyzes what thoughts involve; the analysis leads from value to *Bedeutung* and from the latter to truth value. The argument is as follows: to ascribe to a thought a value is to ascribe to it a *Bedeutung*; to ascribe to a thought a *Bedeutung* is to ascribe to it an object that is a truth-value, that is, either "the True" or "the False." Frege can then introduce the summary formulation, "Judgments can be regarded as advances from a thought to a truth value" (35/159). This argument is quite peculiar. *Bedeutung* is understood in the sense of "object," the latter as truth-value, which for its part is taken to be an "object." Why? Why must the truth value be understood as an "object"? | 253

To show that Frege's position is ultimately incoherent, it suffices to introduce a second famous thesis that he formulates explicitly. In his essay "Thought," he asks, "What is a fact?" And he answers: "A fact is a thought that is true" (1918: 74/342). This thesis, according to which it is the *sense* of the sentence that is either true or false, does not square with the thesis that the *Bedeutung* of a sentence is its truth-value. If a true thought is a fact, then that means that the truth-value of the thought is the display of an entity in the world, namely, the fact (and not the "object," "the True"). In this case, the truth value of the sentence, "Boston is in Massachusetts," is not "the True," but instead the fact that Boston is in Massachusetts.

It is extremely important to note that Frege arrives at his peculiar first thesis by *explicitly* relying on the *compositional* character of subject-predicate sentences: "We have seen that the *Bedeutung* of a sentence may always be sought, whenever the *Bedeutung* of its components is involved. . ." (1892a: 33/157). This "whenever" clause presumably means: whenever what is involved are the components of a subject-predicate sentence and their compositionality. The total divergence of positions noted above with respect to the determination of what is to be understood by the "semantic value" of the (indicative) sentence clearly stems from Frege: he characterizes the semantic value of the sentence as its sense, more specifically as the thought (thus: proposition) that it expresses, *and* as its *Bedeutung*, which he sometimes understands as its truth-value, but at other times as the fact it displays. It appears to be obvious that Frege does not succeed in formulating a coherent conception.

[4] An example of a purely *extensionalist* definition of semantically compositional structures is introduced above. The definition $\langle \mathfrak{A} = A, \mathfrak{a} \rangle$, as it is there given, is only a *partial* definition of such structures, because it does not include the central semantic unit: the sentence. Moreover, it is far too global and undifferentiated. The domain A is said to be a "domain of objects" or "universe," but how is that to be understood more precisely? Because a language of first-order predicate logic is presupposed, the domain of objects consists literally of the set of all *objects*; this is clearly expressed in the definition of the designation function for the set of constants: (b_3) $\mathfrak{a}|_{Co_s}: Co_s \to A$. The question arises | 254

of how to interpret more precisely such a conception or definition of semantic structures. In formal semantics, the usual practice is to understand all "entities" like semantic structures as "abstract" entities; the relation to ontology is generally left wholly unclear. Nevertheless, the usual interpretation involves a clearly ontological import in that the domain is characterized as "domain of objects" or as "universe." If such a conception is to be of interest philosophically—as it is indeed presented as being—it must be taken to presuppose or entail a substance ontology. "Objects" are then nothing other than substances in the sense of substrata: i.e., each is an X that has properties and stands in relations to other such objects. The following section subjects this ontology to a thorough critical analysis.

3.2.2.3.2 Critique of Compositional Semantics and Ontology: The Unacceptability of Substance as Fundamental Ontological Category

It is shown above that compositional semantics finds its articulation in first-order predicate logic and language and thus presupposes or entails a substance ontology. What is so widely termed "object" is thus only another version of the traditional category *substance*. This category is characterized most clearly as follows: it designates an X of which properties and relations can be predicated (and of which states of affairs can be maintained). But how is the X to be understood? The following account shows that this X, as a putative entity, is unintelligible and that it must therefore be rejected.[16]

3.2.2.3.2.1 Substance Ontology and Its Alternatives in Contemporary Philosophy

[1] In contemporary philosophy, the concept *substance* is not understood and determined in a single manner. On the whole, three chief positions can be distinguished. According to those who take the first of these positions, substances are substrata within which properties (and relations) inhere. Thus, the substratum is supposed to be an entity distinct from another entity, the attribute (property or relation), with which however it is correlated. The concrete particular, the individual, is taken to be constituted by those two entities. This substratum has rightly been called a "bare particular," because it is devoid of all determinacy. There are numerous conceptual problems with this concept of the substratum.[17] The root problem of this conception, along with the correlated concept *universals*—which are what bare particulars are supposed to instantiate—is presented below.

Advocates of the second position reject the idea of the bare particular, but not the idea of the "subject." The key concept introduced by these philosophers is *kind*, which is intended to explain what they understand by substances as "concrete particulars." Michael Loux, for example, asserts the following:

> What a concrete particular is, on this view, is simply an instance of its proper kind; and Aristotelians argue that to be an instance of a kind is simply to exhibit the form of being that is the kind. Since that form of being is irreducibly unified, the things that exhibit it are

---

[16] The following account is largely adapted from Puntel (2002).

[17] See Denkel (1996) for a good account of these problems.

themselves irreducibly unified entities, things that cannot be construed as constructions out of more basic entities. (1998: 121)

These authors, however, do not entirely reject the idea of the (ontological) subject. They claim that the substances or concrete particulars themselves have the status of subjects with respect to all of their attributes, but they hasten to add that one should distinguish between the attributes that are essential to their bearers and those that are merely accidental to them. They understand the subject, as bearer of accidental attributes, as a subject whose essence or core does not of necessity include those attributes. The essential attributes, however, *are* necessarily borne by the subject (the substance). These authors understand this to mean that the substance or the concrete, individual thing has also the status of being a subject (a bearer) with respect to its *kind*. Loux explains this view as follows:

> Socrates is also the subject for the kind *human being*. Socrates and not some constituent in him is the thing that is human; but the kind *human being* is what marks out Socrates as *what* he is, so in this case our subject is not something with an identity independent of the universal for which it is subject. Take the *man* away from Socrates and there is nothing left that could be a subject for anything. (127)

Proponents of this view make considerable efforts to eliminate the obscurity of the concept *substratum*, but it is at best doubtful that they have succeeded in solving the problem. If the concrete, single thing is simply identified with its own kind, then the concepts *instance, instantiation*, and *exemplification* lose all explanatory significance. And when it is said, "our subject is not something with an identity independent of the universal for which it is subject," then it is difficult to understand what that really means; how can an X be a subject (substratum) for a universal U if its identity is not independent of U? Perhaps one could say that it is a limiting case of the concept *instantiation*. In philosophy, however, limiting cases of this sort are extraordinarily problematic; generally, they indicate the necessity of introducing a different and more adequate conceptual scheme that is at least better able to articulate the questionable intuition.

    Those taking the third position eschew the concept *substratum* (and *subject*), introducing instead the concept *independence* as the criterion that must be satisfied by any substance (see esp. Hoffman and Rosenkrantz 1994, Ch. 4, and Lowe 1998, Ch. 6). Descartes, Spinoza, and many other philosophers hold this position—that substances are independent in the sense of being capable of existing all by themselves, self-sufficiently, by their own powers. Those who defend versions of this position have strongly diverging conceptions of the relevant independence, but the main problem with this view, common to all versions, is quite simple: independence is only a necessary but not a sufficient condition for being a substance. Existing independently from other entities is only an external characteristic; it does not determine the internal structure of the putatively substantial entity.

    [2] Ontological theories that do not accept the traditional idea of substance explain concrete, individual things (and every kind of complex entity) as bundles of some determinate kinds of entities. Such theories are in general called "bundle theories" (for reasons given below, this book always uses the terms "configuration"

256

and "configuration theory" when characterizing the related theory that it presents). There are various divergent bundle theories; the divergences always relate either to one or the other of the two following factors, or to both: first, the kind(s) of entities that are taken to be bundled, and second, the more exact sense that is associated with the term "bundle."

Three main versions of bundle theories are especially worth mentioning. The first version, "trope theory," is a radically revisionist theory that not only rejects the concepts *substratum* and *subject* but also, and above all, calls into question the concept *universals*. Philosophers who hold this theory accept a new entity or category they call—following Donald Williams (1953/1996)—"trope," characterizing it as an abstract particular or as a particularized or concretized property (or relation). According to this view, tropes are the fundamental elements of being and/or of beings, from which all else can be constructed; more exactly, *trope* is the sole fundamental category, so the complete ontology built on tropes is understood as a unicategorial ontology. Concrete particulars or individuals are explained as bundles of tropes; the entity traditionally called "universal" is reinterpreted as a collection of tropes bundled together by the relation of resemblance or similarity (see Campbell 1990, Williams 1953/1966).

This conception is an interesting new development as regards the concept of ontological category, but it faces many deep difficulties that have been pointed out by various authors (see, among others, Daly 1994, Simons 1994). The chief problem is that trope theory has not succeeded in making clear how it can dispense with the relation of instantiation or exemplification (see especially Daly 1994: 250–60), a relation that presupposes the very concept of universals that the trope theory purports to reject. This difficulty is apparent even in the terminology used by trope theorists, when they say, for example, that tropes are "abstract particulars," "particularized properties (and relations)," or even "instances of properties (and relations)" (see, e.g., Campbell 1990).

This difficulty in its turn is rooted in the systematic deficiency of trope theory: although this theory relies on a valuable intuition, it entirely lacks the semantics needed to express this intuition. Trope theory relies, usually implicitly, on a semantics that in one respect is a (negative) function of traditional substance ontology and in another respect gives rise to an alternative ontology, as shown below. Trope theorists simply understand the entity *trope*, semantically, as the referent of expressions like "Napoleon's posture," "the yellow patch in the lower right of the painting," etc. But they do not ask what kind of ontology is presupposed or implied by the other expressions of natural or ordinary language. The old ideas of universals and particulars are not (semantically) eliminated. If a genuine revision of substance ontology is to be accomplished, the linguistic framework presupposed by this ontology must first be scrupulously examined. The structural-systematic ontology is the result of the attempt to develop systematically what this book accepts as the correct and valuable intuition underlying trope theory. But this ontology requires the introduction of an alternative terminology, one deriving from this intuition, but one that is also in conflict with fundamental assumptions and theses of trope theorists.

A second version of bundle theory takes concrete, particular things or particulars to be bundles of qualities tied together by the relation of compresence. A third version considers concrete particulars to be bundles of immanent universals, which, "by contrast with Platonic universals, are as fully present in space and time as their bearers."[18] Universals so understood are, according to this version, the sole fundamental constituents of the universe.

Some authors defend a fourth version of trope theory, wherein tropes are recognized but are not considered to be items or elements bundled together in the sense of the bundle theory; according to these authors, tropes, not universals, are combined with substrata to constitute concrete particulars or individual things. One variant of this fourth version, the "nuclear theory," distinguishes between two stages: the first stage involves a tight bundle of tropes that form the essential kernel or nucleus or essential nature of the concrete particular, and the second stage adds non-essential tropes that may be replaced without the nucleus ceasing to exist. This second stage is dependent on the nucleus as its bearer: "The nucleus is thus itself a tight bundle that serves as the substratum to the looser bundle of accidental tropes, and accounts for their all being together" (Simons 1994: 568). This version does not, therefore, appear entirely to reject substrata; it appears only to dispense with any ultimately absolute, basal substratum.

3.2.2.3.2.2 The Root Problem with All Conceptions of Substance
The root problem with all conceptions of substance concerns the semantic-ontological framework that all unavoidably presuppose and from which they arise. This is the framework that is characteristic of the natural or ordinary languages of the Indo-European tradition. The syntax and semantics of these languages and substance ontology are two sides of the same coin. More precisely, these are two sides of the same coin within the semantic-ontological framework worked out and accepted by philosophers who use natural language and theorize about it. Natural language as such (i.e., as it is used by concrete (non-philosophical) speakers) should be scrupulously distinguished from natural language as thematized and interpreted by theoreticians. Moreover, there are as a matter of fact many quite divergent philosophical understandings of natural language. But in what follows the term "natural language" is understood as it normally is in philosophy—as fundamentally structured by a first-order predicate language.

To bring the root problem clearly into view, two features of the semantic-ontological framework that as a rule underlie philosophical interpretations and theories of natural language must be made explicit. [i] Even if ontological categories other than the category of substance are introduced and accepted (event, process, etc.), this does not overcome or avoid the basic substantialistic framework; on the contrary, any other such categories presuppose the category of substance and the dyadic concepts that more precisely describe and characterize the category of substance (e.g., subject—universals and subject—attributes [whereby both properties and relations qualify as attributes]). This thesis is based on the indisputable fact that the other ontological categories (event, process, etc.) are explained as entities that have properties and stand in relations to other entities. The basic substantialist vocabulary is thereby retained. When philosophers explain such matters precisely, as a rule they rely upon the language of first-order

259

---

[18] O'Leary-Hawthorne and Cover (1998: 205). These authors call their theory "the Bundle Theory of substance." They simply accept the term "substance" as synonymous with "concrete particular" or "individual." This terminology is at best ambiguous, and engenders fundamental misunderstandings.

predicate logic. According to the standard interpretation, this language or logic has precisely the semantic structure that corresponds to the dyadic framework of subject and universals.

[ii] As is shown in Section 3.2.2.3.1, the semantics of first-order predicate language relies fundamentally and entirely on the principle of compositionality, which, as applied to sentences, is as follows:

(CPP$_S$)   The semantic value of a sentence is a function of the semantic values of its subsentential components.

This principle requires at least a relative independence of the semantic values of the subsentential components. Thus, the singular term has its own referent, the denotatum, and the predicate—at least in a realistic semantics—also has its own designatum, the attribute. In analytic philosophy, the referent or denotatum of the singular term (or noun) is generally and without more precise specification called "object." But because this entity is determined by having properties and relations attributed to it, it indubitably plays the role of the old category of substance, the latter understood as being constituted by a subject (substratum) and universals.

The root problem can now be formulated as the problem posed by predication on the level of first-order predicate language. The simplest and most fundamental form of predication of this kind is "Fa": the ascription of the predicate F to the item a (in quantified form, $(\exists x)(Fx)$). An entity (the subject or substratum) to which a corresponds or that is the value of the bound variable x is simply and absolutely presupposed. The vital problem is then this: such a presupposed entity is not intelligible. *Ex hypothesi*, it must be presupposed as that of (or to) which universals or attributes of any and all sorts—usually, properties and relations—can be predicated (or attributed).[19] But then the question arises of what this presupposed entity is. If all the attributes (properties and relations) and whatever else this presupposed entity may be involved in or within are taken away (and one must be able to take them away, because the entity in question must be a distinct entity), nothing determinate remains: the entity itself is intrinsically determined in no way whatsoever—it is completely empty. Such an entity is not intelligible and therefore must be rejected.

Attempts to rescue the idea of such a subject have been unsuccessful; the reason for their failure is that they rely on the assumption that the entity corresponding to a or x is a subject that in some sense is already determined. But what could be the sense? *Ex hypothesi*, this alleged determinacy of the subject must be predicated of this very subject, but then the following is clear: if every putatively remaining determinacy of the subject must (again) be *predicable* of the subject, then a subject—the same subject? another subject?—must (again) be presupposed, because otherwise this remaining predication would not be possible. But then there would be an infinite regress.

---

[19] According to the substance ontology criticized here, other entities are (often) introduced in conjunction with substances or concrete, individual things; of particular importance are facts. Within the framework of this ontology, facts are the *composite entities* that result from predications. Every fact involves a substance and one or more of its determinations. They are modes of determinate substances.

This attempted solution thus misses the central point: the question concerning the conditions under which predication is at all possible. In other words, what should be   261 explained is the ontological constitution of such a subject.

3.2.2.3.2.3 Quine's Procedure for the Elimination of Singular Terms: An Insufficient Means for Accomplishing a Philosophical Revolution

The necessity, indicated in various passages in the preceding sections, of a radical revision of traditional (compositional) semantics and ontology is chiefly based on the insight that the structure of sentences that qualify as philosophically adequate cannot have the subject-predicate form. This insight can also be formulated as follows: the basic principle of semantics is not the principle of compositionality, but instead the context principle. Quine (1981: 69) aptly articulates this basic point: "The primary vehicle of meaning is seen no longer as the word, but as the sentence." Quine also correctly maintains that this change is a revolution comparable to the Copernican revolution in astronomy. The central questions are, how is the change to be understood, and how is the revolution to be accomplished?

[1] This section analyzes an extraordinarily interesting example of one of the ways a great philosopher, Quine himself, attempts to extract important philosophical consequences from the revolutionary idea of the semantic primacy of the sentence. The example is his procedure, already a philosophical classic, for the elimination of singular terms. If, as is argued above, the subject-predicate structure of the sentence should be radically revised, then it should be revised in such a way that the components of sentences with this structure—subject (singular term) and predicate—are eliminated. Quine's procedure for eliminating the singular term is on the one hand extraordinarily rich in consequences, but on the other in a fundamental respect insufficient. Analysis of the procedure can yield much that is helpful with respect to the development of a coherent semantics and ontology based on the context principle, thus on the primacy of the sentence.

Quine's technical procedure consists essentially of maneuvering singular terms into a standard position "$= a$" that, taken as a whole, is a predicate or general term (general terms are not affected by the problems raised by singular terms). Quine (1960: 179) explains as follows:

> The equation '$x = a$' is reparsed in effect as a predication '$x = a$' where '$= a$' is the verb, the 'F' of 'Fx'. Or look at it as follows. What was in words 'x is Socrates' and in symbols '$x =$ Socrates' is now in words still 'x is Socrates,' but the 'is' ceases to be treated as a separate relative term '$=$.' The 'is' is now treated as a copula which, as in 'is mortal' and 'is a man',   262 serves merely to give a general term the form of a verb and so suit it to predicative position. 'Socrates' becomes a general term that is true of just one object, but general in being treated henceforward as grammatically admissible in predicative position and not in positions suitable for variables. It comes to play the role of the 'F' of 'Fa' and ceases to play that of the 'a.'

Quine develops his technical procedure for the elimination of singular terms in the context of the regimentation of scientific language by means of the standard first-order predicate logic that he takes to be "the adopted form, for better or worse, of scientific theory" (Quine 1985: 170). And he holds that predicate logic gains the expressive strength required for scientific purposes only through reification. He illustrates what this amounts to by means of an example. He considers the sentences that he calls "observation sentences"—sentences "that are directly and firmly associated with our stimulations" (1992a: 2ff)—and that elicit from us agreement or rejection. These are sentences such as "It's raining," "It's getting dark," "That's a rabbit," etc. What is the semantic status of such sentences? Quine's answer to this question is richly consequential, because it immediately brings to light some fundamental theses of his philosophy and philosophical questions of fundamental significance. He responds to the question by reformulating a specific observation sentence (29):

A white cat is facing a dog and bristling.

Quine distinguishes two utterly different readings of this sentence; they are articulated as transformative rephrasings of it. The first "masks" the referential function of the sentence:

It's catting whitely, bristlingly, and dogwardly.

According to Quine, the sentence—reformulated in this manner—has a *non-referential* status; we utter sentences of this form "without meaning to refer to any object" (1985: 169). This putatively non-referential status can be interpreted as follows: the semantic status of the sentence is independent of ontology. Given the presupposition that the world consists of objects that have properties and stand in relations to one another, it is clear that the sentence, in the reformulation, indeed has a non-referential character.

263

On the basis of the assumption—one that he takes to be non-problematic because obvious—that the world is structured in this manner, Quine presents a second trans-forming rephrasing of the example sentence, one aiming to articulate the sentence as *referential*. He achieves this by subjecting the sentence that serves as the original example to a *regimentation*. This rephrasing yields the following:

$\exists x (x$ is a cat and x is white and x is bristling and x is dogward$)$.

[2] This procedure of Quine's is an immensely revealing and highly inspiring example of a semantic correction of ordinary language that is necessary from the philosophical perspective developed and relied upon in this book, although Quine's procedure, as he interprets and applies it, remains incomplete, inconsequent, and thereby ultimately superficial. This is now to be shown in detail.

[i] Quine's procedure must be interpreted as an attempt to overcome the subject-predicate sentential form. Yet this attempt remains only partial and therefore incomplete. The reason for this lies in the fact that Quine introduces the procedure only to solve a single problem, that of the non-referentiality of some singular terms. This leads him to understand his procedure as a purely logical-semantical device.

Within the framework of classical first-order predicate logic, singular terms have real referents (i.e., they denote existing objects). There are, however, singular terms (e.g., "Pegasus") that do not denote existing objects. To solve the problem introduced by such terms, Quine "translates" or "transforms" the sentences in which singular terms appear into sentences preceded by a quantifier that binds the individual variables.

What results are sentences in which the singular terms have been transformed into general terms, as shown above. In these sentences, the (earlier) singular terms do not appear *as such*, but the quantified sentences within which the general terms appear as predicates do articulate what would have been the function of the singular terms had they not been eliminated. The problem of the non-referentiality of some singular terms thus solves itself in the following manner: sentences can be true or false; the quantified sentences into which the sentences containing non-referential singular terms have been translated are simply qualified as false. Consider the example, "Pegasus flies," formalized as *Fa*. The corresponding quantified sentence, in which the (earlier) singular term ("Pegasus") appears as a general term or as a verb and thus as a predicate, has the form: $(\exists x)(x = a \land Fx)$. The sentence is false, but the problem of the non-referentiality of "Pegasus" has been solved. An example of such a sentence that is true is, "Socrates is a philosopher."

264

What are the ontological consequences of the elimination of singular terms, as Quine understands it? The answer is fully clear: there are none. Quine himself insists upon this, as clearly as could be wished for:

> Mistaken critical remarks keep one reminded that there are those who fancy the mathematical phrase "values of the variables" to mean "singular terms substitutable for the variables." It is rather the object designated by such a term that counts as a value of the variable; and the objects stay on as values of the variables though the singular terms be swept away. (1960: 192n1)

This passage makes it fully clear that the elimination of singular terms has, for Quine, *no genuinely ontological* consequences. The world of "objects" remains just as it had been conceived of *independently of and prior to* the elimination of singular terms. Quine's technique of elimination is simply a relocation of the semantic function of language from the level of singular terms to the level of variables. The objects remain the same, but are simply articulated as values (of bound variables) rather than as denotata (of singular terms). For Quine, then, semantic relations to objects are accomplished only when the latter appear as values of bound variables.

If one has philosophical reasons for developing a wholly different ontology, then it is immediately obvious that Quine's procedure, understood as Quine understands it, is in no way helpful. But the procedure can in principle be understood and developed differently, such that the ontological consequences appear as absolutely extensive with respect to the development of a different semantics and a different ontology. This chapter accomplishes this task.

[ii] This possibility for further development can be actualized by means of a precise analysis of Quine's treatment of the example introduced above, "A white cat is facing a dog and bristling." Quine's treatment contains a subtle incoherence and relies on an unfounded but comprehensive thesis.

[a] As shown above, Quine transforms singular terms into general terms. The singular term "a" (or "Pegasus") then appears *only in the position* " = a" (or " = Pegasus"). These terms are then general terms or predicates. They are the "*P*" of "*Px*." How are they to be understood more precisely? Quine reparses " = Pegasus" as the predicate "is-Pegasus" or "pegasizes" (see 1953/1980: 8). He further notes, "The noun 'Pegasus' could then be treated as derivative, and identified after all with a description: 'the thing

265

that is-Pegasus' or 'the thing that pegasizes.'" With explicit reference to Russell,[20] but from a different starting point (1960: 184n4), Quine also eliminates the definite description "the thing that pegasizes" in that he reduces the name "Pegasus" to a general term or to a predicate that cannot be said of any object in such a manner that a true sentence results. In the case of the true sentence "Socrates is a philosopher," the name "Socrates" is reduced to a general term or to a predicate that can be said of only a single object, such that the resulting sentence is true.

This position gives rise to many questions, including the following: how can the presupposition on which this position is based be considered to be satisfied? The presupposition is this: to attach a determinate predicate (e.g., "is Socrates" or "Socrateses") to a single "thing" (i.e., to the value of the bound variable x), this "thing," this value of x, must be *identifiable*. But how can this happen, how can it be achieved, if there is absolutely nothing determinate with respect to the value of x? The talk of a "thing" that is "identifiable" only in being "determined" as the ontological value of the bound variable x is talk that is empty: what could "determined" mean here? If such an identification can enter at all into the framework of Quine's assertion, then it would be necessary to explain the extensive ontological space within which the (*ontological!*) value of the bound variable x is situated. Such an ontological space is, after all, *explicitly* presupposed and thus assumed in that a quantifier is used. It does not suffice, for the identification of the value of x, simply to indicate this ontological space or framework. This question is not, however, further considered here.

266

A subtle incoherence or inconsequence in Quine's treatment of the example sentence introduced above is now to be shown. When Quine develops his procedure for the reduction of singular terms or names, why does he exclude the possibility that such terms be reduced to or translated into sentences? He himself shows that this is syntactically possible, by first reformulating the example sentence as "It's catting whitely, bristlingly, and dogwardly." Why should the singular term "(a) cat" not be accepted as an abbreviation of the sentence "It's catting" and the name "Pegasus" as an abbreviation of "It's pegasizing" and "Socrates" of "It's socratesing"? According to Quine, "It's catting" is—and "It's socratesing" would be—without ontological import. But how can he ground this contention? Does the sentence "It's raining," when true, have no ontological import? Yet this sentence is of the same sort as "It's catting" and "It's socratesing." Why then a reduction of singular terms or names *only* to *predicates*? This thesis appears to be inconsequent.

[b] Yet there is, for Quine, an omnipresent albeit generally implicit justification for his *restrictive* interpretation and treatment of the procedure of eliminating singular terms; it involves the exclusion of the possibility of reducing singular terms or names

---

[20] To avoid the ambiguities of the surface structure of ordinary language, Russell develops his famous theory of definite descriptions (1905/1956). He assumes that names are abbreviated definite descriptions. A definite description, such as "the King of France," can be formalized as follows: $\iota x(Fx)$ ("the x such that x is F"). This expression can be used in place of a name, as in the formula $G(\iota x\colon Fx)$ ("the x such that x is F is G," e.g., "The King of France is old"). But Russell shows that although the definite description is (often) treated and used as a name, it actually is not a genuine name. Instead, the definite description $G(\iota x\colon Fx)$ must be understood as an abbreviation of the formula $\exists y(\forall x(Fx \leftrightarrow x = y) \wedge Gy)$, which is precisely formulated, informally: "There is one and only one thing such that it is the King of France and it is (also) old".

to sentences of the form just introduced. The justification is Quine's grand assumption or thesis briefly introduced above in conjunction with his second reformulation of the example sentence. Quine's striking passage reads as follows:

> I am concerned...with scientific theory and observation as going concerns, and speculating on the function of reference in the linking of whole observation sentences with whole theoretical sentences. *I mean predicate logic* not as the initial or inevitable pattern of human thought, but *as the adopted form, for better or worse, of scientific theory.* (1985: 169–70; emphasis added)

This thesis is one of the central pillars of Quine's philosophy. Because of its comprehensive character, it cannot be adequately criticized in only a few sentences. Here, only two concluding remarks need be added. First, the systematic conception defended in this book can in one respect be read and interpreted as a constant and likewise comprehensive rejection of Quine's omnipresent thesis. Particularly relevant is Section 4.2.1.1, which shows that this thesis cannot be brought into harmony with Quine's other theses concerning ontology. Second, the consideration of Quine given above, subtle though its details may be, should sufficiently clarify the following central point: a single thesis (e.g., the thesis concerning the eliminability of singular terms) can be subjected to serious philosophical analysis and evaluation only if it is precisely situated within the comprehensive systematic framework within which it appears. If any single thesis is considered in isolation, there are not sufficient grounds either for its acceptance or for its rejection.

3.2.2.4 Basic Features of a Semantics Based on a Strong Version of the Context Principle

If philosophical languages cannot rely on compositional semantics, precisely because of the ontological implications of such semantics (i.e., substance ontology), then a different semantics must be developed, one that excludes the elements of compositional semantics criticized above. The alternative semantics must radically exclude any and every analogue of *substance*. This semantics must therefore completely avoid subject-predicate sentences, because these are precisely the sentences that presuppose or imply, and express, some form of substance ontology. This basic insight can be developed by means of the introduction of and adherence to a different basic semantic principle. There is such a principle; it is termed the context principle.

3.2.2.4.1 *A Strong Version of the Semantic Context Principle*

The term "context principle" (CTP) appears to have been introduced by Michael Dummett to designate a principle that, in *The Foundations of Arithmetic*, Frege formulates as follows:

(CTP_F)   "Only in the context of a sentence [*Satz*] do words mean something."[21]

---

[21] 1884: §62; translation altered. Other formulations are found in the same book (see §60, §106).

There is no consensus among Frege scholars concerning either the exact meaning of this principle in the *Foundations* or its fate in Frege's later works. More recently, the CTP—often under the designation "semantic primacy of the sentence"—has been advocated by such philosophers as Quine and Davidson, who understand it to accord to the question of truth semantic priority over questions of meaning and reference. So understood, the CTP is opposed to the "analytic conception of language" (Romanos 1983: 165), whose starting point is considered to be Tarski's truth theory.

Authors who accept the CTP generally defend the CPP as well; the question whether the two principles are compatible is rarely raised. Some authors are led by their acceptance of both principles to theses that without the presupposition of their compatibility would be neither compelling nor even intelligible; this holds in Quine's case for the thesis of the inscrutability of reference (e.g., 1969: 2, 20, 30, 34, 45, 103), and in Davidson's for the thesis concerning the dilemma of reference, from which emerges the additional thesis proclaimed by the slogan, "reality without reference" (1977). The following account largely dispenses with closer consideration of the positions of Frege, Quine, Davidson, and other potentially relevant authors; it seeks instead *systematically* to develop a *strong version* of the CTP and to articulate its semantic and ontological implications.[22]

3.2.2.4.1.1 Incompatibility of the Context Principle and the Principle of Compositionality

[1] The envisaged strong version of the CTP can best be developed and grounded on the basis of a comparison between the CTP and the CPP. The comparison grounds the thesis of the radical incompatibility of the two principles.

To proceed to the fullest extent possible without problematic presuppositions, the formulations rely on the already-introduced neutral term "semantic value." The two principles are then formulated as follows:

(CTP)   Only within the context of a sentence do linguistic terms have semantic values.

(CPP)   The semantic value of a linguistic expression is functionally dependent only on the semantic values of its logically relevant components.

(CPP) understands the CPP *functionally*.[23] A well-known and illuminating example of a functionally understood compositionality is *truth-functional* compositionality. That

---

[22] For a thorough presentation and grounding of the CTP, see Puntel (1990, esp. Section 3.3).

[23] The term "functionally" (and "functionality," etc.) is not used in the same way by all advocates of the CPP, and it is generally used quite vaguely. Dummett attempts to reduce Frege's anything but unequivocal claims about the "decomposition" (and thus the "composition(ality)") of the thought—and, correspondingly, of the sentence expressing the thought—to *two models*: the whole-part model and the function-value model. Dummett himself holds these models to be mutually incompatible (see Dummett 1984: 220). If one accepts the whole-parts model as fundamental, then whether—or in what sense—it makes sense to speak of a semantic primacy of the sentence depends on how the whole-parts relation is determined. Be that as it may, the functional understanding of the CPP is the more widely held. If, however, "function(ally) dependent" is understood not vaguely (as it usually is), but in the strict mathematical sense (as argument-value

the so-called connectives are determined as truth-functional operators means that they establish the truth-values of complex sentences purely functionally: if the truth-values of the individual sentences (i.e., the component sentences) are given, then the truth-value of the complex expression can be determined by means of a mechanical procedure, because it is nothing other than the value of a function whose arguments are the truth-values of the component sentences. Of decisive importance here is that the truth-values of the components must be given *in advance*.

This truth functional compositionality is perfectly reconcilable with the CTP, because it has as components the truth values of sentences. But the compositionality that has as components the subsentential parts of the sentence is not, because in this case the CPP presupposes that the semantic values of the components can or must be determined *before* (and indeed *so that*) the semantic value of the sentence can be or is determined. But this, as shown, is not reconcilable with the CTP as applied to these sentences.

One could attempt to reconcile the two principles by postulating an equiprimordiality or reciprocal interdependence between the semantic values s_1 und s_2—the two subsentential components of the sentence—on the one hand, and the semantic value s_3 of the sentence on the other (see Section 3.2.2.3.1), declaring the semantic values of the components to be "given" in the sense of their being *predicatively* combined with one another. The referent of the singular term would then be "given" only by means of the ascription to it of an attribute (a property or relation). This would succeed in making the semantic value of the sentence be compositionally structured *and* in granting to the components semantic values only within the context of the sentence. Considerations of this sort seem to lie behind the widespread notion that the CPP and the CTP are not only not incompatible but are indeed inseparably interconnected.

This attempted solution nevertheless fails, for two chief reasons. [i] This conception fails to preserve the central idea of the CPP—*functional compositionality*; for this idea, some other as yet unidentified conception would have to be substituted. [ii] The conception leaves fully unclear how *predication* is to be understood or how it could take place, because predication, determined as ascription of an attribute to an object, presupposes that the object can be *identified in advance* (and thus that it be accorded semantic priority); otherwise, how predication could occur is unintelligible. Every other explanation of predication appears to be beyond the framework of the possibilities available given the assumption of the CPP.

270

3.2.2.4.1.2 Basic Features of and Requirements for a Strong Version of the CTP

[1] The designation "strong version" can be initially clarified, briefly, both negatively and positively (the latter here only programmatically): *negatively*, it differs from any version of the CTP that could be compatible with the CPP; positively, it takes radically seriously the semantic primacy of the sentence. But what does that involve?

The strong version develops from the thesis that the semantic value of the sentence determines the semantic values of the sentence's components: the latter must be explained on the basis of the sentence as a whole (i.e., on the basis of its semantic value).

---

relation relative to a function), it becomes clear that the CPP presupposes that the semantic values of the components of a given sentence can and must be identified *before* (and indeed *so that*) the semantic value of the sentence is (or can be) determined.

If—as above—one rejects as insufficient the contention that the semantic values of the sentence's components and of the sentence itself are equiprimordial or interdependent, it might appear that the yet more radical project of establishing a strong version of the CTP is senseless and indeed hopeless. The following paragraphs demonstrate that this project is not merely wishful thinking, that instead it points toward a real possibility and is thus an important task for philosophy.

With respect to the strong version of the CTP, the most important thing to be clarified is what it means to say that the components of the sentence (in the case of an atomic sentence, the singular term and the predicate) are explained on the basis of the sentence as a whole. It is here assumed without argument (but see Puntel 1990: 3,4) that a given sentence can be fully explained only if it *has a semantic information-content* in the sense—at this point, no matter how further determined—of expressing a proposition. The requisite clarification must therefore be this: both the singular term and the predicate must be explained strictly *propositionally*: each must be interpreted as a proposition (or configuration of propositions). Because, however, in syntactic-grammatical terms neither singular terms nor predicates are kinds of expressions of which it can sensibly be said that they express propositions, they must be eliminated in the sense of being explained to be *abbreviations of sentences*. This requirement results from the rejection of the only alternative (i.e., that—as the CPP insists—these components have fully determined semantic values on their own, independently of their occurrences in sentences).

The preceding paragraph presupposes that the language in question uses sentences with the subject-predicate form. This holds above all for natural (Indo-European) languages. It has, however, already been shown that these languages are ultimately inadequate for theoretical and thus for philosophical purposes. Of course, philosophers cannot be prohibited from using natural languages. But that is in no way necessary, because what is solely and exclusively decisive is not the syntactic form of sentences (particularly, whether their syntactic form is subject-predicate), but the semantic question: how are these sentences to be interpreted *with respect to philosophy and for philosophical purposes*? The structural-systematic thesis is that, when such subject-predicate sentences are used, philosophy should understand them as *convenient abbreviations* of great numbers of sentences that do *not* have subject-predicate structures.

Such sentences, within which there are neither singular terms nor predicates in the proper sense (i.e., which include such items only as abbreviations), are here termed *primary sentences*. An example from ordinary language is "It's raining." The general, semi-formalized form of such sentences is "It's φ-ing." "φ" is then not a predicate, but instead, so to speak, what is left of the predicate from a subject-predicate sentence of the form *Fa* when the subject is eliminated. In "It's φ-ing," the "old" predicate "F" appears but as transformed (i.e., no longer as a universal in the strict—traditional—sense). It appears no longer as a universal and unchangeably self-identical entity that is applied to subjects or substrata. The *F* instead has the form of a primary sentence: "It's φ-ing." In this form, the dimension of the universal is retained, although it, too, is transformed.

Proceeding further on the basis of the fundamental *semantic* assumptions made here: every primary sentence has a determinate *expressum*; that is, it has a determinate informational content, it expresses a determinate proposition. This semantic entity constitutes an absolutely indispensable item within the structural-systematic philosophy; it is termed *primary proposition.*[24] Henceforth, "proposition" is used in this book, whenever it relates to the position defended here, only as a convenient abbreviation for this technical term.

[2] With respect to the broader scope of the structural-systematic philosophy, it is important to note that the problems now being considered are not merely semantic; to the contrary, they also have decisive *ontological* aspects. The question concerning the ontological import of the sentence or of the proposition it expresses is doubtless one of the truly central questions of philosophy, because how this question is answered determines what ontology is presupposed or is explicitly maintained. It is shown above that the ontology presupposed by or following from any semantics developing from the CPP is a substantialist ontology (object/individual = composite of substratum and universals). The question is then, what ontological conception corresponds to the CTP? At the outset, what is clear is only that it cannot be substantialist.

3.2.2.4.1.3 The Problem of Identity Conditions for Primary Propositions (and Primary Facts)

[1] This section treats a problem that has played and continues to play a central role in the discussion about propositions: the problem of the individuation of propositions or of conditions of identity for propositions. Many philosophers reject propositions on the basis of the argument that propositions are pseudo-entities because no identity conditions can be provided for them. These philosophers rely upon the slogan *no entity without identity.*

This much-discussed question cannot be treated here in relation to the already extensive literature concerning it.[25] What follows is only a brief defense of a systematic thesis that emerges within the structural-systematic philosophy presented in this book.

First, careful consideration must be given to what is to be understood by *proposition.* The two major approaches in the literature to the determination of this entity are the

---

[24] What term is used to designate what is expressed by primary sentences is not of fundamental importance. Ultimately important for the conception developed here is the fundamental fact that the *expressum* of the primary sentence is an entity that has a quite peculiar, irreducible character. This becomes clear if one compares the *expressum* of a primary sentence with that of a subject-predicate sentence. In the latter case, there is an entity that is a composite (of "substratum" plus property or relation); in the former case, the relevant entity is, to put it one way, primordial rather than composite. To designate the *expressum* of the indicative sentence, the term normally used is "proposition" (and/or "state of affairs"). This designation is also used in this book, but with the indispensable qualification, "primary."

Although primary propositions are not composite entities they are, in the terminology used in this book, not always simple propositions; instead, primary propositions can be either simple or configured. A configuration (of primary propositions) or a configured primary proposition is understood to be a *complex* primary proposition (i.e., one that is a specific form of combination of simple primary propositions). The concept *simple* is thus used as a correlative to *complex*, not to *composite* (in the sense introduced above).

Also to be emphasized is that primary propositions can be of various sorts (e.g., static, dynamic, etc.) or, to use well-known substantives, they can be events, processes, values, etc.

[25] See the thorough treatment in Puntel (1990: 211–19). Moser (1984) presents an astute critique of Quine's position.

272

273

following: first, the proposition is the *expressum* of the indicative sentence; second, the proposition is the object of so-called propositional or intentional attitudes ("Person S believes that *p*"). All more specific determinations of propositions are specifications of one or the other of these approaches. Different ways of specifying the first approach are the following: the proposition is identified as a truth-bearer, or as a translational constant (i.e., as what remains constant or identical when a specific sentence from one language is translated into a sentence in another language), or as a set of possible worlds, etc.

In accordance with the structural-systematic semantics developed in this book, propositions are understood as the *expressa* of primary indicative sentences (i.e., sentences without subject-predicate structure). The *expressum* of a given sentence is conceived of as the informational content that the sentence presents. Because only primary sentences are recognized here, the propositions they express are determined as *primary* propositions.

The question of how primary propositions are to be individuated is misleading; within the framework of the conception presented here it is pointless, because it pre-supposes, in one way or another, that primary propositions must be something like universals that must be *individuated* by concrete cases (see Puntel 1990: 212f.). Here, however, universals are not recognized. The problem of conditions of identity for propositions is, however, different. This is a real problem, also for the semantic concep-tion presented here.

[2] One must distinguish between *semantic* and *ontological* conditions of identity. The semantic conditions of identity are decisive, because the ontological conditions of identity emerge directly from the semantics defended here. The provision of semantic conditions of identity answers a twofold question. First, on the basis of what crite-rion is it the case that two (or more) primary sentences express the same primary proposition? Second, under which conditions are two (or more) primary propositions identical?

274   [i] The first of these questions has, in the semantics defended here, a brief and conclusive answer: there is no such criterion, because the question is based on a false or groundless presupposition—the assumption that it is or must be the case that two (or more) sentences can express the same primary proposition. The semantics developed in this book excludes any such assumption: the structural-systematic phi-losophy rejects so loose a relation between language and world (universe, being) as that assumed by the question. In opposition, according to this semantics *every single sentence token* expresses *its own unique primary proposition*. A (written or spoken or however presented) sentence token is always something individual in the strong sense of being unique.

To understand this thesis correctly, one must consider three aspects in particular. [a] The first is the clarification of a problem that immediately arises. It appears at first glance extremely implausible that every tokening of a given primary sentence expresses its own primary proposition, distinct from all others. One might perhaps note that the fact that language users understand one another supports the opposed thesis that two or more tokenings of a given sentence express one and the same proposition—how else could understanding be possible? Such an assumption or course of argumentation is not, however, sufficiently deep, as is revealed by Quine's considerations of his thesis of the indeterminacy of translation (see Quine 1970a). Moreover, one can question the argument concerning understanding. Understanding in no way presupposes identical propositions (in the strict sense of "identity").

A different explanation is available: the propositions that are exchanged in the process of communicative understanding by means of grammatically/syntactically identical sentences are not identical in the strict sense; they are instead related by means of (extremely high degrees of) *family resemblance*. The latter concept, which stems from Wittgenstein, can explain such phenomena as understanding and both the tight agreements and the obvious differences that unavoidably accompany translations, because this concept allows many degrees that fall short of strict identity. Even exchangeability *salva veritate* is not excluded by this concept; that would be the highest degree of family resemblance, the degree at which the difference is so minimal that—at least in general, and particularly for theoretical purposes—it can non-problematically be ignored. In the case of propositions that are expressed by two or more tokenings of a given grammatically/syntactically identical sentence, one is presumably dealing with the highest degree of resemblance that is possible in this domain, a resemblance that differs so minimally from identity that it can, at least usually, likewise simply be ignored. Things are different with sentences from the same language that are grammatically/syntactically different, but are equivalent or are formed from synonymous expressions. Here, the degree of family resemblance is lesser, so the difference from identity greater. A yet lower degree of resemblance holds between sentences of different languages that are taken exactly to correspond (e.g., "It's raining" and "*Es regnet*"). The propositions expressed by these sentences are not simply identical, but they are so similar that, within the contexts of translations, they can be viewed as equivalent and as pragmatically identical.[26] The immeasurable domain of degrees, gradations, etc., that the concept of family resemblance contains and opens could be analyzed in great detail. Such analysis is, however, neither possible nor necessary here; for the current question, that of how the thesis that every sentence-tokening expresses its own primary proposition is to be understood precisely, what is said above should suffice.

[b] The second aspect considered here is the question of grounding. Why should one accept so radical a thesis? Within the framework of this book, the answer is obvious: the thesis is a direct consequence of one of this book's basic theses: that there is a maximally internal relation between language and the world (the universe, being)—that the two are inseparably connected, as is thoroughly shown particularly in Chapter 5. This thesis itself follows from the yet more fundamental thesis, introduced in Section 2.2.5, that the world (the universe, being) possesses the immanent characteristic of *universal expressibility*. This thesis, too, is thoroughly systematically presented and grounded in Chapter 5. Here, it need only be mentioned within the context of the issue under consideration. Expressibility, however, has a converse relation to a factor that accomplishes the expressing, and this must be language in a comprehensive sense. From this it follows that everything that occurs within the domain of language (even the

---

[26] It is peculiar that a philosopher such as Quine, who decisively criticizes the argument that propositions must be accepted because they are needed as "vehicles of translation" (which presumably means: because they are "identical"), does not arrive at the thought that the purposes of translation do not require unconditionally identical propositions—that instead propositions with high degrees of family resemblance thoroughly suffice.

276   slightest alteration) corresponds immediately and directly to something in the domain of the world (of the universe, of being)—that it articulates or indeed expresses that to which it corresponds.

[c] The third aspect is a small but important specification. There is talk above of "grammatically/syntactically identical sentences." This is the normal case within the framework of a clarified and perfectly developed philosophical language. The philosophical language postulated or assumed here contains no sentences of the subject-predicate form, but only *primary* sentences. But the language that is normally used even in philosophical presentations (e.g., English, as in this book) does contain subject-predicate sentences. If the components of such sentences (and thus the sentences themselves) are, as they must be, eliminated, this does not require the philosopher to cease using English. Here, the difference between the grammatical-syntactical level and the semantic level comes into play. The elimination of subject-predicate sentences occurs *exclusively* on the *semantic* level. Sentences of ordinary language, which as a rule are subject-predicate sentences, must be understood as *abbreviations* of specific collections of primary sentences. With respect to the first question concerning conditions of identity, this has the consequence that the question can unobjectionably be clarified only following the development of the quite complicated procedure for the elimination (i.e., the transformation) of subject-predicate sentences.

[ii] The second question concerning conditions of identity that requires clarification is, under what conditions are two (or more) primary propositions identical? Unlike the first question, this one is genuine and thus one that requires a positive answer.

To clarify this aspect of semantic conditions of identity, one must distinguish between simple and complex primary propositions. Concerning simple primary propositions, the essentials have already been said. What remains is the treatment of complex primary propositions, or configurations of simple primary propositions. This treatment will also clarify the question of *ontological* conditions of identity because a primary proposition, if true, is identical with a fact in the world (see Section 3.3). Because this is so, the following account does not distinguish the question of semantic conditions of identify from that of ontological conditions of identity; for the most part, talk is of "semantic-ontological conditions of identity."

As far as these conditions of identity are concerned, one could refer to Leibniz's
277   famous principle of the identity of indiscernibles or the principle of the indiscernibility of identicals.[27] To be sure, these principles would have to be adapted to the semantics defended in this book, because Leibniz's principles are based on a semantics that recognizes sentences of the subject-predicate form, with the consequence that quantified individual variables and predicates, and/or their semantic values—objects and attributes—are accepted. Such a semantics is here rejected, because here only primary indicative sentences and primary propositions are accepted. The task of adapting Leibniz's principles is not undertaken here. Instead, a different and direct semantic path is followed.

The semantic-ontological conditions of identity are conjuncts of material (contentual) and formal conditions.

Two (or more) configurations of propositions/facts are identical if and only if

---

[27] Formalized, the two principles are the following: $(\forall x)(\forall y)[\forall F(Fx \leftrightarrow Fy) \to x = y]$ and $(\forall x)(\forall y)[x = y \to (\forall F)(Fx \leftrightarrow Fy)]$.

1. each of the two configurations contains all and only the individual simple primary propositions (thus primary facts) contained in the other configuration, and

2. in both configurations, the modes of configuration of the simple primary propositions (thus facts) are precisely the same in all respects (thus, when the two configurations include within themselves precisely the same series of simple primary propositions/facts, and the same forms of their combination).

3.2.2.4.2 The Concept of Contextual Semantic Structure: Primary Propositions as Semantic Primary Structures

This section presents more precisely and in part formally the semantic conception based on the strong version of the context principle.

A given primary sentence expresses a primary proposition or a primary F-ing. Because primary sentences have no subjects (no singular terms) and no predicates (in the proper sense), they are the only relevant linguistic signs that can constitute the set of symbols to be used in the definition of contextual-semantic structures. The general definition can be presented as follows:

A semantic structure for the set of symbols S of primary sentences is a pair:

$\mathfrak{B} = \langle B, \mathfrak{b} \rangle$, such that:

(1) For every constant c in S, $\mathfrak{b}(c)$ is an element in B.

(2) B is the universe or domain of $\mathfrak{B}$. The meaning of the central concept of the universe or domain B is considered in detail below.   278

If one adds *sentence variables* to the set of symbols S, thus if the entire vocabulary of the language in question consists of S $\cup$ *Var*, then a function must be introduced that assigns values from the set B to the variables (i.e., μ: *Var* $\rightarrow$ B). The resulting *complete* semantic structure is thus a pair $\mathfrak{B}^* = \langle B, \mathfrak{b}^* \rangle$, such that

(1) $\mathfrak{b}^*$ is the union of the two functions $\mathfrak{b}$ and μ, thus $\mathfrak{b}^* = \mathfrak{b} \cup \mu$, where $\mathfrak{b}$ is the function of expressing in the sense of (1) of the preceding definition (and indeed for the structure $\langle B, \mathfrak{b}^* \rangle$), and μ is a function that assigns values from B to the variables.

A more precise version of (1) is (1*):

(1*) $\mathfrak{b}^*$ is a function such that

(a) $D_1(\mathfrak{b}^*) = S \cup Var$;

(b) $\langle B, \mathfrak{b}^*|_S \rangle$ is a normal or constant semantic structure;

(c) $\mu := \mathfrak{b}^*|_{Var}$ is a function that assigns values from B to the variables.

(2) B is a non-empty set, the universe or domain, as in the previous definition.

What, precisely, is this universe or domain B, with respect to its semantic structures, within the framework of the semantics developed here? B cannot simply be the *real* universe in a naive or immediate sense; instead, B is the universe as the totality of *primary propositions*. These are entities whose status remains un- or underdetermined. As what follows in this chapter (see Section 3.3) demonstrates thoroughly, the *truth-concept*—better, the *truth-operator*—accomplishes the transition from the semantic structures *qua* un- or underdetermined propositions to propositions that are fully determined. Fully determined primary propositions (semantic structures) are then *ontological structures* designated as *primary facts*.

3.2.3   *Ontological Structures*

3.2.3.1   Definition of Ontological Primary Structures (Primary Facts)

[1] In connection with the preceding definitions, the following provides the simple, universal definition of the ontological primary structure (whereby the boldface letters designate the ontological dimension):

279

Def. A *contextual-ontological primary structure* $\mathfrak{C}$ is a triple $\langle \mathbf{F}, (\mathbf{R}_i^{\mathfrak{C}})_{i \in I}, (\mathbf{G}_j^{\mathfrak{C}})_{j \in J} \rangle$, such that

  i. $\mathbf{F}$ is a non-empty set of primary facts;

  ii. $(\mathbf{R}_i^{\mathfrak{C}})_{i \in I}$ is a (possibly empty) set of ontologically interpreted relations in $\mathbf{F}$;

  iii. $(\mathbf{G}_j^{\mathfrak{C}})_{j \in J}$ is a family of ontologically interpreted finite-ary functions (operations) on $\mathbf{F}$.

One can also provide a fourth (possibly empty) component, a family $(\mathbf{c}^{\mathfrak{C}})_{c \in F}$ of constants (i.e., elements of $\mathbf{F}$). In this case the structure is understood as a quadruple. Allowable simplifications of $(\mathbf{R}_i^{\mathfrak{C}})_{i \in I}$ include both $\mathbf{R}_1^{\mathfrak{C}}, \ldots, \mathbf{R}_n^{\mathfrak{C}}$ and simply $\mathbf{R}$; comparable simplifications are allowable for the other components.

[2] How is this definition to be understood more precisely? "$\mathbf{F}$" designates the totality of primary facts. All the "entities" that populate the world are configurations of primary facts. To be emphasized once again is that *primary* facts are not identical with *simple* facts, so the concept *complex fact* is not opposed to the concept *primary fact*. Primary facts can be simple or complex. Configurations of (simple or complex) primary facts are for their part complex primary facts. As an illustration, consider the simple case of a *conjunction* of primary sentences (or propositions or facts): the conjunction itself is primary, but also complex.

There are thus simple and complex primary facts; configurations of primary facts are themselves primary facts. If one uses the term/concept *ontological category*, then "primary fact" is the only ontological category. Simple primary facts are to be identified and thus understood on the basis of the concepts of the simple primary sentence and the simple primary proposition.

How are complex primary facts to be understood? Initially and in principle, the answer is analogous to the one just given with respect to simple primary facts: complex primary facts are to be identified and understood on the basis of the concepts of the complex primary sentence and the complex primary proposition. But here, there are additional and more specific issues and factors to consider. These are best revealed not by further reliance on the concept *ontological category*, but by the introduction and specification of the concept *ontological structure*, which explains and makes precise the concept of the ontological category.

That complex primary facts are termed ontological primary structures should—given the assumptions made here—not be surprising.

3.2.3.2 Simple Primary Facts as Simple Ontological Primary Structures   280

Given the concept *ontological primary structure*, how are simple (non-complex) primary facts to be understood precisely? At first glance, it could appear that they could not themselves be ontological primary structures in the proper sense and that they would have to be some sort of non-structured *urelements*.[28] If they were, would it not then follow that the *only* ontological category, the primary fact, could not simply be understood as an ontological primary structure? The reason for this would be that there would be a distinction between simple (non-complex) and complex primary facts, and it would then appear that only complex primary facts could be ontological primary structures.

Regardless of the initial plausibility of the just-sketched course of argumentation, it is in no way compelling because its basic premise is false. It can be shown that even non-complex primary facts are to be understood as ontological primary structures in the proper sense.

The basic idea can be formulated as follows: on the basis of what is indicated above, simple primary facts are not to be thought of as isolated or—in the absolutely literal and negative sense—atomistic entities, not as ones that are "windowlessly" enclosed within themselves; instead, each is structured in the sense that each is *determined* by a network of relations and/or functions. More precisely, each *is* such a network of relations and/or functions. Because of this determination—or, indeed, structurality—of simple primary facts, it can and must be said that they themselves are (ontological) primary structures. "Structure," in this case, is to be understood not as *abstract* structure but as *concrete* structure: as the simple primary fact determined by a network of relations and/or functions/operators.[29] It must of course be explained how such facts/structures are to be understood more precisely. Here, there are three possibilities.

[1] First, simple primary facts could be explained as those whose domains or universes were empty. The simple primary facts would then be understood as ontological primary structures on, so to speak, the null level (0 level). They would constitute   281 starting points of series, scales, etc. One could compare the concept of the null set. At first glance, so considering primary facts appears not to be nonsensical. If however one understands simple primary facts as null structures in this sense, one cannot then overlook the fundamental differences between the null set and null structures with respect to form—above all the fact that there is only *one* null set, whereas there is not only one null structure.

---

[28] In set theory, *urelements* are items that are not themselves sets.

[29] Section 1.2.3 explains the distinction between structures in the pure or abstract and in the specific or concrete sense. Pure or abstract structures abstract from the "things" or "entities" ("objects," for Kleene) that are structured by them; specific or concrete structures are the entire entities that include both the pure or abstract structures and the things/entities structured by them.

In part because of this formal difference, the question arises of whether the concept of the (ontological) null structure is a *philosophically* sensible concept. What, more precisely, would it involve? So understood, a simple primary fact would be an ontological structure in the following sense: it would be a triple, $\langle \mathbf{F}, (\mathbf{R}_i^{\mathfrak{A}})_{i \in I}, (\mathbf{G}_j^{\mathfrak{A}})_{j \in J} \rangle$, whereby in this case the domain $\mathbf{F}$ would be empty and either the set $(\mathbf{R}_i^{\mathfrak{A}})_{i \in I}$ of relations or the set $(\mathbf{G}_j^{\mathfrak{A}})_{j \in J}$ of functions, or both, would *not* be empty. In the case of a single simple ontological primary structure, one would have to think of the set $\mathbf{F}$ as a singleton, thus as $\{\mathbf{f^*}\}$, consisting of the single primary fact. But what could it then mean to say that $\{\mathbf{f^*}\}$ is empty? That would mean that one was concerned with a simple ontological primary structure as a *purely abstract* structure, in the sense explained in Chapter 1 (Section 1.2.3). But what would such an "abstract ontological primary structure" be? Would it be an individual, determinate ontological primary structure? It is not clear how it could be.

One could attempt to overcome this fundamental difficulty by conceiving of the set of relations or the set of functions, or both, as not empty in a quite specific sense (i.e., as always already individuated and determined (or concretized) relations and/or operations). This conception would correspond to one of the basic theses of the ontological theory of tropes, when it uses such terms as "abstract particulars" (see Campbell 1990) and the like (see Section 3.2.2.3.2.1[2]). But there is a basic difference: trope theory accepts, as primitive entities or individuated *contentual* properties that it calls "tropes," and only on the basis of such primitively contentual entities are relations and functions introduced and explained. According to the explanation of simple ontological primary structures currently under consideration, however, basally contentual entities would, on the basis of the assumption of an empty domain, disappear. It therefore does not seem possible to accept null structures, in the sense currently under consideration, as appropriate specifications of the entities that this book terms "simple ontological primary structures."

In this context, it is appropriate to introduce an interesting state of affairs in the domain of predicate logic. 0-place predicate constants (e.g., P^0) are understood simply as *sentence constants*. In the strict sense, examples of such sentences in ordinary language are restricted to sentences without the subject-predicate structure, e.g., "It's raining." From the perspective of predicate logic, what happens here is the following: because the predicate P^0 is 0-place, it is not a predicate in the genuine sense; *instead*, it is a *sentence* constant, more precisely, the constant of a sentence without subject-predicate structure (see Link 1979: 69f). To be sure, it thereby remains unintelligible why P^0 is to be taken as a *sentence constant*. This assumption presumably results from the fact that within the framework of first-order predicate logic, sentence *variables* are not accepted. But the real problem does not concern the question whether one accepts sentence variables or only sentence constants, it concerns instead the question of how this procedure is actually to be interpreted. Here, predicate logic clearly reaches its limits. In any case, however, this phenomenon appears to be quite instructive: if a predicate is reduced to null-placedness, it is no longer genuinely a predicate, but is instead a sentence.

That is, to be sure, a precise formal-syntactic description. But what does it mean semantically? The fact that P^0 is still a symbol that is indexed by means of an ordinal number must be interpreted to mean that here there is a coincidence or identification of the two levels indicated by the grammatical-syntactic categories "predicate" and "singular term." From that coincidence there results, syntactically viewed, a primary sentence, thus a sentence without subject-predicate structure, and, semantically viewed, a primary proposition, thus an entity that is *not* composed of an object and an attribute. The level of the predicate is filled or saturated, whereas the other level, that of the argument (the subject or the singular term), is empty.

Despite the emptiness of the second level, P^0 is considered to be intelligible. This phenomenon, which is accorded no further significance in the contemporary literature on logic and semantics, is remarkable in many respects. One remarkable feature points toward the innovative semantics developed in this chapter, which accepts only primary sentences (i.e., sentences without the subject-predicate structure). Another respect concerns the concept thematized in the present context of ontological primary structures as null structures. Just as the concept of a predicate with an empty universe or domain is sensible, so also, in principle, is that of a structure with an empty universe or domain.[30] But the null predicate then transforms itself into a sentence without the subject-predicate structure, whereas a structure with an empty universe is termed a purely abstract structure (see Section 1.2.3). Although *purely abstract structure*, considered in isolation, is a sensible concept and/or entity, ontological primary structures cannot be understood as pure or abstract structures, as pure or abstract structures are characterized in this book.

[2] The second way to attain clarity concerning the determination of simple primary facts as ontological primary structures involves introducing a relation or a function that applies only to a single entity—here, a single simple primary fact. This is the relation of the identity of the simple ontological primary structure with itself—thus, the relation of self-identity. This can be articulated by the logical relation of identity or by the identity function.[31] The identity function is usually defined as the function $f: A \to A$, such that $f = \{\langle x, x \rangle \mid x \in A\}$. It maps every element from (the set) A to itself; it is written as id_A. If A is a singleton (i.e., if A consists of a single element), then the mapping of this element to itself is the same as the mapping of A to itself. Differently put: if A is the set $\{a_1\}$, then $\{a_1\}: = \{x \mid x$ is identical with $a1\}$ or, quite simply, $id(a_1) = a_1$. This yields the following: if the simple primary fact is conceived of as *determined*, then it is understood as—at least— identical with itself. Here, the definition of the concept *structure* is realized more clearly: the domain A is limited to the singleton consisting of the primary fact in question; the set of relations (i.e., the relations that are not functions) is empty. This would suffice to determine simple primary facts as ontological structures in the genuine sense.

[3] The third possibility of understanding primary facts as simple ontological primary structures arises from the analogous application of an elegant procedure developed by Quine to answer the question how "$y \in z$" is to be interpreted when z is an individual (see Quine 1963: 30ff). Quine first rejects as not "convenient" a widespread interpretation according to which "$y \in z$" is in this case false for all y. In addition, the interpretation that relies upon acceptance of a single primitive predicate (i.e., that of being an individual or a class) would be "an unwelcome sacrifice of elegance"

---

[30] See Ebbinghaus (1977/2003: 54): "In opposition to usual linguistic practice we accept structures with the empty universe [*Trägermenge*]."

[31] Identity, in the strict sense, can be understood either as the logical relation of identity (as the two-placed predicate constant that articulates the equality of two individual terms or objects) or as the logical identity function (as it is clarified in the text).

(31). If one insists that "$y \in z$" (when z is an individual) be either true or false, then, according to Quine, the solution is extraordinarily simple: whether the sentence is true or false depends upon whether $y = z$ or $y \neq z$. "The problem of applying the law of extensionality to individuals y and z then vanishes; where y and z are individuals and '$\in$' before individuals is given the force of '$=$', the law comes out true." Quine takes a further step, posing the question of what happens if y is an individual and z a class with y as its only element; thus, $z = \{y\}$. Here, too, the just described interpretation of the element symbol "$\in$" applies. As noted above, "$x \in y$" is true if and only if x is the individual that y is—thus, if and only if $x = y$. Quine next determines that "$x \in z$," too, is true if and only if x is the individual y. The following therefore holds: $\forall x(x \in y \leftrightarrow x \in z)$, so $y = z$. In conclusion, Quine determines that this result at first appears unacceptable, because y is an individual and z a class. In fact, however, the result is harmless:

> Individuals are what rated as nonclasses until we decided to give '$\in$' the force of '$=$' before them; now they are best counted as classes. Everything comes to count as a class; still, individuals remain marked off from other classes in being their sole members. (32)

How does an analogous application of this procedure to the concept of structure in the case of simple ontological primary structures look? The answer is evident: in the case of a structure, the question concerns not the symbol "$\in$," but the relation between the two members of a pair (for the sake of simplicity, structures in this context are considered to be pairs, not triples or quadruples). This relation is indicated not by its own special symbol, but by the notation "$\langle \ldots, \ldots \rangle$." It is the relation of an established ordering, of a fixed, unambiguous serial order. In the case of the simple ontological primary structure, *this* relation must be understood as an *identity relation*. Assume a simple primary fact $\mathbf{f^*}$. If one understands it as a simple ontological primary structure, and understands by R^* the set of real relations over $\mathbf{f^*}$, it is to be understood as the pair $\langle \mathbf{f^*}, R^* \rangle$, and indeed in such a way that it holds that $\mathbf{f^*} = R^*$.

[4] The second and third explanations must be clarified and specified in one quite important respect. A *simple* (i.e., a non-complex) ontological primary structure $\mathbf{f^*}$ can be articulated in various ways. Simplifying, it can be articulated (in the sense of determined) minimally or partially or instead maximally, to mention only the lowest, one intermediate, and the highest degrees on the scale of the possible determinations of the simple ontological primary structure in question. Which articulation or determination comes forth depends upon or results from how the second or third component (thus, the set of relations or of functions/operators) is determined more precisely. What this book calls a "simple ontological primary structure" is the *most minimal* case of determination of an ontological primary structure. The determination reduces itself to this simple entity (i.e., the primary structure) itself, without other simple or complex entities (primary structures) being considered. This point is of enormous importance for the ontological conception developed in this book. If one wanted to speak of a relation (or of relations) of one simple primary structure to another such structure (or other such structures), this would be a somewhat loose manner of speaking; more precisely, one would have to say that in such a case the simple ontological primary structure was *embedded* in a complex ontological primary structure—in this book's terminology, in a structured configuration of simple ontological primary structures. This *embedding* of simple ontological primary structures

in so-conceived ontological structural configurations then attains ever more complex degrees. The additional structural configurations consist for their part of increasingly complex ontological primary structures (i.e., from increasingly complex configurations of structures, etc.).

If a given simple ontological primary structure can be formally expressed by the ordered pair $\langle \mathbf{f^*}, \mathbf{R^*} \rangle$, its second set of components consists of the set of relations $\mathbf{R^*}$. The only determination of this simple primary structure is, however, a relation that does not involve other simple or complex primary structures, but only itself. This relation, as shown above, is the relation of identity. It therefore holds that $\mathbf{R^*}|_{\mathrm{Id}}$ (i.e., the set of relations $\mathbf{R^*}$ is restricted to the identity relation). Correspondingly, according to the third explanatory possibility introduced above, the primary fact $\mathbf{f^*}$ stands in or under the relation $\mathbf{R^*}|_{\mathrm{Id}}$—thus, in or under the relation $\mathbf{f^*} = \mathbf{f^*}$. More precisely, the primary fact $\mathbf{f^*}$ is identical with the relation $\mathbf{f^*} = \mathbf{f^*}$. Formally, $\mathbf{f^*} = (\mathbf{f^*} = \mathbf{f^*})$.

If however the second component of the simple ontological primary structure is understood not as a relation but as a function, the formula for it would be, in notation corresponding to that introduced above, $\langle \mathbf{f^*}, \mathbf{G^*} \rangle$. In this case, the formula would be interpreted in accordance with the second explanatory possibility introduced above— that is, the set of functions or operations $\mathbf{G^*}$ would have to be restricted to the likewise above-explained identity function (i.e., as $\mathbf{G^*}|_{\mathrm{Id}}$).

[5] These matters can be clarified by means of an example. For this example, it is however requisite to anticipate or presuppose some central semantic and ontological theses that are presented only in later sections of this chapter. The theses themselves should be sufficiently intelligible already at this point.   286

The sentence "It's redding" (understood as "It is the case that it's redding"), here presupposed to be true, expresses a fully determined simple primary proposition that, because it is true, is identical with a simple[32] primary fact.[33] How is such a primary fact to be understood? To begin with, it must be emphasized that, according to the central theses defended in this book, the primary fact is not to be understood as the result of an instantiation or exemplification, by means of entities of whatever type (substrata, bare particulars, locations, etc.) of a universal *redding* that remains absolutely self-identical. In this book, an entity such as *redding* is also understood as something *generic*, but in a *wholly different sense*, not in the sense of an instantiable or exemplifiable entity, but in the sense that it has many segments; only the *totality* of all the segments constitutes the entity *redding*. This totality can be understood as the conjunction of all the primary facts that can be expressed by the sentence "It is the case that it's redding (s_1, t_1) & . . . it's redding (s_2, t_2) & . . . it's redding (s_n, t_n) . . ." (whereby s stands for place and t for time). One can thus understand *redding* as a determinate *dimension* of the world, as a dimensional entity that is dispersed throughout the entire world.

---

[32] For the sake of simplicity, it is here presupposed that a determinate red "point" is a *simple* fact. This assumption is made only to clarify the concept of ontological structure by means of an example from the everyday life-world. Whether a thorough analysis of the phenomenon of color, taking into consideration natural-scientific theories, would support the assumption that a red-colored point is a *simple* fact is a question not to be clarified here. Only a fully developed ontology could answer it.

[33] How this fully determined status is to be understood precisely is shown in Section 3.3.

These various occurrences (understood in the just explained sense of segments, not as instantiations or exemplifications) of reddings first form that unity that constitutes the entity *redding*. This unity is thus not to be understood in the sense of a self-multiplying universal property "red" that remains self-identical; instead, it is constituted by the relation of *resemblance* or *similarity*. Its unity is fundamentally a *family resemblance*. No occurrence of redding is identical with any other; in these occurrences, redding is not an identical entity, but a plurality of *similar* entities.

These briefly considered matters are now to be made more precise in relation to the determination of the *simple* ontological primary structure expressed by the simple primary sentence "It's redding"; this primary structure is indicated by "$\mathbf{f^*}$." To be sure, it is a *determined* primary structure, but only in a *minimal* sense. It articulates something ontological, but still in complete abstraction from other simple and complex ontological primary structures: it abstracts, that is, from other reddings (in the sense explained above), from other simple ontological primary structures $\mathbf{f^*_1}, \mathbf{f^*_2}, \ldots, \mathbf{f^*_n}$, from spatial and temporal locations, etc. The simple ontological primary structure $\mathbf{f^*}$ is determined only by *self-identity*. This simple determination is a *minimal determination*.

One mode of *partial* determination of the simple ontological primary structure $\mathbf{f^*}$ is given when $\mathbf{f^*}$ is put into relation to (the) other simple ontological primary structures that are reddings. Thereby, $\mathbf{f^*}$ is placed in connection with an $\mathbf{f^*_i}$ and thereby *embedded* in a sequence $\mathbf{f^*_1}, \mathbf{f^*_2}, \ldots, \mathbf{f^*_n}$. This sequence, however, is a *complex* ontological primary structure, the conjunction $\mathbf{f^*_i}$. That is to say, the simple ontological primary structure $\mathbf{f^*}$ is embedded in the generic dimensional entity *redding* as one of the segments of this entity. *Another* mode enters when the simple ontological primary structure is put into relation to spatial and temporal locations. This involves an embedding of the simple primary sentence "It's redding" in the conjunctive configuration of simple primary sentences "It's redding (s_1, t_1) & it's redding (s_2, t_2) & . . . & it's redding (s_n, t_n)" or placing the simple ontological primary structure $\mathbf{f^*}$ into the conjunctive complex ontological primary structure or structural configuration $\mathbf{f^*}(s_1, t_1)$ & $\mathbf{f^*}(s_2, t_2)$ & . . . & $\mathbf{f^*}(s_n, t_n)$ of simple ontological primary structures situated with respect to specific spatial and temporal locations, whereby it holds that $\mathbf{f^*}(s_1, t_1) \neq \mathbf{f^*}(s_2, t_2) \neq \ldots \neq \mathbf{f^*}(s_n, t_n)$. Among these primary structures the relation is one only of *similarity* or *resemblance*.

Maximal determination is attained when the simple ontological primary structure $\mathbf{f^*}$ is situated within the universe, within being as a whole. This involves complete embedding within the universal and thus absolutely maximal ontological primary structure. To be emphasized is that the scale of ever more comprehensive determinations of simple ontological primary structures, from the level of minimal determination to that of maximal determination, exhibits an immeasurable complexity. To put the matter briefly: the actual (in the sense of complete or maximal) determination of a simple ontological primary structure coincides with its embedding within the comprehensive structure of the universe—more precisely, within the universe or being as a whole *as* the comprehensive ontological primary structure. This presupposes that the entire universe is somehow articulated or described.

3.2.3.3 Forms of Configuration as Ontological Structures

Up to this point, the emphasis has been on providing more precise determination for the first elements within the concept *ontological structure*. These elements themselves prove to be ontological structures of a wholly determinate sort. In language adopted from set theory, this ontological conception assumes no urelements—no elements that are not themselves structures. Even simple primary facts are ontological primary structures in the genuine sense.

As is shown above (see Section 1.2.3), one can and must distinguish between concrete and abstract structures. Any concrete or determinate structure is a pair or triple consisting of a domain or universe and one or more sets of what may be termed *configuring factors* (relations and operations/functions). Abstract ontological structures are constituted by these configuring factors alone; abstract structures are thus the forms of configuration of concrete ontological structures.

How abstract structures are to be understood precisely, in relation to concrete structures, is a difficult and complex problem. This book does not claim to present a response to this problem that is worked out in detail. Instead, it develops only the basis for such a conception. Particularly important is the brief characterization of the many options for this broad field.

The question is thus, how are *configuring factor(s)* or *abstract structure(s)* to be understood?

3.2.3.3.1 *On the Relation Between Logical/Mathematical Structures and Ontological Structures*

An unconditional presupposition for providing a grounded answer to the question posed at the end of the preceding section is the clarification of another (more fundamental) question—the question indicated by the title of this section. The history of Western philosophy is stamped, virtually throughout, by a wholly determinate understanding of this relation, one according to which logical/mathematical structures and ontological structures constitute two wholly separate domains. According to this view, logical/mathematical structures develop within the domain of language or intellect; they are concerns of the domain of language, of the formulation of theories in the sense of abstract models of actuality (of the world); they do not directly concern the world itself. When such an indirect relation of logical/mathematical structures to the world (more precisely, in this book's terminology, to ontological structures) has been or is assumed, however, it remains wholly unclear what this relation could possibly be.

In the structural conception, logical/mathematical structures are themselves ontological, immediately and in a strict sense. More precisely: they have a strictly ontological status *when* they structure primary facts. They themselves then have the status of being ontological (abstract) structures. The reason for this emerges from the radical questioning here undertaken of traditional ontology, the ontology that provides the basis for the traditional understanding of the relation between logic/mathematics and ontology. If the world is not understood as the totality of objects in the sense of substances/substrata that have properties and stand in relations to one another, the basis for the traditional understanding of the relation between logic/mathematics and ontology largely vanishes. If one instead understands the world as the totality of primary facts, configured with great variety and complexity, there arises a completely different conception of this relation. Logical/mathematical structures are then the factors that configure the primary facts. To understand this conception precisely, one must take a much broader view.

3.2.3.3.2 *Configurations and Propositional Logic*

If one accepts primary facts as the only ontological structures, there immediately follow revolutionary consequences with respect to the question of what abstract structures must be. The first and most fundamental of these consequences concerns elementary logic, i.e., propositional and predicate logic.[34] Quantificational logic is generally considered to be a part—indeed the central part—of predicate logic, but that is not wholly correct, because there can be and indeed is a quantificational logic without either predicates or individual variables of the sort found in predicate logic. Thus, one can of course quantify over the "entities" that are found in set theory or set language. Normally, one also terms such entities, including the sets themselves, "objects." But that is nothing more than one of those customs that, despite their virtually universal acceptance, bring with them immense ontological obscurities. In addition, quantification over such entities (elements of sets and/or sets themselves) is not accomplished by means of recourse to predication in the proper sense. Neither the symbol for being an element ("$\in$") nor any other of the symbols used in set theory ("$\subseteq$,"" $\wp$,""$\cap$,""$\cup$,""$\emptyset$," etc.) articulates a proper predication. To be sure, the general tendency—based on the characteristic habituation arising from the current dominance of the language of first-order predicate logic—is to read the "relation" of being an element simply as a predication, in that, for example, one reads the formula "$y \in Y$" as "y is an element of Y." But this convenient way of reading proves to be wholly misleading, if from it are drawn determinate (meta-)logical or philosophical conclusions. Whether—and if so, how—first-order predicate logic can be taken into account by the semantics developed here is considered below.

With respect to the structural-systematic conception, propositional logic is the most interesting case in that it can be shown clearly that its logical structures—its connectives—can have an immediately ontological status. Support for this contention is provided by an essay by Quine from the year 1934, an essay that, despite the fame and expertise of its author, appears to have been virtually forgotten. The essay in question is "Ontological Remarks on the Propositional Calculus." With remarkable clarity, Quine articulates several fundamental points. To begin, he strictly distinguishes between "sentence" and "proposition": "it is the proposition as the denotation of the sentence, i.e., as the entity, if any, whereof the sentence is a symbol, that is the present concern"

---

[34] Propositional logic is often considered to be a subsystem of first-order predicate logic. The reason for this—if one is given—is the following: if one accepts as arguments for connectives only sentences, then one can conceive of the set S of sentences as the set $S_0 := Con_p^0$ of 0-place predicate constants P ('*Con*' for 'constant'; see Section 3.2.3.2[1]). But that is nothing other than a pure artifice that proves to be problematic. The sentence constants or sentence letters of propositional logic—$p, q, r \ldots$—stand for *genuine sentences* of the language in question. It is clear, however, that languages that consist of subject-predicate sentences contain in a great many cases—not to say in most cases—sentences with unary, binary, . . ., n-ary predicates. The sentence constants p and q can for example stand for the sentences represented in predicate logic as Fa and Gb, respectively, such that the conjunction, which would ordinarily be represented as $p \wedge q$, could also be represented, more precisely, as the conjunction $Fa \wedge Gb$. But Fa and Gb contain unary predicates, not 0-place ones. For reasons that are explicit in what follows, the subordination of propositional logic to predicate logic plays no role in this book. It is therefore here ignored.

(1934/1979: 265). He then makes clear that if one accepts propositions, then they have to count as the arguments of the connectives:

> Once we postulate entities whereof sentences are symbols, the logical principles for manipulating sentences become principles *concerning* the entities—propositions— which the sentences denote. Insofar the theory of deduction becomes a calculus of propositions. . . .(266)

To be sure, such a calculus remains in this respect, as Quine notes correctly, "a very partial calculus. . ., since its only principles are those which governed the manipu- lation of sentences antecedently to the notion that sentences were names of any- thing." The task thus remains of explaining and explicating "the residual character of propositions"—a task obviously necessary given that this "character" of propo- sitions is not explained if all that is said is that they can function as arguments of connectives.

291

Quine attains complete clarity by bringing into play the concept of the "total logis- tical system." If the "theory of deduction" (propositional logic) is developed only on the level of sentences, then, according to Quine,

> [It] remains unchanged in structure, but ceases to be a system in the usual sense. The usual sort of system treats of some manner of elements, say cardinal numbers or geometrical points, which are denoted ambiguously by variables; operative upon these elements are certain operations or relations, appropriately expressed within the language of the system. The theory of deduction, when construed as a calculus of propositions, is a system of this kind; its elements are propositions denoted by the variables "p," "q," etc., and its operations are the propositional operations of denial, alternation, material implication, etc., denoted by prefixture or interfixture of the signs, "~," "∨," "⊃,"etc. When, on the other hand, the theory of deduction is reconstrued in the foregoing manner as a mere organon of sentences, it ceases to be concerned with elements subject to operations; the former proposition-variables "p," "q," etc., become ambiguous sentences, symbols *of* nothing, and the signs "~," "∨," "⊃,"etc., become connectives of sentences, innocent of operational correlates in the realm of denotations. (268)

What Quine develops here is governed by his restriction: "Once we postulate entities whereof sentences are symbols . . ." (i.e., propositions). It is of course well known that Quine rejects "propositions"; this makes it the more remarkable that he makes the relevant points with a precision exhibited by no other philosopher or logician. For present purposes, it is wholly unimportant that Quine later changes his position (e.g., then speaking no longer of the "sentence variables 'p', 'q', etc.," but only of "sentence letters 'p', 'q', etc.").

In this book propositions are accepted, but they are understood in a manner wholly different, in a fundamental respect, from any found anywhere else in the philosophical literature (including, of course, in the work of Quine). Here, propositions are *primary* propositions in the sense explained above. But Quine's point concerning propositional logic, articulated in the passage quoted above, remains wholly unaffected by this change in the understanding of "proposition." The reason for this is that all that matters to Quine is the connection between "sentences" and the "entities whereof sentences are symbols" (or, as it is put more often, the entities that are expressed by sentences).

In conjunction with the understanding of the primary proposition developed in this chapter, Quine's point concerning propositional logic has ontological consequences—consequences extending far beyond the relation between sentences and propositions. As demonstrated thoroughly in Section 3.3, every true primary proposition is (in the strict sense of identity) nothing other than a primary fact. From this it immediately follows that the connectives can have primary facts as arguments. In Quine's terminology, the "elements" of the *complete* or *fully determined* "logistical system" are *not only* primary propositions but are *also* primary facts. This makes it clear that the connectives (can) have an immediate and genuine ontological status. The word "can" is here in parentheses because one can, for methodological reasons, restrict the consideration of the entities that are expressed by sentences (here, the primary propositions) to the level of primary propositions in the general (and thereby, as is shown below, underdetermined) sense. When however these entities, the primary propositions, are *true*, then they are identical with primary facts in or of the world; the connectives that have true primary propositions and thus primary facts as arguments thereby attain an immediately ontological status.

3.2.3.3.3 *Configurations and First-Order Predicate Logic*

At first glance, it appears that first-order predicate logic (PL1) cannot be relevant to the contextual semantics and ontology advocated here, given that it is based on the assumption that the sentences of the language to which it is to be applied have subject-predicate form. But such a first glance is misleading. Everything depends upon the semantics that one combines with this logic. Normally, one associates this logic with the so-called standard semantics now to be considered in additional detail; in this case, first-order predicate logic is indeed incompatible with contextual semantics. But there are also non-standard semantics for PL1; whether this logic can have a non-standard, contextual semantics as one of its non-standard semantics is what must be determined.

The flexibility of PL1 with respect to the semantics question can be clarified through consideration of a position that is characteristic of the analytic tradition: the non-standard semantics presented by Gary Legenhausen in his essay "New Semantics for the Lower Predicate Calculus" (1985). Legenhausen relies upon the conception of the relation between logic and ontology accepted by most analytic philosophers, according to which (formal) logic is ontologically and metaphysically neutral. He thereby understands logic as a formal system equipped with semantic rules. Legenhausen distinguishes two wholly different ways of understanding the semantic rules, terming them, respectively, "formal semantics" and "theory of reference." According to Legenhausen, a given system of formal semantics contains an interpretation function for the non-logical terms of the formal system. As he understands it, this function maps these terms to the components of another abstract structure, such as set theory, algebraic theory, etc. The value of the interpretation function for any term is the extension of the term. The task of a theory of reference, on the other hand, is (for Legenhausen) to clarify the relation between linguistic terms and the objects to which they relate. Having introduced this distinction, he then uses the term "semantics" to mean formal semantics.

Legenhausen takes the thus understood distinction between formal semantics and theory of reference as fundamental. The ontological and metaphysical neutrality of logic—which he accepts without restriction—is violated, he argues, if that fundamental distinction is not recognized; its non-recognition leads to an illegitimate identification of extension and reference. From this Legenhausen draws the basic thesis that the acceptance or application of a system of formal semantics with a determinate structure does not have as a consequence that this structure must also be ascribed to the world. Such systems are therefore to be understood instrumentally, not realistically or onto-logically (see 325).

Nevertheless, Legenhausen significantly notes that systems of formal semantics can be taken as *illustrations* (he also speaks of analogies) of metaphysical and ontological theses. This is scarcely comprehensible for a number of reasons, including the following two: [i] How such an "illustration" or "analogy" can be explained remains a riddle. Is it a kind of accident? Or is it the expression of some deeper relation that Legenhausen does not articulate? [ii] If Legenhausen's duality between formal semantics and theory of reference is accepted, the question arises whether it makes sense to develop a formal semantics. If the relation between language and world can be "explained" only by a theory of reference, it would appear that one could simply not bother with formal semantics. In fact, however, Legenhausen's characterization of standard semantics and his development of two non-standard semantics for PL1 constitute nothing other than a self-refutation of both the thesis of the ontological/metaphysical neutrality of logic and the distinction between formal semantics and theory of reference—and thus, of Legenhausen's basic conception as a whole.

Legenhausen presents two non-standard semantics for PL1. The first, S^P, interprets monadic predicates as terms that have extensions defined purely set-theoretically. A given singular term has as its extension a set of the elements of the extensions of predicates. An atomic sentence Pa is true "if and only if the extension of the predicate, 'P', is a member of the extension of the individual term, 'a'" (323). Formally, this non-standard semantics S^P is based decisively on the concept of property ('P' stands for 'property'). A model for S^P (328) "is a pair $\langle \mathbf{P}, J \rangle$, where $\mathbf{P}$ is a non-empty set whose members are extensions of monadic predicates" that are understood as properties. Legenhausen emphasizes, how-ever, that S^P does not entail the existence of universals or of entities of any other sort. The interpretation function J is defined as follows (here only the basics)[b]:

294

1. $J(\mathbf{P})$ is a subset of the power set of $\mathbf{P}$, minus the null set, and [thus provides] the domain of discourse for $\langle \mathbf{P}, J \rangle$. . . . . The domain of discourse is a non-empty subset of the set of all non-empty sets of properties. . . .

2. If a is an individual term, i.e. an individual constant or an individual variable, $J(a) \in J(\mathbf{P})$. [The extensions of singular terms are thus sets of properties that are themselves members of $J(\mathbf{P})$.]

3. If P is a monadic predicate, $J(P) \in \mathbf{P}$.

4. If P is an n-place predicate, $n > 1$, $J(\mathbf{P}) \subseteq (J(\mathbf{P}))^n$, that is, $J(\mathbf{P})$ will be a set of n-tuples of members of the domain of discourse. . . .

5. If A is a [well-formed formula] of the form Pa, so that P is a monadic predicate, $J(A)$ [is true] iff $J(P) \in J(a)$.

   [Etc.]

---

[b] The passages enclosed in square brackets in the following blocked quotation (beginning "1" and ending "[Etc.]") are translations from *Struktur und Sein*; the remainder of the passage comes verbatim from Legenhausen (328–29).

This semantics S^P is non-standard in that it involves, as Legenhausen puts it, "a simple reversal of the standard semantics with regard to the interpretations of individual terms and monadic predicates" (323). The well-known non-substantially oriented so-called bundle theory of the individual could have an (extensionally oriented) articulation in S^P.

The second non-standard semantics Legenhausen develops for PL1 is of far greater significance with respect to structural-systematic semantics and ontology; it is therefore considered here in more detail. The semantics S^f ('f' for 'facts') begins with the assumption that "both individual terms and monadic predicates take as their extensions sets of set-theoretical elements" (325). An atomic sentence 'Pa' is true just when the intersection of the set that forms the extension of the term 'a' and the set that exhibits the extension of the predicate 'P' is a singleton (i.e., when the two sets have exactly one common member). This semantics is non-standard in that "individual terms and monadic predicates have extensions of the same set-theoretic type." Following Frank Ramsey, Legenhausen interprets this assumption in such a way that both individual terms and monadic predicates can be conceived as propositions. The connection between these two sets can be modeled by means of the intersection of the two sets, from which there results a unique proposition that Legenhausen terms a "state of affairs" (see 326).

This can be illustrated with an example. Consider the sentence "Socrates is wise." The predicate "is wise" can, Legenhausen notes, be interpreted as a set containing facts; it would include the facts that Socrates is wise, Plato is wise, etc. Individual terms are then interpreted in the same manner. "Socrates" would be interpreted as a set including the facts that Socrates is wise, that Socrates is a Greek, etc. The connection of Socrates and wisdom, which makes true the sentence "Socrates is wise," is that the fact that Socrates is wise is contained both in the set *Socrates* and in the set *is wise*.

The connection between the unique proposition Legenhausen labels "state of affairs" and the facts of which it is the intersection is not clear. But this unclarity results—as is now to be shown—from the basic ontological unclarity that characterizes his position as a whole.

To present S^f somewhat more precisely, the essential parts of Legenhausen's formalism are now introduced. Let 'μ' be a model structure (i.e., for Legenhausen, a set of models [334]) whose models are based on facts in the sense that facts are the elements of the basic set. A model for S^f in μ is an ordered pair, $\langle F, H \rangle$ where F is a non-empty set whose elements are facts, and H is defined as follows (again: here only the basics)[c]:

1. H(**F**) is a non-empty subset of the set of all non-empty subsets of **F**. H(**F**) is thus the domain of discourse for $\langle F, H \rangle$. The individuals over which the quantifiers range are represented as sets of facts.

2. If a is an individual term H(a) $\in$ H(**F**).

---

[c] The passages enclosed in square brackets in the following blocked quotation (beginning with "1" and ending with "[Etc.]" are translations from *Struktur and Being*; the remainder of the passage comes verbatim from Legenhausen (335–36).

3. If P is an n-place predicate and $a_1, \ldots, a_n$ are individual terms, then $H(Pa_1, \ldots, a_n) \subseteq \mathbf{F}$.[35]

4. If A is a [well-formed formula] of the form $Pa_1, \ldots, a_n$, $H(A)$ [is true if and only if] for each i, $1 \leq i \leq n$, there is an f which is a member of $\mathbf{F}$ such that the intersection of $H(a_i)$ with $H(Pa_1, \ldots, a_{i-1})$ is $\{f\}$.

[Etc.]

296

Legenhausen emphasizes quite clearly that this second non-standard semantics for PL1, S^f, is a "fact-based semantics" (e.g., 326), but denies that this either entails or presupposes any "ontology of facts":

> [I]t [the S^f semantics] remains metaphysically neutral. What it requires is no more than a willingness [!!] to allow that the extensions of individual terms and the extensions of monadic predicates are sets. The members of these sets may be *most naturally* understood as propositions, states of affairs, or facts, but there is no more need to accept an ontology of such entities in order to employ S^f, than there is a need to accept the ontology of substances in order to utilize the standard semantics. (327; emphasis added)

The ambiguity, indeed the incoherence, of such claims is shown above. Here, an additional critical question must be raised: how can the members of the set in question be understood "most naturally" as facts? How is the degree of "naturalness" to be measured if this semantics is indeed ontologically and metaphysically neutral? That aside, it seems miraculous that precisely the assumed set can have facts as its elements. How can this be explained? The fundamental objection to the putative ontological and metaphysical neutrality of PL1 (as of every logic) is that logic would thereby be reduced to purely formal syntax. A semantics for a logic must provide an interpretation for the non-logical symbols. But what is such an interpretation other than precisely an assigning of determinate values to these symbols? Because however the non-logical symbols are determinate terms from ordinary language, the only values than can sensibly come into question are ones that, in ordinary language, are associated with such terms.

In addition, examination of S^f reveals it as an example of a far too common practice: Legenhausen uses formal instruments that might appear to make fundamental and central philosophical terms and concepts more precise, but any such appearance is mere semblance—he says nothing significant about those terms and concepts. In S^f, the term and concept "fact" remain wholly unexplained. At the most, one could say that S^f projects a completely general formal framework that makes possible the development of a propositional semantics and an ontology of facts along with the retention of a language that, syntactically, contains subject-predicate sentences. But then it is unconditionally necessary that independent accounts be provided of how "proposition" and "fact" are to be understood. This book understands singular terms in languages that use subject-predicate sentences as abbreviations of great numbers of primary sentences.

---

[35] Legenhausen here introduces an undefined metalinguistic symbol "∗" that he calls a "plug"; it makes possible a particular treatment of relations. Legenhausen notes that this special treatment of relations is not an essential component of S^f. For this reason—and because of the complexity of the consequences of its introduction—this treatment of relations is ignored here. For present purposes, it is unimportant.

It would make sense to consider whether it might not be preferable to understand them as sets of primary sentences. But here too it is clear that without radical semantic clarification of what sentences are and of what form sentences have or should have (for philosophical purposes), one simply goes around in circles.

3.2.3.3.4 Forms of Configurations: Expansions of Classical Logic and the Multiplicity of Mathematical Structures

In addition to classical elementary logic, the expansions of classical logic must be considered, especially modal logic, which is extremely important for philosophy, and its many specific applications or concretizations, such as temporal logic, dynamic logic, deontic logic, etc. This is an extremely broad area, and the concept of possible worlds is an essential component of many of these modally structured logics. But that is also quite a controversial area—many authors, such as Quine, wholly reject modal logics.[36]

No less important for the structural-systematic philosophy are the alternatives to classical logic, particularly multi-valued logic, relevance logic, intuitionistic logic, free logic, quantum logic, etc. (see Gabbay and Guenthner 1986).

As far as mathematical structures are concerned, their immense multiplicity opens up a great variety of perspectives for the structural-systematic philosophy developed in this book. But precisely this multiplicity presents systematic philosophers with a difficult task. They must attempt to discover *which* mathematical structures best articulate the relatively maximal structuration of the ontological dimension. In contemporary philosophy, there are remarkable new initiatives in this area.[37] This task remains one of the most important ones for the philosophy of the future.

3.3 Theory of Truth as Explication (Articulation) of the Fully Determinate Connections Among Fundamental Structures

The preceding sections of this chapter present sketches of the three grand dimensions of structures: the fundamental formal, semantic, and ontological structures. The question now to be raised is, how do they interrelate? This question concerns chiefly, although not exclusively, the relation between semantic and ontological structures. The concept that provides the answer to this question is the concept of truth.

Section 2.4 of Chapter 2 develops a starting point for a systematic theory of truth. That chapter thematizes the dimension of theoreticity; it shows that this dimension attains fully determined status only with the application of the concept of truth. It formulates a first, global characterization of the truth-concept, but not a sufficiently detailed and precise account. The reason for this is that central elements required for the definition of the truth-concept are presented only in this chapter; these are the elements of the systematics of structure. It can now be said and shown that the function

---

[36] Concerning the central problem of the (logical and philosophical) interpretation of modal logic, Ballarin (2004) is highly instructive.

[37] On this point, see Goldblatt (1979/1984); Bacon (1995); Mormann (1995, 1996); Dipert (1997); and Schneider (1999, 2001, 2002, 2005, 2006a).

of the truth operator is that of fully determining the status of the interconnections of the three types of fundamental structures.

The theory of truth unquestionably belongs among the most intensively treated issues in contemporary philosophy, particularly in recent decades. Despite the great efforts made toward clarifying this issue, however, it cannot be said that even a partial consensus has been attained. A theory of truth that aspires to be acceptable at present could be developed only by taking into consideration the immense truth-theoretical debate currently ongoing, and in opposition to the various other approaches and positions taken in it. Such a discussion would be out of place in this book, whose orientation is systematic. Relevant to this issue are the author's numerous works on theories of truth, wherein he treats the entire spectrum of problems at issue in contemporary debates concerning truth.[38] The quite concise account that follows can be understood and evaluated with full adequacy only in light of the many treatments—not only critical but also concerning the history of the concept and of the problem—to be found in those works. The fact that this theory of truth is presented here, for the first time, within a comprehensive systematic framework has as a consequence that it involves a few—chiefly terminological—differences from the earlier accounts; the most significant difference concerns the term and concept *structure*.

3.3.1 *A More Precise Characterization of the Basic Idea of Truth*

299

[1] Further specification of the informal-intuitive characterization of the basic idea of truth provided at the end of Chapter 2 can develop by comparing it to a similar thought of Frege's. The Fregean thought in question is not, however, an element of Frege's truth-theory—if one can even speak of such—but is instead one of the theses that recur in various of Frege's writings in conjunction with his widely dispersed comments concerning truth. In passages cited in Section 3.1.2.1, Frege describes judgments as "advances from a thought to a truth value" (1892a: 35/159) and as "a passing over from a thought to a truth value" (1891b: 97/150; translation altered), and speaks of the necessity that logic take "the step from thought to truth value" (1892b: 133/178). Frege's terminology ("advance," "passing over," "step") corresponds only in a certain respect to the term used in Chapter 2—the "transition" from the semantic to the ontological dimension—as must now be shown. It must here be kept in mind that, as also noted above, Frege's ontology and semantics are in part idiosyncratic. He understands (1892a: 34/158) the *Bedeutung* of a sentence as its truth-value, and truth-value as an "object". He recognizes two truth-values and thus two relevant objects: the True and the False.

This book relies on a semantics and an ontology wholly different from Frege's. Nevertheless, it is remarkable that Frege articulates the idea of an "advance" from one level to another. He does not, of course, characterize the truth-concept as involving such an advance; for Frege, truth is undefinable.

Whereas Frege speaks of a "progressing" or "passing over" from thought or proposition to truth-value, here the talk is of the "transformation" or "transition" of a sentence or proposition from semantically un- or underdetermined to fully determined status. In a certain respect, transformation or transition also involves passing over or progressing, but the latter terms do not contain the former in its proper sense. Progressing suggests

---

[38] See the relevant items listed in the Works Cited.

that there are two different points or places or entities such that, with the progressing from the one to the other, neither is changed or determined more precisely. In opposition to this, the concept of transformation or transition entails a modification of the condition or status of one and the same thing—here, a sentence or the proposition it expresses. Frege's analysis and terminology are deficient because they fail to recognize or to thematize this essential moment. One must understand this transformation/transition as a qualifying, indeed as a determining, of the sentence or proposition.

That the truth operator "It is true that" is applied to sentences/propositions has both a presupposition and a consequence. The presupposition is that arguments of the operator, which is a determiner, are determinable by it, and thus are less than fully determined. Specifically, the sentence or proposition that is the argument has an incompletely determined semantic status. The effect of the operator is, negatively, the elimination of the un- or underdetermined status of the argument; positively, the effect is that the argument attains a fully determined semantic status (i.e., becomes fully semantically determined, as shown in Section 3.3.3.2.2).

[2] In several preceding sections, this book uses the concept of the fully determined status of sentences and propositions (and, generalizing, of language). But how is this central concept to be understood? Responding clearly to this question is a task that is decisive for the project undertaken here. At this point, it can be said programmatically that the full determination of a language is the complete articulation of its ontological dimension. The following paragraphs in this section show that this thought can be made more precise as follows: the fully determined status of language involves, in the final analysis, the fully explicated interconnecting of the three fundamental dimensions of structure.

Preliminary clarification may be provided by a brief consideration of the "disquotation theory" championed by Quine.[39] The thesis that "true/truth" is a function of disquotation corresponds to the Tarskian truth schema:

(T)   'p' is true $\leftrightarrow p$.

Despite his general acceptance of this schema, Quine—at least in some of his writings—fully recognizes the ontological dimension of truth. His remarks on this topic are unique in analytic philosophy in that they—possibly despite his own intentions—show with crystal clarity just wherein lies the basic error common to all deflationistically oriented theories of truth. One of the most interesting passages is from his *Philosophy of Logic*:

> [T]ruth should hinge on reality, and it does. No sentence is true but reality makes it so. The sentence "Snow is white" is true, as Tarski has taught us, if and only if *real* snow is *really* white. The same can be said of the sentence "*Der Schnee ist weiss*"; language is not the point. In speaking of the truth of a given sentence there is only indirection; we do better simply to say the sentence and so speak not about language but about the world. So long as we are speaking only of the truth of singly given sentences, the perfect theory is what Wilfrid Sellars has called the disappearance theory of truth. (1970b: 10–11; emphasis added)

---

[39] E.g., "Ascription of truth just cancels the quotation marks. Truth is disquotation" (1992a: 80).

This highly interesting text contains a magnificent and deeply correct insight and at the same time a fateful error. The correct insight is that to say "'Snow is white' is true" is to say "*Real* snow is *really* white." More generally, truth involves a relation of language to the world, to the ontological dimension. The error consists in Quine's maintaining that this understanding of truth is the genuinely correct interpretation of Tarski's truth-schema and in his taking this schema, so understood, to be the adequate articulation of—or last word on—truth. It may be—and indeed it appears probable—that Quine's "explanation" agrees with what Tarski had in mind, perhaps consciously, but the important question is whether this explanation agrees with an interpretation of the T-schema that is *correct with respect to the issue at hand*. As can be shown easily, it does not.

The understanding of truth that Quine articulates for the T-schema is accurate with respect to the issue at hand only if it meets the following condition: this understanding of truth must also be articulated or made explicit in the Tarskian T-schema (such a *minimal* condition must be satisfied by any interpretation that claims to be genuinely correct). Tarski and Quine, however, fail to satisfy this condition. The sentence on the right-hand side of the equivalence symbol meaning "if and only if" (thus, "Snow is white," or "p"), without further qualification (i.e., determination), cannot be taken as a sentence that says anything about the world. That should be fully clear from what is said in Chapter 2. But Quine—and with him the majority of truth-deflationists—simply assumes that this sentence is fully determined in the sense of speaking fully determinately about the world. In the correct insight introduced above, Quine does make this explicit by articulating it explicitly, as does Tarski himself, in a remarkably similar formulation introduced below, in [3].

One way this condition can be satisfied is by introducing, on the right-hand side of the equivalence sign, the operator that Quine, in his "interpretation," makes explicit: "really." The result would be as follows:

(T′) It is true that snow is white ↔ really: snow is white.
(T″) "p" is true ↔ really: p.

302

Another possibility would be the introduction of a special notation on the basis of a specific convention. Thus, for example, in some of his earlier works the author has printed the sentence on the right-hand side of the corrected or reinterpreted Tarski-equivalence in bold-faced type to indicate the *fully determined status* of that sentence and of the proposition it expresses:

(T‴) It is true that snow is white ↔ **Snow is white.**
(T⁗) "p" is true ↔ **p.**

Quine's "really" corresponds exactly to the factor that this book terms fully determined status.

[3] An additional point is important to understanding the historical context. The often-named error, just pointed out in Quine, stems from Tarski himself. His informal characterization of the intuitive understanding of truth, cited above more than

once, is the following: "A true sentence is one which says that the state of affairs is so and so, and the state of affairs is *indeed* so and so" (1933/1956: 155; emphasis added). The little word "indeed" is the decisive factor in this characterization, because without the "indeed," what the true sentence is supposed to "say" would be a senseless tautology or repetition of the same formulation (i.e., ". . . says that the state of affairs is so and so, and the state of affairs is so and so"). Tarski's "indeed" is the counterpart to Quine's "really." But in his semi-formalized truth-schema, Tarski in no way articulates this decisive factor; he does not make it explicit, despite his claim that the truth-schema articulates correctly and clearly his accurate formulation of the intuitive understanding of truth.[40] This omission by Tarski is the source of the basic error made in all deflationistically oriented truth-theories.

303

3.3.2 *The So-Called "Truth-Bearers" and the Fundamental Structures*

Contemporary truth-theoreticians regularly speak of "truth-bearers," but there is no consensus concerning their number or their interrelations. Almost all such theoreticians, however, consider sentences to be truth-bearers; some add propositions, and some others also events of utterance. Philosophers before the time of analytic philosophy generally assume that judgments are the proper truth-bearers; the accepted axiom is "*veritas est in iudicio. . . .*"

Talk of truth-bearers arises when and because "truth" is conceived of as a predicate designating a property that, for its part, would require or presuppose a "bearer." In ordinary language the proposition is generally viewed as the genuine or primary truth-bearer, as shown by examples like, "That person A committed crime C is true." Here, the predicate "is true" is applied to the nominalized sentence, "That person A committed crime C"; because this sentence expresses a proposition, the predicate "is true" designates a property of the proposition.

The sentence is generally designated as a truth-bearer only in a more elevated, educated, and indeed scientific form of talk: "(The sentence) 'Person A committed crime C' is true." Since the time of Tarski's epochal work on the truth-concept, most truth-theoreticians have accepted this form as the proper and primary (often as the only relevant) form of truth-talk. As indicated in Section 2.5.1.4, however, talk of "truth-bearers" loses its proper sense when "true" is explained not as a predicate,

---

[40] In the later essay "The Semantic Conception of Truth and the Foundations of Semantics" (1944), Tarski surprisingly asserts that the "indeed" (there in the form of "in fact") "does not occur in the original formulation." The original Polish version of Tarski's monograph, however, contains a word that is accurately translated in English as "indeed" and in German as "*eben*"; this is the word "*wlasnie*" (see Puntel 1990: 43n29). In his essay of 1944, Tarski correctly rejects the *epistemological interpretation* of this "indeed" that was being promulgated at that time, incorrectly, by a number of authors. This rejection is correct because the "indeed" answers not a question of justification, but instead one of intelligibility: Precisely how is the sentence "The state of affairs is *indeed* so and so," which appears in the informal characterization of truth cited in the text, to be understood? This is a semantic question, not an epistemological one. In his later, negative position with respect to the "indeed," Tarski is concerned exclusively with showing *epistemological interpretations* of his T-schema to be fundamental *misinterpretations*. See Puntel (1990: 41ff).

but—as in this work—as an operator ("it is true that . . ."). An operator taking sentences and/or what they express (propositions) as arguments has no "bearer" in the proper sense. Nevertheless, one can use the term "truth-bearer," in a broad sense, to name the entity that the truth-operator determines or qualifies more precisely: the   304 sentence and the proposition.

In the terms used above, the sentence proves to be a purely derivative truth-bearer, because the definition of the truth of the sentence involves nothing other than reference to the truth of the proposition: a given sentence S is true if and only if the proposition that it expresses is true. The crux of the matter, therefore, is the truth of the proposition.

Within the current systematic framework, an additional qualification may be added: sentence and proposition are truth-bearers (in the very broad sense) only because they are, in the final analysis, structures: purely linguistic structures (sentences) and (linguistic-)semantic structures (propositions). Purely linguistic structures in the sense of purely syntactic structures are not analyzed as such in this book, but only considered within the framework of analyses of semantic structures. From what has just been said it follows that one can say not only "The proposition P is true," but also "The structure P is true." The latter formulation is to be sure extremely unusual, but it is fully consequent and compelling within the framework of the systematic theory of truth presented here.

3.3.3 Truth as Composition of Three Functions: The Tristructural Syntactic-Semantic-Ontological Connection

3.3.3.1 The Syntactic-Semantic Dimension: A "Cataphoric" Theory

The task now is gradually to put the preceding contentions concerning the truth-concept into a more precise form, and ultimately into the form of a theory, albeit one that is only sketched programmatically.

[1] The structural-systematic account has in one respect a certain similarity to the *prosentential theory of truth* initially developed by Dorothy Grover, J. I. Camp, Jr., and Nuel Belnap (1975), and modified by Robert Brandom (1994). The central idea of this theory is that "true" is not a predicate, but a fragment of a prosentence. The prosentence is analogous to the pronoun—to a term that has an *anaphoric* relation to another (already used) term, its antecedent. Those who hold this theory rely on what they call the deep structure of language. Sentences containing the truth-term (in any of its various forms) are interpreted as sentences referring back to sentences that have already appeared (generally without the truth-term). Linguistically viewed, these authors contend, every occurrence of the truth-term has the form "That is true," i.e., the form of a prosentence whose sole meaning and function are to refer to a sentence that has already appeared. According to this anaphoric conception, the famous example "'Snow is white' is true" is   305 to be analyzed as "Consider: snow is white. That is true" (Grover et al. 1975: 100).

Robert Brandom has made this conception significantly simpler and more intelligible by interpreting ". . . is true" not as a syncategorematic fragment of a semantically atomic sentence, but instead as a prosentence-forming operator:

> [This operator] applies to a term that is a sentence nominalization or that refers to or picks out a sentence tokening. It yields a prosentence that has that tokening as its anaphoric antecedent. (1994: 305)

The prosentence generated by the operator has an *anaphoric character*: it points back to or retrieves the sentence that is the argument of the operator. It thus holds that "It is true that snow is white" is a prosentence whose status consists in its retrieving the sentence—presupposed already to have appeared—"Snow is white." In addition to being anaphoric, Brandom's theory is deflationary: "'It is true that snow is white' expresses just the same fact that 'snow is white' expresses" (328).

[2] In opposition to Brandom, the structural-systematic theory of truth is *cataphoric*. καταφορά is the conceptual counterpart to ἀναφορά. But the two conceptions share one fundamental element, which makes the difference the more apparent. Common to the two is the assumption that the truth-term is not a predicate but an operator, or should be so interpreted. According to the structural-systematic conception, however, what this operator initially forms are not *prosentences*, but what may be termed *persentences*. The artificial term "persentence" is formed from the Latin *"perficere"* (to finish, to accomplish, to perfect, to determine fully). The "per" in "persentence" is to be understood as abbreviating the gerundive form; "persentence" is then *"sententia perficienda,"* "sentence to be completed or perfected." This reading is meant to indicate that the formation of the persentence is not the only accomplishment of the truth operator and thus that the operator is a composite function.

To make clearer what is meant by "persentence" (and/or "perproposition"), a comparison with the central thesis of the *anaphoric* theory of truth is helpful. According to this theory, the truth operator is one that, so to speak, *points backward*: the value of the operator is the same sentence that is its argument, which means that the sentence in the place of the value of the operator is simply identified with an (earlier) tokening of precisely this sentence. The truth operator, interpreted *anaphorically*, thus articulates a retrospective relation to a tokening of the sentence. One can interpret this anaphoric relation more precisely as follows: the sentence, as argument of the truth operator, is both *syntactically* and *semantically identical* with an earlier tokening of precisely the same sentence; "truth" thus means nothing other than the articulation of a backward-pointing or retrospective retrieval of this syntactically *and* semantically identical item. This presupposes that the semantic status of the sentence is precisely the same in the two cases: the sentence is assumed to be and understood to be *fully semantically determined* in advance, independently of whether it appears as complete in isolation or as the argument or as the value of the truth operator. This assumption or presupposition finds its most interesting expression in the above-described *anaphoric deflationism*.

Cataphorically interpreted, the truth operator articulates a relation that is, in some respects, precisely the opposite. To be sure, here too there is a *syntactic* sameness or identity between the sentence *qua* argument of the truth operator and the sentence *qua* value of the operator. However, the *semantic status* of the sentence, as argument and as value, is *not the same, not identical*: *qua* argument of the truth operator, the sentence has, in the first phase of the explication of the truth operator, an un- or underdetermined semantic status, whereas in the second phase it has a semantic status to be determined, and in the third phase, a fully determined semantic status. The three phases here introduced are formally to be interpreted as *three functions* that explain the cataphorically understood truth operator. As a whole, the truth operator thus appears as *the composite of the three functions* that are explained and defined in detail below.

306

At this point, the aim is only to clarify the *cataphoric* character of the thus explained truth character. It is clear that the cataphorically explained truth operator is *forward-pointing* in the sense that it always points toward a or the more complete semantic determination—up to the level of full semantic determination—of the sentence that qualifies as true; at that level, the sentence is interpreted ontologically. "Persentence" is the designation of the sentence *qua* value of the first of the three functions defining the truth operator: the sentence *qua* value is a persentence in the sense that it is a *sententia perficienda*, a sentence that is *to be determined*.

To simplify the presentation, the following account explicitly considers, for the most part, only the sentence, and indeed as simple (not yet more precisely determined) sentence, as persentence, and as fully determined sentence. The concern is exclusively with indicative sentences, here understood as *primary* sentences. Because however the conception defended here includes the central thesis that every indicative sentence in the presupposed language has an *expressum*, termed the proposition (here understood as *primary* proposition), the explanation of the sentence in the three forms just introduced applies, correspondingly and indeed essentially, also to the proposition, likewise in the three forms: as simple (not yet more precisely determined) proposition, as perproposition, and as fully determined proposition. The three functions that explain the operator "It is true that" thus have as arguments both the three forms of the sentence introduced above and the three corresponding forms of the proposition. 307

[3] This may appear to be quite complicated. It would be simplest and most illuminating to use the reinterpreted or reconstructed Tarskian truth-schema introduced above:

(T_R) It is true that p ↔ **p**.

Yet if one reduces the presentation of this truth theory to this simple and illuminating schema, essential moments remain unexplained, with the usual result that spurious questions and problems arise, and basic presuppositions and entailments remain unexamined and thus appear problematic. It is therefore better to complete a meticulous, indeed pedantic explication of the truth-concept, one that makes explicit each of the individual moments of the conceptual determination of "true/truth."

When all the points made thus far are taken into consideration, the truth operator—T—appears as a *composite* (three-step) function. Symbols for the three functions are, respectively, T^*, T^+, and T^x.

Before each of the three functions is defined precisely, an explanation of the underlying intuitive idea is required. The so-called truth-concept is here understood as a truth operator of the form, "it is true that. . . ." This operator, T, is completely explained only by means of distinctions among and the articulation of *three stages* of its application; viewed formally, these three stages are three functions that form a composite function; the truth operator is explained precisely as this function. Careful attention must however be paid to *how* the *threefold* application of the operator is to be understood more precisely. Negatively, the application is not to be understood such that the truth operator in the form or with the reading "it is true that" would be simply mechanically repeated three times by being applied to three (different) arguments. Positively, the application is to be understood as the complete explanation of the operator, in three stages. The more precise linguistic form in which the truth operator appears or is to be read at each stage can be made clear by quite artificial formulations (such formulations being allowable in the interest of clarity).

308   A quite different means of presentation can also be used to explain the fundamental idea upon which the three-staged truth operator T is based—a chain of explicative equivalences.[41] Paraphrasing an example by exclusively propositional-logical means, the explanation of T* has two alternative formulations:

> It is true that snow is white ↔ it is semantically to be qualified or determined that snow is white.

Or:

> It is true that snow is white ↔ it is semantically qualifiable or determinable that snow is white.

What is expressed here by the laboriously formulated operator "it is semantically to be qualified or determined that" (or "it is semantically qualifiable or determinable that") corresponds exactly to the concept of the persentence/perproposition introduced in Section 3.3.3.1[2].

The second operator, T+, can likewise be articulated informally by a paraphrasing of the example:

> It is established that it is semantically to be qualified or determined that snow is white ↔ it is semantically fully determined that snow is white.

Or:

> It is established that it is semantically qualifiable/determinable that snow is white ↔ it is semantically fully determined that snow is white.

Finally, the paraphrasing of the example in the case of Tx is as follows:

> It is semantically fully determined that snow is white ↔ it is a fact that snow is white.

The complete explanation of the truth concept can thus be formulated as a *chain of explicative equivalences*:

> It is true that snow is white ↔
> it is semantically to be qualified or determined (or: it is semantically qualifiable/
> determinable) that snow is white ↔
> it is semantically fully determined that snow is white ↔
> it is a fact that snow is white.

---

[41] The following account does not explain what a mathematical function, in the strictly formal sense, is. Instead, propositional-logical means are used to articulate the steps that must be taken to explain the truth-concept.

[4] The three equivalences are now presented and defined formally as three functions.

The first function expresses what happens *initially* or *in a first step* when the truth-operator in the form "it is true that" is applied to a sentence or a proposition.

(T_1) $T^*: X \to Y$

   $T^*: p \mapsto p_{PER} \in Y$

where 'X' is the set of fully syntactically determined but incompletely semantically determined sentences or propositions 'p' to which the operator in the form 'it is true that' is applied, and 'Y,' the set of persentences or perpropositions 'p_{PER}' resulting from the application of the operator.

The second function articulates the way in which the (to be completed or perfected) persentence or perproposition is in fact completed or perfected (i.e., such that it attains a fully determined semantic status):

(T_2) $T^+: Y \to \mathbf{Z}$

   $T^+: p_{PER} \mapsto \mathbf{p} \in \mathbf{Z}$

where 'Y' is the set of persentences or perpropositions, and '$\mathbf{Z}$' the set, which results *cataphorically* from the persentences, of fully semantically determined sentences and propositions '$\mathbf{p}$.'

The third function, $T^\times$, which articulates the ontological import of the truth operation, must be analyzed more scrupulously; its analysis is the task of the following section.

3.3.3.2 The Semantic-Ontological Dimension: The Identity Thesis

3.3.3.2.1 *The Fully Determined Semantic Status of Language and the Ontological Dimension*

It is incontestable that, in the entire history of philosophy, the relation to actuality—to the ontological dimension—has been considered to be an essential constituent of what is true. Throughout that history, the utterly natural articulation of this insight is the correspondence theory of truth (which is still accepted in many philosophical circles). With few exceptions, even the authors who reject this theory do not reject the relation to reality; they see it, if not as intrinsic to truth, at least as an indispensable constituent of talk about truth. To provide merely one contemporary reference: deflationists about truth do not reject this ontological feature in every respect, but they fail to explain it adequately. The characteristic case of Quine is introduced above. Other radical deflationists also attempt to avoid altogether excluding from their truth-theories the relation to reality. Paul Horwich provides a symptomatic example of such an attempt. He maintains this of minimalism, his brand of deflationism:

> It does not deny that truths *do* correspond—in *some* sense—to the facts; it acknowledges that statements owe their truth to the nature of reality.... It is indeed undeniable that whenever a proposition or an utterance is true, it is true *because* something in the world is a certain way. (1998: 104)

Horwich's "because" conceals fundamental problems. His contentions reduce the ontological dimension to a factor that is not an internal ingredient of truth and whose status is extremely ambiguous (for a detailed critique, see Puntel 2001: 222ff).

Ultimately, the deflationists presuppose that any given indicative sentence, in isolation, has a fully determined semantic status and that the ontological import of truth is thereby somehow "given." That such a position involves various confusions is demonstrated thoroughly in Chapter 2. In the current context it is important only to emphasized that this shows that even the deflationists acknowledge, albeit in an inadequate fashion, the ontological import of truth.

3.3.3.2.2 *The Ontological Import of Truth as Identity of Proposition and Fact (the Identity Thesis)*

[1] How is the ontological import of truth to be explained precisely? To explain it by means of the traditional correspondence theory of truth is to encounter enormous problems: this theory involves obscurities and difficulties that appear to be insuperable. Here the correspondence theory is not defended, although its basic idea is in a certain respect retained. Considerations of space preclude complete presentation and grounding of this book's conception of truth. Because this book presents only a partial and thus a limited concretization of the theoretical framework of the structural-systematic philosophy, the conception defended here is not developed in every respect.

The primary and fundamental arguments for the operator "It is true that" are primary propositions expressed by primary sentences. Thus, the definitions of the truth of the primary sentence and of the utterance are dependent on the definition of truth of the primary proposition expressed by the sentence or intended by the utterance. A given primary sentence is true just when it expresses a true proposition (the true utterance would be explained in comparable fashion, in relation to the true sentence, but is not here considered). But what is a true proposition? The ontological perspective is central to answering this question.

The idea of an "identity" of thinking and being (reality) has an age-old tradition in philosophy. But this idea has always been rather vague. In 1989, Stewart Candlish uses the expression "Identity Theory of Truth" to designate F. H. Bradley's thesis that truth is identical with reality.[42] There ensues an intensive discussion about this expression. It has been asked whether such authors as Hegel, Bradley, Moore, Russell, and Frege, among others, hold identity theories of truth.[43]

In 1990, the author (Puntel 1990: 325) introduces an identity *thesis*: a true proposition "*is* nothing other, nothing less and nothing more, than a constituent of the actual world." If one calls the constituents of the world articulated by true propositions "facts," then the true proposition and the fact are one and the same entity. That is an identity thesis with a precise content (see also Puntel 1999b).

The structural-systematic conception cannot be designated simply an *identity theory of truth*, because the identity (between true primary proposition and primary

---

[42] See Candlish (1989: 338). J. N. Findlay asserts already in 1933 that "Meinong's theory of truth is . . . a theory of identity or coincidence" (88).

[43] See, among others, Baldwin (1991), Dodd and Hornsby (1992), Stern (1993), Dodd (1995, 1999), and Hornsby (2001).

311

fact) is only one of its moments: it articulates only one—and indeed the third—of the functions constituting the truth operator as composite. Its truth-concept cannot therefore simply be identified with the identity thesis. The various versions of the identity theory ignore the other aspects—more precisely, the other functions—which are, however, essential ingredients of the truth concept.

[2] Frege formulates the identity thesis with maximal concision in a passage cited above more than once: "What is a fact? A fact is a thought that is true" (to be noted is that Frege generally uses "thought" as synonymous with "proposition"; contexts make clear when this is not the case). It is clear that the "is" in the formulation "A fact *is* a thought . . ." is to be read in the sense of identity. It is symptomatic that Frege's formulation has provoked quite controversial discussions that continue into the present. The chief point of contention concerns the concept of the fact. Julian Dodd (1995: 161n44) distinguishes a "robust" from a "modest" sense of fact and attributes to Frege the modest sense. In Dodd's robust sense of identity, facts are entities in the world, whereas in his modest sense they are not. Correspondingly, Dodd distinguishes between a "robust" and a "modest" identity theory of truth. But this discussion suffers from helplessness precisely with respect to the ontological dimension. The structural-systematic theory avoids this helplessness, presenting facts not as merely *among* the entities in the world, but as the *only* entities in the world, as precisely the entities of which the world consists.[44]

312

---

[44] The ontological helplessness just mentioned sometimes bears strange flowers. To show that Frege's "facts" are to be understood "modestly" rather than as "worldly entities," Dodd argues as follows: "Frege famously identified facts with true Thoughts. . . . But because Thoughts, and hence facts, have senses, and not objects and properties, as constituents, he did not take facts to be worldly things. For Frege, facts are true Thoughts rather than occupants of the world" (1995: 161–62).

The identification of Frege's true thoughts with facts in the world would, according to Dodd, yield the following incoherence: Fregean Thoughts are (for Dodd) "things with senses as constituents. This, however, prompts the following problem: facts (if worldly) and Thoughts are in quite different categories, and so the identification cannot be made good" (163).

As noted above, Frege does indeed distinguish between "sense" and "*Bedeutung*." But from that it does not follow that they have nothing to do with each other. Quite to the contrary, senses are meaningful only in relation to the corresponding *Bedeutungen*. Sections 3.1.2.1[2] and 3.2.2.3.1[2] introduce various passages in which Frege articulates the necessity of the step from sense to *Bedeutung*. To be sure, one cannot simply identify the Fregean sense with the Fregean *Bedeutung*, and Frege of course does not do that. But Frege identifies *true* thoughts with facts. A *true* thought is a sense for which the step to *Bedeutung* is realized. The thought as pure sense is thereby surpassed—precisely in that it is true. Of course, the step from sense to *Bedeutung*, in the case of the sentence, is understood by Frege as a step to a truth-value, not to a fact. But, as shown above, that raises the question of the coherence of Frege's position as a whole. What are truth-values, according to Frege? This issue cannot be considered here in further detail.

An additional complication is that it appears scarcely possible to bring Dodd's interpretation of facts as non-worldly entities into harmony with the main stream of Frege's thought. The sense of a sentence, the thought, gains the qualification "true," such that the so-qualified thought is identified with a fact. If Dodd's interpretation were correct, that would mean that the determination "true," with the identification with fact resulting from it, would be *immanent* to the dimension of sense. But how could anything of this sort be conceived? One would have to introduce

313   [3] The identity of true primary proposition or true semantic primary structure and primary fact (as worldly entity) can be related to the thought of correspondence. Throughout the history of the correspondence theory, the concept of correspondence has always been understood as a relation between two *non-identical relata*; the structural-systematic theory's identity thesis is *not* a correspondence theory in *this* sense. Nevertheless, one must say that the identity thesis completely preserves the indispensable core of the correspondence theory, but without inheriting its problems. This core is the relation to the world as an absolutely essential ingredient of truth. This core is preserved in that identity can be understood as a *limiting case*—one might also say "highest" or "perfect" case—of correspondence. That this core is preserved is an invaluable advantage of the structural truth-theory.

As the considerations introduced so far show, the two functions T^* and T^+ do not provide, even in conjunction, a complete explanation of the truth-concept. The factor introduced by the identity thesis must be added as a *third function* ($T^\times$). This third function can be formulated as follows:

$$(T_3) \quad T^\times : \mathbf{Z} \to \mathbf{F}$$
$$T^\times : \mathbf{p} \mapsto \mathbf{f} \in \mathbf{F}$$

Here, '$\mathbf{Z}$' is the set of fully semantically determined sentences or propositions and '$\mathbf{F}$' the set of facts, such that

$$\forall \mathbf{p} \in \mathbf{Z}, \forall \mathbf{f} \in \mathbf{F} \ (\mathbf{f} = T^\times(\mathbf{p}) \leftrightarrow \mathbf{p} = \mathbf{f})$$

The complete truth-concept (CTC) is thus:

$$(CTC) \quad T = T^\times \circ T^+ \circ T^*$$

The final step reveals that the structural theory preserves not only the key insight of correspondence theorists but also, in a certain respect, that of the disquotationalists: the final step performed in the (self-)explication of the truth operator is disquota-
314   tional in that it is the explicit identification of an item within the semantic dimension with an item within the ontological dimension. For the disquotationalists, of course, what results from disquotation remains a *linguistic* item; the idea here that is related to the disquotationalists' idea—the idea of the identity of true primary proposition and primary fact in the world—explicitly articulates the relation between language and world.

---

into the sense-dimension a distinction as great as that between pure thoughts, true thoughts, and thoughts as facts (but not of a worldly nature). In addition, there is the question whether true thoughts *qua* non-worldly entities could at all involve a relation to the world, to the ontological dimension, and if they could, how such a relation could be determined. Such a multiplication of entities appears to be wholly alien to Frege's way of thinking.

3.3.3.2.3 *The Ontology of Primary Facts as the Ontology Appropriate to the Structural Truth-Theory*

Talk of an identity between a true primary proposition or semantic primary structure and a primary fact (in the sense of a worldly entity) is in one respect quite strange and in another quite illuminating. It is strange because it seems to contradict many ideas often held dear. These are the ideas according to which propositions are strictly mental or ideal entities and facts are worldly entities, such that they belong to two utterly different domains. The identity talk is nonetheless illuminating in that it provides the most precise articulation of truth's relation to the world. The genuine philosophical problems come into view only if one looks more closely. They arise with the question, how is the identity of true primary proposition and primary-fact-in-the-world to be thought *coherently*?

Authors who defend identity *theories* of truth presuppose the traditional ontology—substance ontology—according to which the world consists of objects, properties, relations, *and* facts. Facts are understood vaguely as states of affairs involving objects, properties, and relations. How these facts are to be understood more precisely is left unexplained. As a rule, "facts" are spoken of in a purely intuitive sense. The presupposed ontology remains utterly opaque. It is thus wholly understandable that within such a framework the question arises whether facts can possibly be entities in the world. Only in quite recent years has analytic philosophy begun to seek an adequate ontology. The search has brought the concepts "state of affairs" and "fact," *within the framework of substance ontology*, radically into question.[45] But the new ontological initiatives have so far had few if any effects on work on theories of truth.[46] And this is a deep and vital problem,   315

316

---

[45] See, e.g., Dodd (1999) and Vallicella (2000). For a brief critique of Dodd's position and argumentation, see Puntel (2001: 252n7) and note 44 above.

A noteworthy attempt to clarify the problematic of the concept of the fact is undertaken by Stephen Neale in his book *Facing Facts* (2001). He treats fundamental aspects of this concept with respect to epistemology, semantics, logic, and metaphysics. He shows that it is indeed possible to avoid the putative "collapse of facts" that many analytic philosophers (above all Davidson) maintain on the basis of formal-logical and semantic lines of argumentation, but he also shows that this avoidance presupposes radical clarifications and corrections concerning how predicates and descriptions are understood. Nevertheless, Neale himself does not put into question either the ontology—termed in this book *substance* ontology—defended and/or presupposed by analytic champions and opponents of facts—or the semantics, based in the principle of composition, that is usually associated with this ontology. He therefore misses the decisive point with respect to clarifying the problematic of the concept of the fact. The phrase "collapse of facts" generally refers to a famous argument that, because of its brevity, is termed the "slingshot argument" (see Barwise and Perry 1981; the detailed reconstruction and analysis of this argument in Puntel 1990: 163–74; and Neale 2001). Here, it need only be noted briefly that this argument is based on premises that presuppose a semantics based on the principle of composition and thus on the acceptance of sentences having the subject-predicate form, and of the utilization of a corresponding first-order predicate logic. The famous argument thus has no force against the non-compositional semantics defended in this book. Whether or not the argument is conclusive with respect to compositional semantics is thus of no relevance here.

[46] Many authors defend the correspondence theory of truth on the basis of a "metaphysically thin" conception of facts. A characteristic example is Searle's (1995: Ch. 9) attempt to defend a

because *any truth-theory that accords truth an ontological import must remain vague and unconvincing until the ontology it maintains or presupposes is made explicit and investigated critically.*

Fully grounded at this point is the thesis that the ontology sketched in this chapter is precisely the ontology required by the structural-systematic conception of truth. It is, however, to be emphasized that this ontology develops not only or even primarily in conjunction with a clarification of the truth-concept; instead, it is a self-sufficient discipline, albeit one that must be seen as connected with all other disciplines, and thus also with the theory of truth.

The central thesis of this ontology is formulated clearly in traditional terminology in the second sentence of Wittgenstein's *Tractatus* (1.1): "The world is the totality of facts, not of things." An improved formulation is this: the world is the totality not of things but of expressible primary facts that are ontological primary structures. This formulation makes explicit that this ontology dovetails perfectly with the structural-systematic theory of truth.

3.3.4 *Three Concluding Questions*

The truth-theory sketched above seeks only to clarify the fundamental idea of truth. A fully developed theory of truth would have to treat an extensive array of additional questions, and the results would have to be integrated into a comprehensive theory. This challenging task cannot be undertaken here. Nevertheless, three questions must be introduced that are of particular importance to the comprehensively systematic purposes of this book: the question of falsity, the question of the ontological import of the truth-concept with respect to its use in the domains of the formal sciences

---

"substantive" correspondence theory putatively able to avoid the notorious problems of other forms of the theory. According to Searle, a "substantive" correspondence theory maintains that sentences are "made true" by the relation of correspondence to facts. Although he himself conceives of facts as non-linguistic entities, he maintains that his correspondence theory in no way requires a "thick" notion of fact; instead, it is wholly sufficient to understand facts as "*conditions—* specifically, conditions in the world that satisfy the truth conditions expressed by statements" (211–12). He then concludes that "for every true statement there is a corresponding fact, because that is how these words are defined" (214). He introduces a similarly thin concept of correspondence: "'Correspondence to fact' is just a shorthand for the variety of ways in which a statement can accurately represent how things are, and that variety is the same as the variety of statements, or more strictly speaking, the variety of assertive speech acts" (213). The questions to be raised to all this are, what can these "thin" conceptions achieve and why should they be taken seriously? If facts are understood as "conditions in the world," one would like to know what, precisely, they are. The simple assumption of a distinction between "thin" and "thick" entities in the world says little, indeed nothing, and remains empty until it is precisely explained and justified.

A similar criticism should be addressed to Mellor, who has written a book (1995) with the title *The Facts of Causation* that has only this to say about the concept *fact*: "Actual states of affairs, corresponding to true statements, I shall call facts, like the fact that Don falls, which exists iff 'Don falls' is true" (8). Later (166), he characterizes "facts" as "entities trivially entailed by truths." One wonders what is hereby gained. The question immediately presses: what exactly *are* "entities trivially entailed by truths"?

of logic and mathematics, and the question whether this truth-theory entails a specific form of relativism with respect to truth. These questions must be treated in detail sufficient to make this book's theory of truth intelligible and to reveal its coherence.

3.3.4.1 Starting Points for a Theory of Falsity

Are there false primary sentences? If so, do they express false primary propositions? Are false primary propositions identical with anything ontological? Are false primary sentences or propositions fully determined? Such questions emerge quite evidently from the truth-theory presented above. (In the following treatments of them, the epithet "primary" is often omitted, particularly when the concern is not with the specific semantic conception of this book.)

Falsity is normally considered to be a truth-value, a negative truth-value. But mustn't one assume that every sentence or proposition having a positive or negative truth-value has a fully determined status? The highest qualification that a sentence or proposition can have is that of its truth-value. If this is so, how is the *negative* truth-value *falsity* to be conceived?

The acceptance of anything like negative facts can scarcely be considered because negative entities of this sort appear to be unintelligible. Given that, three ways of determining falsity are available.

[1] The first option would be to accept unrestrictedly the thesis that false sentences are fully determined sentences that express fully determined propositions (because the proposition is the crucial factor, false sentences are for the most part not explicitly considered.) To the question, within the framework of this option, of how fully determined but false propositions are to be understood, the answer is at best quite vague: a false proposition would be, say, an abstract proposition that would not be identical with any fact in the (actual) world. Here, the status of the false proposition would be explicitly characterized only negatively, through its non-identity with any fact in the (actual) world. But the question arises of what this abstract entity, characterized only by a negation, is, considered in itself. Simple answers, such as that the so conceived false proposition is a pure idea, a pure construct of our language, etc., no matter how popular and generally intuitively understandable they might be, are not serious philosophical answers, not at least within the framework of the systematic philosophy presented here. The only option would be to provide no answer to the question, and thus to make do with the negative "determination" of false propositions. That would, however, not only be wholly unsatisfying, it would also be a simple failure to make a serious philosophical effort.

[2] A second possible option would be to accept only a restrictedly full determination of false propositions. It would rely on a fundamental distinction: false propositions would be fully determined propositions in a *negative* respect, but would be un- or underdetermined in any *positive* respect. In favor of this option is that a proposition determined to be false would then say *in a wholly determined manner* that it was *not identical* with a fact in the world: negation in the sense of exclusion of the false proposition from the world is fully unambiguous. Positively, the false proposition would however be un- or underdetermined, because nothing determinate would be said about it.

To this option it is to be said that un- or underdetermined entities cannot be made intelligible. Moreover, the mixed status—a mixture of negative and positive—of false propositions is a philosophically weak, indeed problematic and therefore insufficient characterization. To be sure, the distinction as such is clear, but the following question arises: how can any entity that, considered in isolation, has an un- or underdetermined status, be fully determined negatively? Can something that is un- or underdetermined fully exclude anything at all? The notion that it could appears not to be intelligible.

[3] The third possibility, like the first, contends that the status of false propositions is that of unrestrictedly full determination but, unlike the first, the third completely explains this status. To do this, however, it must make an ontological-metaphysical assumption considered risky by many analytic philosophers: it must rely on a distinction between actual and possible worlds, and one that is supported by a realistically oriented semantics and therefore ontology of possible worlds.[47] It then determines the false proposition as a fully determined proposition that is not identical with any fact in the actual world, but is identical with a fact in some possible world. It thus exists in some possible world. The negative moment of falsity is thus understood by means of the disjunction between the actual world and some possible world, whereas the positive moment is interpreted as the identity of the proposition with an entity in some possible world.

How, within the framework of this solution, is what is often termed the *necessary falsehood* of some propositions to be explained? If a given proposition were necessarily false, then according to the current account there would be no world, in the sense of no even possible world, wherein there would be an entity identical to this proposition. From this it would follow that the necessarily false proposition would not be identical with any entity of any sort in any world of any sort. To the fully determined status of any proposition, however, belongs an identity function with respect either to the actual or to some possible world. The only consequent solution would appear to require denying fully determined status to necessarily false propositions.

The genuine solution must, however, have more to say. The concept of an incompletely determined proposition presupposes analytically that the proposition can, in principle, *be completely determined*. With necessarily false propositions, however, this is not the case. In fact, such propositions are—and are better termed—*contradictions*. These must be interpreted as *pseudo*-propositions in the following sense: any such "proposition" purports to combine two propositions or states of affairs (it is here irrelevant whether one speaks of two propositions, two concepts, two entities, etc.). Each of these putative combinata is, for its part, intelligible (this is here assumed)—each can be identified with an entity or fact at least in some possible world. The *combination* of the two is not, however, realizable in any world. What is contradictory about pseudo-propositions means precisely what is described above: the non-realization and indeed non-realizability of the *combination* of two states of affairs in any possible world. But the combination of two states of affairs that is not realizable in any world has by definition no fully determined status or, more precisely, it *necessarily* has no fully determined status. "Propositions," however, that are necessarily undetermined are not genuine propositions and can therefore be designated *only* as *pseudo-propositions*.[48]

319

---

[47] Chapter 5 treats the semantics and ontology of possible worlds.

[48] Such "propositions" are pseudo-propositions rather than non-propositions (1) because the sentences expressing them are syntactically correct and (2) because their analysis reveals that, as

To be noted in conclusion is that the fully developed structural-systematic theory 320 of truth would have to introduce additional elements, including *indices*, to clarify fully the problem of falsity. These additional elements should reveal the precise sense in which—here, with respect to what "world"—a proposition qualified as true is identical to an entity in that world.

3.3.4.2 On the Ontological Import of the Truth of Formal (Logical and Mathematical) Propositions or Structures

The explanation provided above of the third function in the definition of the truth-concept treats the fully determined status of propositions that qualify as true, and interprets this status as the identity of the true proposition with a fact in or of the world. That section speaks of "the world" in a quite intuitive sense: the world of concrete things or the world as the real universe, again in a generally intuitive sense. The sciences and philosophy alike are concerned with such a world. But there are also formal (logical and mathematical) propositions, structures, and facts, and these have no obvious status within this intuitively understood world. At the same time, the structural-systematic truth theory appears to entail that logical and mathematical truths also have an ontological status and thus a relation to "the world," to "facts," etc. How is this relation to be understood? This is doubtless a central question with which this philosophy must deal. The following account introduces only the basics.

The starting point is the ontological character of formal propositions or structures as that character is presented in this chapter (Section 3.2.1) and in the chapters that precede it. Thus far, this character has emerged only with respect to logical and mathematical propositions or structures as, so to speak, *applied*—thus, only insofar as they are indispensable constituents of semantic and ontological structures. The ontological character of formal propositions or structures, understood in this manner—as 321 applied—is understandable more easily in that they appear not only as themselves, but in their interplay with semantic and ontological structures. Here, however, the question concerns the ontological character of these pure structures when they are *not* applied. The structures are encountered in this manner when logical or mathematical sentences, propositions, or structures are *themselves* designated as true. The concern is thus with the concept of truth for pure mathematics or pure logic.

---

configurations, they include individual propositions that are or could be true either in the actual world or in some possible world—thus, propositions that are automatically identical with facts in such a world or in such worlds. For example, the sentence "Fred drew a round square" could be analyzed as follows: "What Fred drew was round" and "What Fred drew was a square." Each of these individual sentences expresses a proposition that is or at least could be true in this actual world (and in other possible worlds); they thus express facts (in such worlds), but the conjunction of the two sentences does not express a proposition that is or could be true in this or in any possible world; it thus does not express a proposition that could be identical with a fact in this or in any possible world. (To be sure, it must be added that in accordance with the technical terminology used in this book neither of the two components of the conjunction is a sentence in the genuine sense; instead, each is an abbreviation for a great number of primary sentences, each of which expresses a primary proposition. But this is only a technical complication that must be noted to prevent misunderstandings.)

From what has already been argued it follows immediately that if logical and mathematical propositions or structures have an ontological character *when applied*, their pure forms must also have an ontological character, for otherwise their attainment of ontological status, when applied, would be some sort of miracle. The question, then, is this: how is the ontological status of pure logical and mathematical structures to be understood?

In this context, it is useful and meaningful to use the term and concept "logical/mathematical world." For the application of the truth-concept to logical/mathematical structures it suffices in principle to assume that every logical/mathematical structure that qualifies as true is identical with a fact—an entity—in this logical/mathematical world, however one may conceive of this world (as an abstract "Platonic" world, as a constructed conceptual world, etc.). Every logical/mathematical proposition that qualifies as true *is* a logical/mathematical fact in this logical/mathematical world. It is interesting to note that the term "logical fact" is currently used non-problematically to designate what are here also termed logical *structures*.

Another question is whether this logical/mathematical world can be understood adequately as something like an intellectual construct. This book decisively rejects any such understanding. Here provided are only what can be termed the *framework* reasons for this position. This book seeks to develop a *coherent* and comprehensive grasp of the entire universe. This cannot be accomplished on any purely materialistic-physicalistic basis, because no such basis can make intelligible, among other things, what it means that the immense logical/mathematical dimension belongs to this universe, in the literal sense of being a dimension of *this* universe. All interpretations of the logical/mathematical world that are not somehow realistically oriented are ultimately characterized by a radically immanent dichotomy that involves an inner incoherence.

322

One indication of this inner incoherence is the following significant fact: the extent of the logical/mathematical world is far broader than that of the physical world, because only a quite narrow segment of logical/mathematical structures are instantiated or realized in the physical world. From this it follows that the physical world is, so to speak, contained or embedded within the broader formal world, and not vice versa.

A characteristic example of the alternative but incoherent position is Quine's, as expressed in the following (1986: 400):

> Pure mathematics, in my view, is firmly embedded as an integral part of our system of the world. Thus my view of pure mathematics is oriented strictly to application in empirical sciences. Parsons[49] has remarked, against this attitude, that pure mathematics extravagantly exceeds the needs of application. It does indeed, but I see these excesses as a simplistic matter of rounding out. We have a modest example of the process already in the irrational numbers: no measurement could be too accurate to be accommodated by a rational number, but we admit the extras to simplify our computations and generalizations. Higher set theory is more of the same. I recognize indenumerable infinites only because they are forced on me by the simplest known systematizations of more welcome matters.

---

[49] Quine refers to Charles Parsons (1986).

Magnitudes in excess of such demands, e.g., $\beth_\omega{}^{50}$ or inaccessible numbers, I look upon only as mathematical recreation and without ontological rights. Sets that are compatible with 'V = L'[51] in the sense of Gödel's monograph afford a convenient cut-off.

Quine thus considers the greater extent or the greater "excess" of the logical/mathematical world over against the empirically accessible world to be a "simplistic manner of rounding out," to be "mathematical recreation, without ontological rights." This totally reductionistic view is not at all well grounded, because it immediately suggests fundamental questions for which it can provide no answers. How is this putative mathematical "extravagance" to be explained? How is the excess of the mathematical world *forced on us* by the "simplest known systematizations of more welcome matters"? How is it that mathematical recreation is "forced on" us? *Is* it forced on us? Or is it a mere pastime? If the *entire* mathematical (and logical) world, with all its "innumerable infinites," were not something that *unavoidably forced itself upon us from out of itself, by virtue of what it is*—and indeed independently of any "application"—then the entire mathematical-physical process that essentially characterizes natural science would be utterly unintelligible.

323

The conclusion is that the logical/mathematical dimension must be accorded a genuine ontological self-sufficiency.

3.3.4.3 A Moderate Relativism with Respect to Truth

The final question to be considered here arises wholly naturally from the fact that the truth-theory just presented must be put into relation to two of the central theses of this book: first, the thesis that every theoretical sentence (and thus every *true* theoretical sentence) arises within *and in accordance with* some determinate theoretical framework, and second, the thesis that there is a plurality of theoretical frameworks. From these theses it follows conclusively that the truth of every sentence and of every proposition involves an intrinsic *relativity* to the theoretical framework wherein it emerges. This relativity raises the well-known and much discussed problem of relativism with respect to truth. Is truth essentially relative? Are all truths then merely relative truths? A positive answer to this question, not differentiated further, characterizes the position that is familiar as total or radical relativism. Likewise familiar is the fact that this position involves an unavoidable antinomy. If all truths are relative, then so too is the truth of the thesis of radical relativism—but it thereby undermines itself.

It is, however, possible to develop a form of relativism that avoids this antinomy and thus the inconsistency of the thesis of radical relativism.[52] This would be a

---

[50] $\beth_\omega$ designates the infinite cardinality of the ordinal numbers.

[51] 'V=L' is an axiom of set theory that Gödel suggests in 1938 (and that he later rejects as false). 'V' is the universe of all sets. 'L' designates the class of all constructible sets (i.e., the class of all sets that can be defined through a specific constructive procedure presupposing that we had access to enough names for all the ordinal numbers). The axiom thus says that all sets are constructible. (On this, see Putnam 1983: 425.)

[52] On this topic, see the excellent Hales (1997), which provides decisive inspiration for the following account.

moderate form of relativism with respect to truth. The decisive point in its development is the following: the paradox or self-contradictoriness of the unrestrictedly relativistic thesis must be investigated more thoroughly than it usually is. It then quickly becomes clear—and it is easy to show—that the self-contradictoriness arises only when the non-relative, thus the absolute character of truth that the non-relativist holds to be indispensable and that the relativist rejects, is understood in a quite specific manner: a sentence or proposition is absolutely true if and only if it is true *independently from any and every theoretical framework* (or any and every *perspective*, etc.) *whatsoever.* This understanding of "absolute truth," however, encounters insoluble problems on various levels.

Of decisive significance is the fact that this understanding of the absoluteness of truth is *in no way compelling.* Indeed, nothing speaks against the understanding of absolute truth *not* as truth that is simply independent from any and every theoretical framework (perspective, etc.), but instead as *truth within all theoretical frameworks* (within all perspectives, etc.). This understanding of absolute truth preserves the relativity of truth to theoretical frameworks and at the same time avoids the antinomy generated by undifferentiated relativism with respect to truth.

The consistent, moderate relativism with respect to truth that is defended here maintains that all truths are relative to some (at least to one) theoretical framework. This does not, however, exclude the possibility that there are truths relative to or within *all* theoretical frameworks, or that these truths, in accordance with the formulated definition, are absolute. This moderate relativism with respect to truth indeed assumes that there are at least some truths that are absolute in this sense. This point is easily clarified by means of a comparison with the (usual) semantics of possible worlds. If one assumes such a semantics, it is clear that the concept of necessary truth is not defined in abstraction from all (possible) worlds, but instead in relation to all possible worlds: a necessary truth is then a truth that exists in all possible worlds.

In principle, the form of this moderate relativism with respect to truth should be clear. Not less important, however, than its solution of the problem of avoiding the antinomy referred to above are additional questions that it engenders. Two of the most important of these are now considered briefly.

[1] The first question concerns the concept of absolute truth as truth within—and thus relative to—all theoretical frameworks (or perspectives, etc.). Here, there are various more specific questions. One of these concerns talk about *all* theoretical frameworks (or perspectives, etc.). Talk about totalities brings with it problems, particularly with respect to semantics, logic, and mathematics. This is a difficult problem requiring extensive treatment. Section 5.2.2 is devoted to such a treatment.

A second question is the following: is such a concept not a fantastic or even illusory one? This question can be posed in the form of a dilemma: either the concept of truth, as defined above, is wholly unintelligible, and therefore senseless and unusable, or it is intelligible but not applicable (realizable), and thus equally senseless and unusable. The answer or solution lies in the rejection of the dilemma: the concept of absolute truth, as defined above, is both intelligible and applicable (realizable). That we somehow understand the concept cannot be seriously doubted. We understand what the formulation "all (theoretical frameworks)" means, and we can precisely indicate (also formally) what it means, as shown in Section 5.2.2. Presumably, the critic would have to mean that this concept of absolute truth is "too big," therefore "inaccessible," or something of the sort. But even quite "big" concepts can be both understood and indeed defined. An example from mathematics (set theory) is the concept of uncountable sets. This concept is well defined and therefore intelligible.

Not so easy to clarify is the assumption or thesis that the concept of absolute truth, as understood here, is applicable or realizable. Here, there are both a problem in principle and a specific problem. The problem in principle arises from the fact that so extensive a concept cannot be "concretized"—stated differently, that it can find no "place" within our theoretical undertaking that is at all determined. The problem is sharpened by the following line of argumentation: for the concept of truth in *all* theoretical frameworks (or perspectives, etc.) to be at all theoretically applicable or interesting, we would have to be in the position of sufficiently grasping and articulating *all individual* theoretical frameworks (or perspectives, etc.), but that is something we will never be able to do.

The answer or solution to the problem just introduced is seen in the fact that the above-defined concept of absolute truth is a *limiting concept* in the sense of a *regulative idea*. The formulation "regulative idea" is, as is well known, introduced by Kant. This book adopts the formulation, but not—at least not in every respect—the concept that Kant connects with it. The application of the concept of absolute truth is similar to that of the concept of uncountable sets in mathematics. It is indisputable that the latter concept has a precisely determined and indispensable "place" in mathematical theories. It serves correctly and precisely to explain fundamental issues within the mathematical domain. Similarly, the concept of absolute truth serves in philosophy to illuminate grand philosophical issues in the sense that it makes possible the determination of the systematic "places" or locations of articulated topics and presented sentences. In this sense, the concept of absolute truth is one of those that "regulate" the entire theoretical undertaking.

The *specific* problem with respect to the applicability or realizability of the concept of absolute truth arises from the decisive importance of the concept of the theoretical framework within the conception presented in this book. Because this book presupposes and/or brings to application a quite determinate theoretical framework (i.e., the structural-systematic one), the question arises of how, *within this philosophy*, absolute truths in the relevant sense can be coherently accepted and articulated. 326

Providing a grounded answer to this question requires that one take into consideration the treatments of immanent and external metasystematics in Chapter 6 and those of the *openness* and the *incompleteness*—to be understood in a sense analogous to that of Gödel's famous incompleteness theorem (see esp. 5.2.2[4]–[6])—of the structural-systematic philosophy. At this point, the following can be said: in that the structural-systematic theoretical framework here thematized and targeted is understood as *one* theoretical framework among many, it is always already beyond itself; it has always already transcended itself. It is understood as embedded within a broader, more comprehensive theoretical metaframework—more precisely, within the totality of actual and possible theoretical metaframeworks. This is clear from the fact that *within this philosophy*, the claim is made that it is based on *one* theoretical framework among many others.

To be sure, this claim must be understood correctly. When it is said that *within this philosophy* the claim is made that it is based on *one* (determinate) theoretical framework, and that this framework is understood as embedded within other theoretical frameworks and indeed within the totality of theoretical frameworks, this formulation, taken strictly and precisely, is already a *meta*-formulation; it does not arise in any compelling manner *within* the structural-systematic theoretical framework itself. For this reason it is, strictly speaking, inaccurate and inadequate. Strictly speaking, the following must be said: this formulation is the expression of the fact that the *theoretician* has access to a capacity for intelligibility that has a scope greater than any formulation that is made possible and attainable by the theoretical framework actually used by the theoretician. Theoreticians always already find themselves within a metadimension with respect to their own theoretical frameworks (i.e., to whatever theoretical frameworks they may be thematizing and utilizing at any given time). Theoreticians' capacities for intelligibility *are not exhausted* by *any* of their theoretical frameworks.

This characterization of the theoretical status of the structural-systematic theoretical framework and of the systematic philosophy developed on its basis is nothing other than the scrupulous interpretation of what happens, philosophically, when, with respect to a philosophy, such claims are made as "This philosophy is based on *a* theoretical framework," and such questions are posed as, "Is this philosophy superior to other philosophies?" "Can more adequate forms of philosophy be developed?" etc.

It is now clear that the characterization of a truth as an *absolute* truth (in the sense defined above) has an unambiguous *metastatus* with respect to the theoretical framework that in fact is presupposed and applied. It is also clear that this metastatus is one that is absolutely comprehensive in the sense that it coincides with that dimension that is *absolute* in that it includes *all* theoretical frameworks. It is the expression and result of the taking into consideration of the *entire* intellectual capacity of the theoretician. If the status of the structural-systematic theoretical framework and of the structural-systematic philosophy is understood in light of what has just been said, the thesis that the relativism with respect to truth defended by the structural-systematic philosophy in no way excludes the acceptance of absolute truths is one that is fully and utterly coherent.

Two additional questions arise. How are absolute truths grasped or articulated? And what are the absolute truths in question?

To clarify the how question, it is important to note that truths that qualify as absolute (in the sense presupposed here) do *not*, strictly speaking, belong to the genuine corpus of the structural-systematic philosophy; instead, they belong to the *metadimension* of this philosophy. That they do follows directly from what has just been said. From that it also follows immediately that one cannot characterize absolute truths *as such* as truths of the structural-systematic philosophy and cannot treat, grasp, or articulate them as such truths. For example, absolute truths are not, strictly speaking, axioms of the structural-systematic philosophy, although *without the qualifier "absolute,"* these *same* truths *can* be axioms of the philosophy and can be so termed.

It is possible to identify a decisive factor that clarifies how absolute truths are grasped and articulated. On the whole, this occurs *intuitively-prospectively-holistically*. Somewhat more precisely, one can put this as follows: on the basis of the activation of their entire capacities for intelligibility, theoreticians can take intuitive-prospective-holistic stances that are concretized in the form of a likewise intuitively-prospectively-

327

holistically determined theoretical (meta-)framework. *This* framework is neither precisely explained nor is it, as such, explicable; its theoretical status is therefore largely underdetermined, precisely in that it is intuitive-prospective-holistic. *Within* this theoretical framework *and* conditioned by its uniqueness, sentences with the (explicit or implicit) qualifier "absolutely true" can be presented. Both the understanding and the grounding of such truths proceed on the intuitive-prospective-holistic level. ₃₂₈

Which are the absolute truths in question? In general, they are comprehensive logical and metaphysical truths. Examples are the principle of non-contradiction, the principle of universal coherence, the principle of the unity of all of actuality (under whatever designation whatsoever), the principle of sufficient reason (if this principle is understood in accordance with this book's basic theses concerning theoretical frameworks), and various other absolute truths.

[2] A second question is important in this context and therefore treated briefly: what are the consequences of this relativism with respect to truth for the philosophy of science and for ontology?

A few indications should suffice here, because this thematic is treated in this book in various contexts. With respect to the philosophy of science, this relativism with respect to truth has the consequence that theories that are abandoned or rejected are not to be designated purely and simply as false. Instead, the reasons for their abandonment or rejection must be investigated scrupulously. If the reasons are that the theoretical framework within which they emerge has been rejected, and when it can be ascertained that the theoretical framework had at least a reasonably defensible scientific status, then it may be concluded that the theories developed in accordance with this theoretical framework are not simply false. They articulate truths *relative to this framework*. The genuine—and extremely difficult—task, in this case, is to compare this theoretical framework with other theoretical frameworks and quite particularly with the successor theoretical framework in order to determine *what sorts of truths* the corresponding theories articulate.

This evaluation concerning the philosophy of science leads immediately to the question concerning the *ontological* consequences of this relativism with respect to truth. What sort of truth is a "relative truth," or how does the truth that is relative to some specific theoretical framework share—what part does it play—in truth *simpliciter*? Because truth has an essential ontological import, the question must be raised concerning what the relativity of truth implies with respect to ontology. That is no doubt one of the most important questions for the philosophy of science and for philosophy in general; for this reason, the quite thorough Section 5.1.5 of this book is devoted to it.

CHAPTER FOUR

World-Systematics: Theory of the Dimensions of the World

The preceding chapters present, in various ways, the dimension named in the title of this book by the word "structure." They thus treat one of the two dimensions whose reciprocal suffusion constitutes the subject matter of the systematic philosophy here developed and defended. The second dimension is named in the title by the word "being." Schematically, one can say that Chapters 4 and 5 treat this second dimension, but this quite global characterization can be misleading because there is a fundamental asymmetry between the two dimensions. To a significant extent, the dimension of *structures* can be presented purely as such, as it is in the preceding chapters. The dimension of *being*, on the other hand, cannot be articulated even minimally independently from the dimension of structure, because its articulation (or explication, conceptualization, etc.) is nothing other than the grasping and presenting of *structures of being* or of *being in its structuration*.

Because of this asymmetry, it would be more accurate to say that in Chapters 4 and 5 the dimension of structure(s) is "applied" to the dimension of being—but this formulation, too, is less than perfect, mainly because of various misleading connotations of "application," chief among them that application is a purely external relation. With respect to the dimensions of structure and being, the "application" is *not* external. Instead, each of the two dimensions, taken in isolation, is purely *abstract*, in a quite determinate sense. Without the dimension of being, the structures remain in a certain respect empty—they lack what makes them meaningful. Without structures, on the other hand, being remains an amorphous mass. For purposes of clarity and of presentation, the two dimensions must be distinguished, but it must be made clear that the central philosophical task is that of revealing their mutual suffusion or interpenetration. The task is accomplished when the *saturated* or *concretized structure(s)* and *structured being* are understood and presented in a manner that reveals their fundamental unity.

From clarification of what is meant by *world* emerges clarification of the organization of this chapter on *world-systematics*. Providing this clarification is the   330
task of Section 4.1.

4.1 The Concept of World

4.1.1 *World, Universe of Discourse, and Being as a Whole*

The word "world" is used in contemporary philosophy—particularly in analytic philosophy—absolutely universally, in virtually all domains treated within philosophy. Only rarely, however, is there any indication of just what meaning is ascribed to the term or precisely what concept is associated with it. As noted in the Introduction, in the first three chapters of the book the term is used as synonymous with the terms/concepts "unrestricted universe of discourse," "universe," "being as a whole," etc. The book's final three chapters, however, require that these terms/concepts be differentiated precisely.

In philosophy, three concepts are associated with the word "world." Each concept is associated with a specific philosophical orientation. The first concept is characteristic of all philosophical positions that in any way assume the basic perspective of classical (or traditional) Christian metaphysics. There, the world is understood as the totality of finite, created things or entities. Thereby presupposed is *absolute being* or an *absolute* (*highest, first*, etc.) *being* that is in no sense whatsoever contained within the world. This being or entity ("God," ordinarily) is, in one way or another, *world-transcendent*.

A second concept of world, one that has been and continues to be widely influential, can be designated the *transcendental* concept. This is the concept of the world as exclusively and completely determined by transcendental subjectivity. Two aspects of this concept are particularly to be emphasized. The first is that this concept, or the dimension that it designates, is one side of a radical and absolutely insuperable dichotomy of which the other side is transcendental subjectivity. The transcendental I is defined precisely, albeit negatively, as *non-worldly*. "World" is the other of or the counterpole to transcendental subjectivity. The second aspect to be emphasized concerns a second dichotomy involving the dimension of the world: "world" is both the "totality of appearances [*der Inbegriff aller Erscheinungen*]" (B391) and the totality of things in themselves,   331
which to us is unknowable.[1]

A third concept of world is the one in most widespread use in contemporary philosophy, particularly analytic philosophy. This third concept simply identifies the

---

[1] Kant generally terms the world in this second sense "intelligible world" (see, e.g., *CPR*: B313, 413, 836). Kant determines more precisely the concept of the (merely) intelligible (or of the *noumenon*) (concretely: of a merely intelligible object and, generalizing, of those totalities that Kant terms "concepts of reason" or "ideas," thus of the intelligible world) in three ways: (1) the intelligible X (an object, the world) is understood in the sense of the "in itself"; (2) the in itself is thought "as merely intelligible, (i.e., as given to the understanding alone and not to the senses at all" (B313); (3) the in itself can be adequately understood only as "an **object of a non-sensible intuition** . . . namely intellectual intuition, which, however, is not our own, and the possibility of which we cannot understand, and this would be the noumenon in a **positive** sense" (B307). On the whole, Kant names the dimension of the "in itself" "thing in itself." For a detailed critique of Kant's distinction, see Puntel (1983).

world with *reality* or *being*: the world encompasses everything that is or exists (such that *being* and *existence* are understood as synonymous). In contrast with the first concept, however, this third one is anything other than unequivocal and precisely determined. With respect to the use (analytic) philosophers in fact make of it, two perspectives, above all, must be distinguished. In general, it is understood as an absolutely universal concept; it is then synonymous with such expressions as "being as a whole," "(unrestricted) universe of discourse," etc. But there is also a *more specific* sense of this third concept, one that is unambiguously restrictive and reductive in the following sense: the constituents of the world are limited to those entities that satisfy specific criteria. Opinions diverge widely with respect to what these criteria should be; the status accorded to the sciences plays a role in their determination. Fully divergent conceptions of the world result: for some, the world consists only of physical entities, for others, it includes mental entities as well, and yet others add abstract (so-called Platonic) entities, etc. The question of the inner (or specific) constitution of the world is in many respects the central question of analytic philosophy. In particular, the domains of the mental and of the abstract (logic and mathematics) are discussed at length.

For the purposes of this book, this specific sense of the third concept is of central significance, as this chapter shows. Because, however, it is less problematic than the general sense, that is the one considered at this point. The question is, to what extent is this third concept of world, independently of its specification, a genuinely universal concept that can be the focus of a genuinely radical philosophical questioning? The answer: the "universal" concept of world as it is understood and used in analytic philosophy is ambiguous and underdetermined. It *can* be understood as universal in every respect, thus as synonymous with such concepts as (*unrestricted*) *universe of discourse, being* (*as a whole*), etc.; so understood, it would also encompass—or at least not exclude—the absolute (God). *This* understanding would, however, require a terminological stipulation; it is *not* the understanding that is—at least generally—associated with "world" in the philosophical tradition and at present. To show that it is not is not a task for this book.

Another fundamental factor must be introduced here. In much of contemporary analytic philosophy, *possible worlds* are accepted. To be sure, there is no consensus concerning how these possible worlds are to be understood, but in at least one important sector of analytic philosophy, they are understood *ontologically*. The most strongly ontological interpretation is presumably David Lewis's (1986) *modal realism*, which posits a plurality of existing possible worlds. According to Lewis (2–3), our world does not differ from others in its "manner of existing," and the things in it differ from other existing things only "in location or lack of it." This makes it fully clear that Lewis's concept of world is not absolutely universal; the acceptance of a plurality of possible worlds brings with it the question of whether—and if so, how—a unity of these worlds is to be conceived of, and how the *more comprehensive* concept corresponding to that unity is to be characterized.[2]

In this book, the term/concept *world*, when used in its genuine or specific sense, is *not* understood as absolutely universal; it is thus not synonymous with (*unrestricted*) *universe of discourse* or *being* (*as a whole*), etc. Instead, it designates that dimension of the universe of discourse, or of being as a whole, that is treated in this chapter. To emphasize this restricted signification of the term, the qualified term "actual world" is

332

---

[2] This question is treated in detail in Chapter 5 (Section 5.2.3).

used. How the actual world, so conceived, is understood in relation to the unrestricted universe of discourse, or being as a whole, is a central issue for systematic philosophy; it is treated thoroughly in Chapter 5.

4.1.2 *The Most Important Domains or Subdimensions of the Actual World*

If one understands "(the) world" in the indicated manner, the next question to arise is that of how its content and innerworldly structuration are to be determined. To be emphasized at the outset is that the content of the world cannot in any way be derived; the German Idealists, above all, attempt such derivations, but utterly fail. The alternative is to accord a—indeed *the*—decisive role, with respect to the determination of the content of the world, to experience. The world presents itself to philosophical view as a *grand datum* (although to be sure not as *the* grand datum) to be comprehended, i.e., here, related to the systematics of structure developed in the preceding chapters. Precisely how the world is to be understood as a *grand datum* must first be articulated carefully.

[1] At the outset, a possible misunderstanding that would have severe consequences must be identified, so that it can be avoided; it is introduced above, but must now be treated in more detail. If one distinguishes, as in this book (see Chapter 3), between the systematics of structure and the world as grand datum, the distinction is literally an *abstract* one: the distinction *abstracts* fully from the fact that the distinguished dimensions are *in actuality* always already related to each other, that they are two sides of the same coin. The structures presented in the systematics of structure are fully explained or fully determined only when they are understood and presented as *structures of the world* (and, as shown below, of being as a whole); they are always already *ontological*. Similarly, that the world can be comprehended or understood at all reveals that it presents itself as structured in some manner or other.

A consequence of the difference between the two dimensions is that the world is always already *given* as somehow structured in some manner or other, and indeed in every respect: with respect to its individual domains and with respect to all the entities constituting those domains. This givenness of the world has two fundamental aspects that, in systematic philosophies, must be scrupulously considered and accounted for. The first is the strictly *empirical* factor: the world is accessible to philosophers, at the outset and unavoidably, as prestructured within both everyday frameworks and available scientific frameworks (see pp. 10–11, above). Negatively viewed, this means (as noted above) that it would be absurd to attempt to derive the world and its domains and its components, in no matter what manner and no matter from what. The second aspect is the *linguistic* factor: one cannot avoid beginning with so-called ordinary language, because it is as articulated by this language that the world and its components initially present themselves (or are initially presented).

The initial givenness of the world just described is only the starting point for the philosophical comprehension of the world. This means, among other things, that philosophy must *not* accept life-worldly experience and ordinary language as its measure. Quite to the contrary: it is with this givenness and this language that the work of the philosopher (or of any other theoretician) begins. The theoretician has the task of *comprehending, explaining,* or *articulating* the world-as-given, but this in no way means that the world, in the way it appears as structured in the lifeworld or to or through ordinary language, could or should simply be reproduced. Instead, the so-understood world must be *comprehended* or *explained* in the genuine sense, which here means that it is to be articulated *in accordance with the dimension of structure developed in the preceding chapters.* This leads to the *reconceiving* of the categories and domains given on the level of lifeworldly experience and articulated by means of ordinary language. This reconceiving is accomplished by means of precise reconstructions and reductions (i.e., here, more or less extensive *restructurations,* which involve the introduction of and reliance upon different structures). The goal is thus a different, new, more precise, and more adequate comprehension of the world.

[2] The world as the grand datum given in everyday and available scientific frameworks thus constitutes the subject matter with which the part of systematic philosophy here called *world-systematics* begins. So understood, how does the world look?

Obviously, the task cannot be that of describing this immense subject matter in all its detail; instead it is to name the grand domains of this world. Since its beginning, philosophy has identified these grand domains and has devoted to them specific philosophical disciplines. They have remained basically constant. The most global division distinguishes three such domains: those of the natural, the human, and the historical. If one attends more closely to each of the three, additional divisions emerge; this book also uses those additional divisions as guidelines.

This array of domains or topics or disciplines emerges from both of the above-identified factors: experience and ordinary language. In fact, we tend both in experience and in ordinary language to distinguish these grand domains. Of course, we distinguish far more domains, ones that can be understood as subdomains of these grand domains. But in a book presenting the *framework* for a systematic philosophy it is not possible to go into additional detail.

4.2 The "Natural World"

335

The term "natural world" is used here to emphasize the particular signification that this book associates with the word/concept "nature." As is well known, "nature" has been and continues to be used in a variety of ways; that need not be considered here. It suffices for present purposes to note three quite distinct ways of understanding and using this term/concept. First, "nature" is sometimes used in an absolutely comprehensive sense, thus as synonymous with being or actuality as a whole (e.g., Spinoza: "*Deus, seu Natura* [God or Nature]" (*Ethics,* Part IV, Preface, 382), "*natura naturans* [nature naturing, or nature *qua* producer]"—"*natura naturata* [nature natured, or nature *qua* produced]" (*Ibid.,* Part I, Proposition XXIX, scholium, 132). Second, "nature" is sometimes understood in a quite narrow sense, above all in recent modernity and particularly in the tradition of transcendental thinking, where the distinctions between mind and nature and between social or human sciences and natural sciences become centrally important. Finally, especially at present, it has a variety of senses that are situated between the broadest and the narrowest senses. This section introduces and uses the term *natural world* to designate the domain of the universe that includes the inorganic and all of the organic, save for human beings.[3] To be emphasized in advance is that this terminological stipulation is not meant to indicate or to suggest that there is no meaningful relation between the domains of the human and of what is here termed that of the natural world; there is such a relation, and it is considered in this chapter. Also to be noted is that when there is no danger of misunderstanding, the term *nature* is used, for the sake of brevity, instead of *natural world*.

4.2.1 *Is a Philosophy of the Natural World at All Possible?*

Throughout most of the history of Western philosophy, the *philosophy of nature* has an unquestionably central position; in modernity, however, the development of the natural sciences is accompanied by increasingly radical consideration of whether *all* theoretical questions concerning the natural world are *natural-scientific* questions, leaving nothing, in this domain, for philosophy to investigate. This is a major development within the history of philosophy. At the extreme, it can appear that philosophy has been left with no subject matter whatsoever. Many philosophers who resist this extreme position view the increases in the domains of the natural sciences as regrettable losses for philosophy, but in fact they are not. Instead, the development of the natural sciences contributes to philosophy an invaluable process of clarification, because that development facilitates determination of the theoretical status of the theoretical enterprise that continues to bear the name "philosophy." To think that this process of clarification has already reached its definitive goal would be to make a utopian assumption, but it is scarcely deniable that significant and even decisive progress has been made.

336

At present, it would be simply senseless to pursue philosophy while ignoring the immense significance and indisputable achievements of the natural sciences. It must be acknowledged, simply and straightforwardly, that the time in which philosophers could develop grand philosophies of nature that could be taken seriously—in, for example, the style of Schelling or Hegel—has passed. But this does not mean that philosophy has nothing to say about the natural world. One must, however, consider carefully what, in the age of the domination of the natural sciences, philosophy can and indeed must contribute to the theorization of the natural world.

4.2.1.1 An Instructive Example: The Philosophical Incoherence of Quine's Attempted Reconciliation of "Naturalism" and "Global Ontological Structuralism"

[1] Many remarks of analytic philosophers concerning the natural world prove to be anything other than clear and well founded. To provide an instructive example, this section subjects Quine's "naturalism" to critical analysis. Quine (1981: 21) characterizes this position as "the recognition that it is within [natural] science itself, and not in some prior philosophy, that reality is to be identified and described." This is not an

---

[3] Human beings can and indeed must of course be studied by strictly natural sciences, such as anatomy, physiology, etc. (and by other sciences such as sociology, etc.); as far as philosophy is concerned, it is important only to note that the natural sciences (and those other sciences) cannot explain human beings fully or adequately; for details, see Sections 4.3.1–4.3.2.

unambiguous or coherent position. More closely considered, Quine's position involves two different kinds of theses that are not easily brought into harmony.

On the one hand, he maintains the following categorically:

> I ... expressed ... my unswerving belief in external things—people, nerve endings, sticks, stones. This I reaffirm. I believe also, if less firmly, in atoms and electrons and in classes. (21)

On the other hand, he describes a "barren scene" for semantics and ontology—barren for semantics because of the "inscrutability of reference" (19), and for ontology because
337   of the consequent "possibility of displacements of our ontology through proxy functions" (21). He calls this position "robust realism" in the naturalistic sense provided by the understanding of naturalism presented above. He then raises the question, "how is all this robust realism to be reconciled with the barren scene. . . ?" His answer is (again) "naturalism," as defined above. To see why this answer is incoherent as a response to this question, the "barren scene" must be analyzed more closely.

Quine initially presents this scene as accommodating a fourfold "reductive reinterpretation" (19) of ontology,[4] one made possible by the assumption that all objects are to be viewed as theoretical objects (see 20). Each of his four examples involves an interpretation or reinterpretation of a domain of objects that identifies that domain with (and thus reduces it to) some part of another domain. In the first case, numbers are identified, one way or another, with classes. In the second, physical objects are identified with place-times. Such place-times are identified (in the third case) with specific classes (i.e., with classes of quadruples of numbers). The fourth case or example concerns the famous problem of the dualism of mind and body, a dualism Quine deems "unattractive" (18): "We can just reinterpret the mentalistic terms as denoting these correlated bodily states, and who is to know the difference?" (19).

Quine, however, recognizes in addition another form of ontological reinterpretation, one that he considers to be equally "instructive" as that just described. With this second form, "we save nothing but merely change or seem to change our objects without disturbing either the structure or the empirical support of a scientific theory in the slightest" (19). This procedure requires only a single rule, one assigning, to every object from a given ontology, a specific object of the new ontological sort. Quine names this rule a "proxy function":

> [I]nstead of predicating a general term 'P' of an object x, saying that x is a P, we reinterpret x as a new object and say that it is the f of a P, where 'f' expresses the proxy function. Instead of saying that x is a dog, we say that x is the lifelong filament of space-time taken up by a dog. (19)

The original object is thus replaced and the general term reinterpreted; Quine designates this process as a revision on the one hand of ontology and on the other of ideology,
338   but comments, "Nothing really has changed" (19). The conclusion that he draws is the famous thesis of the "inscrutability of reference."

---

[4] Quine speaks also of "change," "revision" (19), and "transfer of ontology" (17).

This idiosyncratic conception is based on an additional Quinean thesis: the only points of contact between theories and the world are "observation sentences"—sentences that articulate immediate observations: "This is red," "It's raining," etc. Quine understands such sentences to result immediately from stimuli and to have the property of moving the speakers of a given language to agree (3). These sentences provide "sensory evidence," which is the only evidence upon which theories can rely, but they have no directly referential relation to the world. With respect to the observation sentence "A white cat is facing a dog and bristling," Quine (1985: 169) clarifies his conception as follows:

> This observation sentence, true to form, is one that we will directly assent to or dissent from when suitably situated and visually stimulated. It is in its global susceptibility to visual triggering, and not its mention of two creatures, that its observationality consists. Its referential aspect belongs rather to its devious connections with the ethological theory to which it is meant somehow to bear witness.

The referential aspect of language and thus—according to Quine—of theories is given only when the observation sentences are reformulated within the framework of first-order predicate logic, which Quine presents as "the chosen mold of our scientific theory" (169), "the adopted form, for better or worse, of scientific theory" (170).

Whatever is not on the level of observation sentences belongs, according to Quine, to the level of theory—thus, it is theoretical. As passages introduced above indicate, all *objects* are theoretical because they enter only within the framework of a theory (i.e., first-order predicate logic). At this point, Quine makes an additional assumption that he does not formulate explicitly—or at least, that he does not formulate adequately—but that is of simply decisive importance to understanding what he goes on to say. The assumption is that the level of observation sentences is stable in the sense that it cannot be affected *as such* by any theorization and that it is the basic level or dimension for *every* theorization (1974: 39): "No matter that sensations are private, and no matter that men may take radically different views of the environing situation; the observation *sentence* serves nicely to pick out what witnesses can agree on" (i.e., speakers can rely on or refer to such sentences to attain agreement).

[2] From the theses that have now been presented cursorily, Quine draws a twofold conclusion in the form of two additional theses that have become famous and that constitute one of the most-discussed topics in the analytic philosophy of recent decades: the thesis of the underdetermination of theories and, following from it, the thesis of ontological relativity or of the inscrutability of reference (the distinction between these latter two is unimportant in the current context). The first thesis, which has been and continues to be a topic discussed extraordinarily much in the literature concerning Quine, has various aspects and is presented by Quine in various (partial) formulations. Among the formulations that Quine uses most frequently, and one that already contains a wholly determinate interpretation, is the following: with respect to the empirical basis, all theories are equivalent (see, e.g., 1975, 1990). Another formulation, which can be seen as a consequence of the one just introduced, says that no theory can attain the status of being even an incomplete account of the empirical basis, which remains stable—no theory can

claim adequately or completely to comprehend or to articulate the empirical basis, and thus the world.[5]

With respect to the first thesis, the heart of the Quinean conception is the assumption, introduced above, that the data basis is *stable*. Quine understands its stability in a strict and fixed sense: it remains absolutely the same in relation to any and every theory, with the consequence that the theories remain underdetermined in relation to this stable basis. This assumption is, however, highly problematic. It suffices here to note two points. First, the assumption cannot be defended easily: how is one to show that all human beings "perceive" exactly the same data in exactly the same manner? Many philosophers have shown that observation languages are themselves always already theory-laden. Second, Quine's theses entail the extremely implausible and unrealistic assumption that the data basis provides equal support for any and every theory. This is tantamount to a hypostasizing assumption of a data basis that is fixed or fixated once and for all. At the most, one could say that every possible theory is *consistent* with "the" data basis. But *being consistent with* is wholly different from being *equally supported by* (see Papineau 1996: 303).

[3] The second thesis maintains that no theory can claim to comprehend or to articulate ontology.[6] Ontologies are interchangeable; the "real world," the aim of reference, remains inscrutable. In a 1992 essay (1992b) that summarizes and explains this conception, Quine writes the following:

> The conclusion is that there can be no evidence for one ontology as over against another, so long anyway as we can express a one-to-one correlation between them. *Save the structure and you save all.* (8)

Or, yet more briefly and concisely (1981: 20): "Structure is what matters to a theory, and not the choice of its objects." In the 1992 essay just quoted, Quine names his position "global ontological structuralism" (6); this is a position that purports to explain both concrete and abstract objects. "Structure" designates the entire dimension that stands over against the dimension of "observation sentences"; one could term it the entire *theoretical* dimension. Quine (7) identifies two "key principles" that characterize the entire "structural dimension," in his sense: implication and reification. With respect to these two key principles, he (8) characterizes his position as follows:

> The point is that if we transform the range of objects of our science in any one-to-one fashion, by reinterpreting our terms and predicates as applying to the new objects instead of the old ones, the entire evidential support of our science will remain undisturbed. The reason is twofold. First, implication hinges only on logical structure and is independent of what the objects, the values of the variables, may be. Second, the association of

---

[5] On this matter, Quine's writings contain many (partial) formulations, (partial) interpretations, etc., of a great variety of sorts. One quite comprehensive formulation (1990: 101) is the following: "in the case of systems of the world ... one is prepared to believe that reality exceeds the scope of the human apparatus in unspecifiable ways."

[6] "What particular objects there may be is indifferent to the truth of observation sentences, indifferent to the support they lend to the theoretical sentences, indifferent to the success of the theory in its predictions" (Quine 1992a: 31).

observation sentences with ranges of neural input is holophrastic. It is independent of reification, independent of whatever objects the observation sentences or their parts may be taken to refer to as terms.

[4] This more extensive presentation of Quine's "barren scene" provides the basis upon which it is possible to confront the problem that Quine himself introduces: can such a logical-semantical-ontological framework, i.e., his "global ontological structuralism" (9), be made to agree with his "robust realism"? He himself raises this question explicitly and in complete clarity:

> This global ontological structuralism may seem abruptly at odds with realism, let alone with naturalism. It would seem even to undermine the ground on which I rested it: my talk of impacts of light rays and molecules on nerve endings. Are these rays, molecules, and nerve endings themselves not disqualified now as mere figments of an empty structure?

Quine responds to the question with his stereotypical answer: "Naturalism itself is what saves the situation." He continues,

> Naturalism looks only to natural science, however fallible, for an account of what there is and what what there is does. Science ventures its tentative answers in man-made concepts, perforce, couched in man-made language, but we can ask no better. The very notion of object, or of one and many, is indeed as parochially human as the parts of speech; to ask what reality is *really* like, however, apart from human categories, is self-stultifying. It is like asking how long the Nile really is, apart from parochial miles or meters.

341

The question that comes to the fore here is the following: what interprets what? Does philosophy interpret science, or does science interpret philosophy? Quine entangles himself in a circle from which, given his basic assumptions, he cannot escape: to interpret the dimension of structure (in his sense) *correctly*, he looks to "natural science . . . for an account of what there is and what what there is does." But, as in the passage cited just above, in the same breath, *as a philosopher*, he *interprets* the "account of what there is and what what there is does" provided by natural science—and this *philosophical* interpretation drives the concept of a "robust realism" (1981: 21) directly *ad absurdum*. Quine cannot have the cake (the robust realism) that his "barren scene" devours.

The concluding passage of "Structure and Nature" (1992b) provides a final confirmation of the critique just presented:

> My global structuralism should not . . . be seen as a structuralist ontology. To see it thus would be to rise above naturalism and revert to the sin of transcendental metaphysics. My tentative ontology continues to consist of quarks and their compounds, also classes of such things, classes of such classes, and so on, pending evidence to the contrary. My global structuralism is a naturalistic thesis about the mundane human activity, within our world of quarks, of devising theories of quarks and the like in the light of physical impacts on our physical surfaces.

Visible here yet again is the fundamental incoherence: if one does not somehow rise above naturalism, then naturalism has the first and last word about the world. Then, however, Quine's "philosophical" interpretations of what the natural sciences say and do make no sense whatsoever. If they are claimed to make sense, the paradoxical result is that precisely thereby, the point just maintained—that *only* the natural sciences can exclusively and definitively determine the sense and nonsense of sentences about the world—is destroyed.

If one wants to take the discussion of Quine's position to the limit, a remarkable sentence from a passage cited above provides a focal point: "The very notion of object, or of one and many, is indeed as parochially human as the parts of speech; to ask what reality is *really* like, however, apart from human categories, is self-stultifying." The conception developed in this book supports this contention of Quine's, but also reveals it to have no argumentative value with respect to providing support for Quine's own position. Quine correctly notes that it is senseless to ask what "reality" "really or actually" is in utter abstraction from all human categories, but he utterly neglects to ask, and therefore does not even begin to answer, the more fundamental question concerning how "human categories" are to be understood. Quine takes himself to be able to rely non-problematically on the view that "human categories" are things "produced" by human beings in some naive sense of "produced," so that they are purely subjective and in *that* sense "purely human." Precisely this position is philosophically untenable, because it is undermined by its own incoherence. This is sufficiently demonstrated in the first three chapters of this book.

Strangely enough, one here has a conception that—abstracting from terminology and justification—is bafflingly similar to Kant's thesis of the unknowability of things-in-themselves, albeit with the difference that Kant does not simultaneously hold any problematic position comparable to Quine's naturalism, so that for him, no comparable problem of reconciliation could arise.[7]

[5] Quine does, however, take a surprising and quite revealing turn. Strangely enough, he interprets the "barren scene" as something epistemological:

> The semantical considerations that seemed to undermine all this [i.e., the robust realism] were concerned not with assessing reality but with analyzing method and evidence. They belong not to ontology but to the methodology of ontology, and thus to epistemology. (1981: 21)

That specific semantic analyses are simply reduced to the methodology of ontology can be understood, although such a reduction is not unproblematic. To say, however, that the methodology of ontology belongs to epistemology is anything but illuminating. Nevertheless, one could consider that attribution merely to be a matter of terminology.

---

[7] Quine's contention, in the passage quoted above, that to defend a "structuralist ontology" would be "to revert to the sin of transcendental metaphysics," is unintelligible, or indeed false. "Transcendental" characterizes either Kant's position or one of the positions that is determined in one way or another by the Kantian position. Quine should have said, "transcendent metaphysics." A similar confusion in terminology is to be found in a passage from Hilary Putnam (1983: 226), also quoted in Chapter 5:

> Analytic philosophers have always tried to dismiss the transcendental [*sic*] as nonsense, but it does have an eerie way of reappearing. . . . Because one cannot talk about the transcendent [*sic*] or even deny its existence without paradox, one's attitude to it must, perhaps, be the concern of religion rather than of rational philosophy.

What is important is only this: if the methodology of ontology involves the theses of   343
radical ontological relativity and of the inscrutability of reference, an unbridgeable gap
between it and the naturalistic position is immediately visible. And this further supports and makes yet more fundamental the objection introduced above that Quine's
position is incoherent. The gap is clearly visible in the following passage, which immediately follows the one quoted in the previous paragraph:

> Those considerations [i.e., the "semantical considerations" of the preceding paragraph,
> the description of the "barren scene"] showed that I could indeed turn my back on my
> external things and classes and ride the proxy functions to something strange and different
> without doing violence to any evidence. But all ascription of reality must come rather
> from within one's own theory of the world; it is incoherent otherwise. (21)

What is implied by the possibility, the "could," of which Quine here speaks? This possibility is a fundamental aspect of Quine's methodology of ontology, but it makes the following problem unavoidable: how can this "possibility" be made to agree with the thesis
according to which *every* ascription of reality must emerge and in fact does emerge from
a specific theory about the world? Quine's own theory of the world excludes ontological
relativity, in Quine's sense: an electron is an electron and cannot simply be arbitrarily
interchanged with or replaced by some other "object." And this shows that Quine is not
in a position to develop a coherent connection between his methodology of ontology
and the "scientific" ontology he himself affirms.

Quine, however, takes one more step, one that shows that his incoherence is rooted
even more deeply. He writes, "My methodological talk of proxy functions and inscrutability of reference must be seen as naturalistic too; it likewise is no part of a first
philosophy prior to science" (1981: 21). How can Quine's talk of proxy functions and the
inscrutability of reference be understood *naturalistically*, given that his methodology
excludes the possibility of any scientifically founded ontology?

4.2.1.2 The Interdependence of Philosophy and the Natural Sciences

The example of Quine shows that nothing theoretically interesting or important about
the sciences can be said without the question arising of the status of what is thus said. This
status must be philosophical, because no other science does or can consider questions
of this type as falling within the domain of its subject matter. Also, no philosophy that
understands itself as systematic in the sense relied upon in this book can ignore the   344
grand subdomain of being as a whole that is here termed the *natural world*. It is, however,
also the case that the natural world is the proper subject matter of the natural sciences.
The situation at the outset can therefore be characterized as follows: philosophy and the
natural sciences are inseparably interdependent. The vital question is, how is this relation
to be comprehended and determined precisely?

This interdependence is to be determined on the basis of a fundamental thesis
that must first be characterized precisely. This insight emerges directly from the quasi-
definition of the structural-systematic philosophy upon which is based the conception
here under development: this philosophy is the theory of the universal structures
of the unrestricted universe of discourse (of being as a whole). The natural world,

however, is a partial domain, a determinate or particular domain of the unrestricted universe of discourse. In an abstract formulation: philosophy attempts to comprehend the *universal* structures of this subdomain, whereas the natural sciences are concerned with its *particular* or *domain-specific* structures. This formulation, however, does not yet sufficiently determine the relation in question.

Additional progress is made by the indication that philosophy considers the natural world and everything that it includes *from the perspective of universality*, of being as a whole, whereas the perspective of any natural science is always restricted or particular. To show more precisely what this means, one must take a broader view.

Normally, the presentation of a specific science is introduced by a definition (better: a general characterization) of the science. As an example, consider the following descriptive characterization from a physics reference book:

> Formerly called natural philosophy, physics is concerned with those aspects of nature which can be understood in a fundamental way in terms of elementary principles and laws. In the course of time, various specialized sciences broke away from physics to form autonomous fields of investigation. In this process physics retained its original aim of understanding the structure of the natural world and explaining natural phenomena. (Parker 1993: 1032)

The philosopher must ask about and then explain the precise status, presuppositions, and implications of such characterizations. It should be clear at the outset that no such characterization can be a constituent of the science "physics"—this despite the widespread belief that a descriptive characterization of this sort is the first task of the science itself. In speaking of "aspects of nature which can be understood in a fundamental way in terms of elementary principles and laws," the characterization presupposes that there are aspects of nature that cannot be understood in a fundamental way in terms of such principles and laws; it thus presupposes the concept or domain "nature," but it says nothing about what this (more general) domain "nature" is. The level on which these presupposed concepts are situated is a metalevel.

To be sure, one could attempt to formulate and to understand the quasi-definition of a specific science (here, physics) as emerging on the object level rather than on a metalevel. One might say, for example, that physics is that science that is concerned with entities or beings that behave or appear in the following ways, etc., and one would then list the so-called properties of physical things. One would then say that the domain of things that behave in this manner are designated, as a matter of stipulative terminology, as "inanimate matter." Such a quasi-definition could *in part* be located on the object level, but only in part. Aside from the fact that physics itself does not and cannot treat or define such a concept as *science*, there are other central concepts that have an unquestionably metacharacter, such as *entities* (*beings*); the quasi-definition of physics presupposes that there are entities that do not behave in the ways that those designated "physical" do, etc.

This unavoidable metacharacter is *initially* the level that characterizes the collective of natural sciences: physics is quasi-defined in that it is distinguished from the other natural-scientific disciplines, which happens by way of their being separated from that which characterizes the collective of natural sciences. But the metacharacter extends much further, and indeed, in the case of physics, beyond the level of the collective of natural sciences. Concepts such as "entity (being)" are unquestionably universal concepts whose clarification and treatment must be the task of a universal science. This is nothing other than philosophy.

From the preceding considerations may be drawn conclusions that are important to the issue currently under examination: the precise significance of the concept of the universal perspective that is characteristic of philosophy and of the particular perspectives that determine the statuses of the individual natural sciences—or, in brief, the question concerning the relation between the *universal* and the *particular structures*. The universal structures are constitutive of being as a whole *and* of each of its domains. Crucial here, however, is the distinction between *absolutely* and *relatively universal structures*. The absolutely universal structures are constitutive of being as a whole in the genuine and narrower sense (i.e., for being *as being* in accordance with the strictest sense of the qualification "as"). One of these structures is the *universal intelligibility* of being and of every being. In the grand tradition of metaphysics this structure is formulated as follows: *omne ens est verum*, in which the word "*verum*" is interpreted as indicating intelligibility. This tradition knows at least two additional universal structures: *omne ens est unum*—formulable in contemporary terms as the axiom that every entity is self-identical (formalizable as $(\forall x)(x = x)$)—and *omne ens est bonum*. An additional fundamental universal structure is one that is dually modal: every entity is either necessary or contingent (i.e., this dually modal structure characterizes both being as such and every individual being). The most important and fundamental absolutely universal structure is, however, structure itself: being *is* structure, and thus *every entity* is (a) structure. Chapter 5 considers this issue in detail.

Relatively universal structures are those that characterize not being as such, but the relation of each and every individual being to being as a whole. These are the structures that constitute the *systematic locations* of every being and every domain of beings. A well-founded distinction among different domains of being(s) is based on the distinction and development of different structures.

It can now be said that philosophy develops from a *universal perspective*, and indeed one that is both *absolutely* and *relatively* universal; its task is to develop a theory of the absolutely and relatively universal structures.

Every natural science, on the other hand, is concerned with a specific domain of being(s), and develops from a *particular* perspective; each attempts to discover *particular structures*. But what does "particular" mean here, as qualifying "perspective" and "structures"? It may be said at the outset that "particular" means "concerning the individual itself," without consideration of the perspective of universality, even in the relative sense. This formulation begins to bring into view a problem that merits identification as the fundamental problem concerning the relationship between philosophy and the natural sciences. The relatively universal perspective (and thus also the relatively universal structures) and the particular perspectives (and with them the particular structures) cannot be distinguished in what could be termed a chemically pure manner. The abstractly formulated distinction between philosophy and the natural sciences is without doubt correct and clear, but only on the abstract level. In the reality of philosophical and natural-scientific research, *the borders are largely fluid*. There can be no doubt that a great many philosophers have illegitimately treated topics that fall within the domains proper to natural sciences. On the other hand, it is a fact that some natural scientists, some journalists, and some scientifically unsophisticated philosophers extrapolate from putative natural-scientific "truths" in often fantastic—and putatively philosophical—manners.

The reason that there can be no precise distinction in *all* domains of philosophy between philosophy and the natural sciences is presumably that there is no such distinction between the relatively universal perspective and its structures, on the one hand, and particular perspectives, with their structures, on the other. From the particular perspective, what is comprehended of a given being or domain of being(s) is what can be termed its inner structuration: the being or domain is considered, so to speak, "in itself," in isolation, in abstraction from its relations to other beings and other domains of being(s), and to being as a whole. So viewed, however, neither the being nor the domain is viewed or comprehended *adequately*, in its specific character, because an essential aspect of the *genuine status* (thus of the genuinely *comprehensive* structure) of any being or domain of being(s) is that it relates *in and of itself* to other beings and domains of being, and to being as a whole. On the other hand, from the relatively universal perspective *as such*, all that can be gained is what constitutes the explicit relation of a domain of being(s) or the beings within that domain to the other domains and the beings within them, and to being as a whole; what is particular about or specific to the domain or its beings does not thereby come explicitly into view.

As a consequence of the preceding considerations one must say that neither of the two perspectives can, without the contributions of the other, make the claim of *adequately comprehending* any domain of being(s) or any being(s) within any such domain. From this it does not follow that what develops within either perspective is illegitimate or useless. To the contrary, both are unavoidable and indispensable, because only by relying on both can we human beings attain adequate knowledge of the world. And this is why the two perspectives and their various theoretical undertakings are interdependent: both are required when the task is that of *adequately comprehending* domains of being and the beings within them. This fundamental thesis becomes more intelligible and more plausible if one considers the problem of the methodological procedure of the natural sciences.

4.2.2   *Major Tasks and Global Theses of a Philosophy of the Natural World Connected to the Natural Sciences*

On the basis of the preceding section, the major tasks and global theses of a philosophy of the natural world can now be outlined; that these tasks must be seen as intimately bound up with the natural sciences is, as should be clear, the starting point for the concretization of the most important philosophical tasks and theses.

In what follows, two major tasks are introduced, and theses corresponding to them are formulated. The first concerns the categorial-structural constitution of the natural world, the second, the extent and the problematic of the ontological distinctions among different domains of being(s).

4.2.2.1 The Categorial-Structural Constitution of the Natural World

Without doubt, the usually ontologically understood category of *substance*,[8] along with the semantical-ontological field within which it is situated, plays by far the most important role in the history of ontology and of philosophical doctrines of categories, as well as in that of the relation between philosophy and the natural sciences. To be sure, both in later modernity and especially in recent decades other categories, in addition to substance, have been introduced; these include *fact, event, process, state, field* (see Schneider 2006a), etc. There have also been attempts, albeit few in number, to replace the category *substance*, particularly with *process*.[9] The history of philosophical attempts to understand the results of natural-scientific inquiry on the basis of the category *substance* (and the other categories that come with it, such as *accident*) is extraordinarily instructive, not least because it reveals in exemplary fashion how far wrong philosophers can go when they deal with natural-scientific theories and discoveries.

Originally (with Aristotle and in the centuries-long history of the category *substance*), individual human beings and in general individual living beings are deemed to be substances; in addition, "middle-sized objects" are also understood to be substances. With the development of modern physics, however, it becomes clear how problematic the reliance upon this category is. Early on, the level of substantiality appears to be attained with the domain of molecules, then with that of atoms, then with that of elementary particles, etc. But it is clear that this category is simply not usable in understanding the results of contemporary physics.

As noted above, for some time analytic philosophers have for the most part relied on the word "object" rather than "substance," but careful analysis of uses of "object" clearly shows that what is meant thereby is one or another form or variant of the hoary concept *substance*: an object is an *x* that has properties and that stands in relations to other objects. The *x* is thus the foundation for the properties and relations; it is their *bearer*—and the traditional word for the bearer of properties and relations is "substratum."

In purely mathematically oriented presentations of natural-scientific theories, the categories *substance* and *object* are not present in any relevant manner. Nevertheless, scientists themselves generally use traditional substantialist vocabulary in that they speak of objects, properties, relations, etc. And most philosophers who deal with the natural sciences simply use the old categories of substantialist ontology, and not only in paraphrases but also in attempts to comprehend and to conceptualize. This practice regularly leads to the introduction and treatment of questions based on wholly inappropriate ways of thinking. This happens particularly in the domain of philosophy of mind, as shown in Section 4.3.

The theory of ontological structures, whose basic features are presented in Chapter 3, has the *nearly inestimably valuable* advantage of colliding *in no way whatsoever* with

349

---

[8] The category *substance* has not always been understood ontologically. A characteristic example is Kant, in whose philosophy the category of substance plays a fundamental role although one can scarcely maintain that Kant ascribes to that category an explicitly ontological status.

[9] The perhaps best-known and most important example is Whitehead's *Process and Reality* (1929). See also Seibt (2004).

the results of natural-scientific investigations. That is, the fundamental (and indeed sole) ontological structure *primary fact* is absolutely appropriate as an accurate philosophical interpretation of every single sentence that is formulated in mathematically natural-scientific terms. Whether one speaks of elementary particles, of waves, of cells, of genes, of neurons, of forces, of fields, etc. one is always concerned with simple or complex primary facts. As shown in Chapter 3 (Sections 3.2.3 and 3.4), *all* entities are *exclusively* primary facts, and the primary facts are all *ontological structures*. Here, then, a particular form of *structural realism* is presented and defended.

4.2.2.2   The Natural World and the Plurality of Domains of Being(s): The "Ontological Difference"

The fundamental ontological structure, the *primary fact*, is coextensive with being itself, it constitutes being itself; being itself *is structure*, so that one must say *esse et structura convertuntur*, or *ens et structura convertuntur*. *Structure* is thus the fundamental transcendental ("transcendental" here not in the Kantian sense, but in the classical metaphysical sense).

The *second* grand task or global thesis of the philosophy of the natural world concerns what can be termed the *ontological difference*: the distinctions among the different domains of being(s).[10] As understood here, the domain of the natural world encompasses two subdomains that are sometimes termed *inorganic* or *inanimate* and *organic* or *animate* nature. This distinction brings with it serious problems. But the ontological difference also concerns—and in a particular fashion—the distinction between the domain of nature (the natural world) as a whole, on the one hand, and the non-natural domains on the other.[11] To the latter belong (1) the human world taken broadly, with all that belongs to it (including the social world, the dimension of action, the dimension of morality and values, the dimension of culture, the dimension of religion, the dimension of aesthetics, etc.); and (2) the dimension of the ideal world (of semantic structures, of theories, of logical and mathematical structures, etc.). Chapter 3 considers the ideal world in a specific respect; it is considered again, and in a more comprehensive fashion, in Chapter 5. The human world in the broad sense is the topic of the remainder of this chapter.

4.3   The Human World

What is here called "the human world" has been from the beginning and continues to be the actual world's most important domain for philosophical inquiry; it therefore

---

[10] Heidegger introduces the term "ontological difference" to designate the difference or distinction between beings and being. In this book, the term is used differently. Heidegger uses the term to designate *not* a or the distinction between or among domains of beings or of beings, but instead to designate only the distinction between being as such and the totality of beings, understood globally. If however—as in this book—the plurality of domains of beings or of being is articulated explicitly, then the term used for this plurality—"ontological difference"—also designates the relation of all of these domains of being to being, and thus the *expanded* "ontological difference" in the Heideggerian sense.

[11] See the clarification of the uses of the term "(natural) world" in the opening paragraph of Section 4.2.

merits a central position within any systematically oriented philosophy. This domain encompasses an extensive array of specific questions and topics; this section treats only the most important of them. These treatments are not exhaustive; indeed, they are not even extensive. The purposes of this book require only that the section show how, *in principle*, the structural-systematic theoretical framework can and must be concretized with respect to the philosophical conceptualization of the human world.

4.3.1 *Philosophical Anthropology, or Philosophy of Mind* 351

First considered here are two issues that are the basis for all other topics that concern the human world. The first is that of the ontological basis of the categorial or—better—structural constitution of the individual (Section 4.3.1.1). Then, the specific ontological constitution of the individual human being as person must be articulated (Section 4.3.1.2).

4.3.1.1 What Is an Individual, Categorially/Structurally Considered?

Chapter 3 presents the outlines of the structural-systematic ontology. That account modifies or reinterprets the traditional concept *category* by reducing it to the concept that is simply central to the systematic conception developed here: the concept *onto-logical structure*. The account develops, within the framework of an innovative, non-compositional—instead contextual—semantics, the concept of the primary sentence, i.e., the sentence that has neither subject nor predicate. Primary sentences express primary propositions that, when true, are identical with primary facts (in the world). Chapter 3 also shows that primary facts are to be understood as ontological structures, and indeed as the only ontological structures. That chapter introduces various distinctions that must be kept clearly in mind with respect to the primary fact as the sole ontological structure.[12]

Primary facts differ among themselves with respect to *three parameters*: kind, constitution (degree of complexity), and ontological differentiation. The most important kinds of primary facts include those that are static, dynamic, processual, eventual, concrete, abstract, ideal, etc. With respect to constitution or complexity, a given primary fact is either simple (atomistic) or complex, depending upon whether it is constituted by only a single primary fact or by a multiplicity of primary facts. With respect to ontological differentiation, there are as many kinds of primary facts as there are ontological domains: primary facts can be material, physical, biological, mental, logical, mathematical, valuative, moral, ethical, juridical, institutional, social, aesthetic, etc. To be emphasized is that "primary fact" is here a technical term that must be strictly distinguished from the normal term/concept "fact." One consequence of this 352 distinction is the disappearance of the putative distinction, maintained dogmatically by many philosophers, between facts and values. In the terminology used here, based on a completely different ontological conception, values are also primary facts, albeit ones that differ from other facts, such as material and physical ones.

---

[12] As is indicated above more than once, although there are primary sentences, primary propositions, and primary facts, there are no "secondary" sentences, propositions, or facts. Instead, "primary" simply indicates the fundamental rejection of compositional sentences (i.e., sentences with subject-predicate structures) and of compositionally understood propositions and facts.

For the purposes of this chapter and specifically of Section 4.3, the aspect of constitution is of particular and indeed decisive significance, because this aspect makes possible the development, within the framework of this book's ontology, of a theory of the individual (in a narrower or more exact sense). Complex primary facts are *configurations* of primary facts; configurations themselves are primary structures, and indeed structures of *n*-primary structures. Everything thus depends on the concept of complexity and thus of configuration.

As is well known, the word "individual" is used both in a quite broad and in a narrow sense. In the broad sense, every entity, of any kind whatsoever, that is distinct from other entities is an individual. In this sense, every single simple primary fact, as well as every single complex primary fact, is an individual. In this chapter (and indeed in this book), the word "individual," unless otherwise noted, is *not* used in this sense. In the narrow sense, individuals are configurations of primary facts, thus complex primary facts. Every identifiable configuration that is distinct from other configurations is thus an individual in this narrow sense. But configurations can be of vastly different degrees of complexity. A heap is an individual, but so too is the human being as person. It is obvious that there are philosophically significant differences between these two kinds of configurations. To put the point intuitively and quite generally, a heap is a quite loose configuration, whereas a human person is one whose constituents are quite densely interconnected. Configurations whose constituents are densely interwoven are wholly *specific* configurations and thus wholly *specific* individuals. Individuals in this sense are here termed individuals in the *robust* sense, or *robust* individuals. Extensionally, there are both non-minded and minded robust individuals.

A central task of ontology is to clarify these different kinds of ontological configurations. This book tackles this task quite selectively. Explicitly treated here are only those configurations that are *robust individuals*, and of them, only *robust individuals that are minded*—hence, human individuals as persons. The main point to be clarified is how the configuration constituting the human individual is to be understood.

353

Before this task can be tackled, it must be noted that, within the framework of the ontology developed here, familiar, traditional questions like that of the principle of individuation cannot arise. Such questions are pressing within—but *exclusively* within—ontologies that include traditionally understood universals, i.e., entities that, while remaining identical, can be multiply instantiated. The question of how such universals are to be understood leads to the notion of the individuating instance that must involve some kind of principle of individuation. Nothing of this sort is of any relevance to the current account, because universals (in the traditional sense) have no place in this book's ontology.

4.3.1.2 The Individual Human Being as Person

This section considers the general question of how to clarify the concept of configuration, of complex ontological primary structure(s), in the concrete case of the most important of all ontological configurations (thus structures): the individual human being as person. This question is treated specifically within the framework of this book's ontology, with only occasional references to other ontologies and conceptions. No attempt is made to consider fully the current literature on this topic.

The general question contains two more specific questions that must be distinguished and treated separately. The first is how the configuration that defines the human being as person is to be understood more precisely, so that it can qualify as a concept that characterizes and articulates more appropriately the human being as person, particularly given that the person has two fundamental distinguishing characteristics: first, a unique unity, and second, a center or core made manifest by the use, characteristic of the person, of the personal pronoun "I." The second question is the one that is discussed most intensively in contemporary philosophy of mind: whether the ontological constitution of the human being is reducible to a materialistic or physicalistic ontological basis, or whether instead a different ontological framework, one that is not materialistic or physicalistic, must be developed. Before these questions are addressed, the problematic of the adequate formal articulation of the concept of configuration must be considered.

354

4.3.1.2.1 *On the Problematic of the Adequate Formal Articulation of the Concept of Configuration*

[1] As a rule, we consider every individual human being to be a person, understanding thereby, quite generally and intuitively, an entity with a quite particular unity and possessing a center or core that grounds its capacity to speak in the first person, thus to say "I" and to be a subject. Since Aristotle, however, the category most philosophers use to characterize the individual human being as person is the ontological category of substance.[13] Chapter 3 shows, however, that this category cannot intelligibly articulate ontological complexes. To remedy this defect, that chapter introduces the category of configuration.

Configurations, in the sense relevant here, are complex primary facts. But how is this complexity to be understood more precisely? In contexts such as the present one, the term "bundle" has been, since the time of Russell, the one used most often. That term, however—unlike "configuration"—has far too many non-philosophical connotations.

Configurations, as complexes of primary facts, presuppose pluralities of constituents that, according to this book's ontology, are themselves primary facts.[14] Configurations can best be understood as specifications of the general concept *unity*. It is not controversial that there are various forms and degrees of unity; correspondingly, there are a great many forms and degrees of configuration. Heaps, conglomerates, theories, animals, human persons, etc., are configurations, but it is immediately obvious that these are examples of quite distinct forms of configurations.

[2] If one attempts to use formal instruments to define precisely the concept and the forms of configuration, the immense problem arises of whether such instruments

---

[13] The definition of the "person" most often cited and used in the metaphysical tradition is that of Beothius: "A person is an individual substance with a rational nature [*Persona est naturae rationabilis individua substantia*]" (*Contra Eutychen et Nestorium* 1-3; see Schlapkohl 1999).

[14] Contemporary defenders of "bundle theories" of the individual stemming from Russell accept differing types of entities as the elements that are bundled: qualities, properties, "tropes," indeed even universals.

are suited to characterize and to articulate the entire conceptual realm of the intuitive understanding of configuration or of forms of configuration. The following subsections consider three exemplary formal instruments in order to clarify this problem.

[2.1] The first example is set theory. The concept of the set is currently used universally in all formal sciences. Can it be understood as a more precise articulation of the concept of configuration? If one considers even a heap to be a configuration, then it seems clear that the concept of the set *can* serve as a more precise determination or indeed as an exact characterization of this form of configuration, because heaps, in general, are sets (in the ordinary-language sense) of things lying together in a pile, a hill, etc. The ordinary-language term "set" can be understood and articulated relatively precisely by means of the set-theoretical concept *set*. But already in the case of a *conglomeration* it is no longer clear whether the "theoretical" concept of the set adequately articulates the intuitive conceptual content of the term, because—intuitively understood—conglomerates are mixtures of constituents of different types. Theories, too, are configurations. *Formal* theories, at least, as configurations, can be characterized accurately as sets, because a formal theory is a set of sentences (or formulas) that is closed under logical consequence. If however one considers an animal, or the human individual as person, as a configuration, it is immediately obvious that the concept of the set is wholly inadequate as a precise articulation of this configuration and thus of its form of unity.

How is it that the concept of the set is fully suited to some cases of configurations (heaps, theories), but for others (conglomerates, animals, human persons) it appears, intuitively, to be utterly inadequate? To formulate the answer quite generally: for the latter category of cases of configurations, the concept of the set is not sufficiently rich, either expressively or conceptually. Clarification of this general answer requires taking a broader view.

Georg Cantor, the founder of set theory, defines the concept of the set as follows (1895: 7): "By a set we understand every collection into a whole M of determinate, well-distinguished objects m (termed the 'elements' of M) of our intuition or our thought." There is widespread agreement that this is not a definition in the strictly mathematical sense; Cantor's formulation is instead an intuitive characterization of the concept of the set. Even intuitively, the chief concepts used in the *definiens*—*object, collection, whole,* etc.—are insufficiently determined.

A significant additional step is taken by adding the famous *principle of comprehension*; it can be understood as the principle of set construction and also as a criterion for the adequacy for definitions of the concept of the set. This principle reads as follows (see Ebbinghaus 1977/2003: 8):

For every property P there is the set $M_P = \{x \mid x$ has the property $P\}$.

Fully formalized, the (unrestricted) principle of comprehension is as follows:

$$\exists y \forall x (x \in y \leftrightarrow \varphi(x))$$

where $\varphi(x)$ is a formula within which y does not occur as a free variable.

It is a well-known fact that in set theory, the unrestricted principle of comprehension generates Russell's famous antinomy: a set that contains itself as an element. Various strategies have been developed seeking to avoid this antinomy; here, only three need be mentioned briefly (see Ebbinghaus 1977/2003: 9ff). The first of these is Russell's theory of types, which is based on the notion that a set is an object that is "higher" than the elements it contains. Second, Zermelo's axiomatized set theory merits particular attention because it clearly restricts the axiom of comprehension in that it weakens it to the so-called axiom of separation or of subsets, which allows comprehensions only within the domain of a set that is already presupposed:

To every property P and every set x there corresponds the set

$$y_{P,x} = \{z \mid z \in x \text{ and } z \text{ has the property } P\}.$$

Here, the principle of comprehension is clearly restricted.

The third attempted solution is based on the distinction between *classes* and *sets*. Russell's antinomy is generated when a set is constructed that is (supposed to be) an element of itself. To avoid this, the term/concept "class" is introduced. It then appears consequent to distinguish between classes that can be elements of classes—these termed *sets*—and ones that cannot. What results is called "class theory" rather than "set theory." *Sets* are then those classes that can themselves, non-problematically, be elements. Classes that cannot be elements are designated "genuine classes".

For the purposes of this book it is not necessary to consider the details of these attempted solutions or to consider the various problems they bring with them. Important here is only the question of what set theory can achieve ontologically, specifically with respect to ontological structures that are complex primary facts and thus ontological configurations.

Two considerations are important in answering this question. The first is the central issue, introduced above, that is articulated by the principle of comprehension: the correspondence between the concept of the set and the concept of the property. If multiple "things" have a common property, then they constitute a set. This shows that the concept of the set is on the one hand non-problematic, but on the other hand is of only quite limited utility for philosophy (here ignored is the fact that the concept of the property is problematic in that it is a component of the compositional (substance) ontology rejected by this book; that is irrelevant here, because the limitation would remain if one were to speak not of properties but instead of primary facts). The concept of the set is thus quite adequate to articulate configurations whose structuring factors consist of single properties (or relations—or of single simple or complex primary facts). 357

A (formal) theory is thus a set in the genuine and accurate sense; indeed, it is a set of sentences that have a property (or an expressed primary fact)—being groundable—common to all of them. A heap is likewise a set in the genuine and accurate sense, insofar as all the "elements" of the heap share a property or are a primary fact—the property of being piled atop one another or the primary fact "It's heaping."

From this it follows that if a configuration involves not only one primary fact (or, in a different ontology, one property), but instead many primary facts (or properties) of different kinds, then it is no longer clear what it could mean to speak of the configuration as a "set."

The second central consideration is the purely extensional character of sets: a given set is the extension of a predicate: its elements are such that the predicate φ applies to them all. No extension, however, has an inner structure, and having an inner structure is an essential characteristic of many kinds of configurations.

[2.2] The second example is mereology. The controversy concerning how the relation between set theory and mereology is to be understood need not be considered here (see the remarkable Lewis 1991). For the concept *configuration*, mereology can offer the concept *whole*. In mereology, the whole is the sum or fusion of all the objects (in a wholly unspecified sense) that satisfy a determinate predicate Fξ. Introduction of the variable-binding operator σ makes possible the formal notation of this sum or fusion (see Simons 1987: 15):

σ x ⌜Fx⌝ (to be read as the sum/fusion of all objects x that satisfy the condition Fx; the symbols "⌜ ⌝" indicate the scope of the operator).

358   The following formulation provides a general definition of sum/fusion (of the sum/fusion operator) (35):

σ x ⌜Fx⌝ ≈ ιx ∀y ⌜x ∘ y ≡ ∃z ⌜Fz ∧ z ∘ y⌝ ⌝ ("≈" for "is the same as," "ιx" for "that very x such that," "∘" for "overlap"). I.e.: the sum/fusion of all objects x that satisfy condition Fx is the same as that very x such that for all y, x and y overlap (= have a part in common) if and only if there is a z such that z satisfies condition Fz and z overlaps with y. For the concept of overlapping it holds that two objects overlap if and only if they have a part in common.

The concept *part*, which is here decisive, is understood as a primitive concept. This does not prevent theoreticians from providing the following three formal properties of the concept: irreflexivity (nothing is a proper part of itself), asymmetry (if something is a proper part of something else, then the latter is not a proper part of the former), and transitivity (if something is a proper part of something else, and this latter is a proper part of a third thing, then the first is also a proper part of the third).

What would result if one attempted to explicate this book's concept of configuration by means of the mereological conceptual dyad *whole-part*? The only combining factor that could account for the connection and thus the unity of primary facts would be the concept *overlapping*; this factor would have to explain configurations as complex primary facts. Configurations, as complex primary facts, would be defined by this factor. The very least to be said about this is that this factor is far too weak to provide decisive characterization (or definition) of the factor that structures those configurations that can be termed "robust individuals" (e.g., animals, human persons). This is clearly shown in what follows.

[2.3] The third example is propositional-logical *conjunction*. As Chapter 3 (Section 3.2.3.3.2) shows, the logical connectives have, in the logic, semantics, and ontology of this book, an immediately ontological status: the arguments of the connectives are primary propositions that, when true, are identical with primary facts and thus with ontological primary structures. Complex primary facts are ontological configurations.

The simplest and most illuminating form of such a configuration is the *conjunction*. Normally, one speaks of sentences and propositions as arguments of the conjunctor: the conjunction $p_1 \& p_2 \& p_3 \dots$ is true (thus is a complex primary fact or an ontological configuration) just when all the conjuncts are true (are primary facts).

If one understands an individual as a configuration of primary facts, one can most simply explain the inner structuration of this configuration by conceiving of it as a conjunction. The question that then arises is not whether this is correct, but whether it is adequate. For many forms of configuration, it no doubt *is* adequate. A (formal) theory, for example, can be understood simply as a conjunction of provable or proved sentences/propositions (theorems). Heaps, too, can be viewed accurately and adequately as conjunctions of determinate primary facts. If however one thinks of individuals in a robust sense (animals, human persons) and if one conceives of them as conjunctions, that is not false or incorrect, but it is quite evidently inadequate: a conjunction is a structure that is far too loose to provide adequate articulation of the individual in the robust sense. Taken as only conjoined, the primary facts that constitute the individual appear as bound to one another only by the factor of their being grouped together. This in no way corresponds to the extreme structural complexity and interconnectedness of such an individual. At the same time, however, it must be emphasized that a conjunctive combination of primary facts is *always* given when such facts are indeed conjoined.

Beyond the three examples just introduced, additional mathematical structures could be considered as possible means to make formally (more) precise such fundamental concepts as that of the configuration. These would include ordering structures, topological structures, fields, lattices, sheaves, etc. The literature of contemporary ontology already uses some such structures (see Section 3.2.3.3.4).

From the three examples and from consideration of additional mathematical structures, it follows, generalizing, that no formal structure currently known and used appears to be fully adequate to articulate the extremely complex structure of the individual in the robust sense. It could be that in the future formal instruments will be developed that are at least less inadequate for this task. In the present situation it appears best to use the available formal instruments only when they appear adequate for specific cases. No contradiction arises as long as no claim is made that they can be exclusively applied, in all domains, with absolute and complete adequacy. According to the ontology of this book, these instruments themselves constitute extensive segments of the immense dimension of formal structures.

4.3.1.2.2 Is "Configuration" the Adequate Ontological Structure of the Individual Human Being as Person?

This question can be answered in the affirmative only if full consideration is taken of what is established at the beginning of the preceding section—that as a rule we consider every individual human being as a person, generally and intuitively understanding the person to be an entity characterized by a wholly unique unity and having a core that grounds its capacity to speak in the first person, thus to say "I," or to be a subject. These are thus the two essential factors—its unity and its core—to which the concept of configuration must do justice if it is to qualify as a concept adequate for articulating the human being as individual and as person. Beyond doubt, without further specification the concept of configuration is not adequate for this articulation.

For this reason, it is not surprising that many philosophers firmly reject this and similar concepts (e.g., *bundle*).[15] This situation makes unconditionally necessary the clarification and precise specification of the concept of configuration.

4.3.1.2.2.1 A Fundamental Systematic-Methodological Consideration

The topic of this section is the configuration that determines the human individual as human person. Adequate treatment of this topic is facilitated by the introduction, at the outset, of a fundamental consideration that is paradigmatic throughout world-systematics.

[1] In this book, the distinction between the dimension of fundamental structures and the dimension of the datum (or data) plays a simply central role. Without structure(s), the dimension of data is amorphous, and without data the dimension of structure(s) is purely abstract as well as, in another respect, empty. The entire philosophical and indeed scientific enterprise works to bring these two dimensions together. Nevertheless, this fundamental distinction would be wholly misunderstood if one were to view this bringing together as anything like the bridging of a gap or as a somehow subsequent conjoining of what would have been, prior to the conjoining, two fully heterogeneous domains or dimensions.

361   The misunderstanding just identified is visible in the history of philosophy in two completely opposed basic positions; the positions diverge with respect to which of the two dimensions is considered to be a *tabula rasa*. Traditional pre-Kantian metaphysics considers the world to be the dimension that is fully structured in itself, and that makes its impressions on the mind or on thinking, a *tabula rasa*, in something like the manner in which a seal makes its imprint on a piece of wax. The world, as the totality of data, is then the fully structured dimension, and the domain of mind or of thinking is understood as the dimension of the *tabula rasa*. With Kant's transcendental turn, the subject (the mind, thinking) becomes the fully determined dimension, and the data (in the sense relevant here) are understood as the *tabula rasa*.[16] In contemporary

---

[15] See van Cleve (1985), which draws (103) the following conclusion from the "bundle theory": "[S]ince he [a man] would not be identical with any property or any complex of them, he would have to believe that there is nothing with which he is identical—or in other words, that there is no such a thing as himself. . . ." From this van Cleve (105) concludes, "In a word, it [the individual] is *substance*: an individual is something over and above its properties, something that *has* properties without being constituted by them." For a critique of van Cleve's position see Puntel (1990: 241ff).

[16] This concise formulation must be explicated more precisely. As is well known, Kant distinguishes between the dimension of things in themselves and the dimension of the world-as-appearance. What we cognize, according to Kant, is only the world-as-appearance. How Kant's distinction between these two dimensions is best understood is much discussed in the contemporary literature. Yet however one may interpret the dimension of the thing-in-itself (whether more realistically, as the thing thought in abstraction from all its sensory determinations, thus as noumenon, or as a pure product of thought in the sense of a "transcendental something in general," or indeed even as a pure fiction in the sense of an as-if assumption, or—finally—as an "aspect" or a "perspective" or something of the sort), it remains an undeniable fact that Kant introduces a fundamental distinction that plays an absolutely decisive role in his entire philosophy. This distinction marks a border between two dimensions: on the one hand, the dimension of "our" (i.e., of the transcendental subject's) structures (whereby this

analytic philosophy, there is a near-total lack of clarity and indeed even a situation of chaos with respect to this question, as often noted above and considered in detail in Chapter 5.

In contrast, this book defends and explains the thesis that the dimension of data (or of the universal datum) and the dimension of the fundamental structures (as presented in Chapter 3) are *different* dimensions only in an *abstract respect. Concretely*, they constitute a single, fundamental dimension: the dimension of the conceptualized (i.e., structured) comprehensive datum and of fully explicit (fully concretized, fully determined) fundamental structures. Basically and ultimately, the two are a single basic dimension. Chapter 5 explains and grounds this thesis more precisely and thoroughly.

For present purposes, it is of the utmost importance that a specific aspect of this fundamental unity of the two dimensions be thematized and made more concrete and more precise with respect to world-systematics. For the sake of theoretical clarity, it is indisputably advantageous *first* to present the two dimensions separately. To be sure, the dimension of the universal datum cannot be articulated *in complete abstraction from* the dimension of structure, because its articulation requires the introduction and utilization of structures. In this respect, there is an asymmetry between the two dimensions. But in a quite fundamental respect the data themselves are recognized *before* the appropriate structures are clarified. The relevant fundamental respect corresponds in essence to what can be termed *the concept of experience clarified by systematic philosophy.*

In lifeworldly contexts, but also in concrete scientific practice and in concrete or actual philosophizing, one proceeds on a level of acting, of speaking, and even of theoretical discourse that is characterized by an understanding of the world that is *pregiven* in one way or another. From the perspective and in the terminology of the structural-systematic philosophy presented in this book, this pregiven understanding of the world consists of a *readily available,* for the most part unreflected but nonetheless presupposed, interweaving of data and structures, and thus of data that are understood (structured) in determinate ways. Such data can in no way be understood as entities or instances *independent of* structures, theories, languages, etc. To take them to be so would be to accept a typical—and radical—form of the "myth of the given" (see McDowell 1994/1996, Sellars 1956). This is a myth in the literal sense: a fairy tale about a fundamental and putatively unavoidable and inalterable fixed point. The assumption that any such point exists cannot withstand any penetrating analysis of the concept *datum.*

Nevertheless, the problem-situation for philosophy is somewhat different from that for the natural sciences. In the natural sciences, two points above all must be taken into consideration. The first is the fact that in contemporary philosophy of science, the thesis of the theory-ladenness of the empirical data is widely accepted. This is considered in Chapter 2 (Section 2.4.2.2). To be noted here is only that this central thesis directly corresponds to a fundamental thesis of the structural-systematic philosophy. The second point relates to the character of the specific kind of experience

362

---

term designates Kant's *a priori* forms of intuition, the concepts of the understanding, and the ideas of reason); on the other, a dimension that remains definitively inaccessible to the structural dimension (the dimension of the transcendental subject). This second dimension is, in Kant's conception, a *tabula rasa,* and indeed in the quite determinate sense that, for us—for our structures (in the sense indicated above)—it is and must always remain *empty.*

relevant to the natural sciences: the experience is *experimental*. Experiments serve, most importantly, to support (or refute) scientific theories; in a somewhat different respect, experiments have an *inventive* function, in that they make possible and give rise to the development of (new) scientific theories.

363   As Chapter 5 shows, there cannot as a matter of principle be in philosophy any counterpart at all comparable to the scientific experiment. That, however, does not mean that philosophy has no relation to experience. Indeed, it has a quite subtle and wholly exceptional, indeed unique relation to experience, to the data, to the universal datum. Fundamental in this respect is the fact that philosophy proceeds neither purely *a priori* nor on the basis of experiments, but instead takes as its starting point the understanding of the world *and of the self* that is articulated in ordinary language. This basic fact is in no way in tension with the conception presented in this book; to the contrary, it both expands that conception and makes it more precise. This conception (not its presentation) is wholly present and wholly complete from the very beginning, in the sense that, as the guiding idea, it is itself the result of a lengthy procedure. An essential aspect of that procedure is precisely its having started with the ordinary language that is used to articulate the naturally given understanding of the world and of the self.

The understanding of world and self that is articulated in ordinary language is the *pregiven empirical* unity of the dimension of structure and the dimension of data. That philosophy *begins from* this understanding, however, in no way means that it simply accepts it or that it leaves it as it presents itself in this pregivenness. Instead, the task of philosophy (and of the other sciences) is to use criteria of relatively maximal coherence and intelligibility to test, to correct, in some cases to reject, and ultimately maximally to improve upon this understanding. Chapter 5 treats this state of affairs in detail.

[2] Shown here are the consequences of the preceding considerations that are relevant to anthropology; the first of these concerns what significance is to be accorded the *natural self-understanding of human beings*. The decisive factor is that the perspective taken in this book reveals an *asymmetry* between the understanding of the world and human self-understanding. There is no worldly thing with which we are as familiar as we are with ourselves; we do not know any thing as well as we know ourselves (which does not in any way mean that we know ourselves completely, have complete insight into ourselves, etc.). But regardless of whatever superiority this self-understanding may have vis-à-vis any worldly understanding, only with extreme caution can it be viewed as *given*. It is unconditionally necessary that various important distinctions be introduced and attended to; these distinctions concern the implications of this self-understanding for science and philosophy.

This book emphasizes repeatedly that natural language is not a reliable basis for philosophy and the sciences. However, natural language does articulate our natural self-understanding in the manner described in the preceding paragraph. What follows from this? Here, one must distinguish between two wholly distinct kinds of sentences that articulate this self-understanding: *immediate* (categorially unmediated or unqualified) and *mediated* (categorially mediated or qualified) sentences. Examples of immediate sentences include "I'm thinking" and "I'm awake." Sentences of the second type—mediated sentences—arise within one or another framework for what may be termed "natural" or "non-reflective" metaphysics concerning the human being;

examples would include the italicized portions of the following passage from Bertrand Russell's "What is the Soul?": "When I was young we all knew, or thought we knew, that *a man consists of a soul and a body*; that *the body is in time and space*, but *the soul is in time only*" (1928/1996: 203; emphasis added). Clearly, Russell is not using "soul," "body," etc. in any strictly technical senses.

Scientistically oriented philosophers consider all sentences about human beings that are not presented within the framework of what these philosophers consider to be the purest possible natural-scientific theory model to belong within the domain of "folk psychology" and to be, for this reason, irrelevant to philosophy. This position is, however, untenable, particularly because it exhibits a remarkable lack of discrimination. As is shown by the distinction just introduced, there is no adequate basis for considering the immediate and the mediated sentences articulating "natural" human self-understanding to have the same status or to be on the same level. What the immediate sentences express is absolutely indisputable. They are comparable to what Quine terms "observation sentences" (see, e.g., 1960: § 10) and to the sentences uttered by scientists who describe the phenomena that are *immediately visible* or *observed* when they look into microscopes, read dials, etc.

These somewhat lengthy introductory considerations provide a basis for the attempt to answer the following question, on the basis of the central ontological theses of the structural-systematic philosophy: how can one precisely conceptualize the configuration that constitutes the human individual as human person? This question must be answered step by step.

4.3.1.2.2.2 The Elements of the Configuration Constituting the Human Individual

In the first place, the elements of the configuration constituting the human individual must be investigated carefully. As shown often in this book, these elements can be nothing other than simple and complex primary facts. The human individual is obviously a highly complex unity, as is manifest on the basis of this ontology. This entity is complex in two distinct ways. On the one hand, the number of the individual primary facts constituting it is considerable—indeed in principle, as shown in Chapter 5, infinite. On the other hand, the heterogeneity of the constitutive primary facts must not be overlooked. The question of how many primary facts are involved is considered below. At this point, two main aspects of the heterogeneity of elements constituting the human individual—*modal* heterogeneity and *domain-specific* heterogeneity—are briefly described and explained.

[1] Three specific forms of *modal* heterogeneity can be distinguished: the *absolutely essential* form, the *relatively* or *historically essential* form, and the *contingent* form. These primary facts correspond to the entities that, in the usual terminology of substance ontology, are called "predicates" or "properties or relations." "Correspond" here means that the "predicates" are transformed into primary sentences (i.e., sentences without subject-predicate structure). The "predicate" is thus not "predicated" of or in any way applied to an "object" (substance, substratum), but instead has the form of a sentence without a subject. The general form of such sentences is "It's F'ing." The "property" or "relation" (the "attribute") is transformed into a primary proposition expressed by a primary sentence. Such a primary proposition, if true, is identical to a primary fact.

The *absolutely essential* primary facts are those that are absolutely necessary constituents of human being. These are the factors that characterize every human being, independently of any and every historical, geographical, or other sort of situation. The three most important absolutely essential factors are thinking (reasoning and understanding), willing, and being-conscious (which, as is shown below, includes being-conscious-of-oneself). The exact formulations of these primary facts would be, "It's thinking," "It's willing," "It's being-conscious/self-conscious," etc. As should be obvious, these factors are the proper constituents of what is usually referred to as *mind* ("It's minding").

366   The *relatively* or *historically essential* primary facts are those that cannot be missing from any human being but that, unlike the absolutely essential factors, are always given with a concretization (or indexing). To these belong such facts as bodying. Bodying entails being-in-a-specific-place, -at-a-specific-time, etc. Yet more concrete historically essential factors include having-been-born, growing, dwelling, etc. Also to be included among the primary facts are those that, although in no way essential to the human being as such, nevertheless are essential in that they do not cease to be constituents of a given human being if they ever arise. Examples include having-been-given-a-name, having-studied, being-a-parent, having-siblings, etc.

Contingent primary facts are those that arise historically but that can also, historically, cease to obtain. These constitute the majority of the primary facts associated with any human being, such as wearing-a-specific-outfit-on-a-specific-day, etc.

[2] The primary facts constituting the human individual or person must now be considered within a different perspective, i.e., with respect to the fact that, contentually, they belong to *different specific ontological domains*. They include primary facts that are mental, social, sensory, biological, inorganic (purely physical), etc. Within these levels or domains there are additional relevant differentiations. How these heterogeneous forms are interrelated is one of the central problems in contemporary philosophy of mind. Particularly important is the question whether these are irreducibly different types of primary facts or instead some types can or must be reduced to others. Central to current discussions is the question whether mental, sensory, and biological factors can be reduced to purely physical factors. This question is considered in Section 4.3.1.2.3.

[3] It is an indisputable fact that the human individual *experiences* him- or herself as a *highly complex unity* of all the dimensions of human life that are characterized in [1] as the modally different primary facts. To be sure, the human individual does not experience these dimensions as consisting of items that have been clarified reflectively/philosophically; it must be emphasized that, to the contrary, the sentences articulat-

367   ing this experience are *immediate* in the sense introduced in the preceding subsection and therefore cannot be rejected. This does not, however, mean that they cannot be interpreted and explained. But how is their interpretation and explanation to be accomplished?

The purposes of this book require it to articulate the specific character of this *unity* (i.e., of the configuration of the just-identified forms of primary facts that constitutes the human individual as a person). The approach to the problem taken in this book and the proposed solution now sketched briefly are *fundamentally* different from any approaches taken or solutions suggested and discussed in contemporary philosophy of mind. In brief: the unity of the human individual (in common terminology, "personal identity"[17]) is sought neither in a substance (or substratum) nor in determinate phenomena, such as the continuity of the body, of consciousness, of memory, etc., but instead in an absolutely fundamental ontological-systematic factor that is named the "irreducible ontological-systematic unifying point."[18]

4.3.1.2.2.3 The Unifying Point as the Factor Configuring the Configuration

[1] The configuration of primary facts that is the human individual is neither a set nor a sum nor a conjunction, but is a unique totality that is characterized by a *unifying point*. Instead of "unifying point" one could say "center" or "core," although both these terms have connotations that would have to be ignored. That the term "unifying point" is most appropriate is shown in what follows.

There are many forms of unity, in part because there are enormous differences in how tightly elements within unities are interconnected. Moreover, not every form of unity involves what is here called a unifying point. For example, conjunctive unities have no such unifying points. What is meant here is the following: the uniting bond, the configuring factor, is not only—somehow—omnipresent, it not only encompasses and contains all the united elements, but it also has a quite determinate directedness, indeed one emerging from the point from which the unifying is accomplished and by which the elements are made possible and borne. This unifying bond or configuring factor—the unifying point—is a basic concept; in a specific sense it is a primitive concept. More precisely, it is an absolutely singular, unique concept that can be articulated only on the basis of a concrete and penetrating analysis of the phenomenon "experience of personal unity."

[2] In what, more precisely, does this unifying point consist? It articulates itself by saying "I." The I must not be hypostasized as substance or anything of the sort; it would also be fully insufficient philosophically to interpret or understand the I solely on the basis of linguistic configurations within which the word "I" appears.[19] The task is instead that of comprehending the I ontologically. The concept of the unifying point is helpful for this task, but only if explained more fully.

To explain the unifying point more fully, its extent—the range of what it unifies or configures—must be determined by closer analysis of the specific elements that are included within the configuration and of their various forms. Here, only the most important aspects can be treated.

Common formulations used in the attempt to bring into view the human person in all its complexity include, as indicated earlier, "The human being has a body and a soul," "The human being is a minded, spatiotemporal entity," etc. Within the ontology defended here, the concepts used in such formulations must be understood on the basis of the fundamental (and indeed sole) category *primary fact*. If only physical, spatiotemporal primary facts are considered to be components of the

margin: 368

---

[17] More is said below concerning the concepts *unity* and *identity* in the context of this problem.

[18] The word "ontological" is intended clearly to indicate that the position to be developed is not "phenomenal" or "phenomenological" in any way that would be opposed to the ontological. An example of such a phenomenal or phenomenological position is Bertrand Russell's conception of the individual as a "bundle of qualities," which he understands as a "complex of compresence [of qualities]" (1948: 306f).

[19] This is the approach taken in Tugendhat (1997).

comprehensive configuration constituting the human being, the human being is a purely material being, which would mean the following: the human being's *unity* or its *unifying point* would be restricted to the immediately material/physical (although in this case one would no longer speak of a "unifying point" in anything like the sense intended here).

The sense intended here for "unifying point" becomes clearer if one considers the next stages or steps. If one adds the primary facts that constitute the complex "biological" ("living being"), then the being's unity or what its unifying point unifies expands to include an environment. This is a natural phenomenon. Plants and animals always have determinate environments within which they flourish to greater or lesser degrees, within which they find themselves making exchanges. Their environments constitute the quite restricted horizons for such living beings. They are not capable of being aware of or acting upon or within anything beyond their environments. This restrictedness decisively determines how they are to be conceived of as ontological configurations.[20]

Only the human being, as minding and bodying, has a "world" in what this book accepts as the genuine sense: "a world" in this sense is *the* world, as unrestricted. As shown in Section 2.3.2.5, human beings are capable of surpassing all limits not in that or because they are material-biological beings, but in that or because they are minding beings. They can ask about absolutely anything, develop "theories of everything," etc. This is nothing other than a brief description of that absolutely essential primary fact that we call *thinking* and that decisively (co-)defines the configuration of the human being. No one has provided a briefer or apter characterization of this absolutely essential primary fact than Aristotle, with his famous formulation "The mind is in a certain way all things."[21] One can interpret or clarify the "in a certain way" with the concept *intentional*. The sense is then that *the mind is intentionally coextensive with simply everything, with being as a whole, with the unrestricted universe of discourse*. This is explained further and more precisely below (esp. in Section 4.3.1.2.3.2[2], pp. 288–90).

Among the primary facts that are absolutely essential to constituting the human being as configuration are, in addition to thinking, both willing and being-conscious. These, too, are characterized by *intentional coextensionality* with the universe. Section 4.3.1.2.2.4 reconsiders these three absolutely essential primary facts, especially the primary fact being-conscious.

---

[20] Discussions motivated by the question whether animals (or specific kinds of animals) "think" (or have minds, etc.) usually have nonsensical aspects. Without a criterion for thinking, it is nonsensical to ask whether or not a given kind of animal thinks. Here, such a criterion *is* available: the type of "world" that corresponds to the ontological constitution of a given kind of being. If this "world (*Welt*)" is a pure (hence restricted) environment (*Umwelt*)—only an *Umwelt*, hence not, unrestrictedly, the *Welt*—then there is no "intelligence" in what this book accepts as the genuine sense, because this sense is defined as requiring that the "worlds" of beings that think be unrestricted (as is the case for human beings). Of course, if one associates a different concept with the term "thinking," then some kinds of non-human animals may well "think." But then the question is reduced to a purely terminological one.

[21] See the famous passage in *De Anima* Γ431b21: ἡ ψυχὴ τὰ ὄντα πώς ἐστι πάντα. More literally, "The soul is in a certain way all beings."

[3] Now named and preliminarily explained are all the important factors that must be presupposed if the concept of configuration, for the specific case of the human individual as human person, is to be clarified. Absolutely central is the concept, introduced and preliminarily explained above, of the *unifying point*. The current task is to explain this concept more precisely.

If one takes into consideration all the aspects and factors so far introduced and attempts to fit them coherently together, there results the following conception: the 370 configuration that constitutes the human individual as a human person presents itself—in a first approximation—as the *intersection* of all the previously identified primary facts and of the forms into which they fall. "Intersection" here means, initially, that all these primary facts meet *at* or *in* or *as a point*, within or as which consists the unity of the individual and the person. But this point is nothing like a sum or collection of primary facts that is in any way subsequent to those facts. If it were, then saying "I" would no longer be explicable, because the spoken "I" is the expression of a unifying point that is not the result of a unity that arises after the fact, but is instead the ontological starting point for the unity that experiences and understands itself as human individual, thus as human person.

With this, a yet deeper and more fundamental interconnection becomes visible. The unifying point is originary rather than resultant *because* it is a *systematic point* in the sense that it is situated within this system and at the same time self-situating within the system. This means that it is a point that is situated and can situate itself with respect to any- and everything within the system (within being as a whole). Rocks and cats, for example, cannot situate themselves vis-à-vis, for example, the Pythagorean theorem, whereas the human being situates herself even with respect to (for example) whatever she isn't currently aware of simply by so classifying it. More precisely, as *intentionally coextensive* with the so-understood system, the unifying point is an *originary* locating within the whole of the system (*originary* in the sense of being absolutely ultimate, not reducible to anything else), and one whose determinacy is characterized by or given with both its entire concrete spatiotemporal situatedness and its absolutely universal intentional scope. The minding and bodying individual, the human person, is the concrete reality of such a systematic location.

How then does the concept of configuration present itself in the case of the human individual or the human person? To provide at this point a brief, albeit extremely abstract, answer: *the configuring factor in the configuration that ontologically constitutes the human individual or the human person is the systematic-intentional location.* This location is the *systematically self-situating unifying point* explained above: the point within the system from which the configurational accomplishment of the unity of the individual or the person proceeds.

[4] The basic idea just sketched would have to be developed in detail, but its devel-   371 opment is not among the tasks of this book. Here, only the question of *conditions of identity* is considered briefly.

Within the framework of the conception developed here, there is no problem of individuation of the sort that arises on the basis of the assumption of universals. But there does arise a wholly different question concerning conditions of identity. If one accepts Quine's edict "No entity without identity," it is clear that the current account requires the inclusion of identity conditions for human beings.

Quine's demand is satisfied easily. Two configurations constituting two individuals *a* and *b* are identical if and only if they have exactly the same *systematic-intentional location*. This is the case if and only if the following two conditions are satisfied:

(i) The two configurations have exactly the same primary facts of all forms and in all domains (the material condition).

(ii) In the two configurations, all the primary facts are in exactly the same order and are (formally) interconnected in exactly the same ways (the formal condition).

4.3.1.2.2.4 Intentionality and Self-Consciousness

[1] In this context it is appropriate to consider further the interrelations among the three absolutely essential primary facts identified above: thinking, willing, and being-conscious. Thinking and willing are unambiguously *intentionally* structured: both involve directedness or orientation toward…, and ultimately toward anything and everything, toward the universe in the most comprehensive sense—in brief, toward being as a whole. To be sure, the terms "directedness" and "orientation" are quite general ones that not only can but must be more subtly differentiated in order to be explicated more precisely. In *the mode of thinking* the directedness/orientation to the world has as its goal the articulation (the conceptualization) of the world (in the usual philosophical terminology that involves misunderstandings of various sorts, the term "representation" figures prominently); the directedness/orientation in *the mode of willing* has as its goal the *actualization* of something that is aimed at or attempted. These differentiations are not pursued further because doing so is not necessary for this book.

[1.1] Being-conscious appears to be something quite different; for this reason, a more thorough analysis is provided for this primary fact that is absolutely essential to human being. For many theorists, consciousness (being-conscious) has nothing to do with intentionality; they consider these to be two wholly different concepts or entities. Thus, Manfred Frank—who has written two of the best books on consciousness (and, more specifically, on self-consciousness)—writes as follows (1991a: 16n13):

372   I have deliberately omitted intentionality from my sketch of the essential properties of self-consciousness, because it is a characteristic not of *self*-consciousness, but of knowledge [*Erkenntnis*]. It is also not a necessary characteristic of consciousness, because much of consciousness (moods, emotions, feelings, etc.) is non-intentional. Consciousness also has a broader scope than does intentionality, which of course must also be made intelligible on the basis of consciousness and presumably has to do with the spontaneity of the self.

Frank identifies intentionality with representation and therefore restricts this concept to the domain of knowledge. As shown earlier, this is only one of the two aspects of a comprehensive concept of intentionality. What above is termed "directedness" or "orientation" toward the world encompasses more than the mere articulation or representation of the world. To say that moods, emotions, feelings, etc., are *non-intentional* is accurate, if at all, only if intentionality is identified with an *extremely narrow* concept of representation. Within the framework of a comprehensive concept of intentionality— such that intentionality encompasses all of the aspects of mentality that are world-related—moods, emotions, feelings, etc., are all to be conceived of as intentional, albeit in a weaker sense: ultimately, they are caused or occasioned by elements of the world.

Put generally, they have to do with the situation (of elements) of the world and are therefore also world-related.

[1.2] The history of theories of consciousness (and self-consciousness) in modernity is highly instructive. Here, only the most important of the positions developed in this period and defended at present are briefly named and characterized.[22]

The two best-known and most influential positions have been (and are, insofar as they are still defended) the *egological* and the *non-egological*. The grounding thesis of the egological position is that self-consciousness is always the relation of consciousness to an I, such that "self-consciousness" means "I-consciousness" in the sense of consciousness of the I (Kant, Fichte, Neokantianism). The major variant is known as the "representation- or reflection-model of self-consciousness." Reflection is a form of representation; self-consciousness is seen as an auto-reflexivity, a representation of one's self. The I accomplishes a turn to itself such that the result of the process is its consciousness of itself.

Against this egological conception, two main objections have been raised. The first is that it leads to an infinite regress: the I, as representing/reflecting, is presupposed not to be conscious of itself, because it is supposed to attain consciousness of itself only via self-reflection. The goal attained by the self-reflection is consciousness of the pre- or non-reflective I or self, but this goal is not a genuine *self*-reflection or *self*-consciousness in that the *reflecting* I or self is distinct from the I or self *qua* object of the reflection, which first emerges only through the reflective act. Consciousness of the I *qua* reflecting requires an additional reflective step, and this process—according to the objection—is automatically repeated at every new level of self-reflection, infinitely.

The second objection contends that consciousness of itself must already be ascribed to the I before the I accomplishes any kind of self-reflection. The reason is that it would be absolutely counterintuitive or indeed absurd to attribute to something not conscious of itself any such accomplishment (if the I were wholly unconscious of itself, how could it possibly thematize or reflect upon itself?). Self-consciousness must therefore already be presupposed and indeed is in fact always already presupposed before or in order that the self-reflection of the I can take place. This consideration reveals the egologically oriented theories of self-consciousness to be *circular*: they presuppose (and must presuppose) what they seek to demonstrate.

The non-egological models of self-consciousness explain self-consciousness as consciousness not of an I, but of consciousness itself. These models therefore encounter versions of the two problems (regress and circle) that beset the egological models. Moreover, it remains puzzling how the unity of the dimension of consciousness and particularly how the entrance of the I into this dimension can at all be comprehended.

[1.3] In reaction to the problems described above, Dieter Henrich, in particular (along with some of his students), has attempted to develop a wholly new conception

[margin: 373]

---

[22] The presentation is based primarily on two of Manfred Frank's books (1991a, b). Frank's comprehensive, 187-page Afterword to the second book is an excellent critical presentation of the most important theories of self-consciousness within the period treated in the book (from Fichte through Sartre); it also takes into consideration the most recent developments, particularly in contemporary German philosophy.

of consciousness and especially of self-consciousness. His attempt is guided by two considerations: on the one hand the avoidance of the aporias that arise with the models introduced above, and on the other the search for the *original phenomenon of consciousness*. His basic idea (1970) is extremely simple: consciousness is a wholly *relation-free original phenomenon*, an "I-less, anonymous dimension." Henrich immediately concedes that from the anonymity of this dimension there can emerge a genuine subjectivity, an active principle of unification that can be accorded the name "I" or "self." What is decisive, however, is his contention that this principle, the self or I, is not the original or primordial phenomenon of consciousness.

For more precise characterization, Henrich introduces four factors.[23] Consciousness is, first, an *event* that is relationless in the sense that it is "not related either to itself or to anything that itself is without consciousness [*beziehungslos und ohne Relation zu irgend etwas, in dem Bewusstsein nicht vorkommt*]*." Second, although consciousness is itself to be understood as a pure "medium" or an anonymous "dimension," it is an event that makes possible other (derived) events that are synthetically interrelated with one another. Third, consciousness is a *simple* or *exclusive* dimension in the sense that it cannot be the case that multiple occurrences of consciousnesses could overlap and thereby form something like a common space of consciousness. The fourth factor is the most problematic: despite his contention, insistently repeated, that consciousness is *relationless*, Henrich ascribes to consciousness an "acquaintance with itself [*Kenntnis seiner selbst*]" or a "familiarity with itself [*Vertrautheit-mit-sich-selbst*]" that is an internal property of consciousness. Consciousness is a dimension that is *prereflective and immediate*, and that therefore contains no knowledge and is not conceptually mediated. This leads Henrich to produce paradoxical formulations such as the following: "The knowing self-relation that is found in reflection is not a fundamental state of affairs, but an isolating explicating [*ein isolierendes Explizieren*]; it does not presuppose a self-consciousness of any kind whatsoever, but instead an (implicit) selfless consciousness of self" (280).

The chief objection to Henrich's position focuses on the fact that it is critical in the sense that it attempts only negatively to avoid circular "explanations" of self-consciousness. This objection leads Henrich already in 1971 to develop, in a still unpublished text titled "Being-a-Self and Being-Conscious [*Selbstsein und Bewusstsein*]," the basic features of a more detailed theory. A summary of the theses presented in that text appears in 1986 with the title "Self-Consciousness: A Field of Problems with Open Borders." According to the generally new insight or thesis, the fact that consciousness is an "anonymous dimension" does not exclude its being internally structured, if one takes its structuration to be a factual inseparability of moments that are clearly conceptually distinct from one another.

Henrich names three concepts that constitute the internal structuration of self-consciousness: consciousness, being-a-self, and a formal, epistemic self-relation. He continues to understand consciousness as an "anonymous dimension" "within which something comes to be present"; being-a-self is characterized as the "source" of conscious activity (in correspondence to Kant's "spontaneity"); the "abstract self-relation"

---

[23] See Frank's excellent treatment (1991b: 586ff).

is still termed "familiarity [*Vertrautheit*]," but Henrich no longer uses the reflexive pronoun in its designation.

Henrich concludes as follows:

> Self-consciousness has an inner complexion [*Komplexion*[a]], an inner multiplicity, because, to put it summarily, it includes or contains within itself the difference between apprehending and what is apprehended, and that between knowledge and actuality. Nevertheless, common forms of analysis provide us no access to the way in which these aspects are connected with one another within self-consciousness. (7)

At the end of his presentation of Henrich's position, Manfred Frank accurately   375
remarks:

> What the model continues to leave in the dark is the unity of self-consciousness as a phenomenon that, for its part, is conscious. Consciousness constitutes ... an I-less, anonymous "field" within which the I first registers tendencies, rules, and abstractions. But it remains unknown what makes the "inner complexion" of the three moments in fact inseparable. (1991b: 598f)

Here, Frank identifies the decisive problem, which is basically the same problem that arises above with respect to the configuring factor in the configuration that constitutes the human individual. This problem is addressed below, but worth noting before it is addressed is that the discussions briefly described above, concerning theories of self-consciousness, for the most part exhibit clear deficiencies in conceptual-methodological and conceptual-analytical acuity. Three factors are chiefly responsible for these deficiencies: first, the generally naive search for "*the* primordial phenomenon"; second, their attempted reliance on or application of the extremely obscure "procedure" of "deduction" or "derivation" that is simply central to the German Idealists; and third, the (early Romantic) notion of the *innerness* of self-consciousness in total separation from anything that could count as "the world" or "the other."[24] An example of a thinking oriented toward the "primordial" is Heidegger's *Being and Time*, which presents self-consciousness as a merely "derivative mode" (1927/1972: §44; see Frank 1991a: 10ff).

[2] On the basis of the concept developed above of the configuration of the human individual, it is not difficult to show how to understand the absolutely essential primary

---

[a] In using the word *Komplexion*, Henrich presumably seeks to emphasize an *active* character. The English "complexion" serves not in any current ordinary-language sense, but only if understood as the nominal counterpart to a verbal meaning the *OED* lists as obsolete, i.e., "To constitute by combination of various elements; to put together, compose."

[24] The influence of this third factor, which continues to play a decisive role for authors such as Henrich, is the more surprising giving the frequency with which they refer to and indeed decisively rely upon German idealism. Hegel, who is if not the greatest of the German Idealists then certainly one of the greatest, subjects the notion of "pure innerness" to a thoroughgoing critique. Consider one of the famous sentences from the dialectic of self-consciousness in the *Phenomenology of Spirit*: "*Self-consciousness achieves its satisfaction only in another self-consciousness* [*Das Selbstbewußtsein erreicht seine Befriedigung nur in einem anderen Selbstbewußtsein*]" (110/*TW*3:144). Here, "satisfaction" is synonymous with "realization" or "full determination."

facts *being conscious* and *being self-conscious*. The human mind is distinguished as being *intentionally coextensive* with the universe, with being as a whole. *Intentionality* appears here in a comprehensive sense (i.e., it is understood in a comprehensively systematic manner). To explain and to comprehend the individuality of the human individual as minding and bodying, reliance is put upon the experience of saying "I." That reliance leads to the identification of a unifying point in the system, a point that is to be interpreted as the factor that configures the configuration constituting the human individual. *The configuring factor in the configuration constituting the human individual or the human person is the systematic-intentional locating.* This locating or situating is the systematic unifying point: the point in the system from which emerges the achievement that configures the unity of an individual or person.

It is now possible to provide a quite simple interpretation or explanation of the phenomena *consciousness* and *self-consciousness*. The decisive point can be briefly put as follows: the comprehensive intentionality that characterizes the human mind, precisely because it is systematic, extends itself to or encompasses the universe *and everything that it includes—therefore also the unifying point itself, the mind as subject*. With this is given not only consciousness of the *universe* but also consciousness of the *self*. The unifying point itself—and therewith the subject or I—is included within the dimension of consciousness. Consciousness and self-consciousness are thus absolutely essential *ontological factors* within the configuration constituting the human individual as minding and bodying. Self-consciousness is not the result of some reflection that is somehow subsequent; it is instead a *prereflective ontological self-relation of the unifying point itself (i.e., of the subject, the I)*. The configuring factor in the ontological configuration constituting the human person is the systematic-intentional locating. This locating is the systematic unifying point explained above. It thus cannot be said that consciousness or self-consciousness is a primordial phenomenon on the basis of which the subject or I can be explained, nor can it be maintained that either consciousness or self-consciousness is a "derivative mode." The comprehensive phenomenon of the human being as minding and bodying is an extremely complex configuration within which systematic intentionality and consciousness (including self-consciousness) are two sides of one and the same coin.

4.3.1.2.3 *Is the Human Individual or Person Explicable Materialistically/ Physicalistically?*

The question posed by this section's title is the basis for one of the most intensively discussed issues in contemporary philosophy of mind. An exhaustive response to the question cannot be provided in this book. Instead, the account that follows provides first some suggestions aimed at clarifying the entire discussion and then a central argument in support of a negative response to the title's question.

4.3.1.2.3.1 On the Current Discussion

[1] The concepts *materialistic* and *physicalistic* are not identical, but in fact they are explicitly equated by many analytic philosophers and implicitly equated by most. The result is a deeply disturbing confusion. "Materialistic" designates a comprehensive metaphysical conception that identifies the world or the universe or being simply with

matter. "Physicalistic" relates to physics and designates both the method of physics and what this science investigates. Scientistically oriented philosophers are inclined simply to identify the two concepts, as do most of the analytic philosophers who are concerned with the philosophy of mind. But most accomplish this identification in a highly striking manner—they introduce subtle distinctions in the attempt to do justice to the peculiarity of the mind, but they do so in such a way that the ultimate metaphysical basis remains a purely materialistic one that is then simply understood as physicalistic. This leads to a situation characterized by a fundamental unclarity and incoherence. Several observations provide support for this contention.

The following account understands and assesses physicalism exclusively as a *metaphysical* thesis; other uses of this term in the literature are not considered here. The metaphysical thesis defended by physicalism, in its simplest formulation, is the following: all fundamental facts (at least of the actual world) are physical facts. There is currently an extensive variety of types and schools of physicalism; this variety is not considered here (see Stoljar's (n.d.) excellent "Physicalism"). The following account considers more closely only one position, the one that appears currently to be dominant; this is *non-reductive physicalism*, a position that must first briefly be situated within the entire spectrum of views currently being defended.

The three most important positions are eliminativism, reductive physicalism, and non-reductive physicalism. *Eliminativism*, as its name indicates, eliminates (in the sense of "denies the existence of") mental facts. This must, however, be made more precise. Eliminativism maintains that mental language (i.e., here, the language that corresponds to "folk psychology") has no "referential" character: it designates nothing ontological, or yet more precisely, it maintains that the singular terms of this language have no referents and the sentences of this language no truth-values.

Of the many varieties of *reductive physicalism*, here only the most strongly metaphysical version need be introduced. According to this position, all properties that are   378
expressed by predicates of (for example) a psychological theory are identical with the properties that are expressed by the predicates of (for example) a neurological theory (see Stoljar (n.d.): 6). This variant of physicalism is thus essentially identical to "type physicalism" or the "identity theory."

Eliminativism and reductive physicalism are the two most radical variants of physicalism. They are not here subjected to separate critiques, because the anti-physicalist argument presented in Section 4.3.1.2.3.2 aims directly at both of them. Here, it suffices to note to that both forms are untenable not because they are directly or unequivocally incoherent, but for other reasons. In this way, they differ from non-reductive physicalism, which is criticized most often precisely for being incoherent. The following section reveals this criticism to be on target.

[2] The two best-known forms of non-reductive physicalism are Davidson's token physicalism and supervenience physicalism. The first can be characterized as follows: "For every actual particular (object, event, or process) x, there is some physical particular y such that $x = y$" (Stoljar (n.d.): 6). Despite its reliance on this central thesis, token physicalism also maintains that the mental is irreducible. This can also be put as follows: there is no reducibility by definition or on the basis of psychophysical laws (this position is also termed "anomalous physicalism"). Nevertheless, this variant of physicalism understands itself as metaphysical physicalism in the strict sense (i.e., as materialism).

The by far best known version of non-reductive physicalism is *supervenience physicalism*, of which again there are various forms. The thought of supervenience (in the domain of the philosophy of mind) is easily explicable intuitively: two different things (objects, properties, organisms, etc.) cannot be distinct mentally without also being distinct physically. This global thesis includes *three specific* subtheses that characterize the relation between a basis (here, the physical) and the level of supervenience (here, the mental): *covariance* of properties (if two things are indistinguishable in terms of their basic properties, they are also indistinguishable in terms of their supervenient properties), *dependency* (supervenient properties are dependent upon or determined by the basis), and *non-reducibility* (covariance and dependency can hold even when the supervenient properties are not reducible to the basic properties).

379   Stoljar illuminates the ontological difficulties that this concept of supervenience brings with it by means of an example presented by David Lewis. This example is of a picture produced by a matrix of dots:

> A dot-matrix picture has global properties—it is symmetrical, it is cluttered, and whatnot—and yet all there is to the picture is dots and non-dots at each point of the matrix. The global properties are nothing but patterns in the dots. They supervene: no two pictures could differ in their global properties without differing, somewhere, in whether there is or there isn't a dot. (Lewis 1986: 14; see Stoljar n.d.)

Applied to the domains of the mental and the physical, the thesis is that the physical features of the world correspond to the dots and the psychological, biological, social, or mental factors to the global properties of the picture. The latter are nothing other than a pattern of or in the points; correspondingly, the psychological/biological/social/mental factors are nothing other than patterns of or in the physical factors of the world. In the terminology generated by the concept of supervenience, this means that the global factors in the picture are supervenient in relation to the dots; correspondingly, anything and everything non-physical is supervenient in relation to the physical, *if* physicalism is true. Lewis's example makes clear that in the case of the picture, supervenience means that no two pictures can be identical in the arrangement of their dots but at the same time different in their global properties. Correspondingly (if, generalizing, one applies the supervenience concept to worlds), one would have to say that no two possible worlds could be identical in their physical properties but differ, somewhere, with respect to their mental (or social, biological, etc.) properties.

It is now possible precisely to articulate the *ontological* problem that supervenience physicalists do not even pose, much less solve: what *are* the patterns themselves? What is the *ontological status* of the patterns? Clearly they are not identical with the points (or with the basic physical properties). Supervenience physicalism confronts a dilemma: either the patterns are something ontological, or they are not. If not, then the entire analysis or explanation based on the supervenience concept has not the slightest philosophical value; but if so, then supervenience physicalism accepts entities that it can neither thematize nor bring into harmony with its thesis of metaphysical physicalism.

380   What these considerations yield may be put as follows: non-reductive physicalism irreparably lacks a developed, coherent ontology. In a fully obscure and undeveloped manner, a single and comprehensive ontological-metaphysical dimension is accepted; it is designated "matter (material reality)" or "the physical (physical reality)." Then,

however, "non-physical" entities are accepted as well, but are themselves said to be *fundamentally* or *ultimately* understood as "physical."

[3] To illustrate and support the contention with which the preceding paragraph closes, it suffices to consider the example of a philosopher who is well known both as a radical physicalist and as a radical non-reductivist with respect to the mental domain: John Searle. On the one hand, Searle defends (1995: 7) an exalted and comprehensive *physicalist* ontology:

> Here . . . are the bare bones of our ontology: We live in a world made up entirely of *physical* particles in fields of force. Some of these are organized into systems. Some of these systems are living systems and some of these living systems have evolved consciousness. With consciousness comes intentionality, the capacity of the organism to represent objects and states of affairs in the world to itself. Now the question is, how can we account for the existence of social facts within this ontology?

On the other hand, Searle also maintains (1992: 14–15) the following:

> If there is one thesis that I would like to get across in this discussion, it is simply this: The fact that a feature is mental does not imply that it is not physical; the fact that a feature is physical does not imply that it is not mental. . . . When I say that consciousness is a higher-level feature of the brain, the temptation is to hear that as meaning physical-as-opposed-to-mental, as meaning that consciousness should be described *only* in objective behavioral or neurophysiological terms. But what I really mean is consciousness *qua* consciousness, *qua* mental, *qua* subjective, *qua* qualitative *is physical*, and physical *because* mental.

To the question formulated in the first Searle extract—"Now the question is, how can we account for the existence of social facts within this ontology?"—Searle answers concisely (1995: 228): "From dollar bills to cathedrals, and from football games to nation-states, we are constantly encountering new social facts where the facts exceed the *physical* features of the underlying *physical* reality" (emphasis added).

This is peculiar. If "we live in a world made up entirely of physical particles in fields of force," how can there be in this world "social facts" that "exceed the *physical* features of the *underlying physical* reality"? To be emphasized is that "physical" appears twice in Searle's answer, first as characterizing the ultimate metaphysical dimension or reality (such that the "features" of this reality are also—quite consequently—characterized as "physical"), but *then* entities ("social facts") are introduced that *exceed* the "physical features" of the "underlying physical reality," and that are therefore *non-physical*. The incoherence is obvious: ultimately, what—for Searle—is "the physical"? If *anything and everything is physical* ("We live in a world made up entirely of physical particles in fields of force"), what can be meant by the contention that this world ("made up entirely of physical particles in fields of force") also contains non-physical entities (non-physical features)?

In all probability, Searle is guided by the notion that David Lewis presents in the example of the dot-matrix picture, introduced above. But then he is confronted with the question posed above: what *are* the patterns themselves? What is the *ontological status* of the patterns? In Searle's terminology: what *are* the "higher-level" features "that exceed the physical features"? What do "higher-level" and "exceed" mean here? Searle does not even pose these questions, and thus of course does not answer them.

Instead, Searle appears to take a different path in the attempt to avoid the obvious difficulties his position involves. This path is briefly mentioned in one sentence from one of the passages quoted above: "The temptation is to hear that [i.e., that consciousness is a higher-level physical feature of the brain] as meaning physical-as-opposed-to-mental, as meaning that consciousness should be described *only* in objective behavioral or neurophysiological terms." This path would also involve the distinction between different *descriptions* (of one and the same *physical* reality?). But here again Searle does not thematize the problem that would arise. The distinction introduces an entire domain of the physical (i.e., the mental) that is not accessible to any physical description. But how can that be? Something physical that is not accessible to a physical description? What then does "physical" mean?

Here, he is confronted by a dilemma: he assumes either that "descriptions" of whatever is to be conceptualized (the relevant ontological domain or situation) are fully external, or that descriptions, when they are correct, precisely articulate whatever it is that they are descriptions of. If he assumes the former, then his enterprise loses all scientific or philosophical significance, because it then makes no difference whether one uses a physicalistic or a mentalistic description. If he assumes the latter, then he is forced to recognize and to distinguish among *many specific subdomains* within the framework of the comprehensive or basic dimension that he names the physical (it is here designated $physical_0$; the additional domains can be designated $physical_1$, $physical_2$, . . . $physical_n$). But then the question presses: how do these subdomains relate to $physical_0$? And above all, how is $physical_0$ itself to be understood *and described*?

A *coherent* conception could *not* maintain the thesis that there are subdomains of $physical_0$ (concretely, subdomains such as *the mental, the social*, etc) that would "exceed" the "features" of the fundamental domain $physical_0$, because then these subdomains would be *non-physical* precisely in the sense just explained (i.e., that of being non-$physical_0$ domains). Nevertheless, this is precisely the thesis that Searle does maintain. Actually, Searle maintains two theses that contradict each other: (1) $physical_0 \neq physical_1$ (e.g., the mental subdomain) $\neq physical_2$ (e.g., the social subdomain) $\neq \ldots \neq physical_n$; and (2) $physical_0 = physical_1$ (e.g., the mental subdomain) $= physical_2$ (e.g., the social subdomain) $= \ldots = physical_n$. Searle can avoid the contradiction only by precisely explaining the concept *physical*. If a difference between *physical* and *non-physical* is introduced and accepted, then these can only be two different subdomains of what would have to be presupposed as a distinct domain that would be comprehensive or fundamental. If one then uses or wants once again to use the term "physical" for this comprehensive or fundamental domain, one must *specify it in complete clarity* in order to distinguish it clearly from the mutually distinct subdomains of the "physical" and the "non-physical." One could then write something like the following: "$physical_F$" ("F" for "fundamental domain"). If, however, Searle were to do this, the problem would become particularly acute, because the question would arise concerning what could be meant by "(the dimension of the) $physical_F$," given that it could be neither understood *nor described* as $physical_0$ or as $physical_1$, etc. With Searle, the case is as follows: his grand thesis, "we live in a world made up entirely of *physical* particles in fields of force," is utterly ambiguous, indeed incoherent, and thus says nothing. He maintains both thesis (1) and thesis (2) in that he completely neglects to ask about and thus to clarify the presuppositions and implications bound up with his confused and contradictory assertions. Comparable and comparably fundamental ambiguities and incoherences characterize all variants of "non-reductive physicalism."

4.3.1.2.3.2 An Argument Against Physicalism

[1] In the current literature, there are several much-discussed arguments against physicalism. The most important are those presented by Thomas Nagel, Frank Jackson, and David Chalmers. In the widely discussed essay "What Is It Like To Be a Bat?," Nagel (1974) attempts to show that every physicalistic explanation of the universe, because of its essentially objective character, does and indeed must leave the subjectivity of perspectives out of consideration. On this basis, he develops a thought experiment, arguing that this lack of consideration manifests itself in the following fact: if we knew all   383 the facts about the physiology of beings in a species wholly different from ours, we still would not know "what it is like to be them."

Frank Jackson and David Chalmers, particularly, have reformulated and developed this argument in various respects. Jackson's (1982) so-called knowledge argument is also based on a thought experiment: one considers the case of a super neuroscientist, Mary, who, enclosed in a black-and-white room, has learned everything that can be learned about her visual system by means of books and by watching a black-and-white television. Jackson maintains that Mary would nevertheless learn something new—a new fact—the first time she perceived real colors. She would learn, for example, what the perceptual experience of "red" really is. From this Jackson concludes that physicalism must be false, because we could know all the physical facts without knowing, even in principle, the totality of facts.

Chalmers's (1996) "conceivability argument" draws a conclusion from the thought that we can think of or imagine a universe that would be identical to ours in everything except the following: our counterparts in the other universe would have no conscious mental lives. They would be subjective zombies. Chalmers argues that such a universe is thinkable or conceivable and also logically possible. He takes this to demonstrate the falsity of physicalism in that he shows that the facts about qualitative consciousness are additional facts that are not adequately determined by the physical facts.

More recently, Gregg Rosenberg (2002) has presented an argument based on an analogy between physical facts and facts about cell automata. A cell automaton consists of points (or cells) in an abstract space, such that all these points or cells have certain sorts of causal properties (see Rosenberg's Chapter 2). Rosenberg describes an example of such an automaton that he considers to be a model of a miniature physical world. On this basis, Rosenberg develops the following argument: (1) From facts about cell automata, facts about phenomenal consciousness cannot be derived. (2) If facts about phenomenal consciousness cannot be derived from facts about cell automata, then facts about phenomenal consciousness cannot be derived from facts about the purely   384 physical world. (3) Therefore, facts about phenomenal consciousness cannot be derived from facts about the purely physical world.

Rosenberg's strategy is to use the physics corresponding to the cell automata to develop the categorial structure of physical theories in general; that structure identifies the kind of information that physical theories provide and exposes the kinds of conditions that make physical properties the kinds of properties that they are.

With such (and similar) thought experiments and/or methods, anti-physicalistically oriented philosophers attempt to demonstrate that there is no *a priori* entailment between physical facts and "phenomenal" or "experiential" facts—thus, that experiential facts (facts of experience) cannot be derived from purely physical facts on the basis of *a priori* reasoning. These authors conclude that physicalism is false.

This book does not attempt to determine whether or not these arguments are conclusive. Instead, it presents an argument that can be designated, in a certain respect, a direct argument (direct in the sense that it does not rely on thought experiments, analogies, etc.). It could be termed an argument from knowledge (in a sense fully different from Jackson's), or from the truth of scientific theories, or from intentionality.

[2] Section 4.3.1.2.2.3[2] (p. 276) introduces, but does not fully explain, the basic thought of the *intentional coextensivity of the mind with the whole* (*with the universe, with being*). The concept *intentionality* is central to the philosophy of mind, but no precise explanation of it has attained widespread acceptance. Only the concept's central aspect is considered here; it provides the basis of the argument to follow. Quite generally, intentionality is the directedness of mental acts (thus acts of the mind or of thinking) to the world, to reality. It is highly controversial just what kind of directedness it is. Its more precise determination depends in part on the semantics being relied on. If the semantics is what this book terms *compositional* semantics, then singular terms play a central role in the determination of the concept of intentionality. The situation is quite different if instead a contextual semantics, which recognizes no singular terms as referential expressions, is used. But in every semantics the truth-concept plays the decisive role, and rightly so, because truth, however this concept may be determined more precisely, involves in one way or another an *ontological import*, and indeed the one that is absolutely definitive. One can therefore say that the directedness of the mind or of thinking to reality (the universe, being as a whole) is fully determined only when the mind or thinking entertains or articulates one or more true sentences. Generally and simply formulated: intentionality attains its maximal realization when there is *knowledge*. (The will and phenomena related to it are also relevant here, but are not explicitly included in order to simplify the presentation.)

What, more precisely, does the just-mentioned ontological import involve? Given present purposes, the answer can be formulated as follows: with truth and/or knowledge, we articulate or grasp or reach *the things themselves*: reality (the universe, being as a whole) or a segment of the world or of reality. There is a tendency to take terms such as "grasp" and "reach," used in this context, to be mere metaphors. The reason for this tendency is presumably that since Kant, the thesis that there is a deep gap between the knowing subject and the world or reality has been elevated to a universal dogma in philosophy. According to this dogma, the knowing subject is something that does not really belong to the world or to reality, so that the relation between the two dimensions involves an unbridgeable gap and thus becomes a mystery. Everything changes, however, if one understands the subject to be an *integrating component* of the world or of reality, because then the relation between the two dimensions is itself innerworldly. Terms such as "grasp" and "reach" are then no longer merely metaphorical; instead, they express innerworldly relations.

There can be knowledge of all possible objects or subject matters, because in principle absolutely nothing evades the intentional scope of the human mind. The concrete, the abstract, the physical, the theoretical, the practical, etc.—knowledge is

385

possible within all of these domains. At least as first glance, the problem of physically explaining all these varieties of knowledge appears to be insuperable. The following account thematizes only one instance of knowledge and of truth that is particularly central to science; this instance shows unambiguously that no physical explanation of knowledge and thus of the mental is possible.

A basic component of modern cosmological physics, not seriously doubted by anyone, is that there are systems of stars that are many light-years distant from the Earth: from 160,000 light-years (the distance from the Earth to the galaxy known as the Great Magellan Cloud) to around 13 billion light-years (the distance from the Earth to a "thin" galaxy discovered in 2004).[25] The question arises: how is human knowledge of such a galaxy to be explained *physically*? Is a physical explanation even at all thinkable? The answer is fully clear: contemporary physics provides no basis for a physical explanation of such knowledge. This is briefly shown in what follows.

That we know true scientific sentences about such systems of stars means that we grasp or reach the *real galaxies*. We thus enter into a relation with these heavenly bodies, just as, in every case of knowledge that attains truth, the knower enters into a relation with what is known. How this "enter into a relation" is designated is a secondary matter, one merely of terminology. Important here is only the question of how is it possible. However this entering-into-a-relation may be explained (by means of light impulses, signals, etc.), in no case can these "physical bearers" (as they may be termed here) of the knowledge-relation, and thus the entering-into-a-relation with the relevant heavenly bodies, be faster than the speed of light, because according to the theory of relativity the universal constant, the speed of light ($c_0 \approx 300,000$ km/s in a vacuum[26]) is the natural upper limit for all possible speeds at which matter or energy in any form, thus also as signal, can expand or move.

What follows from this, with respect to the issue in question? On the basis of an explanation of knowledge that was purely physical (according to contemporary physics), we could never be in the position of coming up with sensible, true sentences about objects like the galaxies introduced above. We would never reach them. Our so highly treasured scientific sentences would be merely subjective games or flights of fancy without any objective (i.e., true and real) content. There thus can be no physicalistic explanation of the human mind.

The question here arises, how can philosophers have ignored and continue to ignore the situation just described? The answer is twofold. [i] These philosophers have never coherently considered even the minimal ontological implications of the truth concept. Because they have not done that and continue not to do so, they fail to raise and to consider fundamental questions, above all the one introduced above: how, on any physicalist basis, is it possible to explain knowledge of immense and comprehensive cosmological complexes? [ii] Despite the fact that these philosophers are scientistically oriented, they have never seriously attempted to understand the dimension of mind or of thinking *as* a genuine dimension of or in the world. They continue to attempt to

386

---

[25] See *The New York Times*, Feb. 16, 2004. Of known heavenly bodies, this galaxy is by far the one most distant from the Earth. This galaxy is "thin" or "small" because it has a diameter of only 2,000 light-years (the diameter of the Milky Way is around 100,000 light-years).

[26] Precisely: $c_0 = 299,792,458$ m/s.

387   *reduce* the mental, thinking, etc., to the physical; they appear not to have noticed that they thereby totally miss what is characteristic of this dimension. Their procedure fails to think through this dimension *as such* as a dimension of or in the world; it yields only a caricature of the mental and of thinking.

[3] Against the argument just sketched, an objection worth discussing could be raised. One could claim that the only "real" contact we could have with the heavenly bodies introduced above would be physical contact (i.e., that light reaches us following an immense journey); this would be taken to be fully sufficient for scientific theories and for the true sentences they contain. To explain knowledge of distant heavenly bodies, there would then be no need to consider the knowing subject as somehow leaping over distances measured in light-years.

This argument correctly points out a fundamental fact, but draws from it the wrong conclusions. The genuinely important and decisive point is that science introduces various sentences concerning these galaxies, their interrelations, their development, etc., as they are *today*, however this "today" may be interpreted. Science speaks about real galaxies, not about sensory material, sense data, or phenomena in instruments. But already the simple fact that the light that reaches us is *interpreted* or *explained* as light sent from galaxies many light-years distant from the Earth sufficiently shows that the human mind is capable of overcoming, in an instant, the immense physical distances between these galaxies and our Earth. This is a fact that shows all physicalistic attempts to explain the human mind to be utter failures.

4.3.2 *Moral Action and Moral Values (Ethics)*[27]

Along with anthropology in the narrower sense, the practical dimension is a central part of the human world. In the Aristotelian and Kantian traditions, this dimension provides the subject matter for so-called "practical philosophy," and includes, in addition to the domain of morality, the domains of social philosophy, political philosophy, philosophy of law, etc. This book treats only the first of these domains, and uses the term "ethics" to designate its philosophical theory that has that domain, morality, as its

388   subject matter.[28] What is said here about ethics and the moral domain is also centrally relevant to theoretical inquiry thematizing the other domains of "practical philosophy" but, as emphasized often above, the world-systematics developed in this chapter makes no attempt to be a comprehensive and exhaustive presentation of all the (not even all the important) domains of the world, and thus of course of the countless problems that arise in the extensive domain of the practical. The goal of this chapter—and this holds in exemplary fashion for this section—is to show in specific instances and to a limited degree how to understand and accomplish the development or concretization of the structural-systematic theoretical framework.

For two reasons, the current topic must be seen and treated in close connection with the truth-concept. The first reason is that, as Chapter 3 shows, the truth-concept is

---

[27] Most of this section is taken relatively directly from Puntel (2004).

[28] "Practical" and "ethics/ethical" are rooted in Greek, "morality/moral" in Latin, and "*sittlich*," here translated as "practical," in medieval German. In the philosophical tradition, the terms are used and interrelated in various mutually inconsistent ways that need not be considered here.

absolutely central to concretizing the structural dimension. The second reason is that the ethics developed here has a *cognitive* status; it is, as indicated above and explained more fully below, a determinate theory. Decisive in this respect is that ethical sentences are sentences that can be true, and thus can express propositions that are identical to facts.

Whether ethical sentences[29] can be true is a question with which philosophy has always been concerned in one way or another. At present, this question is discussed particularly intensively (see, e.g., Hooker 1996) because of a remarkable convergence of discussion within the two relevant disciplines: ethics and truth-theory. On the one hand, considerations of the status and the basis of cognitively conceived ethics lead to the explicit thematization of the applicability of the truth-concept in the domain of ethics; on the other, the search for a truth-concept that is maximally adequate and comprehensive brings with it the tendency, almost to be designated as "natural," to extend this concept to ever more distant fields and to clarify the semantic problems that arise from these extensions (see, exemplarily, Wright 1996). But the precise sense in which the truth-concept can and must be applied in ethics cannot be considered to have been clarified even minimally. This fact results primarily from the hopeless lack of consensus with respect to the truth-concept itself; the centrally problematic point concerns its ontological import. If truth involves directly or indirectly an ontological import, the question arises of how to understand the world or the ontological dimension, in the case of ethical truth. 389

4.3.2.1 On the Theoretical Character of Ethical Sentences

At first glance, the "practical" status of ethical sentences introduces the particular difficulty involved in relating them to the truth-concept. "Practical" sentences are considered to belong to *practical philosophy*, which is traditionally distinguished from *theoretical philosophy*. For the most part, this distinction—which since Aristotle and particularly today is taken to be something like a continental divide in the philosophical landscape—is assumed to be more or less obvious. Nevertheless, the attempt to characterize more precisely the understanding that is thereby presupposed brings with it problems that show clearly and quickly that the widespread consensus is relatively empty. Is ethics—to which discipline the problematic of "practical philosophy" is here restricted—not a *theoretical* discipline? Is ethics not a philosophical theory?

4.3.2.1.1 *The Ambiguity of "Practical Philosophy" and of "Normative Ethics"*

In the entire history of philosophy—or, more precisely, since Aristotle—there has been a fundamental ambiguity concerning the precise status of ethics. If however one does not ascribe to ethics some kind of non-theoretical or even anti-theoretical status, then

---

[29] Here, there is a strong terminological distinction between ethical sentences and moral sentences. Ethical sentences are sentences that are presented within the framework of the discipline *ethics*. Moral sentences are ones that are or can be expressed in all sorts of lifeworldly and linguistic contexts, and that thus relate to the domain of morality. In a specific respect, ethical sentences are metasentences concerning moral sentences.

it must be distinguished from all the activities that are or are termed "practical" in any *direct manner* (e.g., advice, therapy, education, instructions or invocations concerning specific ways of acting, etc.). But what remains other than to understand ethics as a discipline in the strict sense, thus, as *theoretical*?

The confused status of ethics as the central discipline of *practical* philosophy stems from an ambiguity in the determination "practical science" that is due primarily to Aristotle.[30] Practical philosophy and with it ethics have always been somehow Janus-faced in that they have been and continue to be determined by two factors whose relation to each other, and with respect to the definition of ethics, remains murky. On the one hand, ethics is determined by its subject matter, the domain of the practical in the sense of moral action; it is thereby distinguished not in the slightest from any other science, in that all sciences have subject matters. As the study of this subject matter, ethics would be a theory, the theory of practical (moral) action. On the other hand, however, another factor is always introduced as at least *in fact* a codefiniens: a specific goal (i.e., the practical intent). Aristotle brings the goal that determines a specific intent into the characterization of the science itself: a science whose goal is itself, thus one pursued purely for the sake of knowledge, is a "theoretical" science; a science that is not its own goal, one having a goal not identical to itself and thus one not pursued purely for the sake of knowledge, is not theoretical; if its goal is action, then it is a "practical science."

This Janus-faced determination of practical philosophy and thus of ethics yields no coherent conception. On the one hand it is clear that all possible goals can be associated with any science traditionally classified as "theoretical"; knowledge or truth for its own sake is only one of these goals. Currently, indeed, the actual goal of science is not generally considered to be truth or knowledge itself, but instead technical mastery of the world. On the other hand, the development of ethics need not be connected to action, for example in the form of contributing to the development of a more human world (whatever that might be). As far as philosophy, including ethics, is concerned, various other goals can also be named: humanistic ones, national-cultural ones, and yet many more, including quite primitive ones (e.g., earning money, avoiding boredom, etc.). A philosopher can, however, also of course develop an ethics purely for the sake of the truth.

The preceding considerations reveal that the provision of a goal cannot define the status of a science. Instead, sciences are fundamentally determinable only by their subject matters and by their specific conceptual, argumentative, logical, etc., structures. The traditional confusion concerning the status of ethics comes most clearly into view with the problematic of *normativity*. Because the domain of the practical, which essentially includes the phenomenon of the normative, is incontestably the subject matter of ethics, ethics of course involves practical sentences in the sense of normative sentences. The question is, how is this involvement to be understood precisely? And this is exactly where the traditional ambiguity arises.

There are two essentially different kinds of practical sentences. If one articulates the relation of sentences to the practical from the perspective of positing goals or intents, as introduced above, then the status of the sentences is practical in a *primary sense*; such

---

[30] Aristotle does not use the designation "practical philosophy." See Bien (1971).

sentences are therefore termed *primarily practical sentences*. These are sentences that directly make requests or demands. The strongest form of such sentences is the *imperative* form, such as (politely) "You ought to do X" or (directly), "Do X!" Kant's famous formulations of the categorical imperative are typical examples of such sentences. But if one articulates the relation of sentences to the practical instead from the perspective of the subject matter of ethics, also introduced above, then the sentences have practical contents, but not the status of being primarily practical sentences; they can therefore be designated *secondarily* or *objectively practical* sentences. As the following account shows in detail, there are two quite different varieties of such sentences. Sentences of both varieties have the status of being *not primarily practical*, but instead *theoretical-practical*.

When there is talk of normative ethics, as a rule it is not clear what precisely is meant by "normativity." That ethics can and indeed must consider primarily practical sentences as the *subject matter* for investigation or theorization is obvious. But can ethics *generate* primarily practical sentences as *its own* sentences, thus as sentences within the theory itself? This is the central and decisive point with respect to the ambiguity of ethics as "normative science." The position presented in this book excludes the possibility of ethics itself generating such sentences. A science that would be "normative" in the strong sense would be a simply self-contradictory endeavor, because it belongs essentially to science to investigate what is the case (either empirically or universally, the latter meaning, in the context of this book's treatment of ethics, "as a matter of universally valid ethical principle"). The first step toward developing an ethics that is a genuine science requires overcoming the ambiguity that plagues the traditional understanding of "practical philosophy"; that first step is taken in the following subsection.

4.3.2.1.2 *Primarily Practical, Theoretical-Deontic, and Theoretical-Valuative Sentences*

The general tendency in contemporary ethics is to develop not (only) metaethics but also normative ethics. Metaethics investigates only the meanings of the terms used in ethics; normative ethics seeks in addition to show what is ethically right or wrong—thus what are the moral norms, and indeed not only purely abstractly, but concretely and in detail. Briefly, normative ethics seeks to show *which moral norms* are valid. But how is this concept *normative* to be determined more precisely? Instead of speaking of the concept of the normative, one can also—and, in many respects, more adequately—consider or formulate normative sentences; these are generally characterized as practical sentences, but the precise meaning of "practical" is usually left unclear and underdetermined.   392

The concept of the normative (or of normative or practical sentences) is usually explicated by general verbal paraphrases (e.g., "concerning a norm"). When attempts are made to say more, two extensive groups of normative concepts (or normative or practical sentences) are introduced: *deontic* concepts (sentences) and *valuative* concepts (sentences). Valuative concepts are also termed concepts of *values* (or sentences concerning values; see, e.g., Kutschera 1982b: 1ff).

The central deontic concepts are the obligatory, the forbidden (or prohibited), and the allowable (or permissible). Just how unclear the precise meanings of these concepts are is shown by the ways in which deontic sentences are understood more precisely.

[1] Deontic concepts can appear in sentences both as predicates ("This act is obligatory") and as sentence operators ("It is obligatory that A," where "A" is a sentence). Either of these complex sentences can be symbolized as "O(A)," but these formulations are anything but clear.

[i] O(A) can mean (can be read as), "It is obligatory that A," or "The proposition A, expressed by the sentence A, ought to be done" (as a rule, A is a specific act). An example for sentence A: "It is obligatory that the basic rights of every human being are respected." In this example, "It is obligatory that" is the operator, and the sentence, "the basic rights of every human being are respected," serves as the argument of the operator. The proposition A expressed by this sentence can, in the usual mode of expression, be articulated as, "(the state of affairs or proposition that) the basic rights of every human being are respected"; simplified, the proposition can be articulated as, "(the act of) respecting the basic rights of every human being." One can therefore, quite easily, present the *content* of the sentence in question—"It is obligatory that the basic rights of every human being are respected"—as follows: "It is obligatory that A is accomplished" (i.e., that the basic rights of every human being are indeed respected).

So interpreted, O(A) is an indirect form of an *imperative* sentence: "Every person ought to do A" or, directly and in a specific case: "Do A." Such a sentence is here termed a *primarily practical* sentence, or a *practical-deontic* sentence; strictly speaking, the qualifier "deontic" is superfluous, because in the terminology used here every primarily practical sentence is a deontic sentence. But—as shown shortly—not every deontic sentence is a primarily practical sentence; for this reason, clarity is served by the use of "deontic" as an additional qualifier, such that "practical-deontic sentences" are distinguished from "theoretical-deontic sentences."

393 [ii] O(A) can, however, also be read and interpreted quite differently: as an *indicative* and thus as a *theoretical* sentence. There are two ways in which it can be so read and interpreted.

[ii-i] The first is a *theoretical-empirical* reading: "It is empirically the case that it is obligatory (or it is established that it is obligatory) that A, or that A is actualized." The operator "It is empirically the case that" (or "it is established that") indicates an empirical situation, a circumstance that immediately affects the second operator: "it is obligatory that." Making this reading explicit yields a paraphrase of the following sort: "It is empirically the case (or it is established) that in the (social, cultural, historical, geographical, etc.) situation s it is obligatory that A, or that A is actualized," or "It is empirically (i.e., in the social, historical, cultural, etc. situation s) the case that it is obligatory that A, or that A is done."

[ii-ii] The second form or possibility is a *theoretical-universal* reading, one in which not an empirical, but instead a universal operator is presupposed: "It is universally the case that it is obligatory that A, or that A is actualized." The content of the sentence can also be expressed as, "It is universally obligatory that the basic rights of every human being are respected." Or, again more simply: "It is universally obligatory that A is actualized."

Sentences of the forms introduced in [ii-i] and [ii-ii] are *theoretical-deontic* sentences.

To attain complete clarity concerning the statuses of these sentences or of the three readings introduced for the sentence O(A), it is helpful to provide each initial operator with its own symbol. The symbol "Ⓟ" ("Ⓟ" for "primarily practical") designates the *practical-deontic operator*; it thus characterizes the status of *primarily practical* sentences in the form of imperatives in the broader sense. ⓅA is to be read or paraphrased for the three forms of the deontic operator as, "It is ethically obligatory or forbidden or allowable that A (or that **A** is actualized)" or, in the direct form, "Do **A**" (as a matter of ethical duty), "Do not do **A**" (because **A** is morally forbidden), "Do **A** or do not do **A**" (because both are morally allowable)).

The symbol "Ⓣ" designates the *theoretical* operator: "It is the case that. . . ." The theoretical operator can be the determinative operator even in the domain of the practical. Practical sentences governed by it are *theoretical-deontic* sentences; for the theoretical-deontic operator, the general symbol "Ⓣ$_D$" is used. But this operator is incompletely   394 determined, because the validity it introduces can be either empirical or universal, and these yield two fully different kinds of sentences. The difference is indicated by the two symbols "Ⓣ$_{DE}$" ("It is deontically-empirically the case that. . .") and "Ⓣ$_{DU}$" ("It is deontically-universally the case that. . ."). If one of the three specific forms of the deontic operator ("obligatory," "forbidden," "allowable") is intended (as it must be in every specific case), this form must be made explicit. Formalizing, the "$_{DU}$" or "$_{DE}$" should be given the appropriate secondary index (thus, "$_{DU/O}$," "$_{DU/F}$," "$_{DU/A}$," and similarly for "$_{DE}$"). There are then formalizations such as "Ⓣ$_{DU/O}$(A)," read as, "It is deontically-universally the case that it is obligatory that A." Such details are not considered further here.

The three readings—those of [i], [ii-i], and [ii-ii]—of the apparently so simple and clear sentence O(A) yield the three sentential forms "Ⓟ$_O$(A)," "Ⓣ$_{DE/O}$(A)" and "Ⓣ$_{DU/O}$(A)." There are corresponding formulations for the specifications "forbidden" and "allowable" or "permissible." Sentences of the general forms "Ⓣ$_{DE}$(A)" and "Ⓣ$_{DU}$(A)" can also be termed "theoretical-practical" sentences, whereby "theoretical" indicates the operator determining the entire status of the sentence, and "practical," the domain to which the sentence relates (i.e., the practical domain).

[2] The second extensive group of normative/practical concepts is that of concepts of *values*; to them correspond *valuative sentences*. Examples of these sentences are "This act is good," "Keeping promises is good," etc. It is of course not possible in this book to consider all the specific aspects of this class of concepts or sentences (see, e.g., Kutschera 1982b: 10ff). To be noted is only the following: such concepts (e.g., *good*) are used not only both predicatively and attributively but also as operators: "It is good that promises are kept." The latter use has significant advantages with respect to ontology and is therefore preferred here.

Valuative sentences clearly have the syntactic form of (descriptive-)theoretical sentences. Non-cognitive ethicists do not deny this, but attempt to reinterpret these sentences, making of them for example expressions of preference, of disinclination, etc. This book does not consider this position, although it is held by more than a few   395 philosophers; as noted above, the attempt here is to develop a coherent and tenable *cognitivist* conception. To accord cognitive status to valuative sentences means to interpret them as fundamentally indicative and thus as fundamentally theoretical sentences. To be emphasized, however, is that they are theoretical-*valuative* sentences. To indicate them notationally, the symbol "V" (for "valuative") can be used. Theoretical-deontic sentences formed with the operator Ⓣ$_D$ and theoretical-valuative sentences containing the operator Ⓣ$_V$ are the two forms of theoretical-practical sentences. Here again empirical and universal sentences must be distinguished; this requires the introduction of the operators $\textcircled{T}_{VE}$ and $\textcircled{T}_{VU}$. In the valuative domain, there is no counterpart to the primarily practical sentences introduced above, because valuative sentences are always *theoretical*-practical sentences. There is thus no complete symmetry between the deontic and the valuative domains.

At this point, the thesis formulated above, that ethics cannot itself contain purely or primarily practical sentences, can be made more precise: ethics presents theoretical-practical sentences—thus, sentences determined by the operators $\textcircled{T}_{DU}$ and $\textcircled{T}_{VU}$. Every ethical sentence is a sentence with one of these two operators. (In ethics, theoretical-deontic-empirical and theoretical-valuative-empirical sentences (thus sentences of the forms "$\textcircled{T}_{DE}(A)$" and "$\textcircled{T}_{VE}(A)$") occur only incidentally, e.g., to clarify the status of theoretical-practical sentences; the same holds for all theoretical-empirical sentences.)

Ethical sentences, thus understood, are indicative-theoretical sentences and thus sentences that *can be true*; they are sentences with which the question of truth can *and must* be raised. This thesis brings with it a number of serious problems. Difficult as it is to develop a truth-concept about which there could be a consensus, it is the more difficult to show the precise sense in which ethical sentences are or can be true. The following subsection provides an introductory clarification of basic aspects of this issue.

4.3.2.2 The Ontological Dimension of Ethical Truth: Ontological Values

[1] Philosophers who at present are concerned directly or indirectly with the issue of truth in ethics can be divided into two groups. Those in the first group maintain that ethical sentences are true (or false), but do not consider explicitly or sufficiently the precise meaning of truth in this context. Generally, they rely upon a mostly general or purely intuitive understanding of truth, such that truth is generally understood in the sense of one of the many—as a rule unanalyzed—variants of the correspondence theory; some scarcely distinguish between truth and justifiability (see, e.g., Kutschera 1982b,[31] McGinn 1997: Ch. 3,[32] Ricken 1993/1998: 46f, etc.).

Those in the second group, which includes most of the philosophers working in this area, maintain an explicitly *epistemic* truth-concept: truth is reduced to something like objectivity, intersubjectivity, ideally justifiable assertibility, etc. (see, e.g., Wiggins 1996: 46ff). The following account introduces two characteristic examples.

---

[31] Kutschera affirms the truth-character of ethical sentences; as far as the truth-concept is concerned, he simply refers, in a footnote, to a section in his book about epistemology (see 1982b: 47n11, which refers to 1982a: section 1.6). The latter book defines truth as follows: "**TK:** *a sentence is true just when the state of affairs that it expresses exists.*" Kutschera explains, "A sentence A is called true just when the case in actuality is just as A presents it" (46). In his *Grundlagen der Ethik* (1982a), Kutschera does not at all explain what it does or can mean to say that a moral state of affairs "exists," or what in the case of a true ethical sentence A it can or does mean to say that the case in actuality is just as A presents it. That it is not obvious what this can or should mean is made evident by the current debates on the issue.

[32] Particularly consequential is Colin McGinn's conception. Referring to his (1997) position, he maintains, "In my own view . . . truth applies equally and univocally to moral and scientific statements" (2003: 72n2). The cited book contains a spirited defense of the ontological character of moral norms, but no explanation worth mentioning of the truth-concept used in ethics or of the ontology that corresponds to it.

Crispin Wright (1992) has developed a much-discussed theory of truth that is known as the "minimalist superassertibility theory of truth." According to Wright, a sentence is true just when it is superassertible, which he understands as follows: a sentence is superassertible just when it is assertible in a specific state of information and then remains assertible regardless of any new information that may become available. This concept of truth, Wright maintains, stands "in compliance" with a number of "principles" that—because of their putative obviousness—he terms "platitudes." He formulates one of these platitudes as, "[T]o be true is to correspond to the facts" (34). His strategy is to interpret the platitudes in such a way that they are emptied of meaning and content. He aptly characterizes this strategy (12) as follows:

> It is ... a platitude that a statement is true if and only if it corresponds to the facts. But it is so only in so far as we understand a statement's correspondence to fact to involve no more than that matters stand as it affirms. For reflect that if 'p' says that p, then matters will stand as 'p' affirms if and only if p. Since by the Disquotational Scheme, 'p' is true if and only if p, it follows that matters stand as 'p' affirms just in case 'p' is true—essentially the Correspondence Platitude. What this simple argument brings out ... [is] that the phraseology of correspondence may embody much less of a *metaphysical* commitment than realism supposes.

397

The truth of a sentence is thus reduced to "matters stand as it [the statement] affirms." But what does it mean to say "matters stand"? On the whole, Wright reduces "truth" to a diffusely understood "objectivity," which he says is attained in the domain of morality by means of "convergence." Nevertheless, at the end of his treatment of truth in ethics, he makes a remarkable concession, indeed, a deeply revealing admission or confession: "How much, and what kinds of moral appraisal may ... contain the seeds of such convergence seems to me a great—perhaps the greatest—unresolved question in moral philosophy" (18). However many paraphrasings and phraseologies one may use, one cannot escape the ontological question, as is shown clearly by the presentation of a quite different theory of truth in Chapter 2 and especially Chapter 3 (Section 3.3). Moreover, the argument against Habermas sketched in the following paragraphs applies also to Wright.

With respect to the topic under current consideration, Jürgen Habermas has become quite well known because, with a terminological artifice, he claims to have "solved" the problem of truth in ethics. The artifice is to distinguish "truth" from "correctness," such that truth is understood somehow in the sense of the correspondence theory. In distinction from truth, "the sense of 'correctness' is limited to justified acceptability" (1999: 285). Ethical sentences are thus not true or untrue, but only correct or incorrect. Correctness in Habermas's sense is precisely what is maintained by one of the best-known variants of the epistemic theory of truth. This makes clear that what Habermas achieves is only a terminological relocation of the problem.

A thoroughgoing confrontation with Habermas would far exceed the purposes of this book, but those purposes are served by the brief introduction of a fundamental aporia found in Habermas's conception. How would an "ideal justification/acceptability" at all be understood? In any case, not as one that would be itself supported by any consensus of any kind. Otherwise, this question would arise, again and then yet again: on what is the (new, deeper) consensus based? Indeed, upon what must a Habermasian "ideally rational" participant in discourse rely? How is rationality to be measured? The empty appeal to "reason" is a fateful (self-)misunderstanding of the putatively so enlightened modern man. In actuality, this is an appeal to a hypostatized abstraction, for how is "reason" at all to be determined? Does it even have a measure? If so, where is it to be found? As long as these questions remain unanswered, the appeal to ideal rationality or ideally justified acceptability exactly resembles what Nietzsche, in *Beyond Good and Evil*, says brilliantly (albeit wrongly) about the freedom of the will: "The desire for 'freedom of the will' in the superlative metaphysical sense, which still holds sway, unfortunately, in the heads of the half-educated . . . involves nothing less than to be precisely this *causa sui* and, with more than Münchhausen's audacity, to pull oneself by one's own hair out of the swamp of nothingness into existence" (§21; translation altered). However one may twist and turn, one cannot avoid confronting the *ontological level*: there is a basis or a measure only if an ontological basis is accepted. A given justification is ideal only if it ties or links the sentence to be justified directly to the ontological domain to which the sentence relates.

Some forms of *moral realism* attempt to do justice to the ontological dimension of the moral domain (see Brink 1989, Rescher 1991). Even in those, however, it as a rule remains unclear how the corresponding ontology is conceived of and in what more precise sense the truth-concept plays a role. In current discussions of truth in ethics, however, the centrally problematic point is precisely the question concerning the onto-logical dimension of ethical sentences that qualify as true.

[2] How is the truth-status of *theoretical* ethical sentences to be understood? When a theoretical-deontic or a theoretical-valuative sentence (or the proposition expressed by such a sentence) qualifies as true, that means, according to the semantic-ontological truth-conception presented in this book, that the sentence or proposition is a fully determined sentence or proposition, which it was not prior to its qualifying as true. But then it must be asked: what is this fully determined status in the present case (i.e., in the ethical domain)?

There is no need to repeat the presentation of the truth-concept provided earlier, particularly in Section 3.3. It suffices to consider here more carefully the third func-tion, T^x, in the domain of ethical truth. As the preceding account shows, the onto-logical dimension of the truth-concept is the central problem in the case of the ethical truth-concept.

4.3.2.3 The Distinction Between Basal-Ontological Values and Moral-Ontological Values

Theoretical-valuative and theoretical-deontic sentences express propositions that can qualify as true (or false). How are such ethical propositions and facts to be understood? As an example, consider the ambiguously valuative sentence, "It is bad to execute inno-cent human beings." Understood as theoretical-valuative, this sentence expresses the proposition that it is universally the case that it is bad for any innocent human being that that human being is executed. For reasons presented in Section 4.3.2.4, this proposi-tion is a specification of the more general proposition that it is universally the case that it hinders (or, in this extreme case, prevents) the increased self-perfection of any human being (including of course those who are innocent) that that human being is killed (whether by execution or otherwise). Understood as theoretical-deontic, on the other hand, the sentence expresses the proposition that it is universally the case that it is forbidden that innocent human beings are executed. These are quite complex propositions.

[1] If one wanted to develop an analysis of the theoretical-valuative sentence (or the proposition expressed by it) on the basis of the semantics of a predicate logic and of the corresponding ontology of objects (substances, events, processes), properties, and relations, one would come up with the following: the "object" or the "event" or indeed the "action" *executing innocent human beings* would have the (first-order) "value-ontological property" of being bad and the entity determined by it (executing innocent human beings *plus* the property of being bad) would have the (second-order) property of being universally the case. This is fundamentally inadequate. An alternative analysis is based on the semantics and ontology developed in this book, which relies only on operators and primary sentences, primary propositions, and primary facts.

Such an analysis—developed here not in detail, but only as far as present purposes require—yields, from the example, a basal sentence ("Innocent human beings are executed") and two operators ("it is bad that" and "it is universally the case that"), such that the basal sentence is the argument of the first operator and the sentence resulting from its application the argument of the second operator. With the aid of a quite unnatural paraphrasing and of various brackets, the result of the analysis can be presented as follows:

{It is universally the case that [it is bad that (innocent human beings are executed)]}

Semantic-ontological analysis of the theoretical-deontic sentence yields a wholly analogous result: 400

{It is universally the case that [it is forbidden that (innocent human beings are executed)]}

[2] The ontology on which these sentences are based is in each case articulated by the second operator and by the sentence determined by it. In both cases, there is at first a non-qualified proposition or state of affairs that is normally termed an act: "an innocent human being is executed" ("innocent" is here taken in a quite neutral sense—an innocent human being is one who did not commit a specific act). The state of affairs or act of execution is then qualified in the first example by the qualifier or operator, "(it is) bad (that)," and in the second example by the operator, "(it is) forbidden (that)." The operators or qualifiers are on the one hand related or similar, but also, on the other hand, different. They are similar in that each articulates a *value* (or a *valuation*), but fully different in that the first qualification (the first operator) designates a value as such (simply "(being) bad"), whereas the second qualification (the second operator) connects a practical way of acting with the determination of the value as such: being forbidden is connected to (being) bad. This yields immediately, first, that both of the sentences analyzed above articulate valuational-semantic propositions or states of affairs, and second, that the first proposition or state of affairs serves as the *basis* for the second. To indicate both the similarity and the difference terminologically, the following designations are used: the state of affairs expressed by the first sentence

is a *basal-semantic valuational state of affairs,* that by the second, a *moral-semantic valuational state of affairs.*

If these theoretical-valuative and theoretical-deontic sentences qualify as *true,* this means, according to the third function that defines the truth-concept (thus, the identity thesis), that these states of affairs are *identical to* facts in the world. This immediately yields the following: every true theoretical-valuative sentence expresses a basal-valuational proposition (or state of affairs) that *is* (in the sense of identity!) a *fact in the world* and thus a *basal-valuational fact* or simply a *basal-ontological value.* And correspondingly, every true theoretical-deontic sentence expresses a moral-valuational proposition (or state of affairs) that *is* (in the sense of identity!) a *fact in the world* and thus a *moral-valuational fact* or simply a *moral-ontological value.* The following sections basically consider only the explicitly *ontological* level.

401

4.3.2.4 The Ontological Status of Basal-Ontological Values

Basal-ontological values are value facts, thus entities in the world, that do not (yet) involve anything prescriptive. They are termed "basal" because they provide the basis for *moral* values in the genuine or proper sense, as is now to be shown. How are they to be understood? The concept of basal-ontological values can be developed in two ways that provide a twofold grounding of the thesis that—and in what sense— basal-ontological values are to be accepted. The first way can be termed the *general-metaphysical,* the second, the *metaphysical-anthropological.*

4.3.2.4.1 *The General-Metaphysical Perspective*

The general-metaphysical perspective has a long history within the framework of the tradition of Christian metaphysics, in which Thomas Aquinas is of particular importance. It is there articulated as the thought of *perfection (perfectio).*[33]

The world, or being as a whole, is not an amorphous mass of somehow fully formless or unstructured things or entities; instead, it presents itself as a well-structured whole consisting of well-structured individual beings. Every item within this whole has its own constitution, an ontological constitution. Every such item has a location or situation within the whole that fits it—that corresponds to its ontological constitution. Every being is truly itself to the degree that it occupies this location or situation, and therein develops and actualizes its capacities.

This abstract metaphysical thought can be illustrated quite concretely. As the relevant sciences demonstrate, of subhuman beings one says wholly correctly that every such being (e.g. a (type of) plant or of animal (or animal species), etc.) has its own environment as the arena within which it can live, the one that corresponds to its ontological constitution. For example, life is possible only when water is available. Here, the entire history of life in the plant and animal kingdoms can be treated in detail. And if one then considers human beings, then the thought of the perfection of their being (i.e., of their

402

---

[33] See such formulations as the following: "a thing is good from the fact of its being perfect [*Bonitas ... uniuscuiusque est perfectio ipsius*]" (*ScG*: 82 [Bk. I Ch. 37]); "the goodness of a thing consists in its being well disposed according to the mode of its nature [*in hoc enim consistit uniuscuiusque rei bonitas, quod convenienter se habeat secundum modum suae naturae*]" (*STh*: FS Q 71 A 1 Obj 3)].

ontological constitution as minding and bodying) attains simply universal dimensions, as shown in what follows.

This brief consideration shows four things.

[i] *Value* is fundamentally the *measure*, determined or established by the ontological constitution of any entity, for the capacities of actualization of that entity. If in place of "value" one uses the more traditional "good," then it is immediately clear that this thought is articulated in the classical dictum, *Ens et bonum convertuntur*. That entities do not "occur" or are not "there" or "present" once and for all as fixed and finished means that they have ontological constitutions characterized by arenas within which their capacities can be actualized. Every entity itself establishes such a measure by means of its ontological constitution. This is the fundamental meaning of the formulation, *every being has a value*. This "value" is not yet to be identified with anything like a *moral* value, because it has not yet been shown in what manner the measure (i.e., the value of a given being) is or is to be actualized.

[ii] Every being has a value. This follows directly from the fact that every being, thanks to its own ontological constitution, determines or establishes the measure for the actualization of its capacities. Negating this thesis would require that there be a being that had no ontological constitution—and that would be a self-contradictory non-thing.

[iii] This briefly sketched line of thought also shows that and how different beings each with their own values form *domains of values*. This thesis, too, follows directly from what is shown above. The question is, how are these domains of values to be understood more precisely?

[iv] The preceding line of thought yields, finally, that these values are *ontological* values in the genuine and proper sense. Here one could object that a distinction must be introduced between the ontological constitution of a being and the measure for the actualization of that being's capacities: the value, conceived as such a measure, would not be (according to the objection) simply identical to the ontological structurality of the being, but instead something "supervenient" upon this structure; the value would then be something "non-natural." But such an objection would miss its mark. "Value" does not here mean "moral value," but only "basal-ontological value." This latter, understood as a measure in the sense explained above, emerges from the analysis of the ontological constitution of a given being. To be sure, this ontological constitution is not to be understood naively as anything purely atomistic or purely static—in a word, purely positivistic—but instead philosophically, as something holistically metaphysical. The being is considered insofar as it is an item or element within being as a whole, with all the consequences that thereby result. It is said above that the ontological structurality of the being "establishes" or "determines" the relevant measure. This formulation should not be mistaken as suggesting that there are two different "entities" (for example, in accordance with the model, object-property); instead, the "relation" between ontological constitution and measure/value is not truly a *relation*, because measure/value introduces nothing beyond the fully determined or fully analyzed ontological constitution itself.

4.3.2.4.2 *The Metaphysical-Anthropological Perspective*

[1] The second perspective is in a fundamental respect only a specification of the just-presented general-metaphysical considerations to the case of that being that stands

out from all others: the human being. It is called a second perspective chiefly for methodological reasons.

What does it mean for a conception of being as a whole or simply of the universe that there are human beings as minding-bodying, as thinking, willing, freely acting, and here particularly as moral beings?

There is a widespread tendency, particularly in analytic philosophy, to accept—most often only implicitly—a conception that can be termed an "absolutely abstract-metaphysical realism." According to this conception, the universe is actuality "in itself"—and that here means, without human beings as factors that have a significance worth mentioning with respect to the determination of what the universe or actuality truly is. This position is fundamentally a radically *materialistic* position. Because even defenders of this position are aware that there are human beings, they must deal with them, as a phenomenon, in one way or another. There are three basic ways of so dealing. [i] One simply ignores human beings *as factors to be taken seriously ontologically*; if at all, human beings are considered only as "social" or "psychological" or "historical" factors, whereby the question is not raised concerning what *ontological status* the "social-psychological-historical" dimension has. [ii] One considers human beings as factors to be taken seriously ontologically, but attempts to *reduce* them to purely materialistic-physicalistic entities or processes. [iii] One maintains that the mental is not reducible to purely physical processes, but continues to uphold a comprehensively materialistic (metaphysical) conception. This last position is so obviously incoherent that it is amazing that there are (analytic) philosophers who advocate it.

It is not surprising that within the frameworks of positions [i] and [ii] something like an ontological value can only appear only as a "queerness" (Mackie 1977: 43; see Section 4.3.2.5, below) or indeed as an "ontological monstrosity" (Greimann 2000: 139). The reason is clear: this purely materialistic-physicalistic ontology can neither understand human beings *as such* nor integrate them into the universe. Within the frameworks of positions [i] and [ii], human beings play, with respect to the determination of the universe, no ontological role worth mentioning: with human beings, the universe remains ontologically exactly what it had been or would be without human beings. Only position [iii] accords to human beings an important place in the determination of the universe, but this position, as shown above (Section 4.3.1.2.3.1[2], pp. 283–87), is not only ambiguous but is also incoherent. If it could free itself from its fundamental materialism, it would at least be pointing in the right direction.

[2] If one takes human beings seriously, with all their ontological dimensions and facets and in their genuine and proper ontological constitution, then it is immediately clear that the universe or being as a whole cannot at all be conceived of in the way in which absolutely abstract-metaphysical realism attempts to conceive of it. The universe or being contains beings other than the ones that fit into the framework of any positivistic-materialistic ontology. Frege speaks of a "third realm [of thoughts]," in addition to the (first) realm of things in the external world and the (second) realm of representations (1918: 337/69). But far, far more dimensions must be acknowledged (e.g., the dimension of logical/mathematical structures). The domain of values is one of those dimensions of being as a whole that no positivistic-materialistic position can acknowledge.

As far as the dimension of ontological values as the basis for moral values is concerned, consideration of the status or placement of human beings within the universe is of decisive significance. Human beings stand out from all other beings in the manner in which they are embedded within webs of values. Their ontological uniqueness is such that the status within the universe *adequate* to their ontological constitution is—in the traditional (Kantian) formulation—the status of being *absolute ends in themselves*. To say that human beings are absolute ends in themselves is not yet to introduce a moral thesis, but still only an ontological one: it contradicts the ontological constitution of the human being if, *qua* human being, the human being is not seen and treated as end in itself. On the basis of their intelligence, their wills, and their freedom, human beings are, as minding, intentionally coextensive with the universe: none can be reduced to the status of serving as means for other items or elements within the universe. In other words: within the universe, every human being is an absolute point of reference. To consider and to treat the human being as a means is thus tantamount to an ontological degradation, because it directly contradicts the ontological constitution of the human being. This is perhaps even clearer in light of the identification in Section 4.3.1.2.2.3[3] (p.277) of the human individual as *self-situating* within being as a whole. If such a being is used as a mere means, it is *situated* within the whole by whomever so uses it; *being situated* by someone else clearly contradicts *situating oneself*.

End in itself is thus a basal-ontological value. To be emphasized is that this is a constitution or determination that is genuinely *ontological* in every respect. Anything else would be a self-contradictory subjectivism. But in what sense is *end in itself* a genuinely ontological entity? The answer is given above: *"end in itself" designates the fully determined or fully analyzed ontological structure or constitution of the human being*. This is nothing like a "supervenience" in relation to any kind of already presupposed ontological constitution; instead, the ontological constitution *is* end in itself; this constitution *includes* the basal-ontological value *end in itself*.

4.3.2.5 The Ontological Status of Moral-Ontological Values

Moral-ontological values are moral norms, prescriptive values. The most universal and highest moral norm is, in traditional terminology, *Bonum est faciendum—Malum est vitandum* (The good is to be done, the bad is to be avoided). How is this *prescribed bonum* (and this *proscribed malum*) to be understood? Can or must pre- and proscribed values, as moral values or norms, be understood ontologically?

[1] Particularly against such values the objection is raised that they are "queer" or "monstrous" (see Section 4.3.2.4.2[1]). Stegmüller, in his commentated summary of Mackie's book containing the argument directed against the assumption of real moral values or "facts"—the one asserting the "queerness" of such entities—elaborates this "queerness" more precisely by introducing what he takes to be a fantastic scenario. This is clearly shown by the passage in which Stegmüller maintains (1989: 175) that whoever believes in objective values and norms "must populate his world with peculiar beings like *ought-to-be-done, ought-to-be-avoided*, or *values equipped with authoritative prescriptivity*. Plato's forms provide an intuitive and at the same time dramatic image of these entities." Despite its popularity in many philosophical circles, this "argument" presents an extreme and provocative thesis. What is its argumentative value?

Everything is decided by what one understands by *actuality*. For one who has no inkling of modern physics or thinks nothing of this fundamental science, the entities that physics introduces, describes, explains, uses, etc., are simply "queer." To say that (for example) a table, a solid, reliable table that one sits at, uses, touches, etc., every day, is nothing other than a heap of molecules, elementary particles, etc., would be for this person simply "queer." Mackie defends a radically empiricist-materialist conception of "actuality"; it is therefore not surprising that he views the likes of "moral facts" as nothing but "queerness." The real problem however lies not in these specifically moral entities, but in the more general question of what notion of actuality a given philosopher has. The mention just made of the entities of physics should be expanded through the introduction of other concepts and/or entities, such as logical/mathematical structures, numbers, sets, modalities, etc. Mackie himself mentions this point. His position with respect to these entities is one of self-expressed helplessness, and is highly symptomatic:

> I can only state my belief that satisfactory accounts of most of these can be given in empirical terms. If some supposed metaphysical necessities or essences resist such treatment, then they too should be included, along with objective values, among the targets of the argument from queerness. (1977:45)

As the final sentence shows, Mackie's procedure or position is quite problematic. Every X that is inaccessible to empirical "explanation" is simply "explained" as "metaphysical" and "queer." In this manner one can indeed explain problems away, but problems that are explained away are not problems that are solved.[34]

[2] This section briefly shows that moral norms in the genuine sense are to be understood ontologically. Shown above, without difficulty, is that basal-ontological values have an ontological status. Perhaps less obvious is the thesis that moral norms or values likewise have an ontological status. How is this to be explained and supported more precisely?

To answer this question, a distinction must be introduced between *first-order* and *second-order* ontological status (although to be emphasized is that nothing depends upon the terminology). This distinction does not agree completely with that between primary and secondary qualities, but there is a significant analogy between the two distinctions.

The distinction between first-order and second-order ontological status concerns in the first instance those beings that are designated as minding and bodying, and as persons. These are human beings. Here presupposed is that human beings as moral beings are free in the strong sense.[35] First-order ontological status characterizes those structures of human beings that belong to their constitution as minding and bodying. Second-order ontological status designates everything that emerges from the acts of human beings as minding and bodying or is connected with those acts; briefly, it is the domain of entities that are produced by human beings. This includes the domain that Hegel calls "objective spirit," social facts, the domains of right, of morality, of

---

[34] For a critique of Mackie's argument from the "queerness" of the acceptance of ontological values, see the similar considerations in McGinn (1997, esp. 19f). To be sure, these considerations are quite general; for example, no distinction is made between basal-ontological and moral-ontological values.

[35] "Free in the strong sense" can be characterized briefly as follows: freedom, so understood, excludes any compatibility with any form of determinism.

institutions, etc. Moral norms have this second-order ontological status. It is clear that, if human beings did not exist, neither would entities with this second-order ontological status. But given the existence of human beings, entities with this second-order status are constituents of the universe just as are colors, etc.

Here, the question of how the moral-ontological values *emerge* from the basal-ontological values plays a central role, because clarifying this is an essential presupposition for making intelligible the ontological status of moral norms. This issue cannot be pursued here; it suffices to indicate that the relation between basal-ontological and moral-ontological values is not analytic: there is no logical-deductive connection between the two. To maintain such a connection would be to commit a naturalistic fallacy. The central point here is the following: moral-ontological values (thus norms in the strongly prescriptive sense) are, without the inclusion and mediation of willing, inconceivable. How such willing and the connection between the two dimensions of   408
value that it mediates are to be thought is not shown here.

As shown in Section 2.2.3.1, practical and practical-deontic sentences are also meaningful in that they articulate or express something or other; in a word, they also have *expressa*. The *expressum* of such a sentence is a *practical* or a *practical-deontic* proposition. But practical and practical-deontic sentences differ semantically from theoretical-deontic sentences in their *mode* of expression or articulation. In practical-deontic sentences, this mode is that of *demand*, not the *theoretical* mode made explicit by the operator, "It is the case that ..." Because practical-deontic propositions, thus practical norms, commands, etc., are grounded in basal-ontological values (in the sense explained above) and arise from them, in consequent fashion—not deductively, but only as mediated by willing—they are second-order ontological entities, in the sense explained above.

In sum: a given sentence is true just when it expresses a true proposition. A proposition is true just when it is identical with a fact (in or of the world). Ethical truth is a specification of the universal concept of truth. A given ethical-valuative sentence is true just when it expresses a true ethical-valuative proposition; an ethical-valuative proposition is true just when it is identical with a basal-ontological value, thus with a basal-ontological value *fact*. A given ethical-deontic sentence is true just when it expresses a true ethical-deontic proposition; an ethical-deontic proposition is true just when it is identical with a moral-ontological value, thus with a moral-ontological value *fact* (a second-order entity). Within the domain of morality, genuinely practical sentences, thus primarily practical or practical-deontic sentences, and the propositions they express, are neither true nor false, because only sentences and propositions with theoretical statuses have truth-values. To be sure, however, practical-deontic sentences (and the corresponding practical-deontic propositions) are "correct," "rational," "justified," etc., when they are grounded in basal-ontological values.

4.4 The Aesthetic World

From its first beginnings, philosophy has always been concerned with the dimension of the aesthetic and of art, and this dimension continues today to be central to philosophy. It could be considered to be a subdimension of the human world, but this book takes the human world in a narrow sense. The reason it does so is systematic: human beings,   409
with all that relates essentially to them, do not provide *the* perspective from which

everything else is to be considered; instead, human beings are understood as integrating parts of the world. "World" here is not simply synonymous with "universe" or "being as a whole." "World" is not the greatest or most comprehensive totality within which human beings are included. Nevertheless, "the world" is a determinate totality that is not understood adequately if viewed anthropologically (i.e., purely from the perspective of human beings). To indicate this negative moment, "the aesthetic world" of this section and "the world as a whole" of the following are not considered as subdimensions of the human world.

Given the purposes of this book and particularly of this chapter, introduced and explained above, the task here cannot be that of treating so extensive a domain as that of the aesthetic with anything approaching adequacy. Only how this domain can, in principle, be thematized within the framework of the structural-systematic philosophy is shown here.

There are remarkable similarities between the domain of the aesthetic and the just-treated domain of the ethical. Just as the concept of the practical sentence plays an absolutely central role in the domain of the ethical, significant treatments of the aesthetic must attend closely to the concept of the *aesthetic judgment*. But the analogy does not end here. In the case of the ethical, the concept of the practical sentence proves to be deeply unclear and easily misunderstood; the same holds for the aesthetic sentence. The first task is therefore that of clarifying this concept. Only after that is done is it possible to present and examine the central contentual questions of a general philosophical aesthetics and of a theory of art. This also explains the organization of this section on aesthetics.

4.4.1 *The Three Central Logical-Semantical Forms of Aesthetic Sentences*

[1] Two points must be made at the outset. First, the term "judgment," despite its long and significant history, is misleading and inadequate because it is drawn from the vocabulary of the mental domain and thus accords to that domain an unacceptably privileged status. For a philosophy that accords central significance to language and thereby also to semantics, this is reason enough not to use the term. Instead, in systematic contexts the talk here is, as a rule, of aesthetic *sentences*.

410   Second, the centrality in aesthetics of the concept of the aesthetic sentence is as a rule seen and treated only quite one-sidedly. The correct question is posed of what an aesthetic sentence is, but the concept of the aesthetic sentence is accorded a peculiar form of *univocity*. It is, however, easily established that this concept is anything other than univocal. Indeed, one must distinguish among at least three fundamentally different forms of aesthetic sentence.

As is the case in the domain of ethics, in that of aesthetics, too, the three forms of aesthetic sentence are best identified and explained with the aid of operators: the *theoretical operator* $\textcircled{T}$, and the aesthetic operator $\textcircled{Æ}$. In the domain of the former operator there are both purely aesthetic sentences $S_{Æ}$ and aesthetic sentences $S_{Æ}$ preceded by the aesthetic operator $\textcircled{Æ}$. There are thus the following sentence forms: "$\textcircled{T}(S_{Æ})$" and "$\textcircled{T}(\textcircled{Æ}(S_{Æ}))$" and "$\textcircled{Æ}(S_{Æ})$."

Aesthetic sentences $S_{Æ}$ are sentences with aesthetic contents; more precisely (in accordance with the semantics developed in this book), they are sentences expressing primary propositions that are in the aesthetic domain. How "aesthetic domain" is to be understood is shown shortly below. For the considerations that follow immediately, it is important that close attention be paid to the strict distinctions between operators and sentences and, in the case of operators, between the theoretical operator and the specifically aesthetic operator.

[2] The three forms of aesthetic sentence must be clarified and made precise logically and semantically.

In accordance with the semantics developed in this book, the domain of arguments for the sentence forms introduced above—"$\text{T}(S_{\!Æ})$," "$\text{T}(Æ(S_{\!Æ}))$," and "$Æ(S_{\!Æ})$"—must be interpreted *semantically* in such a way that they articulate or express *aesthetic propositions* belonging within the aesthetic domain. The more precise clarification of these sentence forms that follows is based on this fundamental thesis.

[2.1] Sentences of the form "$\text{T}(S_{\!Æ})$" can—somewhat awkwardly—be read as follows: "It is the case that (or: there is a theoretical presentation such that) the primary proposition expressed by the sentence $S_{\!Æ}$ is articulated." Ordinary-language examples of such aesthetic sentences are, "The Dolomites are beautiful mountains," "Beethoven's Ninth Symphony is magnificent," etc. To be sure, to be noted is that if such sentences are not made precise by means of formalization, they can be understood and interpreted differently. The ordinary-language sentences can be preceded by the purely theoretical operator T as well as by the combination of the theoretical operator T with the aesthetic operator (so, $\text{T}(Æ...)$), or by the aesthetic operator $Æ$ alone; the status of the informal sentence $S_{\!Æ}$ is fundamentally different in each of the three cases.

[2.2] Sentences wherein the theoretical operator T is the main operator but that also contain the aesthetic operator $Æ$ as a second operator, within the scope of the first operator, and such that the domain for their arguments is restricted to aesthetic sentences $S_{\!Æ}$—thus, sentences of the form, "$\text{T}(Æ(S_{\!Æ}))$"—can be read as follows: "It is the case that there is an aesthetic presentation such that the aesthetic primary proposition expressed by the sentence $S_{\!Æ}$ is articulated." Or, more simply, "It is the case that the aesthetic primary proposition expressed by the sentence $S_{\!Æ}$ is aesthetically presented."

[2.3] Finally, sentences that contain not the theoretical operator, but only the aesthetic operator $Æ$, thus those of the form "$Æ(S_{\!Æ})$," are to be read as, "There is an aesthetic presentation such that the aesthetic primary proposition expressed by the sentence $S_{\!Æ}$ is articulated," or, more simply, "The aesthetic primary proposition expressed by the sentence $S_{\!Æ}$ is aesthetically presented." This third form of aesthetic sentence is the one generally meant when, in philosophical aesthetics (particularly since Kant), there is talk of "aesthetic judgments." The fact that this form has generally been considered to be the *only* form of aesthetic sentence has given rise to various fundamental misunderstandings, confusions, and one-sidednesses.

The most important and decisive distinction is that between aesthetic sentences that are theoretical and those that are not. A philosophical aesthetics can produce and contain only *theoretical sentences*. Aesthetic sentences of the third form cannot be generated by any philosophical aesthetics; they can only be parts of the *subject matter* of philosophical aesthetics. How the theoretical operator T is understood is explained above. The aesthetic operator $Æ$ requires a detailed explanation.

411

[3] The three fundamental operators that articulate the relation of the mind to the world are the theoretical, the practical, and the aesthetic. The dimensions of theoreticity, of practicity, and of aestheticity are the three fundamental dimensions of this relation. Considered more closely, these three are the dimensions of *the presentation of the world*. They are equiprimordial in the precise sense that no one of them is reducible to any other. This is fundamental, as Kant, in his terminology and within the framework of his transcendental philosophy but in connection with basic assumptions that date from Aristotle's metaphysical psychology, recognizes clearly:

412   [A]ll faculties or capacities of the soul can be reduced to the three that cannot be further derived from a common ground: the **faculty of cognition**, the **feeling of pleasure and displeasure**, and the **faculty of desire**. For the faculty of cognition only the understanding is legislative, if (as must be the case if it is considered for itself, without being mixed up with the faculty of desire), it is related as a faculty of a **theoretical cognition** to nature, with regard to which alone (as appearance) it is possible for us to give laws through *a priori* concepts of nature, which are, strictly speaking, pure concepts of the understanding.—For the faculty of desire, as a higher faculty in accordance with the concept of freedom, reason alone (in which alone this concept has its place) is legislative *a priori*.—Now between the faculty of cognition and that of desire there is the feeling of pleasure, just as the power of judgment is contained between the understanding and reason. (*CJ*: 178–79)

What Kant calls "the faculty of desire" is simply what is usually called "will." In a footnote to the passage just quoted, Kant presents his definition of this faculty: "**the faculty for being through one's representations the cause of the reality of the objects of these representations.**" Yet although Kant's distinctions among the three basic dimensions (traditionally: "higher faculties") are basically correct, he does not develop a coherent conception of them. That he does not is strikingly clear from the following lines, which come shortly after the passage cited above:

[E]ven if philosophy can be divided into only two parts, the theoretical and the practical; even if everything that we might have to say about the proper principles of the power of judgment must be counted as belonging to the theoretical part, i.e., to rational cognition in accordance with the concepts of nature; still, the critique of pure reason, which must constitute all this before undertaking that system, for the sake of its possibility, consists of three parts: the critique of the pure understanding, of the pure power of judgment, and of pure reason, which faculties are called pure because they are legislative *a priori*. (*CJ*: 179)

The concepts "theoretical" and "practical" part of philosophy are here manifest in their full ambiguity and incoherence. It is utterly unintelligible how "everything that we might have to say about the proper principles of judgment must be counted as belonging to the theoretical part." The division of philosophy as a whole into two parts and the division of the project of critique into three simply do not fit together. One aspect of this basic difficulty is the problem, extremely controversial and heavily discussed in Kant exegesis, of how to explain the relation between the two parts of the *Critique of Judgment*. The talk of "aesthetic" and "teleological judgment" on the basis of the concept *purposiveness* is obviously artificial and incoherent; its function is purely ad hoc, in that Kant, having produced the first two *Critiques*, later attempts to introduce an artificial coherence into his comprehensive philosophical conception.

It is important here to note that the presumably deepest reason that Kant's position   413
is unclear and aporetic is his failure to recognize the extensive semantic and theoretical implications of his distinction among the three "faculties of the soul." Many of the unclarities and incoherences in his conception are immediately overcome if one distinguishes, explains, and consequently applies the three operators introduced above.

[4] The aesthetic operator Ⓔ is sufficiently characterized by means of the provision of two factors: a basal or indirect factor, and a determining or direct factor.

[4.1] The *basal* factor is the totality of presuppositions that must be satisfied if there is to be talk of a *directly aesthetic* operator. Essentially, this involves a central thesis: that an aesthetic operator is possible only on the basis of a relation of the mind to the world. Briefly, this factor is the *presupposed ontological import* of the aesthetic attitude. It involves a determinate mode of presentation of the world, and indeed one that is comprehensive in a sense to be specified—one in which, implicitly, both the theoretical and the practical modes of presentation play significant roles. As shown below, this is the case both with natural beauty and with the work of art.

To clarify the basal factor with respect to determining the aesthetic operator, it is appropriate to indicate the strongly analogous concept, *basal-ontological value*, that is articulated by means of valuational practical sentences (see Section 4.3.2.1.2). Such sentences are preceded, explicitly or implicitly, by the operator Ⓣ$_V$ ("V" for "value"). These sentences are indicative-theoretical sentences, but have practical-valuational *expressa*. They articulate, in the theoretical mode of presentation, practical-valuational ontological contents. More precisely, they can be characterized as sentences that are determined by the theoretical operator whose domain of arguments contains valuational sentences, such as "This act is good" (interpreted in accordance with the form "Ⓣ$_V$(A)," this sentence can be paraphrased as, "It is the case that this act is good").

The practical basal-ontological values provide the basis for the moral-ontological values (see Sections 4.3.2.3–4.3.2.5). The moral-ontological values are expressed by practical sentences—thus, by sentences governed by the practical operator Ⓟ. The practical operator reveals itself to be a determinate mode of presentation of the world, precisely the practical mode, which however has as its presupposition an (implicitly) theoretical mode of presentation.

In complete analogy, the aesthetic operator is the articulation of a determinate   414
mode of presentation of the world, but one that presupposes both the theoretical and the practical modes of presentation of the world. Therein consists the *basal* factor of the aesthetic operator.

This fundamental circumstance explains the well-known fact that many philosophers accord to the aesthetic dimension, and quite particularly to art, *also* both a *cognitive* and a *moral aspect*; to be sure, however, these aspects are usually not understood and presented correctly and clearly. At this point, what must be clearly introduced is the following fundamental thesis: the cognitive and moral aspects of the aesthetic dimension are to be understood as a basis, always already presupposed, for the existence of the aesthetic dimension and the accomplishment of aesthetic presentations. From this it follows that the aesthetic dimension is in no way separate from the world, as is maintained by so many forms of aesthetic subjectivism and idealism.

[4.2] The just-explained basal-ontological import of the aesthetic operator is only a presupposition for the determination of the aesthetic operator. A second factor must also be considered, one that *directly* characterizes the aesthetic operator. Although this operator presupposes or includes a basal ontological import, it is, considered in itself, neither cognitive nor practical; instead, it is a third operator, equiprimordial with the theoretical and the practical operators. How is this to be understood?

The aesthetic operator designates and accomplishes the following: the aesthetic "object" to which the sentence that is the argument of the operator relates is not grasped and articulated "as itself"; instead, it is expressed and articulated from the perspective of the *lived experience*[b] of the speaker/subject of the "object itself." This lived experience is a quite particular condition of the subject, one that fully *determines* the character of the sentence that is uttered. To be sure, in the semantic sense the sentence that is the argument of the aesthetic operator expresses a primary proposition that, if true, is identical with a primary fact in the world, but this semantic process is embedded within the sphere of the lived experience of the speaker/subject. In that the sentence is preceded by the aesthetic operator, the form of presentation or of articulation of the proposition expressed by the sentence reveals itself to be neither theoretical nor practical, but instead aesthetic, in the sense sketched above: it is the mode of presentation/articulation that is proper to the sphere of lived experience.

One can term this mode or form of presentation or articulation the "expressive." This term has here a specific meaning that only at times and even then only in part coincides with the meaning that this word normally has in philosophical aesthetics. Generally, *expressivity* is viewed as a concept central to aesthetics; there is also an *expressive theory of art* (see, e.g., Gardner 1995). There, however, the concept is usually explained by means of the concept of the *emotional condition of the subject*. In contrast, the concept of *aesthetic experience* used here to explain expressivity is a significantly more comprehensive concept, as shown in what follows.

Providing a thorough and genuinely adequate characterization of aesthetic experience in the sense intended here is a difficult task. At least two aspects that are constitutive of this experience can, however, be identified.

The first can be designated the *contentual* aspect; this is an additional characterization of what is introduced above as the ontological import of aesthetic sentences. To clarify negatively, in aesthetic experience the aesthetic "object" is grasped and articulated in a manner that is non-theoretical and non-practical. Positively, the grasping and articulation develop in a *holistic* or *comprehensive* and at the same time in a *global* and therefore *undifferentiated* manner. The aesthetic grasping and articulation of the (aesthetic) object are *complete* or *comprehensive* in that they *include* both the theoretical and the practical levels, such that the object is grasped and articulated in its totality. But the grasping and the articulation are at the same time in a specific sense *wholly global* and thereby wholly *undifferentiated*, not only with respect to the individual elements or aspects that constitute the object but also—and especially—with respect to the *mode* of grasping and/or articulation. In aesthetic experience, the theoretical and practical

---

[b] "Lived experience" translates the German *Erlebnis*, one of two terms—the other being *Erfahrung*—that can translate "experience." The difference relevant here is that whereas in cases of *Erfahrung* the experiencing subject can be disengaged—the experience can, for example, be no more than a transitory perception—in cases of *Erlebnis* the subject is intimately involved. Colloquially: a lived experience (an *Erlebnis*) is one that makes an impression on the subject who undergoes it. In what follows, "aesthetic experience" is aesthetic *Erlebnis*.

moments are not differentiated; instead, the object is grasped and articulated at once, in a kind of continual global apprehension. At the same time, it would be inadequate to characterize aesthetic grasping and articulating as any kind of original intuition or anything of the sort, if one understood by "intuition" something that had to do only with the theoretical dimension. This point becomes somewhat clearer if one takes into consideration on the one hand the second of the aspects characterizing aesthetic experience introduced above and treated below, and on the other the treatment provided below of the central aesthetic concept, the concept of *beauty*.

The second aspect of aesthetic experience is in a specific respect of a *formal* nature. Aesthetic experience brings the mind together, so to speak, in its *integrality*: *all* of its so-called faculties, all its structures, are involved, are at work. Correctly understood, aesthetic experience is the *integral self-consummation* of the mind in its situatedness in or with respect to the world. To make this general characterization more precise, one must iden- 416 tify the specific difference between theoretical and practical presentation or articulation, on the one hand, and aesthetic presentation or articulation, on the other. In the theoretical attitude, the mind is directed completely and exclusively toward the "object" (in general: the world); "object" here can mean any entity, any state of affairs, any fact, in short any being of any sort whatsoever—thus, absolutely anything that can be considered to be a component of the universe of discourse and thus of the world or of being as a whole. The exclusive directedness toward the object in this sense means, among other things, that the subject as such does not appear; only the subject matter in question becomes articulated. In the theoretical attitude, the subject elevates itself to universality; it attains a coincidence with rationality and objectivity and thereby makes itself, *as subject*, superfluous. This is considered in detail in Chapter 2 (see esp. Section 2.3.2.4[4], pp. 113–17).

The attitude that is articulated by means of the *practical* operator is likewise directed at the "object" (thus, in general, at the world), but its directedness differs in two decisive ways from theoretical directedness. The first way is that the practical atti-tude relates only to a specific aspect or dimension of the object (the world): the dimen-sion of ontological values, more precisely, of ontological value-facts (see Sections 4.3.2.3 and 4.3.2.4, above). Because every entity, fundamentally considered, is good or valuable, the practical directedness is indeed also universal, but only in the sense that it concerns *one* specific universal dimensionality. In contrast, the theoretical directedness is simply or unrestrictedly universal in the omnidimensional sense. The second difference arises from the fact that the practical attitude involves the subject in a manner that is wholly different. The practical subject cannot at all be ignored; it is an essential and constitu-tive element of the practical attitude, because it is the subject, in its free decisions, that forms the practical presentation or articulation of the practical dimension of the world, thereby situating itself in a determinate manner within the universe or within being as a whole. But this *practical* self-accomplishment or self-situating of the subject within the world or universe happens on the basis of the *explicit relation* to the ontological dimen-sion of value-facts in the sense that the subject seeks to *realize* an ontological value by means of its self-accomplishment.

This fact also distinguishes the practical attitude essentially from the aesthetic; the latter also has a relation to the ontological dimension, but that relation is implicit. That is, the subject so to speak incorporates this relation into the sphere of its lived experience. 417 The aesthetic directedness toward the (aesthetic) object (generally, toward the aesthetic dimension of the world) is not accomplished *as such*, but is completely determined by the lived experience of the subject. The aesthetic self-accomplishment of the subject, unlike the theoretical and the practical self-accomplishments, is not directed primarily toward the object, but instead toward the subjective accomplishment itself, albeit, to be sure, with the involvement of the ontological import because, as emphasized above, the aesthetic attitude involves the theoretical and the practical.

This involvement has a unique character. The theoretical and practical attitudes are not presupposed in such a manner that they would remain a fixed or firm determining foundation. Instead, they are presupposed in the sense that they are present, but in such a way that they are simultaneously suspended: they undergo a kind of transformation in that they are no longer determinative for the aesthetic attitude, but are instead involved in the play of the *expressive* presentation or articulation of the self and the world that is accomplished by the subject. The aesthetic attitude is the expressivity of the mind in its integrality insofar as it articulates its relatedness to the world as its lived experience. In brief: the aesthetic attitude accomplishes or realizes itself as the expressivity of the aesthetic experience in this sense.

To be noted in this context is that the just-described mode of aesthetic directedness to the object (the world) has two distinct forms, one relating to natural beauty and the other to the work of art. The first form has, to put it one way, a *more prominently* objective or ontological status than does the second. This noteworthy state of affairs is considered more closely in Section 4.4.2.

[5] To make this briefly sketched conception somewhat more intelligible, it may be helpful to compare it to the position taken by Kant in the *Critique of Judgment*. As noted above, Kant does not recognize anything like operators, and thus nothing like an aesthetic operator. If this point is ignored, it can be shown that, in comparison with Kant's position, the position sketched here is on the one hand remarkably similar, but on the other deeply and irreconcilably different. To show this, however, it is important to attend both to the different terminologies and to the incompatibility of most of the basic theses of the two philosophies.

A long passage from the Introduction (VII) of the *Critique of Judgment* can be viewed as a brief and summary presentation of the Kantian position with respect to the issues of current concern:

418   If pleasure is connected with the mere apprehension (*apprehensio*) of the form of an object of intuition without a relation of this to a concept for a determinate cognition, then the representation is thereby related not to the object, but solely to the subject, and the pleasure can express nothing but its suitability to the cognitive faculties that are in play in the reflecting power of judgment, insofar as they are in play, and thus merely a subjective formal purposiveness of the object. For that apprehension of forms in the imagination can never take place without the reflecting power of judgment, even if unintentionally, at least comparing them to its faculty for relating intuitions to concepts. Now if in this comparison the imagination (as the faculty of *a priori* intuitions) is unintentionally brought into accord with the understanding, as the faculty of concepts, through a given representation and a feeling of pleasure is thereby aroused, then the object must be regarded as purposive for the reflecting power of judgment. Such a judgment is an aesthetic judgment on the purposiveness of the object, which is not grounded on any available concept of the object and does not furnish one. That object the form of which (not the material aspect of its

representation, as sensation) in mere reflection on it (without any intention of acquiring a concept from it) is judged as the ground of a pleasure in the representation of such an object—with its representation this pleasure is also judged to be necessarily combined, consequently not merely for the subject who apprehends this form but for everyone who judges at all. The object is then called beautiful; and the faculty for judging through such a pleasure (consequently also with universal validity) is called taste. For since the ground of the pleasure is placed merely in the form of the object for reflection in general, hence not in any sensation of the object and also without relation to a concept that contains any intention, it is only the lawfulness in the empirical use of the power of judgment in general (unity of imagination with the understanding) in the subject with which the representation of the object in reflection, whose *a priori* conditions are universally valid, agrees; and, since this agreement of the object with the faculties of the subjective is contingent, it produces the representation of a purposiveness of the object with regard to the cognitive faculties of the subject. (*CJ*: 189–90)

It is not easy, indeed perhaps scarcely possible, to interpret this text in a way that yields a coherent conception. Many of its concepts remain wholly unclarified and problematic.[36] Here, however, only several other points important for a comparison of this book's position to Kant's need be noted.

What Kant laboriously constructs in the passage just quoted is the result of the attempt to conceive of the aesthetic within the framework of his transcendental architectonic of the "faculties of cognition." The concept central to his conception, the *purposiveness* of the object, is explained by his reference to a "comparison": reflecting judgment compares the forms of the objects of intuition (forms "grasped" in the imagination) with its capacity to relate intuitions to concepts. The object is then to be viewed as "purposive" for reflecting aesthetic judgment when something happens that Kant describes as follows: "if in this comparison the imagination (as the faculty of *a priori* intuitions) is unintentionally brought into accord with the understanding, as the faculty of concepts, through a given representation and a feeling of pleasure is thereby aroused." This is extremely difficult to understand. The "form of the object" is, according to Kant, the object on the one hand without "the material aspect of its representation, as sensation" and on the other hand "without relation to a concept that contains any intention." If this is so, what can the "object" be? Kant speaks of "its [the object's] representation" and emphasizes that such a representation is given without any sensation and without any concept. But what could such a representation of an object be? Kant here plays with empty words and concepts. He then maintains that the imagination "is unintentionally brought into accord with understanding, as the faculty of concepts, *through a given representation*" (emphasis added). If the "representation of the object" is, as shown just above, wholly empty, how can there be any talk of "accord"? And how is it to be understood that from such an unintentional "accord" "a feeling of pleasure is ... aroused"? Kant's attempt to clarify the aesthetic shatters in a fundamental respect because of the complete unclarity of what he says of the "object." Moreover, there arises for him the problem of the incoherence of this attempt

419

---

[36] How, for example, can Kant speak of an "object" on the basis of the assumption that no concept is involved, given his assumption elsewhere that objects can be understood only by means of the concept of the concept? Such details cannot and should not be considered here.

within the framework of theses he elsewhere presents as fundamental—but this is not considered further here.

Despite these insuperable difficulties, some valuable intuitions can be detected in Kant's conception, intuitions that can be incorporated into an utterly different basic philosophical framework. Perhaps the most important is Kant's thought of "a thoroughly interconnected experience" (184) that he can explain within his comprehensive philosophical position only by introducing or postulating a **"principle of purposiveness** for our faculty of cognition" (184). He speaks of "an agreement of nature with our faculty of cognition" (185); he often speaks of "the faculties" in the plural, and not only of *the* faculty of cognition (e.g., "agreement of the object with the faculties of the subjective"; 190). But Kant grasps this purposiveness (or suitability), with respect to aesthetics, as something purely subjective. It is therefore fully consequent for him to go so far as to (re)interpret natural beauty in characteristically transcendental fashion:

420   [A]nd so we can view natural beauty as the presentation of the concept of the formal (merely subjective) and the natural purpose [*Naturzwecke*] as the presentation of the concept of a real (objective) purposiveness. We judge the former by means of taste (aesthetically, with the mediation of the feeling of pleasure), the latter by means of understanding and reason (logically, in accordance with concepts). (*CJ*: 193)

"Real (objective) purposiveness" is the teleology of nature that is treated in Part II of the *Critique of Judgment*. What Kant terms "real" or "objective" is to be understood in the transcendental sense. "Real/objective" are quite misleading terms in Kant's conception and terminology: the "reality/objectivity" intended here is radically enclosed within the limits of the world as the totality of *appearances*. When aesthetic purposiveness is understood as "subjective," the subjectivity in question is of a second order or level: the world as the totality of appearances is already a radically subjective dimension, a first-order subjectivity; the aesthetic purposiveness of nature is once again "subjective" relative to the (fundamental) subjectivity of the world as the totality of appearances, however that is to be understood. It remains remarkable, however, that Kant speaks of an "*agreement* [*Zusammenstimmung*; harmony] with the faculties of the subject." This thought is a transcendental abbreviation of the thought, presented above, of the *integrality of mind* that articulates itself expressively and thereby aesthetically.

4.4.2 *The Universal Aesthetic Dimension: Beauty as Fundamental Concept*

This and the following subsection treat *two fundamental contentual questions* that a philosophical aesthetics must consider. The first concerns what can be designated in general as *the concept that is fundamental* within philosophical aesthetics; the second concerns the specific domain of art.

Chapter 3 (Section 3.1.2.1) shows that what is usually termed "concept" is highly problematic, but also that the term is scarcely avoidable in philosophical presentations because it is an extremely convenient designation. If therefore one does not want to avoid the term "concept," then it should be understood as an abbreviation for "structure." Chapter 3 thoroughly presents the precise signification of "structure." Given this qualification, the "concept" *beauty* can be presented, without danger of misunderstanding, as the central concept of philosophical aesthetics. What is meant by it is that structure that, contentually, is to be interpreted as the *fundamental aesthetically valuative primary proposition or primary fact.*

[1] According to traditional (classical) aesthetics, beauty is, if not *the* fundamental aesthetic category or quality, at least one of the aesthetic categories or qualities that ground the unity of the discipline "philosophical aesthetics" (see Gardner 1995: 585, 590–91, 599–600). Both modern and contemporary aesthetics generally reject this position; one consequence of its rejection is the general absence, from modern and contemporary aesthetics, of a fundamental unity.[37] The history and current status of aesthetics are not critically examined here. Instead, the following account simply presupposes that these aesthetics suffer from the lack of clarification of their strictly systematic-philosophical bases and/or of their situatedness within systematic-philosophical totalities. The following account explains and supports this presupposition only in outline, not in detail. The logical-semantic status of aesthetic sentences, clarified in the preceding section, provides the basis from which it is possible to articulate several basic aspects of the concept of beauty. 421

It would be scarcely possible to provide a direct proof of the thesis that beauty is the central aesthetic category or quality. It is however presumably possible to provide a proof that is *indirect* in the following sense: the basic elements of a coherent and comprehensive conception of aesthetics can be presented. Such a conception can then be compared with other conceptions, thereby demonstrating its superiority. Even independently from such a comparison, however, it becomes clear as the basic features of this book's philosophical aesthetics are presented that the closedness of the systematic conception, within which every philosophical domain can be unambiguously situated, on the one hand presents persuasive grounding for its superiority, and on the other confirms the immanent stringency of the relevant discipline (in the current case, philosophical aesthetics).

[2] The *systematic* explanation of the concept of beauty involves three factors. First, the concept must be understood and explained within a strictly systematic framework; that means, in the context of the structural-systematic philosophy, that the concept must be situated within the comprehensive *theory of being.* Second, the three forms of aesthetic sentences introduced in the preceding section must be taken into account, because they articulate the conceptual content of beauty in different ways. Third, the different domains of being wherein beauty is to be found must be considered explicitly.

[i] Within this book's comprehensive systematic framework, beauty appears as a basal ontological structure; more precisely, it is what Chapter 5 (Section 5.2.3) names *one of the universal, immanent characteristics of being,* and indeed a quite particular one: beauty is the *consonance or harmony (consonantia) of all the (other) universal immanent characteristics of being* (i.e., the consonance or harmony of universal intelligibility, universal coherence, universal expressibility, and the goodness of being as a whole). 422

Precisely insofar as these other four characteristics of being are *immanent* characteristics of being, they form a unity; their interconnection is characterized by a thought that plays a not inconsiderable role in the entire philosophical tradition: the thought

---

[37] Peres (2000) counters, presenting beauty as the central category of aesthetics.

of consonance or harmony (*consonantia*).[38] Thereby, what in the terminology of this book is an aesthetically valuative primary fact is given as a highly complex ontological configuration. The attempt can be made to present the precise inner structurality of this configuration. On the whole, every characteristic of being implies all the others and is implied by all the others. The thought of consonance can be understood more precisely as the highest perfection (*perfectio*), the highest or complete or completed self-presentation or self-manifestation of being and of every individual being.[39]

[ii] The comprehensively systematic determination of the concept of beauty just presented can be termed the *valuative-basal-ontological* concept of beauty. It is assumed or presupposed by all sentences in which "beautiful/beauty" occur; it is presupposed as the *valuative-basal-ontological determination* of beauty. In its pure form (i.e., *only* as valuative-basal-ontological determination), it is the ontological value that is articulated in aesthetic sentences of the first form; these are the aesthetic sentences that have only a single operator, the theoretical one: $\text{Ⓣ}(S_Æ)$. Such sentences *directly*—here, in the purely theoretical respect—articulate fundamental aesthetically valuative primary propositions or primary facts. It is thus clear that the following holds: if any such sentence is *true*, then the aesthetically valuative primary proposition that it expresses is also true and is therefore identical to an aesthetically valuative primary fact in the world.

If there is talk of beauty in aesthetic sentences of the second form (thus the form, $\text{Ⓣ}(\text{Ⓔ}(S_Æ))$), then the basal-ontological aesthetically valuative determination of beauty in such sentences is *presupposed* in the sense of being always already *coexpressed*; beyond this, however, in such sentences beauty is already articulated in a concrete determination. The decisive point is the circumstance that the *aesthetic operator*, as the second operator, explicitly brings in the *subject*. The determination *beauty* is then articulated not only as a basal-ontological determination but also as a determination that co-constitutes the aesthetic self-expressivity of the subject.

---

[38] See, e.g., Thomas Aquinas: "a certain moderate and fitting proportion, and this is what we understand by beauty" (*STh* SS Q 141 A 2 Rp 3); for Thomas, "*proportio*" is synonymous with "*consonantia*" (harmony), as is revealed by his use of the phrase, "*proportio sive consonantia.*" Elsewhere, Thomas characterizes beauty as follows:

> For beauty includes three conditions, "integrity" or "perfection," since those things which are impaired are by the very fact ugly; due "proportion" or "harmony"; and lastly, "brightness" or "clarity," whence things are called beautiful which have a bright color (*STh* FP Q 39 A 8).

These characterizations of beauty are not unconditionally mutually exclusive, because it cannot be assumed that Aquinas wants here to provide an absolutely precise and exhaustive determination of the concept of beauty. One must distinguish among various levels of consideration, as made explicit in the main text.

[39] At first glance Aquinas, the great medieval thinker of being, appears to understand goodness (*bonum, bonitas*) as the perfection (*perfectio*) of every individual being. He thus writes, for example, "a thing is good from the fact of its being perfect" (*ScG*: 82 [I 37]). But Thomas can or indeed must be understood as here having in mind the aspect of *action*; this is made clear by texts like the following: "That which virtue implies 'consequently' is that it is a kind of goodness: because the goodness of a thing *consists in* its being well disposed according to the mode of its nature" (*STh* FS Q 71 A 1; emphasis added).

This factor is well known in the philosophical tradition (which does not mean that it is grasped and exhibited with sufficient clarity), particularly in the way that the perspective of "*placere*,"[40] of "pleasure,"[41] etc., is connected with the determination of beauty. This connection appears at first glance to broaden the concept and the determination *beauty* significantly. On closer consideration, however, Kant's attachment of pleasure to beauty, which continues to be a guideline for many philosophers, proves to narrow the concept in two ways. First, Kant reduces beauty to a merely reflexive structuration of the subject: the aesthetic harmony of the faculties of imagination and understanding. Second, he reduces aesthetic experience, which in fact is characterized by integrality, to a *feeling* of pleasure: a reflexive being-pleased-by the subjective harmony of the imagination and the understanding. In response to this narrowing, Constanze Peres (2000: 156) has developed a broadened and concretized concept of beauty as an "ontosemantic constellation." She makes her expansion of the concept of beauty more precise by presenting actual beauty as occurring as a singular and unique constellation involving a subject who designates an object as beautiful and an object designated as beautiful.[42]

424

---

[40] See Aquinas: "the 'beautiful' is something pleasant to apprehend" (*STh* FS Q 27 A 1 Rp 3; see also *STh* FS Q 5 A 4 Rp 1).

[41] See Kant *CJ*, esp. Introduction VI–VII.

[42] Persuasive as this ontosemantic expansion of the concept of beauty is on the one hand, on the other it is to be criticized on a systematically important point. Peres explains as follows (155–56):

> Presupposing that we do not proceed from an absolute sphere of ideas and values or from "objectively" pre-found values, then every value consists fundamentally in the relation between whatever is valuable and whatever recognizes it to be valuable. Something is beautiful in relation to the one who perceives it and judges it to be beautiful. Without this perceiving and linguistic symbolization, it might or might not exist for itself, but even if it did exist for itself, if we did not judge it to be beautiful we would know nothing about that. *In that* someone evaluates something as "beautiful," thus in and by means of the symbolization, what is judged to be beautiful *acquires* that value. . . . "Beauty" is thus not a property of things or states of affairs. It is however also not a property of a subject, for example in the sense of (post-)Kantian pleasure. Instead, "beautiful" is always "symbolized-as-beautiful-by." It *exists* as an aesthetic relation that is unmistakably symbolized as "*x* is beautiful." In this sense, in the judgment of beauty that to which a relation is taken—namely, this aesthetic relation—is always maintained. Beauty is thus an ontosemantic constellation. *Within* this ontosemantic constellation, that of which "beautiful" is predicated in fact ("objectively") has the property of in fact ("objectively") moving the evaluator to make the judgment of beauty and the evaluator in fact ("objectively") has the property of being attuned in such a way that it ascribes to the object the predicate, "beautiful."

This passage expresses a certain restriction (or reduction) that consists in the attachment of the ontological dimension to the human subject. As is shown in many passages in this book, and demonstrated particularly thoroughly and systematically in Chapter 5, the thesis that the "ontological dimension" is fully independent from language, mind, etc., is untenable. One must, however, see the entire scope of this problem. Many authors are inclined to avoid the so-called metaphysically realistic conception by immediately introducing the human being as the unconditionally necessary perceiving subject. Here, caution is advisable if one is to avoid reducing the world (the universe, being as a whole) to something for which human beings provide the measure. Chapter 5 shows how this danger is to be avoided. Some formulations in the passage cited above are problematic in just this respect ("*In that* someone evaluates something as 'beautiful,' thus in and through the symbolization, what is judged to be beautiful *acquires* that value"; "'Beauty' is thus not a property of things or states of affairs").

In sentences of the third form ($\oplus(S_{\cal E})$), beauty is articulated exclusively such that it is an element—more precisely, the central element—within the sphere of the self-expressivity of the subject. The mind or the subject articulates itself *integrally-expressively as* having a lived experience of beauty. To be sure, the sentence that articulates beauty continues to exist, but it does so precisely as the argument of a single operator, the aesthetic operator. In this case, beauty is not primarily known or willed, but instead experienced; it is what triggers and constitutes the lived experience of the subject. The subject who is engaged aesthetically is here the determining factor or the determining perspective from which beauty is articulated.

425   [iii] The third factor to be taken into consideration with respect to the adequate determination of the concept of beauty concerns the different *domains of being* within which beauty occurs or appears. The two most important domains are those of nature and of art. The following section considers the domain of art. The concept *natural beauty* encompasses an extraordinarily broad spectrum of phenomena that is determined in part by the concept of nature that is relied upon; already at the point of such reliance the investigation of natural beauty can appear to be boundless. For this reason, adequate determination of (*natural*) *beauty* in accordance with the concretely differentiated spectrum of sites of natural beauty lies beyond the scope of this book.

4.4.3 *The Specific Dimension of Art*

[1] In opposition to the traditional (classical) conception, Hegel reduces beauty to artistic beauty; he consequently excludes natural beauty from the philosophical theory of aesthetics, thereby establishing the superiority or priority of art over nature that continues to characterize contemporary aesthetics. The reason for Hegel's turn lies in the basic idealistic premises, characteristic for him, on which his comprehensive philosophical conception rests.[43] Within the framework of Hegelian idealism, the distinction between or opposition of natural beauty and art (or artistic beauty) is a relatively clear thesis. Independently of the idealistic premises, it is quite problematic. Although analytic philosophy is not based on idealistic premises, it nevertheless tends to accord art absolute priority (and indeed at times even exclusivity) and generally to exclude completely the dimension of beauty. This tendency is particularly evident in philosophies that do not develop adequate ontologies, even when they do not in principle exclude ontology. This appears to be the case today in significant parts of analytic philosophy.

    The structural-systematic philosophy developed in this book fundamentally excludes any such tendency or thesis. Because of this exclusion, it requires that

---

[43] See, from his *Aesthetics*:

> Artistic beauty is beauty that is born from spirit and reborn, and to the same degree that spirit and its products stand above nature and its appearances, artistic beauty stands above the beauty of nature. . . .
>
>   The superiority of spirit and its artistic beauty, in comparison with nature, is however not only a relative superiority; instead, spirit is that which is true, that which encompasses everything within itself, so that everything beautiful is truly beautiful only as participating in this superior domain and only as produced by it. In this sense, natural beauty appears only as a reflection of the beauty that belongs to spirit, as an imperfect, incomplete mode, a mode that, in substance, is contained within spirit itself (I:2/*TW* 13:14–15; AW translation).

aesthetics be anchored within the comprehensive philosophical theory and thus within the theory of being. The dimension of art must therefore be situated within the framework of being or of the theory of being; as a consequence, so-called natural beauty must be accorded priority over the dimension of art.   426

Art is an extraordinarily complex domain. The scope of this book wholly excludes even touching on the most important philosophical questions about art. The following paragraphs briefly treat only one of the relevant questions, albeit presumably the most fundamental; this is the question concerning the starting point for a philosophy of art. There is no question here of a definition or of a theory of art.

Of the many theories of art that are today defended and discussed, only four—presumably the most important—need be mentioned briefly in order to situate the approach introduced here within the contemporary theoretical spectrum. The four are the mimetic, the formalistic, the expressive, and the semiotic theory (see Gardner 1995: 612ff). The *mimetic* theory—in the current, quite broad sense—understands art fundamentally as an imitation of nature in the sense of a determinate examination of the world; art is thus seen from the perspective of the relation of art to the world. In opposition, the *formalistic* theory conceives of art as fully independent of the world; only the form—the complex constellation of elements that are constitutive of a specific work of art—has artistic significance. The *expressive* theory considers art to be a means for the communication of feelings. Finally, the *semiotic* theory of art proceeds from the fundamental assumption that works and forms of art should be analyzed within the framework of a logical-semantic system of concepts containing such concepts as meaning, reference, denotation, syntactic and semantic rules, etc.

The approach envisaged here cannot adequately be classified as any one of the four types of theory just introduced. Clarifying how this approach relates to those four types would require fundamental corrections and specifications. Nevertheless, some connections may be indicated. The approach suggested here has a certain proximity to the mimetic theory of art, because that theory is unambiguously ontologically oriented. At the same time, however, both the term and the concept "mimesis" are extremely problematic. The other three theories, too, contain certain insights and elements that cannot be ignored. Details become clear in what follows.

[2] Art is a *wholly determinate form of presentation of the relation of human beings to the world.* Because this form is neither theoretical nor practical, this question arises: what can it be? To be sure, as shown above, the theoretical and practical forms of presentation of the relation of human beings to the world are also involved in the aesthetic relation, but neither one is *the determining factor.* But how then is the all-determining factor, the specifying character of the work of art, to be determined? If everything shown in the preceding paragraphs is taken into consideration, a conception emerges that is here formulated only thetically.   427

Art presents not the actual or real world, but a *transformed* world. *A work of art is a non-theoretical and non-practical, but instead precisely an aesthetic presentation of a specific aspect or "piece" of a transformed world.* Such a transformed world is, however, not definable independently of the actual or real world. That it is not emerges not only analytically from the concept *transformed world* but also from the already introduced fact that the theoretical and practical relations to the actual or real world are always already *involved* in the artistic presentation of the world.

The here-introduced concept of a *transformed world* is envisaged in one way or another in various theories of art, above all when there is concern with a thought that one could term the concept of the *idealization* of the actual or real world.[44] Without scrupulous interpretation, however, this concept remains quite vague; as a consequence, the point that is ultimately decisive with respect to comprehending art is missed. It can, however, be interpreted in a manner such that this point is grasped correctly. The point is the following: an idealized form of the transformed world that is adequately grasped with respect to art is one that presents the *complete* or *perfect* structurality of the world or of one of its segments. The artistically idealized form must not thereby be misunderstood, as in classical theories of art, as oriented with respect to a determinate, somehow objective ideal measure (e.g., harmonious proportion, noble subject, balanced coloration, etc.) in the sense of a normative concept. Likewise, however, it must not be understood as some sort of free or arbitrarily invented or imagined form bound to no measure at all (i.e., one that would be somehow ideal rather than genuinely real). Instead, the form that idealizes the actual world by transforming it must be understood such that it presents the full ontological perfection of the selected aesthetic segment of the actual or real world.

This thought is based on two presuppositions. First, the world is not closed and thus is open. Differently stated, the current reality of the real or actual world and of all its elements always falls short of the complete and perfect structurality that the actual world and its elements are capable of exhibiting and that is adequate to them. "Idealization" indicates what the actual world and its elements *could be*, if they were to attain their maximal form, and what they *ought to be*, if they behaved in accordance with this form. This makes clear that art, with respect to its metaphysics, contains or presupposes not only a theoretical but also a moral component. For this reason, theories of art that have attempted to articulate the cognitive and moral aspects of art have headed in the right direction, but these aspects must be understood correctly (i.e., situated systematically). Second, the openness of the world allows for not only one, but in principle infinitely many possible perfective, idealized forms. For this reason, art is something like the dimension of, in principle, infinitely many formative possibilities for presenting the possible and most adequate completion of the world and of all its elements, and it is accurate to say that art opens not only a but *the* grand domain of human freedom.

[3] It is clear that the just-sketched approach would have to be made more precise and concrete in many ways before there could sensibly be talk of a developed philosophical theory of art. This is a task that this book cannot undertake. In conclusion, only two points need be clarified to put the just-sketched approach into the proper perspective.

[i] First, the approach could suggest the question whether art belongs in the domain of a theory (semantics and ontology) of possible worlds. Possible worlds are, after all, modified or transformed versions of the real or actual world (at least according to the theory called "actualism"). It is also correct that, according to the suggested approach, works of art are presentations of transformed worlds (or, generally, of segments or

---

[44] The concept *artistic idealization* should not be confused with the often-used concept *theoretical idealization*.

elements of such worlds). But one must keep in mind the decisive distinction concerning the *mode of presentation*. A "possible world" in the sense of a philosophical theory (semantics and ontology) of possible worlds is a (modified/transformed) world exclusively *qua* topic of a purely *theoretical undertaking*. In opposition to that, a work of art is not a theoretical, but instead an *aesthetic* presentation of a possible (transformed) world. The distinction thus lies in the mode of presentation.

[ii] Within the framework of the structural-systematic philosophy developed here, the question unavoidably arises: what is the *ontological status* of the work of art? The basic answer to this question emerges from the same basic explanation used as the foundation for the answer to the question concerning the ontological status of moral values (see Section 4.3.2.5). It consists of the introduction of a *twofold ontological status*: a first-order and a second-order ontological status. In the first instance, this distinction concerns those beings designated as minding and bodying and as persons. In the case of ethics, the concept of basal-ontological ethical value is introduced; it has first-order ontological status. In the domain of aesthetics the concept of beauty is introduced as a primary fact with a *basal-valuative ontological status* (see Sections 4.3.2.3 and 4.3.2.4)—thus, likewise with a first-order ontological status. Both the basal-ontological ethical values and the basal-ontological aesthetic (artistic) values are forms of universal, immanent characteristics of being as such—respectively, of goodness and of beauty (these characteristics are presented in Chapter 5, Section 5.2.3). The parallelism between the moral values/norms that are specific to human beings and works of art is illuminating with respect to ontology. Like the moral values/norms, works of art presuppose the existence and the activity of human beings and are therefore second-order entities.

Works of art are products of human activity. If one wants to characterize the ontological status of these products more closely, one must pay the closest attention to the ontology that is presupposed. In the case of moral values, the conception of Mackie, according to which such entities are "queer" and therefore must be rejected, is criticized. Second-order entities are constituents of the concrete world (i.e., the world insofar as it contains human beings). If one terms such entities "mental" or "abstract" or "intensional," etc., these designations quickly prove to be misleading and therefore inappropriate. It is best to term them entities *sui generis* and to take care in grasping and articulating their specific characters.

The status of works of art is analogous to that of theories, insofar as one considers theories in terms of their *status as presentations*. So considered, theories (in accordance with the so-called logical conception of theories) are collections (more precisely, sets) of sentences—thus, linguistic signs in determinate configurations—that present something or other (generally, propositions and/or facts). Works of art also have a status as presentations, but a quite different one: an expressive-integral one, insofar as they are not only and not primarily forms of presentation from the perspective and in service of theoretical activity, but prove to be forms of presentation that, as a matter of principle, involve all the forms of presentation given with the integrality of mind.

4.4.4 *Two Objections*

Many objections can be raised to the approach to a theory of aesthetics and of art sketched above. Further development of the approach would overcome many of these, but such development is not undertaken here. This section concludes instead by introducing two objections that doubtless belong among the most important that can be made against this approach. The first objection concerns the just-sketched approach to a theory of art; it holds that the attempt to understand art fundamentally as a determinate form of presentation of the relation of human beings to the world fails from the beginning, because not every form of art and thus not every work of art can be considered to be a presentation of something relating to the world. The second objection is directed against the thesis that the concept of beauty is the fundamental aesthetic concept; it is supported in particular by such facts as that there can be an "aesthetics of the ugly"; this the title of a book by Karl Rosenkranz (1853/1996), a student of Hegel's, that thematizes "non-beautiful" phenomena. Moreover, beauty plays a decreasingly central role in modern and particularly contemporary art (see Peres 2000: 146ff). The following subsections subject these two objections to critical analysis.

[1] The first objection is supported by putative counterexamples. Thus, for example, it is suggested that abstract paintings do not present anything that relates to the world, that it would be absurd to say that music (especially "absolute" music) presents any such thing, etc. But this objection arises from an analysis of the concept of presentation that restricts presentation to some sort of copying and that is therefore, systematically, superficial. Nothing happens to human beings that does not involve their relation to the world; human beings as individuals or persons are after all systematically definable only in terms of the infinitely many relations that situate them within the world (the universe, being). Products that are putatively purely "abstract," such as certain forms and examples of works of art, present themselves only at first glance as entities that have nothing to do with the relation of human beings to the world. In actuality, already the concept of the *abstract* product or the *abstract* work of art, insofar as it abstracts from the world, is unintelligible if the relation to the world is not involved. If the work of art presents anything at all, it presents it in relation to the world—and it thereby also presents the world, to be sure not the real or actual world, but a world that is transformed in the sense explained above.

The case of ("absolute") music is only apparently more difficult. If beauty is definable by means of the principal concept of *consonantia*, of harmony, and if this concept is not somehow purely subjective, but instead a genuinely ontological primary fact (the harmony of the immanent characteristics of being), then it should be clear that *an exemplarily segmental tonal presentation* of this universal harmony is fully intelligible.

[2] The facts supporting the second objection can be interpreted in a way that prevents them from undermining the fundamental status in aesthetics here accorded to the concept of beauty. It is indeed a fact that there are works of art that present non-beautiful (ugly) or negative phenomena. A suffering face does not appear to be beautiful. Albrecht Dürer's famous picture *Ritter, Tod, und Teufel* ("Rider, Death, and the Devil") presents nothing beautiful—at least, it must be added immediately, nothing that is beautiful in the usual and unspecified sense of the term (i.e., as a counterpart to a subjective experience of pleasure). More precise analysis, however, reveals that this picture and others like it do indeed present beauty, and in a fascinating manner: either *ex negativo*, as a whole, or as an ugly component of a (larger) contextually beautiful whole. A suffering face, presented artistically, is ugly or non-beautiful only if it is considered utterly superficially. In the negative features

of the face, the beauty of the face is visible quite directly.[45] Because it is possible to present something *ex negativo* or as a part of a contextual whole, it is possible in principle artistically to thematize all the superficially negative phenomena that can be presented artistically.

The thesis that the development of art in the modern and contemporary periods exhibits a clear diminution of the significance of the beautiful is ambiguous. It can be understood—and generally is understood—as referring to the *self-interpretation* of artists and to what is said by interpreters of art (art critics, philosophers of art). So understood, the thesis may well be accurate, at least with respect to the majority of artists and interpreters of art.[46] But neither the self-interpretations of artists nor the opinions of (most) interpreters of art can be considered to be the criterion—and certainly not the proof—of the correctness of the thesis, because the phenomenon in question can also be interpreted quite differently, such that the thesis of "the diminution of beauty in modernity" can and indeed must be understood wholly differently.

There is indeed a "diminution" in modernity, but not a diminution *of beauty as such or on the whole*; the diminution instead involves an inadequate concept of beauty. Modern art turns wholly away from a fully static, metaphysically abstract concept of beauty. Drastically oversimplifying, one can say that modern art particularly rejects two characteristics of this "traditional" concept of beauty; these can be termed its "non-dialectical" and its "non-processual" characteristics. On the whole, the traditional concept is "non-dialectical" in that, at least for the most part, it does not sufficiently take into account or put into practice the presentation of beauty *ex negativo*. Instead, beauty is traditionally presented almost exclusively in a *wholly positive* or *affirmative* manner. Thereby, immense and powerfully expressive possibilities of presentation are utterly neglected and indeed ignored.

The second characteristic putatively characterizing beauty as traditionally understood that is put into question by modern art is the following: beautiful objects are virtually exclusively understood as finished objects in the sense of their being objects whose beauty is presented so to speak all at once, in perfected form. In brief: *from the perspective of modernity* objects that are beautiful in the premodern sense are understood to be utterly complete, perfect objects. The entire *process* that characterizes the reality of the coming to be or of the self-manifestation of beauty is often ignored virtually completely. But the beauty of so-called classical art works is also characterized by the processes of their coming to be, and thus by stages and imperfections. Its (at times virtually exclusive and excessively emphasized) attention to processual, partial, and incomplete elements no doubt characterizes modern art and distinguishes it from traditional art, which tends not to speak of or to thematize these fragments and fractures that are so often emphasized in modernity.

45 A picture or symphony that consisted only of beautiful parts or passages (i.e., pleasant ones) would be not only boring (this not an ontological category), but also non-truthful (kitschy) with respect to the ontological structure of the world.

46 Symptomatic of this is, for example, the assertion of the artist Barnett Newman, quoted by Danto (1986: 13): "The impulse of modern art was this desire to destroy beauty . . . by completely denying that art has any concern with the problem of beauty."

4.5 The World as a Whole

The domains of the world treated in the preceding sections undoubtedly belong among the most important topics of concern to contemporary philosophy. To be sure, the preceding accounts do not consider all, nor even all of the most important of the relevant domains and themes. As is regularly emphasized, the goal of Chapter 4 is only to show how the structural-systematic theoretical framework, as developed in the preceding chapters, can and should be developed or concretized with respect to the explicitly *contentual* topics of world-systematics. This book attains this goal *not* in absolute completeness (that would require a simply comprehensive presentation), but only by indicating how, within this framework, some of the most important domains and topics can be treated.

433

One of the domains that belongs among the most important philosophically is *the world as a whole.* "World," as indicated at the beginning of this chapter, is not synonymous with "universe" or "being as a whole"; instead, it is taken in a restricted sense. In this sense, "the *world* as a whole" is a topic to be treated explicitly in this chapter because it is a domain that is more comprehensive than are those treated in the preceding sections; it therefore presents an interesting case for the concretization of the structural-systematic theoretical framework.

Coming to philosophical terms with the world as a whole requires consideration of three extensive topics: first, the natural-scientific enterprise of developing physical/cosmological "theories of everything"; second, the phenomenon of religions and their historical development; and third, the most comprehensive arena of human life, world history. These are three heterogeneous domains; the gap separating natural-scientific theories of everything from religion and world history is particularly wide. Systematic clarification of the constellation of topics arising from the concept of the world as a whole requires considering all three, because each concerns and illuminates a specific facet of the world as a whole.

4.5.1 *Natural-Scientific Cosmology*

In the history of philosophy, the term "cosmology" is used most often to designate a *philosophical* theory of the world (including humanity).[47] Today, however, the term is

---

[47] The designation "cosmology" for a philosophical discipline is common since the work of Christian Wolff (1737). It is understood within the framework of the division of traditional metaphysics into general metaphysics or ontology (*metaphysica generalis* or *ontologia*) on the one hand and special metaphysics (*metaphysica specialis*) on the other; the latter comprises rational theology, rational psychology, and rational cosmology (see Wolff 1728/1996). General metaphysics is concerned with the order of the world as a natural system of physical substances: "General cosmology is the science of the world or the universe in general, the science that considers beings as interconnected and alterable [*Cosmologia generalis est scientia mundi seu universi in genere, quatenus scilicet ens idque compositum atque modificabile est*]" (1737: §1).

It is interesting that Wolff distinguishes between *cosmologia scientifica* and *cosmologia experimentalis*(§4):

General scientific cosmology consists in a general theory of the world that develops on the basis of the principles of ontology. In opposition thereto, experimental cosmology develops the theories it presents

usually understood in the *natural-scientific* sense and used to designate the physical   434
theory of the origin, development, and structure of the comprehensive physical cosmos or universe. For this theory, the extremely misleading designation "theory of
everything" is also used. This section presents some considerations concerning the
fundamental status of this natural-scientific theory from the perspective of philosophy. Because of the immense breadth and difficulty of the topic, the presentation is
restricted to one specific position; it concentrates on the theory (or theories) of the
physicist and mathematician Stephen Hawking.

[1] Natural-scientific cosmology poses the question of the beginning or origin of
the universe. On the basis of theorems that he and Roger Penrose have proved, Stephen
Hawking maintains "that the universe must have had a beginning," but adds that these
theorems "didn't give much information about the nature of that beginning" (2001: 79).
It must be asked, however, how "beginning" is to be understood here. With respect
to this term, there is a deplorable unclarity and confusion that is responsible for the
formulation of questions based on false presuppositions, the presentation of utterly
unfounded contentions, the construction of nonsensical oppositions, etc., by both
theistic and atheistic scientists and philosophers.

According to the standard natural-scientific, physical-cosmological theory, the
universe began with a "big bang." According to the theorems of Hawking and Penrose
alluded to above, 10 to 20 billion years ago the distance between neighboring galaxies
was zero. At this point, both the density of the universe and the curvature of space-time
are calculated to have been infinite. Such a point would be a case of what mathematicians term a *singularity*. Decisive here is the following consequence: because all scientific theories presuppose that "space-time is smooth and nearly flat" (1988: 40), at the
point of singularity all the laws of nature fail to hold. It therefore makes no sense to ask
what there was *before* the singularity, because time itself begins only with the big-bang
singularity. From this it appears to follow that the question concerning the origin or
beginning of the universe receives a univocal natural-scientific answer: the universe
begins with the big bang, and there is no basis for introducing anything like a creator.

[2] In the past two decades, a heated controversy has arisen among natural
scientists, philosophers, and theologians concerning the status of the physical-cosmological theory or theories; the controversy continues today. Many theologians
and philosophers attempt to find in the fact that it has been proved by the natural   435
sciences that the universe has a beginning a direct proof of a creator God; they interpret the big-bang theory *creationistically*. Other philosophers attempt to see in this
theory precisely the opposite: they view the natural-scientifically proven beginning
of the universe as an explanation that makes every other explanation superfluous,
and one that is irreconcilable even with the assumption of the existence of a creator
God; according to them, the origin of the universe is uncaused, as the universe arises

---

or is to present within the scientific domain from observations [*Cosmologia generalis scientifica est,
quae theoriam generalem de mundo ex Ontologiae principiis demonstrat: Contra experimentalis est, quae
theoriam in scientifica stabilitam vel stabiliendam ex observationibus elicit*].

Wolff's division of philosophy and his distinctions were quickly incorporated into many textbooks of philosophy.

spontaneously.[48] But some physicists have interpreted the big-bang theory in a theo-logically creationistic manner; see J. Narlikar, who writes (1977: 136–37) as follows:

> The most fundamental question in cosmology is, "Where did the matter we see around us originate in the first place?" This point has never been dealt with in big-bang cosmologies, in which, at $t = 0$, there occurs a sudden and fantastic violation of the law of conservation of matter and energy. After $t = 0$ there is no such violation. By ignoring the primary creation event most cosmologists turn a blind eye to the above question.

It can be shown that *all* these positions or interpretations rest upon *misunderstandings* because they do not recognize a fundamental factor: the *specific status* of the natural-scientific, physical-cosmological sentences and theories. When *within the framework* within which these sentences/theories arise there is talk of an "origin or beginning of the universe," the talk is deeply misleading and ambiguous. The meanings of the terms used by the cosmological physicists and ambiguously speaking philosophers are determined by the status of the natural-scientific *theoretical framework* or *model* that is relied upon by the physical cosmologies. *Within* this model, as the Hawking-Penrose theorems show, it can be "proved" that the universe has its origin or beginning at a point in the past, the singularity. But this can only mean that, in any such account, the universe as a structured whole can be explained in terms of the familiar laws of nature only from this point. There can be no talk of an origin or creation *ex nihilo*, out of noth-ing, because the entire model presupposes precisely the singularity—and *it is of course not nothing.* The singularity is only something that is not "structured" by the familiar laws of nature. The physical-cosmological big-bang theory consists *only* in this: it shows how *one* state of the physical universe, the big bang, leads to *another* state of the physi-cal universe (ultimately, the current one). In the strict sense, one cannot even say that it shows the current state of the universe to have developed from an earliest state of the *same* universe, because the singularity itself cannot be situated in space-time; according to the model, time first begins with the big bang.

436

Within the physical-cosmological theoretical framework or model, one can neither raise nor respond to any question such as how the singularity itself arose. Hawking puts this point as follows:

> In fact, all our theories of science are formulated on the assumption that space-time is smooth and nearly flat, so they break down at the big bang singularity, where the curvature of space-time is infinite. This means that even if there were events before the big bang, one could not use them to determine what would happen afterward, because predictability would break down at the big bang.
>
> Correspondingly, if, as is the case, we know only what has happened since the big bang, *we could not determine what happened beforehand.* As far as we are concerned, events before the big bang can have no consequences, so *they should not form part of a scientific*

---

[48] This controversy is not documented carefully here. It suffices to introduce two names. For an atheistic interpretation of big-bang theory, see Smith 1991 (65): "If the arguments of this paper are sound, then God does not exist if big bang cosmology, or some relevant similar theory, is true. If this cosmology is true, our universe exists without cause and without explanation." Also of particular importance is Grünbaum (2000).

*model of the universe. We should therefore cut them out of the model and say that time had
a beginning at the big bang.* (1988: 49; emphasis added)

In the physical-cosmological model, the philosophical-metaphysical question concerning the origin or beginning of the universe is not posed—*and indeed cannot be posed*, because that question does not concern simply the relating of one state of the universe to another state of the universe; it concerns instead what—in the Introduction, earlier in this chapter, and in Section 5.3—is termed the *contingent dimension* of being *as contingent*. The question concerns the fundamental status of being or existence, of absoluteness and contingency.

[3] Hawking later changes his position. He comes to maintain that the universe did not arise from a singularity (53). Within the framework of an attempt to combine the two basic physical theories, relativity and quantum mechanics, within a unified quantum theory of gravitation, he includes quantum effects in his considerations; this leads him to the assumption that "the universe has no limits in space and time" (2001: 83). But he and Jim Hartle have discovered that "there is another kind of time, called imaginary time, that is at right angles to the ordinary real time that we feel going by." The histories of the universe in real time and in imaginary time determine each other, although they can be quite different. Hawking and Hartle accept (85) Richard Feynmann's concept of the "multiple histories" of the universe, which are to be thought of as limitless, closed, curved surfaces comparable to the surface of the earth. According to this view, the universe can have neither a beginning nor an end in imaginary time. This position cannot be treated in detail here, but it is to be noted that its implications for a philosophically determined theory of everything can appear to be enormous. Hawking writes,   437

> If the histories of the universe in imaginary time are indeed closed surfaces, as Hartle and I proposed, it would have fundamental implications for philosophy and our picture of where we come from. The universe would be entirely self-contained; it wouldn't need anything outside to wind up the clockwork and set it going. Instead, everything in the universe would be determined by the laws of science and by rolls of the dice within the universe. (85)

In an earlier work, *A Brief History of Time*, Hawking writes as follows:

> So long as the universe had a beginning, we could suppose it had a creator. But if the universe is really completely self-contained, having no boundary or edge, it would have neither beginning nor end; it would simply be. What place, then, for a creator? (1998: 146)

To be noted right away is that Hawking appears to be unfamiliar with the history of the thought of creation. The philosopher and theologian Thomas Aquinas defends clearly and directly the possibility of a creation of an—in the terminology of his time—"eternal world" (i.e., a world without beginning or end). According to Aquinas (see *Eternity*), this involves no contradiction, because creation means only this: the dependence of all contingent being on a (better: the) absolute being. This dependence can be determined more precisely as causality, and indeed as the production of all (contingent) being, not only of the state of being of something or other. Such a dependence or causality in no way implies that the world must have a beginning in time. Aquinas is concerned with dependence and causality only in a metaphysical sense.

Correctly understood, the creation thesis does not contradict the partially ambiguous sentences of Hawking cited above:

> The universe would be entirely self-contained; it wouldn't need anything outside to wind up the clockwork and set it going. Instead, everything in the universe would be determined by the laws of science and by rolls of the dice within the universe. (2001: 85)

438 According to the thought of creation, the *entire* being of the contingent universe *and thus* also the clockwork itself and the entire course of its running (or of the development of the universe) are created. In this sense, there is indeed *no outside* influence that would have to set the clockwork in motion. The creation of the contingent universe could not be anything like the winding of a somehow—but how?—*already present* clockwork. If it were, then it could never explain how the clockwork itself had come to be. If one remains with this terminology, one should say instead that if the contingent universe, understood as a clockwork and the course of its running, is created, then the following holds unrestrictedly: the clockwork is always already wound, the universe is fully enclosed within itself, and it is subject to no outside influence (creation is *not an outside influence*, because it does not concern anything *in* the universe, but instead the universe itself or as such). Hawking's final sentence, in particular, shows just how ambiguous his formulations are: "Instead, everything *in the universe* would be determined by the laws of science and by rolls of the dice *within the universe*" (emphasis added). Hawking's theory concerns everything—but *only* everything—that happens *within* the universe; it does not concern the universe *as such*. Differently put: it does not concern the status of the being of the universe.

Hawking's *additional* clarification (or contention), also cited above—"if the universe is really completely self-contained, having no boundary or edge, it would have neither beginning nor end; it would simply be"—brings to succinct expression a quite widespread way of thinking from which arise notions that obstinantly oppose the idea of the creation of the world. But this way of thinking simply fails to recognize that the "closedness" here in question is determined by the framework of a physical-cosmological model. *Within the framework of this model*, it is completely accurate to say that the universe "would simply be," and in addition that everything in the so-determined universe would be determined "by the laws of science and by rolls of the dice within the universe." But this natural-scientific, physical model in no way exhausts the capacities of questioning and of intelligibility that characterize human understanding. We can—and because we can, we must—ask, how, or in what sense, we can say groundedly or justifiedly that the universe simply *is*, that it is enclosed within itself. We have access to concepts that make what is maintained and understood within the framework of this model appear fully insufficient; these include above all *modal* concepts, which allow us to ask whether the being of this universe is *contingent* or instead *necessary* (or *absolute*). The general tendency of our scientist age is either not to pose or even actively to suppress such questions. In the worst case, they are rejected as senseless. This rejection, however, contradicts the human capacity for intelligibility. This issue is treated thoroughly in Chapter 5.

439 [4] One could acknowledge the modal concepts *necessary* and *contingent* but continue to affirm the thesis of the closedness or simplicity of the universe, understood physically-cosmologically; one would then add the thesis that *the universe itself* is

absolute, indeed *the absolute*. The latter thesis would take the discussion to a new level. How could the physical universe be understood as the absolute? Here, it need only be stressed that, as has already been emphasized, a defensible conception must do justice to *all phenomena*. The physical conception of the universe as *the absolute* does not appear to be coherent.

It is revealing to note that, already within physical cosmology itself, one can find some reflections concerning what is termed just above a *coherent* conception. These reflections concern above all the *anthropic principle* (introduced by the physicist Brandon Carter (1974/1990)). In its strong version, this principle holds that the structurality of the universe (thus, the universe with its natural laws and constants) must be understood in such a way that the universe had to produce life and intelligence at one time or another. This principle is quite controversial within physics, and it raises many questions for philosophers. What does "intelligence" mean here? Is "intelligence" understood here purely physically? Or is "intelligence" understood as Section 4.3.1.2.3.2 shows it to be, as intentional coextensionality with the universe? If the latter, then this question arises: how can "the universe" be understood as producing—indeed, as having to produce—beings that have this physically inexplicable characteristic?

Hawking provides the anthropic principle with an interpretation that, although not univocal, appears to have clearly idealistic features. Concerning Feynmann's thoughts about the multiplicity of histories, he asks,

> What picks out the particular universe that we live in from the set of all possible universes? One point we can notice is that many of the possible histories of the universe won't go through the sequence of forming galaxies and stars that was essential to our own development. . . . Thus, the very fact that we exist as beings who can ask the question "Why is the universe the way it is?" is a restriction on the history we live in. It implies it is one of the minority of histories that have galaxies and stars. This is an example of what is called the anthropic principle. The anthropic principle says that the universe has to be more or less as we see it, because if it were different, there wouldn't be anyone to observe it. (2001, 85–87)

These are thought-provoking formulations. They are based on the implicit premise that there must be someone who can observe the universe. But why should this be so? It sounds like idealistic philosophy. In any case, one would still have to ask, how can beings who can ask why the universe is as it is be explained physically? Such beings, in that they ask this question, reveal themselves to be intentionally coextensive with the universe. Are they then merely or purely physical? To maintain that they were would be scientistic dogmatism (see Section 4.3.1.2.3).   440

4.5.2 *The Phenomenon of the Religious and the Plurality of Religions: The Necessity of a Philosophical Interpretation*

[1] In all societies and cultures of all times, religion has played a central role, and even if Christianity, since the beginning of modernity and particularly at present, has lost the political power it once had in the West, it is not the case that the phenomenon of religion has ceased to influence human life and culture in manifold ways. On the

contrary, even in the West there is at present a new and remarkable perception of the phenomenon of the religious, caused, in part, by the new and unexpected upswing of religion in many parts of the world. This is particularly clear in the case of Islam, a religion with more than a billion adherents that is extremely active particularly in the domain of politics. In India, too, with its population of more than a billion, the many religions are livelier than ever.

Here, this immense domain cannot be considered in detail. Instead, only a few remarks from the perspective of the thematic of this book are necessary and possible.

In a certain respect, every high religion—Christianity, Judaism, Islam, Buddhism, and Hinduism—involves a comprehensive view of the world. More cautiously formulated: it is a fact that the adherents of these religions take themselves to possess comprehensive views of the world. Even if the senses of the comprehensive views of these various religions are not determined more precisely, one should not undervalue the extensive ramifications of this fact.

[2] That there is a phenomenon of the religious cannot be denied, given that the existence of religions has been a fact throughout world history. The first question with which this presents the philosopher is how to interpret this phenomenon. This is a comprehensive question whose clarification requires (or at least is facilitated by) its more precise formulation. One such formulation is the following: where is *a or perhaps the starting point* for the interpretation of the phenomenon of the religious—and thus also of the history and plurality of religions—to be found?

The following paragraphs develop a systematic approach, and do so in two stages. For the sake of clarity, these are designated "approach$_1$" and "approach$_2$." [i] Approach$_1$ emerges directly within the structural-systematic philosophy presented in this book. Negatively viewed, this approach is characterized by proceeding *neither* from the assumption that the religious constitutes its own *a priori* in the sense of a "hidden 'predisposition' of the human spirit" (Otto 1917/1958: 115) *nor* from the assumption that religion is a phenomenon comparable to every other, such that it could or should be interpreted somehow neutrally. Instead, the approach to the religious briefly sketched here understands it as a domain or dimension that is interpretable *only* on the basis of the fundamental characteristic of the human mind: its intentional coextensivity with the universe. The religious is always concerned with the universe, with being as a whole, albeit in a non-theoretical manner.

The phenomenon of the religious (or of religion(s)) is *fundamentally* to be explained on the basis of the fact that the intentional coextensivity with the universe that characterizes the human mind is not, for human beings, some sort of abstract dimension, but is instead something that concretely concerns all of human life in all its facets, particularly with respect to practice. Human beings attempt to understand their place in the universe in their concrete lives and to live their lives in manners appropriate to that place. Given this practical orientation within the context of their coextensionality with the universe, it is fully consequent that human beings not view the universal intentional "space" as structured simply by an ultimate or first principle that is *purely abstract*; instead, they shape their entire lives with specific reference to this comprehensive space insofar as they experience it as determined by an ultimate, highest intelligence and an ultimate, highest will. The religious, or religion, is the expression of this comprehensive experience that is manifested in the entire lives of human beings. This expression takes various forms, some of which are constant among religions, whereas others vary considerably from religion to religion.

All religions designate this highest intelligence and will as "God" or "Gods," depending on whether they are monotheistic or polytheistic. This distinction itself clearly shows that in different religions, ideas about what can be termed the ultimate or highest dimension within the universal intentional space could scarcely diverge more sharply. Indeed, there is probably no other domain within which human beings have developed so many and such disparate forms and contents. These disparities and the history of religions are among philosophy's most obscure and impenetrable topics.   442

[ii] Despite the obscurity and impenetrability of this topic, it is one that philosophy must consider, because it is among the topics that are most central to human life. How can at least minimal fundamental clarity with respect to a defensible approach to its interpretation be attained? The answer lies in the development of the second stage of the systematic approach envisaged here (i.e., in approach₂).

At the beginning, philosophy finds itself confronted with a literally chaotic situation (i.e., here, constellation of issues). It can begin to bring order to this chaos only if it begins by *itself* attaining clarity with respect to the ultimate or highest dimension, which, for religions, is the absolutely central point of reference with respect to what religions call "God." The way in which philosophy interprets this dimension will have a simply decisive influence on the entire philosophical interpretation of religion(s). This is presumably relatively obvious.

There are two distinct ways in which philosophy can develop such an interpretation; one is direct or immediate, the other indirect. The direct way is straightforward examination of the phenomenon of the religious as such, without presuppositions of any sort. As history shows, the attempt to understand the phenomenon of the religious in this direct manner can lead either to a positive or to a negative result: the "God" of religion appears either as an existing being, with all the predicates that religions ascribe (or some specific religion ascribes) to it, or this "God" is considered to be an illusion, a pure projection of the human mind, a mythological construct, or something similar. This direct or immediate approach is extraordinarily problematic, because as a rule it relies implicitly on fundamental but unanalyzed presuppositions both methodological and contentual.

The second, the mediated or indirect way, is the one appropriate to systematic philosophy. Before the phenomenon of the religious is subjected to an analysis of any sort, the basic features of a universal theory, a theory of being as a whole, must be developed. This universal theory then provides the framework within which it is possible to develop a genuine and philosophically founded interpretation of religious talk of "God."

Chapter 5, Section 5.3, develops the basic features of such a universal theory. It includes in particular the thesis that being as a whole is characterized by a metaphysi-   443
cal two-dimensionality: an absolute or necessary and a contingent dimensionality. The former dimension is explained more precisely as absolutely necessary, free, personal being. It is presumably obvious that, on the basis of such a universal theory, the interpretation of the religious, and of religion and religions, can provide quite determinate results.

[3] On the basis of the just-sketched two-stage approach it is now possible fundamentally to clarify a complex of problems in the domain of the religious (quite

especially concerning the high or world religions); the complex belongs within the scope of world-systematics and concerns the assessment of the plurality of religions. As is well known, there are many religions, generally divided into several categories: nature religions, world religions, high religions, polytheistic religions, monotheistic religions, revealed religions, etc. The purposes of this book do not require that it examine in detail the questions arising from this multiplicity. Its purposes do, however, require that it clarify the fundamental question concerning which philosophically transparent *criterion* can decide how these religions are to be assessed, particularly with respect to the question whether one religion can be and indeed must be ascribed a clear and indisputable superiority over the other religions. The remainder of this section sketches a well-founded answer to this question.

Polytheistic religions can simply be excluded, because the assumption of a plurality of gods is not a philosophical possibility. With respect to monotheistic religions, clarification is a complicated task. Here, only the most important points outlining a comprehensive philosophical program need be outlined. The decisive point is that only Christianity has developed a *genuine* theology: one that satisfies the highest demands and challenges of theoreticity. The grand theological tradition of Christianity, which in contemporary philosophy is, unfortunately, largely unknown and/or ignored, should without doubt be included among the greatest achievements of the human mind. Here however it is relevant only to the following question: why is this unique status of Christian theology important and indeed decisive for the development and attainment of philosophically grounded and transparent criteria for deciding the question concerning the assessment of (monotheistic) religions?

The answer should be obvious: because the development of a theology that is genuine in the sense just described shows that the corresponding religion (i.e., Christianity) is prepared to subject its understanding of God (thus, the so-called Christian God) to a *philosophical clarification*; the history of Christian theology shows that Christianity is also *capable of* subjecting its understanding of God to such clarification. This is a factor whose significance for the philosophical assessment of monotheistic religions can scarcely be overestimated, because it yields virtually all of the criteria necessary for providing a rational answer to the question. It is scarcely contestable that within the philosophical perspective developed here, Christianity is the incomparably superior religion.

4.5.3 *World History*

In the two preceding sections, the world as a whole is thematized in two forms: in the form of grand physical-cosmological theories and in the form of the phenomenon of religion(s). There must be talk of two "forms" because in both cases the world as a whole is explicitly thematized, but in two fully different ways. And there is yet a third form of the world as a whole, one that in a certain way encompasses the two already considered along with all the individual domains of the world: *world history*. It is no accident that world history has perpetually been a topic for philosophy and continues to be one today.

World history encompasses all aspects of the phenomenon of history. For the most part, one speaks only of the "philosophy of history," not of the "philosophy of world history." The former designation is not misleading only insofar as in philosophy, at least since Hegel, the most important questions that concern history *as a whole* are seldom treated by philosophers. This section, however, *does* treat the topic of *world* history, albeit within an extremely restricted perspective. It considers only four aspects or issues that clearly belong within a philosophy of world history: (1) the relation of philosophy to world history and to the science of history; (2) the ontology of world history (i.e., determination of the kind of entity that world history is); (3) whether world history has an inner structure; and (4) whether world history has a meaning.

4.5.3.1 Philosophy of World History and the Science of History

This book treats problems concerning the relation between philosophy and the specific sciences in various places (see esp. Section 4.2.1.2); the final two chapters consider additional aspects of this relation. A particularly interesting and difficult case of such a problem is the relation between the philosophy of world history and the science of history. If one considers what historians do, it becomes clear that the borders between philosophical considerations of world history and the work of historians are quite flexible, over an extensive scope.   445

    Among the elementary and absolutely indispensable tasks of the science of history are, most importantly, the following: examining sources, discovering new sources, collecting facts, and sensibly interrelating sources and facts. The first, simplest, and most elementary form of presentation that emerges from this work with sources and facts is the historical *narrative*. On the basis of the sources, discovered facts are presented in the form of a temporal development that appears as a constant and uninterrupted sequence of events. A qualitatively higher level of achievement is attained by the historian who presents an *explanation* of the sequence of events. The explanation presents the events not merely as a temporal sequence, but as linked together, particularly by causal connections.

    Insofar as explanations have more or less local characters, they are unquestionably to be viewed as accomplishments of the science of history, because empirical or specific sciences have the task of conceptualizing and explaining data and their interconnections. The hermeneutic stream in philosophy speaks not of "explanation" but of "understanding." This distinction is often strongly emphasized and even at times seen as insuperable. If however one attempts precisely to explain "explanation" and "understanding" as they apply to the subject matter of history, it quickly becomes clear that the distinction is in no way insuperable. The apparent incompatibility of the two arises only if the two concepts are not clarified. On the one hand, advocates of "explanation" often defend or presuppose a concept of explanation that is far too narrow, such that explanations would have to be understood in terms of the deductive-nomological model or concept of explanation. On the other hand, "understanding" is taken in a purely intuitive sense; it is not subjected to precise critical analysis. More subtly differentiated conceptions overcome these limitations (see Stegmüller *Probleme* I/1969: Ch. VI, VII).

    If however explanations in the domain of theories of history are comprehensive, then the question arises whether they still fit within the domain of the science of history. At the very least, it is scarcely possible to draw a clear line between the science of history and the philosophy of world history. Briefly: a philosophical world history, if it is more than abstract speculation, must be based on the empirical knowledge developed by the science of history, but on the other hand the science of history, precisely because it is a *science*, has the immanent tendency to attempt to explain *all* the facts or phenomena or data, and thus the tendency to expand itself ever more broadly. The following sections, particularly Section 4.5.3.3, consider some of the problems that thereby arise.

4.5.3.2 The Ontology of World History

This section develops the central thought guiding this book's approach to a systematic philosophy of world history. There is talk of ontology because world history is viewed and thematized as a phenomenon within the scope of being as a whole: a systematic philosophical theory of world history must be situated within a universal theory of being.

[1] Correct and adequate understanding and appreciation of this approach require the brief introduction of some of the most important approaches to philosophical conceptions of world history that have arisen in the history of philosophy, up to the present. These can be divided into three approaches that are metaphysically oriented, and three that are not. In what follows, each is characterized briefly.

[i] Classical metaphysics, arising from Plato and Aristotle, does not develop even the starting points for a theory of world history. Only under the influence of Christianity does there develop a grand theological metaphysics that includes various philosophical elements, particularly conceptual ones, whose contental claims are drawn from Christianity. The first comprehensive presentation of this expressly theological metaphysics of history is Augustine's *The City of God*. With respect to content, no other philosophy of world history comes close to comparing with this work. The next most significant philosophical accomplishment in this domain is Hegel's philosophical world history, but even that is largely based on the central contents of classical theological metaphysics.

[ii] A comprehensively metaphysical approach diametrically opposed to Augustine's is the cosmological-materialistic approach, which views world history as an episode in the evolution of the material-physical cosmos. This approach is seldom developed explicitly, but it is presupposed not only by traditional materialists but also—albeit generally only implicitly—by many analytic philosophers. In the Introduction to *Dialectics of Nature*, a book written between 1873 and 1883, Friedrich Engels presents a particularly drastic characterization of this position:

> It is an eternal cycle in which matter moves, a cycle that certainly only completes its orbit in periods of time for which our terrestrial year is no adequate measure, a cycle in which the time of highest development, the time of organic life and still more that of the life of beings conscious of nature and of themselves, is just as narrowly restricted as the space in which life and self-consciousness come into operation; a cycle in which every finite mode of existence of matter, whether it be sun or nebular vapour, single animal or genus of animals, chemical combination or dissociation, is equally transient, and wherein nothing is eternal but eternally changing, eternally moving matter and the laws according to which it moves and changes. But however often, and however relentlessly, this cycle is

completed in time and space, however many millions of suns and earths may arise and pass away, however long it may last before the conditions of organic life develop, however innumerable the organic beings that have to arise and to pass away before animals with a brain capable of thought are developed from their midst, and for a short span of time find conditions suitable for life, only to be exterminated later without mercy, we have the certainty that matter remains eternally the same in all of its transformations, that none of its attributes can ever be lost, and therefore, also, that with the same iron necessity that it will exterminate on earth its highest creation, the thinking mind, it must somewhere else and at another time again produce it. (1873/1940: 24–25)

Purely materialistic metaphysicians of the present of course present the cosmological-physicalist position in different terms, but fundamentally, they retain Engels's basic thesis.

[iii] The third comprehensively metaphysical conception of world history is developed by Hegel. In a certain respect, it is a secularized form of Christian-theological metaphysics. Central to Hegel's conception are the concepts of reason and of world spirit: "The only thought that it [the philosophy of world history] brings with it is . . . the simple thought of reason, that reason rules the world, that world history has therefore proceeded rationally" (*History*: 27/28; translation altered). As concretely ontologically historical, reason is, in Hegel's terms, "spirit." Hegel distinguishes between *absolute spirit*, which he identifies with God, and *world spirit*, which is the principle of world history:

World spirit is the spirit of the world as it reveals itself through human consciousness; the relationship of men to it is that of single parts to the whole that is their substance. And this world spirit corresponds to the divine spirit, which is absolute spirit. Because God is omnipresent, he is present in everyone and appears in everyone's consciousness; and this is world spirit. The particular spirit of a particular people may perish; but it is a link in the chain of the development of world spirit, and this universal spirit cannot perish. (52–53/60; translation altered)

Hegel can thus say, "world history is the record of spirit's efforts to attain **knowledge** of what it is **in itself**" (54/61–62; translation altered). Spirit in itself, according to Hegel, is *freedom*. Thus, "spirit's consciousness of its freedom (which is the precondition for the *reality* of this freedom) has been defined as . . . the ultimate end of the world in general" (55/63). Correspondingly, Hegel "defines" world history as "the progress of the consciousness of freedom" (54/63). **448**

In long disquisitions that are available only in students' transcripts, Hegel attempts to present these grand intuitions in specific detail by recounting the concrete course of world history through its various epochs, peoples, and cultures. The difficulty with this conception lies particularly in the extraordinary vagueness of Hegel's concept *spirit*. Formulations like "spirit's coming together with itself" appear to presuppose that spirit is a mysterious collective quantity.

[iv] Members of the second group of conceptions of world history differ from those just introduced by their strictly non- or anti-metaphysical character. Three are considered here.

The first such view can be termed *transcendental*, not because Kant explicitly presents it, but because it is a kind of *expansion* of Kant's transcendental philosophy to include the topic of history (see Baumgartner 1972, Baumgartner and Anacker 1973).

This position turns specifically against the thought of any objective unity of process or occurrence or development—briefly, against any ontological-metaphysical conception of history. It is a critical theory of history in the sense that it extends the Kantian critique of metaphysical thinking to our knowledge of history. The decisive point is its critique of an understanding of unity that according to this conception produces, by means of the problematic application of the concept of totality, an empty objectivism. One of the consequences of such an objectivism, according to this transcendental position, is that it makes impossible any practical relation between the acting human being and history (see Baumgartner and Anacker 1973: 554).

Despite its critique of historical objectivism, however, this position does not want entirely to abandon unity and coherence. It seeks to retain them in two ways. The first is by taking recourse to linguistic analyses of historical sentences. It assumes that historical sentences have a quite particular structure: such sentences thematize not pure facts and their interconnections under universal laws, but instead interconnections of descriptions of experiences. This means, according to this conception, that historical experiences themselves have the structure of narrative connections. From this, the two authors conclude the following:

> History is founded upon narrative constructions and is thus realized only in the telling of histories. . . . History is thus possible only as the retrospective construction of narratives about past experiences that is in a certain manner arbitrary. The resulting thesis—that neither *the* ultimately valid history of the past nor *the* history as the determinable interconnection of past, present, and future is an idea free from contradiction—appears to lead to the conclusion that it is impossible sensibly to speak of *the* history. (555)

It is clear that Kant's original transcendental view is applied fully here to the case of history.[49]

The second way that this transcendental position attempts to do justice to the unity of history is by its application of the typical Kantian procedure of relying upon the distinction between cognition and (mere) thinking: we cannot *cognize* (or *know*) the unity of history, but we can *think* it. This fundamental Kantian thesis finds its expression in the concept *regulative idea*. The resulting position is the following:

> If one wants to avoid, in contradiction to the inescapable interest in narrating, declaring history, as a particular form of human knowledge, to be trivial or vain and if one is also forced, for reasons of consistency, to reject as contradictory the notion of an objective, universal course of history, then the only possibility is to think of history as a regulative principle that makes possible only retrospective narrative constructions with practical aims. "History" is then the symbol of the idea of totality and unity and retrieves that idea's

---

[49] In the *Critique of Pure Reason*, this view is expressed in such formulations as the following:

> Up to now it has been assumed that all our cognition must conform to the objects; but all attempts to find out something about them *a priori* through concepts that would extend our cognition have, on this presupposition, come to nothing. Hence let us try whether we do not get farther with the problems of metaphysics by assuming that the objects must conform to our cognition. . . . (Bxvi)
>
> [A] light dawned on all those who study nature. They comprehended that reason has insight only into what it itself produces according to its own design. . . . (Bxiii)

transcendental status as an a priori guideline, without which coherent knowledge would be impossible. To be sure, it is a guideline that cannot be thought of as itself a real object or as a factual interconnection of an objective history. (556)

Chapter 5 (see Section 5.1.2) systematically criticizes this position by thoroughly treating the problem of the putative cut or gap between theory (knowledge) and the (objective) system of the world.[50]   450

[v] The second anti- or non-metaphysical position is the one accepted and defended by most analytic philosophers. It consists virtually entirely of the treatment of epistemological problems of historical knowledge; it thereby leaves the distinction between the philosophical theory of (world) history and the science of history for the most part unthematized and thus unexplained. This is not a full-scale systematically philosophical approach. One achievement of the analytic theory of history that should be noted is, however, its thematization of the *relativity* of historical knowledge and thus of philosophical theories of history to implicit metaphysical assumptions and/or pre-suppositions. Thus, in his brief presentation of the philosophy of history, Leon Pompa (1996: 438) writes the following:

[S]uch relativization does not mean that they [i.e., the historian's final accounts] are not warranted ways of thinking of the past. It could only mean this if the historian could escape the temporal and conceptual constraints of his or her situation to achieve some   451
God's-eye viewpoint from which to see the whole. But it is not clear that the idea of such a whole and such a viewpoint is intelligible. The fact that historical knowledge must be from a perspective cannot therefore vitiate it. It does, however, put an end to the old hope of a once-and-for-all universal history, for historical knowledge is not simply aggregative in the way that such a hope requires. . . . The ultimate reason, therefore, why any account as a whole must be thought as revisable lies in the fact that the perspective from which it operates involves revisable metaphysical assumptions.

---

[50] The main text does not attribute the briefly sketched "transcendental conception" of (world) history directly to Kant, presenting it instead as the consequential extension (better: as one of the consequential extensions) of his transcendental position. It is however to be noted that Kant treats world history explicitly in several relatively short essays, particularly an essay of 1784—an essay thus written and published *after* the (1781) appearance of the first edition of the *Critique of Pure Reason* but before the (1787) appearance of the second—"Idea For a Universal World History with a Cosmopolitan Intent." In this essay, Kant presents a conception that appears to be fundamentally independent of transcendental-philosophical assumptions; it is therefore better considered to be an early version of the conception of philosophy that is sketched in the *Opus Postumum*. It is based on nine theses, including the following:

First thesis: All of a creature's natural capacities are meant to develop completely and in conformity with their end. [...]

Second thesis: In man (as the sole rational creature on earth) those natural capacities directed toward the use of his reason are to be completely developed only in the species, not in the individual. [...]

Eighth thesis: One can regard the history of the human species, in the large, as the realization of a hidden plan of nature to bring about an internally and, *for this purpose*, also an externally perfect civil constitution, as the sole state in which all of humanity's natural capacities are developed. (*History*: 18ff/30, 36; translation altered)

This book avoids basic reliance on such terms as "viewpoint" and "perspective"; instead, it uses the fundamental concept of the theoretical framework, which captures and brings to theoretical clarity many intuitions widespread in analytic philosophy (e.g., pluralities of "conceptual schemes" or of "paradigms").

[vi] The third non- or anti-metaphysical position to be considered here is the *hermeneutical* position. Here, only the variant of this position developed by Hans-Georg Gadamer is considered; it is the variant that is at present the most important and influential. At the center of Gadamer's position is the concept of understanding, which Gadamer interprets in an almost excessive manner from the perspective of historicity. According to Gadamer, understanding is always, and essentially, historical. This means, among other things, that the understanding and articulation of history as comprehensive, coherent, and unitary are impossible. The human being, as characterized by understanding, is always already enmeshed within a historical context. Every context determines specific conditions and possibilities of understanding. Gadamer terms the totality of such contextual conditions the *horizon of understanding*, and characterizes it (1960/2003: 304/309) as follows: "The horizon is . . . something into which we move and that moves with us." But the horizons of understanding are not separated from one another:

> In fact the horizon of the present is continually in the process of being formed because we are continually having to test all our prejudices. An important part of this testing occurs in encountering the past and in understanding the tradition from which we come. Hence the horizon of the present cannot be formed without the past. There is no more an isolated horizon of the present in itself than there are historical horizons which have to be acquired. *Rather, understanding is always the fusion of these horizons supposedly existing by themselves.* (306/311)

Elsewhere (290/295) he writes, "*Understanding is to be thought of less as a subjective act than as participating in an event of tradition,* a process of transmission in which past and present are constantly mediated."

From these central assumptions of his theory of understanding Gadamer draws far-reaching consequences with respect to the conception of world history. He summarizes them (299–300/305) as follows:

452   Real historical thinking must take account of its own historicity. Only then will it cease to chase the phantom of a historical object that is the object of progressive research, and learn to view the object as the counterpart of itself and hence understand both. The true historical object is not an object at all, but the unity of the one and the other, a relationship that constitutes both the reality of history and the reality of historical understanding. A hermeneutics adequate to the subject matter would have to demonstrate the reality and efficacy of history within understanding itself. I shall refer to this as "history of effect." *Understanding is, essentially, a historically effected event.*

This view is popular for two reasons. First, it is articulated in relatively ordinary, uncomplicated language that more or less ordinary human beings can understand (or take themselves to understand). Second, it is formulated purely intuitively. No attempt is made precisely to articulate it. The problems to which it leads are thus simply ignored and remain fully unthematized. This is a case of a philosophy in the domain of and governed by the requirements of a highly developed erudition that is far from stringent.

[2] The approach briefly sketched in what follows attempts to do justice both to the empirical-scientific data and to the systematic framework.

[i] Methodologically, world history is not developed on the basis of an absolute (God) or of world spirit (whatever that might be) or from any comparable principle, but begins instead with the empirical-scientific datum that human beings (humanity) come to be at a determinate temporal point in the evolution of the cosmos. This datum is not, however, understood as some kind of *factum brutum*; instead, it is situated within the comprehensive systematic framework. This means that the coming to be of humanity is not viewed and interpreted as a simple physical event like any other, such as the eruption of a volcano, the explosion of a galaxy, etc. Instead, its significance within the cosmos as a whole is recognized explicitly from the outset: the coming to be within the cosmos of the human being, who is *minded*, is the arising within the cosmos of a point that is *intentionally coextensive with the cosmos*. This book considers this fundamental thought in various places. With the introduction into the cosmos of beings intentionally coextensive with it, the cosmic process that is rightly termed *world history* begins.

Many fundamental questions arise here. The most important are the following: How is it possible for such points to come to be? How must the cosmos as a whole be understood for this possibility to be explicable? These and similar questions cannot be avoided. The partial answers provided by cosmological physics are considered in Section 4.5.1. Fascinating though these issues are, they cannot be considered further here.

[ii] The cosmic point just introduced is not a single individual in the usual sense, but instead a *collective*: not a human being, but humanity. The question that this section must pose and answer is the following: what kind of entity is this collective, humanity? The ontological theses developed in this book provide the basis for the following answer: the collective *humanity* is a configuration of configurations; more precisely, it is a configuration of configurations constituting individuals *qua* persons that themselves consist of primary facts. Because individuals or persons are themselves configurations, the collective, humanity, must be determined more precisely as a *third-order* configuration: as a configuration$_3$ of configurations$_2$ of configurations$_1$. This answer is no doubt quite abstract, but in another respect it is concrete in the sense that it characterizes the configuration *humanity* in a genuinely ontological manner: as a particular entity within the universe, within being as a whole.

To be emphasized is that the collective *humanity* is not anything outside or beyond the universe; it is instead a phenomenon—and thus an entity—*within the universe, of the universe*, belonging to the universe, co-constituting the universe. The process of the development of this collective is the process of the development of that extremely complex configuration that has been characterized ontologically as humanity. This process is world history.

This approach differs fundamentally from the cosmological-physical approaches that consider human beings to be ultimately random or accidental products of nature, that do not acknowledge the immense significance of the appearance within the physical

453

universe of minded beings, and that do not thematize this phenomenon. Only if one attends to and acknowledges fully, from the outset, the full constitution of human beings as minded does one have a solid and adequate starting point from which to comprehend world history.

4.5.3.3 Does World History Have an Inner Structure?

[1] The collective *humanity*, as the comprehensive configuration of human beings *qua* individuals/persons, has a univocally ontological structure—that structure that arises from the (already explained in principle) structures of configurations that constitute individual human beings as persons. As such, the collective itself has an inner structure. But how is the factor that configures this collective configuration to be understood more precisely?

454   It is clear from the outset that this configurator cannot be identified with and indeed is highly dissimilar to the configurator that, as shown in Section 4.3.1.2.2, succeeds in configuring human persons as individuals. A collective of individual human persons is not itself an individual person. But what then is it? Answering this question means clarifying what a society (of human beings) is.

Unlike individual persons, collectives of persons are not entities that can themselves think, act, feel, etc. From this it follows that the unity that constitutes such a collective is fundamentally different from the unity that constitutes the individual human person. The configuring factor of the unity of the individual person is termed above the *systematically unifying and self-situating point*," the intersection of the primary facts that define both the concept of the *intentional coextensivity of the human mind with the universe (being as a whole)* and the capacity to say "I." The configurator that configures a collective is of a completely different sort. It further configures those configurations that are already fundamentally constituted by means of the configurators that configure individual persons. The new configurator, which forms a collective, does not have a *systematic status* comparable to that of the configurators that form individual persons. The individual person is definable only *in direct relation to* the entire system (the universe, being as a whole). The collective, on the other hand, is only *indirectly* related to the system—it is related to it only through the elements (the individual persons) who themselves are directly related to the system (the universe, being as a whole).

How is this collective-forming operator or configurator to be understood more precisely? It is understood most adequately and appropriately as the *intersection of all the factors of development* that characterize the individuals/persons. One can proceed on the basis of the assumption that *development* is a factor fundamental to the cosmos and to all of its elements. Individual persons are thus likewise determined by this factor: every individual person possesses the immanent ontological determination of the capacity and necessity of *development*. This *factor of development* is, in this book's terminology, a central *complex primary fact* that relates to—here, (co)determines—*all the other* primary facts constituting the individual person. The configurator that configures the collective can be understood as the *intersection* of the developmental factors constituting the immanent determinations of the individual persons. Formulated somewhat loosely and more generally intelligibly: what produces a collective (a society) and holds it together is the intersection of the necessities and capacities of development of the individual persons within it.

[2] It is now possible to formulate an *ontological determination of world history.*   455
World history is that process that concerns, encompasses, and presents *all* collectives of
human beings, thus *all* societies, *in every respect.* This process is for its part a third-order
configuration: a configuration$_3$ of configurations$_2$ of configurations$_1$. Configurations$_2$
are human collectives and thus societies of all sorts; configurations$_1$ are individual
human persons. World history as a configuration$_3$ is thus a highly complex *processual
primary fact.* How is the configurator that configures history to be determined more
precisely? Purely abstractly, the answer is as follows: the history-configurator is the
intersection of all collective-configurators. But an additional question arises immediately: how is this history-operator more precisely to be determined *contentually?*

[3] The preceding, relatively abstract considerations are now to be concretized.
This concretization must take several forms. One central question, stated somewhat
loosely, is the following: is there anything like one or more "driving forces," or some
comprehensive principle, that determines and structures the process that is characterized as world history? More precisely: how does the history-configurator present itself
contentually and concretely as the intersection of all the collective-configurations that
determine every individual human person and thus all developmental factors?

If one considers the ontological structure of the individual human being as person
and assumes in addition that—as is scarcely deniable—human beings are constituents
of a cosmic process of *evolution*, then it appears to be correct to consider the first thesis
of Kant's "Ideas Concerning a Universal World History with a Cosmopolitan Intent,"
cited above, as an accurate formulation of the fundamental aspect of the developmental factor just introduced: "All of a creature's natural capacities are meant to develop
completely and in conformity with their end" (30/18)—although instead of the easily
misunderstood "natural capacities," it would be better to say, "ontological constitution
or structure." Applied to individual human persons, this means that they configure
themselves both *collectively* and *historically.* Human beings can develop all of their
capacities only as socially organized (only as collectively configured).

The history-configurator is, as shown above, the intersection or focal point of all
the already collectively configured capacities of human beings. Two of the most fundamental components of the history-configurator are the following: knowledge/science
and striving for freedom. These two factors are two fundamental driving forces that   456
lead already collectively configured human beings to form those additional configurations that present themselves as the global process of world history. Indeed, the overall
development of world history, as the process of the "development" (Kant) of all the
ontological capacities of the ontological constitution of human beings, leads to a progressive increase in and refinement of *knowledge*, particularly *scientific knowledge*, and
to what Hegel characterizes accurately as "progress in the consciousness of freedom."
Other, related concepts can also be used to characterize this process, including *enlightenment, independence, self-sufficiency,* etc.

To be sure, these theses remain largely abstract. They require concretization in
many respects, but their wide-ranging consequentiality should not be underestimated,
because they articulate what can be termed the universal and fundamental guidelines
of world history.

[4] A concrete example that illustrates quite clearly this entire state of affairs and
problematic is the much-discussed and now famous book of Francis Fukuyama, *The*

End of History and the Last Man (1992). To be noted at the outset, however, is that the title of the book contains a gross error that is the more to be criticized because it has triggered in many readers utterly false notions and connotations. By "end of history," Fukuyama means only this: the end of the *ideological evolution* of humanity, an evolution that is characterized by strife and tension or, in short, by a clash of competing ideologies. According to Fukuyama, world history is directional: its goal is capitalistic, liberal democracy. He identifies two chief factors—forces—that drive the historical process. The first is natural science, together with the technology it makes possible; together, these lead to increasingly homogeneous cultures. The second motor of history is a thought that Fukuyama adopts from Hegel: the striving for recognition (and thus freedom), which leads to innovation and to personal achievements. The end of history, for Fukuyama, is the culminating point of history, which (again) he understands as the ideological evolution of humanity.

This leads to the question, what comes next? In the second part of his book, Fukuyama considers various aspects of this problem under a rubric that he borrows from Nietzsche: "the last man." He asks whether, in this "last" phase of world history, we all are or will be "last men," complacently self-absorbed. At this point, Fukuyama replaces his previously Hegelian perspective with a Nietzschean one.

457   Fukuyama relies on Hegel in clarifying how the end of history (in his sense) is to be understood; he then uses Nietzsche to show how unsatisfactory this endpoint of the historical process is. What happens when all human beings mutually recognize one another as equal and every struggle except the struggle to attain purely material goods ceases? Fukuyama leaves this question open.

4.5.3.4 Does World History Have a Meaning?

The issue with which the preceding section ends is in effect the question posed by this section's title. "Have a meaning" is generally understood in two ways: first in the sense of having a direction or a goal and second in the sense of being meaningful as a whole, of having a meaning that encompasses both the process (the direction) and the result of the process—the goal or endpoint. If one describes only the goal or endpoint, the additional question arises of how to evaluate such a goal or endpoint. Fukuyama is concerned with the latter question when he introduces the concept of the last man.

4.5.3.4.1 *Preliminary Clarifications*

First, the concept *meaning* must be clarified more precisely in this context. One could understand "meaning" as whatever is understandable, conceptualizable, or explainable (in a broad sense). But what does "understandable or conceptualizable or explainable" mean? The short answer is that what is understandable or conceptualizable or explainable is anything that, in whatever manner, exhibits a *coherence*, and thus involves and manifests a relation to something or, ultimately, to everything else. But that does not sufficiently explicate the sense intended with the use of "meaning" in the question currently under consideration, because there is a *second* essential component of the concept *meaning*: the understandable or conceptualizable or explainable (whatever it is that is coherent) must prove to be something *univocally positive*, in the sense of something that is to be affirmed, something that is to be striven for. This second component is often simply identified as a "goal," but one can use the term "goal," in this context, only quite cautiously—in such a way that it designates *not* the *endpoint* of a process *in independence from* all the elements or stages of the process, but instead the endpoint *along with* all the elements and stages, thus, briefly, together with the ways or means   458 to the end. "Meaning" in this sense concerns the totality of some development, some process, etc. Does history have such a meaning?

It should be clear that Fukuyama's conception, briefly presented above, can in no way be seen as an answer to this section's question as explained in the preceding paragraph, no matter how many individual structures, regularities, etc., it may illuminate within world history. In addition, various other attempts to explain individual interconnections, predict future events, etc., are sensible and interesting, even if they prove to be not particularly accurate. But the question concerning the meaning of (world) history concerns history *as such*, history *as a whole*.

[2] One can distinguish between two quite different types of theories about the meaning of world history: *restricted* and *comprehensively systematic* (hence *unrestricted*) theories.

Restricted theories of world history consider world history as a self-contained whole that bears its meaning within itself, independently of whatever status it may have within being as a whole (the universe, the cosmos). One such theory is the Marxian theory of *historical* (*not* dialectical) *materialism*, which understands history as the process leading to and arriving at the stage of the classless society that is a socialist paradise (to be sure, this "historical-materialistic" conception is usually presented together with the philosophy termed "dialectical materialism," because dialectical materialism is a comprehensively systematic materialistic theory through which historical materialism—and thus also the historical-materialistic theory of world history—attains a comprehensively systematic status). Generally, philosophical theories of world history—and especially analytically oriented ones—when they are developed at all, remain theories that are restricted in this sense.

Comprehensively systematic theories of world history at least implicitly involve one or another comprehensive metaphysics: world history is understood and thematized as an occurrence within the framework of being as a whole or of the universe. An example of such a theory is the dialectical-materialistic one formulated by Engels (see the passage quoted on pp. 334-35). Every materialistic conception of actuality implies a theory of world history of this type, although it is to be noted that scarcely any analytic philosopher in the materialist camp formulates even the starting points for such a theory. Scientists (cosmological physicists) who extrapolate philosophically from their   459 particular physical theories likewise fail to articulate, in the comprehensively metaphysical sense, the universal frameworks of their theories of history.

An interesting example of such a physically-cosmologically oriented theory of world history is the one presented by Stephen Hawking. According to Hawking, physical cosmology strives to become a "Theory of Everything" that "will govern the universe and everything that it contains" (2001: 175). Moreover, Hawking asserts, "we may have already identified the Theory of Everything (ToE) as M-theory." According to Hawking, the M-theory is not a single theory, but consists of "a network of apparently different theories that all seem to be approximations to the same underlying fundamental theory in different limits." The M-theory unites the various string theories within a superordinate theoretical framework. It appears to have eleven spatiotemporal dimensions. Hawking compares the M-theory to a puzzle whose outside pieces, which are small with respect to the specific parameters of the theory, we already know relatively well. What we do not know is what is in the middle of the M-theory puzzle; there, "there is still a gaping hole."[51]

Hawking's assertions are the strongest, the most ambitious, but also the most problematic that any serious and widely respected natural scientist has ever made. The question arises concerning how they are to be evaluated philosophically. Taken literally, many of his claims are scarcely defensible philosophically, as shown in Section 4.5.1. Many of his formulations are simply utterly ambiguous—for example, M-theory "will govern the universe and everything it contains." What does "universe" mean here? Is it the physical universe, and only the physical universe? Hawking appears to include not only the physical domain and not even only the domain of life, but even the domain of mind (of thinking, of intelligence) when he says "and everything it contains." He appears to understand "life" and "intelligence" simply as extremely complex physical entities. The preceding sections of this chapter reveal, however, that this is a deep misunderstanding.

If we presupposed *hypothetically* that the universe, understood as the totality of what is usually termed inorganic matter, life, and mind, could be "governed" by means of the completed M-theory, where would *we* then be? And above all, *how* would *we* then be there?

As a rule, we—we normal human beings, we individuals, politicians, teachers, etc.—daily live in situations within which we have to make decisions on which, at times, a great deal depends, concerning both our own lives and the lives of others. Assume that we understood the comprehensive M-theory and accepted it unrestrictedly. What effect would this have on us in the specific situations in which we would continue to find ourselves? We would then *know* that everything, including the acts that follow our decisions, is always already governed by laws discovered by natural science. We would be faced with decisions, but would "know" that those decisions and the acts ensuing upon them would be "governed" not by us but by purely physical processes. What *sense* would it then make for us to act? If we knew in advance that our acts had complete natural-scientific explanations, then decisions would be pure illusions, albeit illusions that themselves would be explained fully by the M-theory. If we were convinced of

---

[51] Briefly to be noted is that string theory is the best but not the only candidate to be a theory of quantum gravitation. At least one other (less lavish mathematically) model is attracting increasing attention and respect; this involves so-called loop quantum gravities. This topic cannot be considered further here. Hawking's "suggestions," still quite controversial among physicists, are considered here only as an example of a physical-cosmological "theory of everything." Today, Hawking is the unusually popular natural-scientific cosmologist who has made the most comprehensive—but also the most radically mistaken—assertions about the beginning or the origin of the universe. It can *in no way* be the task of philosophy to investigate such assertions *as physical-cosmological* ones or to accept or reject them as assertions *about physical cosmology*. The task of philosophy lies instead in (among other things) pointing out the unclarities, ambiguities, and extrapolations beyond physics that easily creep in when physical-cosmological formulations are used carelessly and without scrupulous differentiation. The account in the main text is devoted to this task.

that—if we knew that—then all talk of freedom and responsibility would be nonsense, mere blather. What this shows is the following: confronted with the unavoidability of deciding to do one thing or another, we would all, in practice, reject or indeed refute, *by our acts*, the thus-understood, putatively all-explaining M-theory.

This would be the case even if one acknowledged that the M-theory did not exclude various indeterminacies (see the reference above to the "particular universe that we live in"), because such indeterminacies would in no way suffice to make intelligible or to do justice to the (usually daily) situations within which we have to make decisions. The reason is that such indeterminacies cannot explain the basic fact that the decisions are *our* decisions, for which we are responsible—they are not the results of chance occur-   461 rences, however these may be understood, because if they were, they simply would not be *ours*.

Non-materialistic theories of world history as a rule contain in various ways, as noted above, *religious* elements. The most important are those inspired by Christianity. Among the purely philosophical non-materialistic theories of world history, Hegel's is unquestionably the most significant.

4.5.3.4.2 *Reasons for the Necessity of a Comprehensively Systematic Theory of World History*

Why is a *restricted* theory of world history, in the sense explained above, insufficient? And why is a comprehensively systematic theory of world history necessary? This section does not present the reasons directly and abstractly; instead, it develops them somewhat slowly, in confrontation with the remarkable reflections of Thomas Nagel, an analytic philosopher who has posed and responded to the question concerning the meaning of life. This approach makes possible a more concrete treatment of the issues at hand.

[1] Nagel devotes the final chapter of his 1987 book *What Does It All Mean?* to the topic, "The Meaning of Life." As is now to be shown, his reflections are directly applicable to the issue of the meaning of world history. His beginning point is the following insight: "The problem is that although there are justifications and explanations for most of the things, big and small, that we do *within* life, none of these explanations explain the point of your life as a whole—the whole of which all these activities, successes and failures, strivings and disappointments are parts" (96). To the objection that we must simply be satisfied with what we do "within life," he responds as follows:

> This is a perfectly good reply. But it only works if you really can avoid setting your sights higher, and asking what the point of the whole thing is. For once you do that, you open yourself to the possibility that your life is meaningless. (97)

The problem Nagel identifies is that being satisfied only with things "within life" leaves a *broader* question unanswered. The broader question is the following: if one considers life as embedded within a broader context, does it have a point, or not? Nagel responds with a course of argumentation that requires consideration in detail, because for the   462 most part it is vague and intuitive in a manner that is characteristic of most human beings *and of most philosophers*. This detailed consideration proves extremely useful with respect to the topic of this section.

Nagel maintains that "If you think about the whole thing, there seems to be no point to it at all" (96). His argument for this contention is a variant of an argument leading to an infinite regress:

> If one's life has a point as a part of something larger, it is still possible to ask about that larger thing, what is the point of *it*? Either there is an answer in terms of something still larger or there isn't. If there is, we simply repeat the question. If there isn't, then our search for a point has come to an end with something which has no point. But if that pointlessness is acceptable for the larger thing of which our life is a part, why shouldn't it be acceptable already for our life taken as a whole? Why isn't it all right for your life to be pointless? And if it isn't acceptable there, why should it be acceptable when we get to the larger context? Why don't we have to go on to ask, "But what is the point of all *that*?" (human history, the succession of the generations, or whatever). (98)

Nagel extends the argument by considering the assumption of an ultimate meaning or ultimate point. His argument continues to apply, he contends, because we can ask,

> "And what is the point of *that*?" It's supposed to be something which is its own point, and can't have a purpose outside itself. But for this very reason it has its own problems. (99)

Strangely, Nagel considers the assumption that life has an ultimate point to involve the attribution of a "religious meaning to life" (98). That involvement is absolutely to be rejected. From the beginnings of philosophy, questions concerning what is ultimate and absolute have *always* been *philosophical* questions. That the word "God," characteristic of religion(s), is sometimes used when these philosophical questions are addressed is based on a deep misunderstanding, as shown in Chapter 5 (Section 5.3). The facile application of the designation "religious" to characterize questions and theses concerning what is ultimate is the more regrettable in that it tends to arouse an extensive series of negative psychological connotations and emotions that as a rule simply hinder rational discussion or even make it impossible. The following discussion of Nagel's position does use the word "God," but only in connection with his formulations. In strict opposition to Nagel, however, the word is here used *purely philosophically*.

Against the assumption of an ultimate meaning (which he terms "God"), Nagel raises two objections. First (99), he proceeds from the assumption that the idea of (a) God "seems to be the idea of something that can explain everything else, without having to be explained itself." Nagel maintains that it is "very hard to understand how there could be such a thing," and then raises the following questions:

> If we ask the question, "Why is the world like this?" and are offered a religious answer, how can we be prevented from asking again, "And why is *that* true?" What kind of answer would bring all our "Why?" questions to a stop, once and for all? And if they can stop there, why couldn't they have stopped earlier?

The second objection is based on a fact that, according to Nagel, leads to the same problem. The fact is that God and his plans or intentions are presented as the absolutely "ultimate explanation of the value and meaning of our lives." Nagel continues as follows:

The idea that our lives fulfil God's purpose is supposed to give them their point, in a way that doesn't require or admit of any further point. One isn't supposed to ask "What is the point of God?" any more that one is supposed to ask, "What is the explanation of God?"
   . . . Can there really be something which gives point to everything else by encompassing it, but which couldn't have, or need, any point itself? Something whose point can't be questioned from outside because there is no outside? (99–100)

[2] The following paragraphs show that Nagel's two arguments, which are the basis of all his other reflections, are based on misunderstandings.

   [i] The first argument is based on the thought that at every point whatsoever, new why questions arise; they would thus arise also at a or the putatively ultimate point, however that point might be termed or understood, but this shows the putatively ultimate point not to be ultimate. This kind of argumentation—which is quite common—depends upon a simple and peculiar lack of understanding and of thematization of the logic, semantics, and ontology of why questions. The argument depends upon the naive assumption that in every context and with respect to every state of affairs (every entity, object, question, concept, thesis, etc.), why questions appear   464
mechanically or can be posed or repeated. To be sure, why questions *can* be generated mechanically everywhere and at all times, but closer analysis shows this to be a senseless abuse because it is incoherent. It is incoherent in that it simply ignores the presuppositions that make such questions possible and sensible. This is now to be shown briefly.

   A question is sensible only if it has a clear target; otherwise, the question is undetermined and in fact asks nothing. Such a target is, however, for its part possible only on the presupposition of a specific context or framework, precisely because what the question asks for is the relation between its target (however specified: object, state of affairs, etc.) and *something else*. In brief, to ask "why?" about something is to ask that that something be related *to something else*, which is a reason or ground. A why "question" that did not ask for this would not be a question but instead an undirected, pointless, and thus senseless attempt to get to something completely indeterminate and empty. Indeed, a genuine and sensible why question about something, as the attempt to discover how that something is related to something else (a reason or ground), presupposes that the questioner has raised the question within a theoretical framework that explicitly contains such a relation. In other words, the question is based on the presupposition that a (theoretical) framework, consisting at least of a logic, a semantics, and an ontology, is already in place.

   From this basic fact there follows something that is highly consequential: questions asking "Why?" cannot be repeated or iterated arbitrarily because they would soon escape the theoretical frameworks within which alone they make sense; they would be undirected and thus senseless, no more than empty plays with words. This case is wholly analogous to that in which demands for grounding or self-grounding are made utterly independently of theoretical frameworks. As this book shows in various contexts, sentences are meaningful only within theoretical frameworks. This holds also for interrogative sentences.

   To Nagel's question, "What kind of answer would bring all our 'Why?' questions to a stop, once and for all?," the answer is the following: the answer that brings all our why questions to an end is the one that understands correctly *all why questions* and interprets them as questions that are meaningful only within specific theoretical frameworks. The case of a why question concerning an ultimate point is then interpreted as follows: if an X is understood as an ultimate point, then the question concerning the why of this ultimate point is utterly empty *if* it is understood as the demand for *something more*, an additional reason or ground to which the putatively ultimate point would be related. This "something more" would be something *ex hypothesi beyond* or *outside of* the presupposed theoretical framework, because there could be nothing *within* the theoretical framework beyond that ultimate point. The introduction of any such thing into the framework would be the introduction of a contradiction. Differently stated: the articulation of an ultimate point on the basis of an accepted framework stops every iterative, mechanical posing of why questions. Why questions make sense only within frameworks. If all of the potentialities of a given framework are taken into consideration by the articulation of an ultimate point, the process of posing why questions comes to an end in the sense that it turns upon itself: the why question concerning the ultimate point is answered by the ultimate point itself. If the why question concerning the ultimate point is not understood in this way, then it simply breaks down, because there is nothing that could be its answer.

This state of affairs is helpfully comparable to the mathematical concept of the *fixed point*, which can, oversimplifying to the extreme, be presented (for the case of ordinal numbers) as follows: let F be an operation that maps ordinal numbers to ordinal numbers. Fixed points of F are ordinal numbers α with $F(\alpha) = \alpha$. The application to the philosophical case is then clear: the operation F corresponds to the concept "why" or "the reason for"; α corresponds to the "ultimate point." $F(\alpha) = \alpha$ is then to be interpreted contentually as indicating that the reason for the ultimate point is the ultimate point itself.

In another respect, the ultimate point can and indeed must be put into question. Concretely, the question "Why this ultimate point?" is not understood as, "What is the ground or basis for this ultimate point itself?" but instead as, "Why *this* ultimate point?" The sense of this question is the following: is the point that has been determined to be ultimate actually the *adequately* understood or determined ultimate point?

This question is completely legitimate. But it then means precisely the following: the question, correctly formulated, problematizes one of the following two possibilities: *either* the question concerning the ultimate point is not posed correctly or clarified *within* the presupposed framework, *or* the presupposed theoretical framework *itself* is inadequate and therefore must be replaced by one that is more adequate or more comprehensive. This second possibility is the one that is genuinely interesting philosophically, because the most fundamental questions and the most decisive divergences arise as a rule from differences between theoretical frameworks.

In order, however, for the question, "Why *this* last point?" not to prove yet again to be an undirected and thus empty play with words, it must be ascertained that in this ultimate case (i.e., in the case that requires the development or the assumption of a broader, more comprehensive theoretical framework), such a broader, more comprehensive theoretical framework is *actually* presupposed *and explicated*, or at least that it *can be explicated*. Then, however, the mechanical procedure of asking "Why?" within this broader, more comprehensive theoretical framework comes to a standstill, as shown above.

If nevertheless the question "And why *this* (additional) last point?" is posed in a *meaningful manner*, the question is actually meaningful only if *a broader, yet more comprehensive* theoretical framework is or can be envisaged *and* presupposed. But can this procedure of envisaging or discovering or projecting ever broader, more comprehensive theoretical frameworks at all be maintained? At first glance, this appears to be impossible. In fact, we human beings cannot simply envisage or discover or project ever more—in principle, endlessly many more—theoretical frameworks *that are meaningful*; the thought that we could is an illusion, an empty notion or contention. Nothing prevents us, as human beings (philosophers, theoreticians), from remaining *open* to additional theoretical frameworks or from attempting to envisage, discover, develop, or project them. Our remaining thus open is in complete accordance with the basic theses defended in this book concerning the central thought of the theoretical framework. As shown in the concluding section of Chapter 3 (Section 3.3.4.3), this remaining open in no way entails an inconsistent relativism. Also to be kept in mind in this connection is that the broader, more comprehensive theoretical framework in question must prove to be *more adequate* than its predecessor. Even if it does, however, that does not entail that the new theoretical framework would have to negate, in every respect, everything characterizing the preceding framework or frameworks; instead, the relation between the frameworks is that of relatively *greater adequacy*. This implies a thoroughgoing and fundamental commonality between the theoretical frameworks with respect to central factors. Thus, *all* theoretical frameworks, both actual and possible, are structured by means of the grand and fundamental thought of coherence, even if "coherence" is not understood in all in precisely the same way. To all belongs as well—and essentially—the perspective of an *ultimate factor,* or a last point.

[ii] On the basis of what has just been said concerning Nagel's first argument or objection, Nagel's second argument or objection is dealt with easily. The objection says that an accepted ultimate point or God, as putatively ultimate explanation for anything and everything, itself requires explanation and thus the introduction of something else. The response to this objection should be obvious: here again, the central importance of the theoretical framework is simply and wholly ignored. The operator "is an explanation for" has, methodologically, precisely the same status as the operators, thoroughly considered above, "Why ... ?" and "The reason for ... is. ...." *Within* a presupposed theoretical framework, the ultimate point explains itself; it is simply a (better: the) point that is fixed systematically. However, the fixed point explains itself precisely in that it explains anything and everything: it is the coherence that constitutes the interconnection of everything, itself included.

Nagel's above-cited question, "Can there really be something which gives point to everything else by encompassing it, but which couldn't have, or need, any point itself?" is ambiguous. As shown above, it is incoherent to say or to assume that the ultimate point within a theoretical framework T requires or is subject to explanation *outside of* the accepted or presupposed theoretical framework T.

To Nagel's additional question, "[Can there really be] something whose point can't be questioned from outside because there is no outside?," the answer is the following: this is not only possible, it is necessary, *presupposing* that one understands "outside" as "outside the theoretical framework being used" *and* that one interprets "because there is no outside" as meaning "there is nothing outside *of the accepted and presupposed theoretical framework, as long as* one accepts, uses, or presupposes *exclusively* this theoretical framework." This thesis is explained and grounded thoroughly above.

[3] From this confrontation with Nagel emerge several insights important to clarifying this question: Isn't a philosophical theory that is restricted, in the sense introduced above, enough? The answer is no, because such a theory would understand world history as an isolated phenomenon within the universe, within being as a whole, one related only to itself. In other words, a restricted theory of world history could develop only a quite limited, relative meaning for world history, and this directly contradicts the principle of the unrestricted intelligibility of all things and thus also of world history, of the universe, of being as a whole. In Nagel's well-targeted formulation, such a theory, and the restricted meaning it would accord to world history, "only works if you really can avoid setting your sights higher, and asking what the point of the whole thing is" (97). Only an internal or external prohibition could lead to the rejection of the attempt to develop a comprehensively systematic theory of world history. Such a prohibition, however, is not an argument. Answering the question concerning the meaning of world history presupposes a comprehensively systematic theory of world history.

To avoid misunderstandings, it must be added that the "relativity," described above, to whatever theoretical framework is being used is in no way in conflict with the philosophical theory of world history being *unrestricted*. The relativity in question brings with it the factor of the *degree of adequacy* of the theoretical framework, whereas the unrestrictedness of the requisite theory of world history concerns the *thematic extent* and the *thematic completeness* of the theory. In brief: unrestricted theories of world history, in the sense intended here, can be developed within more and less adequate theoretical frameworks.

Now, however, the following question arises: Under which (at least minimal) necessary conditions is it possible and sensible to develop a positive, non-materialistic, comprehensively systematic theory of world history? The following section seeks to clarify only this preliminary question.

4.5.3.4.3 *Presuppositions for a Comprehensive Systematic Theory That Clarifies the Meaning of World History*

[1] World history, as a third-order configuration (see Section 4.5.3.2[2], pp. 339–40), exists only as long as there are configurations constituting individuals or persons. A comprehensive systematic theory of world history therefore depends *decisively* upon how one conceives of individuals or persons within the universe as a whole. If within the universe they are purely ephemeral appearances, then world history itself is an ephemeral episode, as Engels describes it impressively in the text quoted above. In this case, world history also has a purely ephemeral meaning or significance.

It must be emphasized that theories of world history that are restricted in the sense introduced above can and must *always* explicitly maintain meanings for world history. But such theories can never coherently accept any non-ephemeral meaning because by definition they lack explicitly metaphysical theoretical frameworks, and such frameworks are unconditionally required if non-ephemeral meanings are to be proposed. An additional qualification is, however, necessary. Restricted theories of world history must be divided into those that either reject the metaphysical question or leave it wholly open, and those that defend or presuppose—implicitly or explicitly—one or another metaphysical theoretical framework, but one within which an unrestricted theory of world history cannot develop. Materialistic theories of all types are examples of the latter. Materialistic theories are metaphysical in the sense presupposed here (i.e., they are theories about the totality of actuality or of being): such theories reduce absolutely everything, one way or another, to matter.[52] Such theories are therefore comprehensive theories, theories that are in no way restricted. *Because of their materialistic character*, however, such theories can accord to world history only purely ephemeral meanings. There is thus, in a certain respect, a paradox: these are metaphysical theories, thus comprehensive theories concerning the totality of actuality, that contain *and entail* only *restricted* theories of world history. To avoid misunderstandings, this must be understood as follows: any comprehensive materialistic theory can present and defend only a restricted meaning for world history. More precisely: materialistic theories are metaphysical, and therefore comprehensive and unrestricted, but ones that must necessarily understand world history as a purely ephemeral episode. Almost all analytically oriented philosophers maintain such theories, implicitly or explicitly, because all, at least implicitly, presuppose or accept ultimately materialistic fundamental frameworks. Until recently, however, the best-known variant of such a materialistic theory has been *dialectical materialism*. One of its main champions, Friedrich Engels, clearly formulates the restricted meaning of world history in question here in the passage cited above, which reads, in part,

> [W]e have the certainty that matter remains eternally the same in all of its transformations, that none of its attributes can ever be lost, and therefore, also, that with the same iron necessity that it will exterminate on earth its highest creation, the thinking mind, it must somewhere else and at another time again produce it. (1873/1940: 25)

The question here must therefore be, are individual human persons absolutely contingent beings in the sense that with death they simply cease to exist in every respect, that they therefore disappear from the world's stage ("world" here in the comprehensive sense: being as a whole) and thus from history? Contemporary philosophy exhibits the peculiar practice of even mentioning this question only in conjunction with religion (see Nagel 1987: 98ff, 2001: Ch. 7). To religion or religions (and quite particularly to Christianity), it ascribes the thesis (or, as is generally said, the belief) that the individual person has a life after death. Thereby wholly ignored or even suppressed is the fact that the theory of the immortality of the soul or the mind is among the very oldest *purely philosophical* theories—it is discussed already in the time of Plato. Contemporary philosophy remains virtually utterly and thus quite strikingly silent on this topic. Analytic philosophy does all that it can to avoid it.[53]

For a philosophy that takes its task and its subject matter seriously, the question is unavoidable. Moreover, for the topic under current consideration—that of the meaning of world history in a comprehensive sense—it is simply fundamental. At the

470

---

[52] It must be noted here that many theories that are currently either explicitly defended or implicitly presupposed are "materialistic" in quite vague senses. This issue cannot be considered further here.

[53] When—as is extremely rare—an analytic philosopher says anything at all on this topic, what is said is as a rule on so low a level that it scarcely deserves designation as "philosophical." An example is provided by Chapter 6 of Bernard Williams's *Problems of the Self* (1972). The chapter is titled, "The Makropulos case; reflections on the tedium of immortality."

same time, to work out a conception of it in detail is an extraordinarily difficult and complex task. Quick-fix solutions accomplish nothing. The specific nature and goals of this book allow for and require presentation only of the kind of framework within which a positive solution to the problem could develop. Some requisite aspects of the relevant kind of framework are sketched by means of the informal presentation of two arguments (better: of the schemata of two arguments); each argument concludes that individual human persons do not, with death, simply cease to be or to exist.

[2] The first argument is not conclusive in the strict sense, but is an inference to the best explanation and thus an exhibition or accomplishment of a better or higher intelligibility and thus coherence. The value of such arguments, in philosophy, should not be underestimated, as should be clear from various passages in this book.[54] The argument brings together the considerations introduced above in the discussion of Thomas Nagel.

471   One of the most important premises is the requirement, demonstrated above, that comprehensive systematic accounts address the question concerning the ultimate meaning of world history. Avoiding this question would require the utter suppression of one of our intellectual capacities, but suppressing that capacity does not make it disappear; it instead confirms its existence. That our minds have this capacity establishes that they have no physical limits, that they are intentionally as vast as the universe itself.

A second premise is the decisive one, but it is here presented only quite cautiously and tentatively. It is scarcely to be conceived that an entity with the intellectual constitution just alluded to should simply cease to exist as an individual entity.

This thought has many aspects. Here, only one is presented explicitly, and that one only briefly (additional aspects are considered below in conjunction with the second argument). With the constitution of the human being as minded, a grand mental dimension opens; this term and concept designate what are often termed the intellectual world (the world of science, of theories, of ideas, of formal structures, etc.), the ethical world (the world of norms and values), the cultural world (the world of art, literature, of human works in general), the social world (the world of society, of the state, of right), the world of religion, etc. In each and all of these worlds, humanity expresses the constitution characteristic of the mind, its intentional coextensionality with the cosmos (the universe, the world as a whole). These worlds do not disappear with the deaths of individual human beings. Would it be genuinely meaningful if the human being, as creator of all these (non-physical!) worlds, simply ceased, with death, to exist? What could the meaning be? If the human being were simply to disappear, the only intelligible interpretation would be that the dimension characteristic of the mind was an immense imaginary, illusory, epiphenomenal one, without any real value: it would be something absolutely ephemeral, as would be the human being. To put it cautiously: this *appears* to yield no meaning.

The plausibility of the assertion just made can be supported in various ways. Here, exemplarily, only one is briefly articulated. If the human being, upon dying, were simply to disappear in every respect, then there would be much that would be inexplicable; particularly, it would be some kind of miracle (or absurd phenomenon, which comes to

---

[54] A particularly interesting example of a quite informative and successful application of this method to a similarly basic and comprehensive philosophical topic is Forrest (1996: esp. 26–35, 41–42, 117–21).

the same thing) that the human being and that humanity have developed and continue to develop so magnificently, that human beings (at least often, but indeed usually) have expended and continue to expend so much creative energy developing the intellectual or spiritual dimension described above. Where one winds up if one does not consider this point, or does not consider it adequately, can be read from the last two pages of the book of Thomas Nagel cited and discussed above, *What Does It All Mean?* Nagel there confronts the phenomenon that he characterizes as "an incurable tendency to take ourselves seriously. We want to matter to ourselves 'from the outside.'" He comments further, "[I]f we can't help taking ourselves so seriously, perhaps we just have to put up with being ridiculous. Life may be not only meaningless but absurd" (101). 472

This is a characteristic example of ignoring the ontological level, the ontological intellectual constitution of human beings. Nagel simply psychologizes an enormously dubious phenomenon. The problem is not psychological; it is ontological.[55]

[3] Unlike the first argument, the second argument *is* conclusive. Its starting point is the more narrowly understood ontological constitution of the human individual or person, and it is a variant of the argument presented against physicalism in Section 4.3.1.2.3.2. This argument can best be viewed as a fully new and fully revised version of the central argument for the so-called immortality of the soul presented most frequently in the grand metaphysical tradition. In its traditional form, the argument is based on the distinction between immaterial and material beings, a distinction traditionally held to be unproblematic. Material beings are traditionally understood as beings that can decompose into their (material) components and that in fact one way or another come in every case to be decomposed. Their disappearance is understood as this decomposition. Because the immaterial being ("the soul," in the terminology of and in accordance with the conception of traditional metaphysics) is characterized by the ontologically fundamental property of *simplicity*, it cannot decompose; therefore it cannot cease to exist. The death of the human being is then explained as the separation of the soul from the body, and grand speculative cogitations are devoted to the "essence" of the "separate soul." This is not the place, however, to consider this conception further. 473

------

[55] Nagel himself is an excellent critic of the conception articulated (or position described) in this passage. In a later book, *The Last Word* (1997), he argues successfully against "evolutionary naturalism," which he considers to result from "the ludicrous overuse of evolutionary biology to explain everything about life, including everything about the human mind" (131). To evolutionarily naturalistic "explanations" of reason, he objects as follows:

> I have to be able to believe that the evolutionary explanation is consistent with the proposition that I follow the rules of logic because they are correct—not *merely* because I am biologically programmed to do so. But to believe that, I have to be justified independently in believing that they *are* correct. And this I cannot be merely on the basis of my contingent psychological disposition, together with the hypothesis that it is the product of natural selection. I can have no justification for trusting reason as a capacity I have as a consequence of natural selection, unless I am justified in trusting it simply in itself—that is, believing what it *tells* me, in virtue of the *content* of the arguments it delivers. (135–36)

This on-target line of argument makes clear that our "taking ourselves seriously" (or not) cannot ultimately be anything like a "psychological disposition," but depends instead exclusively on the standpoint of objectivity (thus, on whether the relevant arguments are valid in terms of content, etc., not on whether "we like them," etc.).

The distinction between material and immaterial beings, as found within the traditional metaphysical conception, today appears to be too simple in that it either presupposes or implies a dualism that is deeply problematic in various respects. That distinction is in no way the basis of the conception, presented in this chapter, of the human being as minded individual or person. Here, the human being is understood instead as a component of the *world* (of the cosmos, of the universe) such that the constitution of the human being *as minded* is itself understood as a component of the world, albeit not a physical one. The argument is similar to the traditional one in that its key premise introduces a fundamental ontological distinction between human beings and all subhuman entities.

In this respect, the human being is that item of or in the world that is essentially characterized as being intentionally coextensive with the entire world, with the universe. This thought arises and is treated in this book in various contexts. If one comprehends this intentional capacity in a genuinely *ontological* manner, it becomes clear that it cannot simply disappear from the world (in the sense specified above); it is not conceivable that this could happen as if the human being were something like a disappearing galaxy or anything else of that sort. No galaxy is in any way coextensive with the world, the universe. What makes the significance of this thought so hard for philosophers to acknowledge and to appreciate is their tendency to understand intentionality (thus knowledge, truth, etc.) as a purely "subjective state" of the brain (in the head, etc.) of the human being. This understanding not only utterly fails to provide an ontological understanding of intentionality but is in addition truly bizarre in that one of its consequences is that it must appear as a *miracle* that we human beings, putatively somehow trapped within our skulls, (can) *actually know* the *actual* things in the world, the *actual* galaxies in the universe, no matter how distant from us they may be; that we *actually know* them entails that we can *reach* them. That we do indeed reach them is on the one hand utterly familiar to us, but on the other quite astonishing, when we reflect upon it. It is a fundamental fact that philosophy must explain, not explain away.

Beings constituted as minded cannot, given that constitution as described above, simply disappear from the cosmos (the universe, being as a whole), cannot simply cease to exist. The minded entity, as a systematically unifying point within the cosmos (in the sense specified in Section 4.3.1.2.2.3), can of course undergo transformations of various sorts. It can, for example, hone its intellectual capacities, increase its knowledge, concretize the dimension of its will by acting, etc. But no such transformation can destroy the proper being or existence of the genuinely individual person, of the being constituted as minded; instead, every such transformation is a *further determination* of that entity, and thus one that presupposes *and* confirms its being or existence. This cannot be said of any merely or purely physical entity.[56]

If it is said that physical entities, too, "do not simply disappear," it is not clear what, in purely physical terms, can thereby *positively* be meant. Be that as it may, one thing should be beyond question: it *cannot be meant* that purely physical entities have self-identical individualities that they *always* retain, that they thus cannot lose; in light of

474

---

[56] Here and in the remainder of this section, any entity that is not constituted as minded is termed, for convenience, a "physical entity"; all subhuman animals are thus, in this sense, physical entities.

the physical sciences, that is simply untenable, because those sciences reveal physical entities as, strictly speaking, *not* being self-identical individualities.[57] The only possible transformations are transformations *within* the physical universe, thus transformations that affect only parts or segments of the physical universe; they do not concern the physical universe as a whole and do not point beyond that universe.

The self-identically individual being constituted as minded *cannot* be transformed *in this purely physical* sense. The decisive point in this second argument can now be articulated as follows: as shown above, the self-identically individual mind is intentionally coextensive with the universe, and indeed not only with the physical universe, but with the entire, utterly unrestricted universe, with being as a whole. Because however this intentionality, as demonstrated above, can be adequately explained only if understood as an *ontological dimension*, the *actuality*—and with it the self-identical individuality—of the being constituted as minded (the human being) must be interpreted as a *universal phenomenon* in the genuinely ontological sense—the same sense in which one usually speaks of the actuality of *physical* phenomena—and thus as a *universal entity*. For this reason, the self-identical being constituted as minded is completely comparable with such physical "phenomena" as the forces of nature or of the physical universe, such as the force of gravity (the gravitational constant, the law of gravity). All physical transformations are transformations *within the framework of these forces*; the forces themselves are not transformed by anything else, nor do they transform themselves; they remain constant, because they are *universal*—because they are coextensive with the physical universe. They are presupposed by every physical transformation. They therefore cannot simply cease to exist or simply disappear (as long as the physical universe exists). In partial analogy to them: because the self-identical being constituted as minded is a *universal phenomenon in the sense of a universal being*, it cannot undergo any transformation such that it would cease to exist *as such* or would simply disappear. And because the *universality* of this being constituted as minded is not restricted to the physical universe, but is instead coextensive with the absolutely unrestricted universe, it excludes *every* possibility—not only the physical possibility—of the cessation of its individual existence.

Of the transformations that the minded, individual human person can (and, in this case, must) undergo within the purely physical sector of the universe, death is the most significant and most radical transformation, because it radically affects—transforms—the spatio-temporality (called "body") that is absolutely essential for the integrity of the individual human person. The daunting philosophical task would be that of comprehending this *radical transformation* of spatio-temporality. Without an

475

---

[57] An example of a contemporary natural-scientific "interpretation" of the physical world can serve here as illustrative. The physicist and mathematician Roger Penrose writes, of the physical phenomenon of quantum entanglement,

> Despite their falling short of providing direct communication, the potential distant ('spooky') effects of quantum entanglement cannot be ignored. So long as these entanglements persist, *one cannot, strictly speaking, consider any object in the universe as something on its own.* In my own opinion, this situation in physical theory is far from satisfactory. There is no real explanation on the basis of standard theory of why, in practice, entanglements *can* be ignored. Why is it not necessary to consider that the universe is just one incredibly complicated quantum-entangled mess that bears no relationship to the classical-like world that we actually observe? (1994: 300; emphasis added)

essential relatedness of the intellectual constitution of the human being to spatio-temporality, the human being would not be the human being with whom we are familiar, not the individual person with whom we are familiar (and whom we have also defined). Although traditional metaphysical theories interpret death as the separation of the soul from the body, this metaphysics never reduces the human being to the purely separated soul (*anima separata*). How the relatedness to spatio-temporality is to be thought, ontologically, as continuing beyond death is the formidable task with which philosophy finds itself confronted. Adequate treatment of this question would require a monumental effort; this book cannot even begin to undertake such a treatment. The task here has been only that of indicating how the question can be approached in one quite specific respect: in the context of a fundamental clarification of the question concerning the comprehensive systematic meaning of world history.

[4] To be able to determine this comprehensive systematic meaning from a non-materialistic systematic perspective, one must presuppose that the intellectually constituted individual person, upon dying, does not simply disappear from the stage of the world and thus from world history. If this is so, then the genuinely grand task becomes visible; it is that of determining the comprehensive meaning of world history. To this task philosophy has thus far contributed virtually nothing. It is instead religion, and particularly Christianity, that has developed an in part extraordinarily extensive eschatology. From that eschatology has arisen a grand Christian, theological, comprehensive theory that is unmatched—which is not, of course, to say that it is acceptable philosophically. The case instead is that from it emerges for philosophy the task of coming seriously to terms with it. If philosophy neglects this task, it does not deserve to be called the universal science.

Comprehensive Systematics: The Theory of the Interconnection of All Structures and Dimensions of Being as Theory of Being as Such and as a Whole

5.1 The Philosophical Status of Comprehensive Systematics

Comprehensive systematics is the theory of being as such and as a whole. Section 5.2.1 explains why the twofold formulation "being as such" and "being as a whole" is used. Preliminarily, it may be said that this chapter thematizes the comprehensive structure or the comprehensive architecture of being. The task here is to show how, more precisely and in detail, this thematization is to be understood and accomplished; this task can be done only step by step. The first step treats the basic problem of whether a comprehensive systematics of this general sort is at all possible or whether instead there are considerations that rule it out in principle. Providing a penetrating and comprehensive response to this problem requires consideration of an extensive array of other central questions, ones usually discussed in quite different contexts. After their consideration, a detailed solution can be provided to the basic problem.

5.1.1 *Comprehensive Systematics as Structural Metaphysics*

The traditional term that would best name what is here termed "comprehensive systematics" is "metaphysics," but this term should be used only with the greatest caution, because it has been used in so many ways that reliance upon it invites calamitous misunderstandings. At the same time, simply to avoid it might well be an overreaction because there are deep similarities between the questions and topics of comprehensive systematics and the explicitly formulated or implicitly effective intentions and basic intuitions that, in one way or another, characterize the "metaphysics" of all times, including those in which metaphysics has been declared to be *passé* or even completely dead.

478   Despite the fact that the early years of the twenty-first century are often taken to be among those times, consideration of only the titles of a great many philosophy books currently being published would suggest, ironically, that metaphysics is flourishing as never before. Yet the publication of these many books does not prove conclusively that metaphysics is in fact a sensible or possible philosophical discipline, much less one that is indispensable. Be that as it may, because of the deep continuity between the metaphysical tradition and the program of comprehensive systematics required for this book, the book does use the word "metaphysics," but only somewhat marginally and in any case with extreme caution. The following paragraphs provide additional explanation of and support for this manner of proceeding.

Given the complexity of the history of the term "metaphysics," the philosopher does well either to avoid it or to explain precisely in what manner and in what sense it is or is to be used. For those opting to explain their usages of the term, there are basically three ways of proceeding further.

The first way is for the philosopher to presuppose or present explicitly a determinate concept of metaphysics familiar from the history of philosophy. Methodologically viewed, this procedure is unobjectionable in that, if it is followed, the reader knows exactly how metaphysics is then to be understood. If that is not made clear to the reader, of course, then accepting or rejecting the account as "metaphysical" is senseless. Nevertheless, most of the current rejections of "metaphysics" do not first adequately determine what it is that is being rejected. Moreover, serious problems arise from the fact that objections have been raised that are claimed to undermine any account that could qualify as metaphysical in any traditional sense. The three best-known anti-metaphysical positions are Kant's transcendental philosophy, the logical positivism of the Vienna School, and Heidegger's "essential or originary thinking." The philosopher adopting the first procedure must somehow deal with these positions.

The second possibility can be termed the "directly systematic." It begins with the philosopher's own definition of the term/concept "metaphysics." This philosopher's reliance upon the term indicates that some important intuition connected to the word "metaphysics" is retained, whereas the idiosyncratic definition reveals that the intuition is articulated in an innovative manner. This presupposes that the articulation has no

479   direct precedents in any of the historically familiar forms of metaphysics. The innovation may, in principle, allow the philosopher using this approach to ignore the traditional critiques of metaphysics.

The third approach is quite different. Here, metaphysics is not identified at the outset with either a specific traditional form or an idiosyncratic definition. Instead, the endeavor is determined, so to speak, *a posteriori* or *ex post facto*: a specific philosophical theory or account is presented, and only thereafter designated "metaphysical." The term "metaphysics" may of course occur in the title of the work presenting such a theory or account, but if it does so, that is merely as a matter of convenience: it anticipates what is made explicit only near the end. The initial answer that such a work provides to the question, "What is metaphysics?," would be something like this: "Metaphysics is, for example, what is done here." Given the chaotic use of "metaphysics," in the past and at present, this procedure has much to recommend it—but only if the precise conceptual or definitional determination anticipated at the outset is actually provided at some point within the account.

This book, in developing its unprecedented *structural* metaphysics, uses a combination of the second and third approaches. The retention of the term "metaphysics" indicates that this account corresponds in a significant manner to central intuitions and problems that, within the history of philosophy, are designated "metaphysical"; the qualifier "unprecedented" indicates that the theory presented here is innovative in the sense of being importantly discontinuous from any of its historical predecessors (thus, of having no direct precedents); the qualifier "structural" indicates the basic feature that makes this metaphysics innovative and unprecedented.

Against a comprehensive systematics that is in part continuous with and in part discontinuous from traditional metaphysics, many objections may be raised. Many such objections that have already been formulated in the history of philosophy can be considered overcome, including Kant's famous objection that metaphysics depends on synthetic *a priori* sentences that prove to be empty because they can have no grounding in experience and can present no direct conditions of possibility of the cognition of empirical objects. The duality of sensible and supersensible world, the distinction—decisive for Kant—between analytic and synthetic judgments, the specific distinction between synthetic *a priori* and synthetic *a posteriori* judgments, along with various other assumptions that for some time played decisive roles in the history of the critique of metaphysics, no longer play such roles. That they do not might appear at first glance to be a bold thesis; the purposes of this book do not require that this thesis be supported in detail, but they do require that it be introduced, because the account provided in this chapter presupposes it.

The situation is similar with respect to analytic philosophy. When language and formal logic become central to philosophical inquiry, anti-metaphysical positions develop—the most famous example is the logical positivism or empiricism of the Vienna School—but these have been overcome within analytic philosophy itself. Currently, metaphysics is recognized, within the framework of analytic philosophy, as a legitimate discipline or set of issues. Despite this fact, however, analytic metaphysics remains, in comparison with its traditional counterpart, extremely limited; in traditional terms, one would classify its inquiries as *metaphysica specialis* (i.e., as focusing on specific domains that are deemed "metaphysical"). Within analytic philosophy, the move toward a universal or comprehensive metaphysics continues to be slow and hesitant. The chief cause of this slowness is the significant problem introduced in the following section; for the most part, this problem lurks in the background, and because it does not appear explicitly, it continues to have insidious effects within all of contemporary philosophy.

5.1.2   *The Primary Obstacle to the Development of a Comprehensive Systematics as Structural Metaphysics*

The obstacle referred to in this section's title is a problem that is a historical legacy of Kant's philosophy. It influences all of post-Kantian philosophy extensively and in a variety of ways. Currently, its most virulent form is in evidence in all the forms of analytic philosophy that can be gathered under the rubric "anti-realism." The problem is generally referred to as that of the "cut" or the "gap."

480

481

5.1.2.1 The Problem of the Gap Putatively Separating the Theorist from Reality as It Is "In Itself"

This problem emerges directly from Kant's so-called transcendental turn. Those taking this turn take themselves to see a deep and unbridgeable gap between the dimension of the subject and that of reality, such that all knowledge, everything conceptual, etc., remains on the subjective side of the putative gap. Kant characterizes the gap as the radical separation of subjectivity, with all that is ascribed to it, from the domain of things in themselves. Kant's entire critical philosophy depends upon this strict dichotomy.

With explicit reference to Kant, Hilary Putnam (1990) places this problem at the center of the philosophical thematic of the middle phase of his philosophical development and describes it in a manner that is definitive with respect to much of the current philosophical discussion. According to Putnam, there is a "gap" or "cut" between, on the one side, subject(ivity)-thinking-mind-language-theory and, on the other, the "system," no matter how the latter is specified further: being (as a whole), actuality, world, universe, etc. The tradition determined by Kant holds this gap to be absolutely unbridgeable.

Putnam introduces two examples that he takes to be both central and highly illuminating. The first relates to the domain of quantum physics. Putnam cites Eugene Wigner, who speaks of a "cut between the system and the observer" (4). Wigner locates the entire apparatus making the measurements that test the theory's predictions on the side of the observer; from this it follows that the theoretician or observer does *not* belong to the system under observation. Putnam concludes that a quantum-mechanical theory of the entire universe (i.e., here, the universe conceived as a whole that includes the theoretician or observer) is utterly impossible, because such a theory would require that the gap or cut between theoretician/observer and system be overcome. Only such overcoming could yield a theory of the entire universe from what Putnam often calls a "God's eye view." Putnam adds the following:

> The dream of a picture of the universe which is so complete that it actually includes the theorist-observer in the act of picturing the universe is the dream of a physics which is also a metaphysics (or of a physics which once and for all makes metaphysics unnecessary). (5)
>
> [W]hat it means to have a cut between the observer and the system is ... that a great dream is given up—the dream of a description of physical reality as it is apart from observers, a description which is objective in the sense of being "from no particular point of view." (11)

482

Putnam's second illustrative example is the so-called liar paradox and the solution to it provided by Tarski. As noted in Chapter 2, every natural language, according to Tarski, is semantically closed—that is, remaining within its own dimension, it can form all possible constructions, including ones that yield paradoxes. These include "The sentence that the reader of this book is currently reading is false," and, schematically,

(1)   Sentence (1) is false.

Independently of the problem of truth-paradoxes, Tarski introduces the "truth-schema"

(TS)   'p' is true if and only if p.

Replacing p in (TS) with (1) yields

(1')   "Sentence (1) is false" is true if and only if sentence (1) is false

and thus

(1")   "Sentence (1) is false" is true if and only if "Sentence (1) is false" is false.

The contradiction is obvious.

Tarski avoids the paradox by introducing the distinction between object language and metalanguage, as well as a hierarchy of (meta)languages. No contradiction arises if the nominalized sentence on the left-hand side of the "if and only if," the sentence qualified as true, belongs to the object language, which is *different* from the language—a metalanguage—in which the same sentence appears in non-nominalized form on the right-hand side of the equivalence indicator ("if and only if"). This solution, although elegant as well as logically unobjectionable, has a price that the systematically thinking philosopher will be loath to pay: in accepting his solution, Tarski abandons the prospect of developing a unitary and universal truth-concept. His solution requires that every language L, L_1, L_2, etc., have *its own* truth-concept. Putnam raises the appropriate question: what is the language spoken by the theoretician who discusses just this state of affairs? To this, Tarski has no answer.

According to Putnam, this second example reveals the same basic phenomenon of "gap" or "cut" between speaker/theoretician and system that is visible in quantum theory, although in a different form (one in which there is no "observer"): the language of the speaker/theoretician (and with it all of its conceptuality, syntactic and semantic structures, etc.) remains *outside* the "whole" (the "system") *about* which the theoretician speaks. Between the two yawns an absolute gap.   483

If, independently of Putnam's examples, one wanted to maintain that there is a gap or cut of the sort just described, would acceptance of the resulting consequences be the last possible word in philosophy? Would "insights" thereby be attained that would warrant designation as definitive for philosophy? The sections that follow in this chapter show not only that this position is not compelling but also that it is based on a series of assumptions, notions, and ways of questioning some of which are false, some insufficiently clarified, and some one-sided.

5.1.2.2 Examples of Failed Attempts to Solve the Problem of the Putative Gap

This is of course not the place to consider all possible or even actual attempts to deal with this problem; the following paragraphs treat only a few particularly characteristic examples of such attempts that are clearly unsatisfactory.

[1] The path of philosophical development that Putnam has followed in considering this problem is characterized by often surprising changes in course; it is instructive to consider the various mutually contradictory conclusions he draws

and then presents as solutions to it. Initially, he represents a position he names (1983) "metaphysical realism":

> What makes the metaphysical realist a *metaphysical* realist is his belief that there is somewhere 'one true theory' (two theories which are true and complete descriptions of the world would be mere notational variants of each other). In company with a correspondence theory of truth, this belief in one true theory requires a *ready-made world* (an expression suggested in this connection by Nelson Goodman): the world itself has to have a 'built-in' structure since otherwise theories with different structures might correctly 'copy' the world (from different perspectives) and truth would lose its absolute (non-perspectival) character. (211)

This position is based on the assumption that the "system" is simply the world or the universe that in itself is fully structured, and is so fully independently of our (human) minds, our language(s), and our theories. According to Putnam's metaphysical realist, of course, the task of philosophy and the sciences is to develop this one true theory that accomplishes the perfect mirroring of the world. The tenability of this project is based on the assumption that we humans are in the position to attain some sort of God's eye view—thus, to attain a standpoint from which we can view comprehensively *both* sides of the gap or cut, so that we can then *compare* theory and world in order to determine whether the two correspond. Putnam consistently notes, as in the passage just cited, that this metaphysical realism presupposes or entails a correspondence theory of truth.

484

Putnam abandons metaphysical realism for three chief reasons. First, he comes to recognize that the assumption that we can attain a God's eye view is a presumption and an irreality.[1] Second, he concludes that the fact that the metaphysically realistic theory is developed by humans, who are not God(s), has a consequence that is untenable for a number of reasons. Among them is that because the gap between theory and system (world, universe) is, *for us, as finite*, unbridgeable, the possibility cannot be excluded that even if we developed an *ideal* theory (i.e., a theory that satisfied all our conditions for theoretical perfection), that theory might nevertheless fail to correspond to the world, and thus be false. Putnam remarks, "The metaphysical realist's claim that even the ideal theory T_1 might be false 'in reality' seems to collapse into unintelligibility" (1980/1983: 433). A third factor, arising from the increasing influence exerted on Putnam by the later Wittgenstein, plays an essential role. As Wittgenstein's influence increases, Putnam (1994: 449) begins to see in positions like metaphysical realism a "naiveté about meaning" that he considers no longer to be tenable.

[2] Rejecting metaphysical realism, Putnam (1994) embraces a form of anti-realism that he terms "internal realism."[2] It is summarized in the well-known claim

---

[1] A characteristic passage: "We don't have notions of the 'existence' of things or of the 'truth' of statements that are independent of the versions we construct and of the procedures and practices that give sense to talk of 'existence' and 'truth' within those versions" (1983: 230).

[2] In his Dewey Lectures, Putnam refers (1994: 461n36) to "internal realism"—a phrase he there encloses in quotation marks, for reasons he provides—as the position "I have been defending in various lectures and publications over the last twenty years." In large part, the Dewey Lectures are devoted to Putnam's revealing "where I feel I went wrong" in defending this position (456).

(1983: 205–28) that because there is no "ready-made world," the only reality (world, universe—hereafter, generally, "reality") is a dimension formed or configured by *us* (by the human intellect, human languages, human conceptual schemes, human theories). The epithet "internal" designates the putative fact that "reality" makes sense, indeed exists at all, only *within* a conceptual scheme that *we* project and apply. The question whether "reality" *in general or on the whole*—i.e., from the perspective of internal realism, as *not* conceptualized by us—does or could exist is avoided in a characteristically obstinate manner; this avoidance is one of the weaknesses of this position.[3] The cut or gap between theory and system is—in complete faithfulness to the Kantian tradition—thereby deepened radically and indeed made permanent in that the objective pole (the "system," reality) itself is radically split once more by a distinction between what is in itself and what is for us. What is in itself is not, to be sure, negated directly and explicitly, but it is left fully in the dark; it serves merely as a foil to prevent our thinking that the reality we posit is exhaustive of reality itself. Beyond this, it is simply set aside on the basis of its making no sense—and of course it must make no sense, if all sense comes from us.

Putnam's internal realism is the result of a peculiar combination of two significant theses, one correct but the other both superficial and incorrect. The correct thesis concerns the relation between intellect/language and world/reality. The assumption that the world (reality) would or could exist in absolute independence of intellect and language (theory, conceptuality, etc.) is unintelligible and therefore untenable. Yet Putnam never works out the true and only compelling reason for this unintelligibility. It cannot suffice merely to announce that a so-conceived world/reality would be a bare or naked reality, or anything of that sort. It would have to be shown in what precise sense the so-conceived world or reality was naked or bare, what would be the implications of the assumption of such a world or reality, etc.

The true and compelling reason can be formulated in preliminary fashion as follows (the issue is considered in more detail in later sections of this chapter): no questions at all concerning the putative gap or cut between thinking/intellect/language and "system"/reality, or concerning positions like realism and anti-realism, can be clarified fundamentally unless one accepts the following simply central *ontological* thesis: whatever else the dimension that is called "system," "world," "universe," "reality," "being," etc., may be, in any case it possesses the following fundamentally *immanent* and *genuinely ontological* structuration: *complete expressibility*. The term "expressibility" is used here in a comprehensive sense, as a blanket term or as an abbreviation for an extensive series of other terms that, in specific contexts, should be used in place of it; these other terms include "intelligibility," "conceivability," "graspability,"

485

486

---

[3] Putnam (1983: 226) writes as follows:

> I am not inclined to scoff at the idea of a noumenal ground behind the dualities of experience, even if all attempts to talk about it lead to antinomies. Analytic philosophers have always tried to dismiss the transcendental [*sic*] as nonsense, but it does have an eerie way of reappearing. . . . Because one cannot talk about the transcendent [*sic*] or even deny its existence without paradox, one's attitude to it must, perhaps, be the concern of religion rather than of rational philosophy.

"understandability," "articulability," "explicability," etc.[4] But expressibility (in the sense relevant here) is unintelligible without language (conceptuality, mind...) because expressibility presupposes the inverse relation of *expressing* and therefore language, as the means by which what is expressible can be expressed. (Language is understood here in the broadest possible sense, as introduced in Chapter 2 and developed below).

Without this central *ontological* thesis, absolutely nothing relating to language (in its descriptive segment), theory, truth, etc., can be made intelligible. Given this thesis, however, the assumption of a "naked" world in the sense of a *world encapsulated within itself and conceived as without any relation to anything like language, mind, conceptuality, etc.*, is a metaphysical impossibility and thus unintelligible.

The second assumption on which internal realism is based—the one designated above as both superficial and incorrect—is the thesis, astonishingly presupposed to be obvious, that "reality" (the world, the universe), as independent of *our* language and *our* minds, would be "naked" in the sense articulated in the preceding paragraph. But we ourselves—and with us our language(s), our theories, etc.—are also taken to be purely contingent things, without which the so-called physical world not only could exist but in fact has existed. But if this were the case, how could humanity (thought as no matter what kind of totality) be taken to be the measure (so to speak) for what the world *is*? That humanity could be such a measure is obviously an extremely implausible assumption. Because, however, the first insight (in its modified and adequately articulated form) is correct, the task arises of answering the following question: how can and must language, mind, etc. be understood in ways that support the contention that a "world" unrelated to language, mind, etc. would be an impossibility? The following sections, which present an *adequate* solution to the gap or cut problem, also develop an answer to this question.

487

[3] Putnam himself comes to see the untenability of his internal realism. In his (to date) most recent developmental phase, he abandons this position in favor of one he terms (1994) "pragmatic" or "direct" or "natural" or "commonsensical" realism.[5] This position results in part from his characterization (1994: 487) as unintelligible of the two forms of realism he had earlier defended. This of course leaves open the question whether Putnam's "natural" realism itself satisfies the criterion of intelligibility. His Wittgensteinian leanings in any case make the "natural" form of realism quite different from the earlier ones. Presumably decisive in this respect is his following Wittgenstein

---

[4] In that various philosophers in the course of the history of philosophy have attempted to articulate this—in this book's terms—immanent structuration, the thesis expressed in the sentence to which this note is appended can reasonably be said to recur throughout that history. Nevertheless, it is crucial *at present* to articulate the thesis in terms adequate to *the present stage* of philosophical development. Formulations of the thesis in ancient philosophy include that of the sameness of thinking and being. In the tradition of Christian metaphysics, the thesis appears in the specific form, *onme ens est verum* (which presumably means nothing other than *omne ens est intelligibile*). More recent versions include Hegel's dialectical notion of the identity of being and "concept" and Gadamer's "Being that can be understood is language" (1960/2003: xxxiv/GW2: 444).

[5] See the detailed presentation in the Dewey Lectures (Putnam 1994). On the terminology, see 454. Putnam particularly singles out William James, Husserl, Wittgenstein, and Austin as those who bring him to see "that progress in philosophy requires a recovery of 'the natural realism of the common man'" (469).

in ascribing to natural language a simply determinative role. The position is utterly dependent upon Wittgenstein's famous dictum, "Meaning is use,"[6] which Putnam interprets as follows:

> [T]he use of words in a language game cannot, in most cases, be described without employing the vocabulary of that game or a vocabulary internally related to the vocabulary of that game. If one wants to describe the use of the sentence "There is a coffee table in front of me," one has to take for granted its internal relations to, among others, facts such as that one perceives coffee tables. By speaking of perceiving coffee tables, what I have in mind is not the minimal sense of "see" or "feel" (the sense in which one might be said to "see" or "feel" a coffee table even if one had not the faintest idea what a coffee table is), I mean the full-achievement sense, the sense in which to see a coffee table is to see that it is a coffee table that is in front of me. (458)

This is a characteristic example of a reduction of all philosophical sentences to sentences in natural or ordinary language, a reduction that is an aspect of what Wittgenstein terms the true, *therapeutic* method that must be applied to philosophical sentences. The philosophical problem that arises immediately is this: what is thereby gained with respect to intelligibility? The paraphrasing of sentences of whatever sort by means of sentences from natural or ordinary language is nothing other than a wholly "natural" reduction of all philosophical questions, insights, and theses to the level of natural or ordinary language. But is natural or ordinary language intelligible? If it were, how could philosophy and science ever have arisen? What reductions of this sort in fact result in is the trivialization of philosophical sentences. The last sentence in the passage just quoted is exemplary: "By speaking of perceiving coffee tables, what I have in mind is not the minimal sense of 'see' or 'feel' (the sense in which one might be said to 'see' or 'feel' a coffee table even if one had not the faintest idea what a coffee table is), I mean the full-achievement sense, the sense in which to see a coffee table is to see that it is a coffee table that is in front of me." "The full-achievement sense" of "see" is hereby articulated by means of a remarkable tautology: "to see a coffee table is to see that it is a coffee table that is in front of me." Reductions of this sort are reductions to trivialities.[7]

[4] Despite the triviality of its reductions, Putnam's "natural" realism involves one factor that is of great significance both for epistemology and for comprehensive systematics. Following such authors as John McDowell, Putnam stresses the necessity of overcoming the thesis, predominant throughout modernity and defended by some still today, that between the epistemic subject and the subject matter to be known there is or must be an intervening medium or interface. Throughout much of modernity, this medium has been taken to be "sense data." The notion is that we as knowers have immediate access only to the medium or interface; the question then becomes, how is

488

---

[6] Wittgenstein's exact formulation (*PI* §43) reads, "For a large class of cases—though not for all—in which we employ the word 'meaning' it can be defined thus: the meaning of a word is its use in the language."

[7] Further evidence that Putnam—surprisingly—takes "to see a coffee table is to see that it is a coffee table that is in front of me" as a serious philosophical thesis, albeit one requiring "a sort of cultivated naiveté," is provided by his additional use of it (1993: 183) in a similar context, and also without further explanation.

the move made from the medium to a dimension that is genuinely *outside* the subject? Whatever is beyond the inputs, whether or not they be "sense data," is on this view supposed to relate to mental processes only *causally*, not *cognitively*. This notion, an assumption fundamental to modern philosophy, has produced peculiar dichotomies and subjectivisms of various sorts.

In opposition to this notion, Putnam writes as follows:

> I agree with [William] James, as well as with [John] McDowell, that the false belief that perception *must* be so analyzed [i.e., "we perceive external things" analyzed as "we are caused to have certain subjective experiences in the appropriate way by those things"] is at the root of all problems with the view of perception that, in one form or another, has dominated Western philosophy since the seventeenth century. James's idea is that the traditional claim that we must conceive of our sensory experiences as *intermediaries* between us and the world has no sound arguments to support it, and, worse, makes it impossible to see how persons can be in genuine cognitive contact with a world at all. (1994: 454)

489

This is no doubt correct, but it does not suffice simply to reject the supposed interface between the knower and the world; what must be shown in addition is precisely how the resulting position is to be understood and explained philosophically. Putnam's natural realism is nothing other than a grand assertion within the framework of ordinary language; it lacks philosophical penetration or suffusion. Here, Wittgenstein's influence reaches its apex:

> I argued that this philosophy of perception [i.e., the "causal" theory of perception] makes it impossible to see how we can so much as *refer* to "external" things. And I argued that we need to revive direct realism (or, as I prefer to call it, natural realism)—more precisely, that we need to revive the spirit of the older view, though without the metaphysical baggage (for example, the mind "becoming" its objects, though only "potentially", or the mind taking on the "form" of the object perceived "without its matter"). (469)

Putnam provides no explanation of how a "natural realism . . . without the metaphysical baggage" is at all possible or how, more precisely, it is to be understood. It may sound good, but assertions that merely sound good are not responsible or defensible philosophical theses. The authors to whom Putnam (467) alludes in the passage just cited (he includes among them Aristotle, along with "Aquinas and other scholastic philosophers") may develop theories that are no longer tenable, but at least *they attempt to make direct realism philosophically comprehensible*; they never resort to making assertions within the framework of an ordinary language that simply sweeps aside all philosophical issues.

Putnam and McDowell appeal programmatically particularly to Wittgenstein, and cite above all §95 of the *Philosophical Investigations*: "When we say, and *mean*, that such-and-such is the case, we—and our *meaning*—do not stop anywhere short of the fact; but we mean: *this—is—so.*" McDowell (1994/1996: 27) interprets this passage as follows:

> [T]here is no ontological gap between the sort of thing one can mean, or generally the sort of thing one can think, and the sort of thing that can be the case. When one thinks truly, what one thinks is what is the case. So since the world is everything that is the case (as [Wittgenstein] himself once wrote), there is no gap between thought, as such, and the world.

This overcoming of the gap or cut between thinking and world (system, reality) could 490 not be maintained more decisively. But is it actually understood and philosophically developed in such a way that it is anything other than a verbal formulation that sounds good? Not by McDowell (any more than by Putnam), as is clear from his commentary on the passage just cited:

> But to say there is no gap between thought, as such, and the world is just to dress up a truism in high-flown language. All the point comes to is that one can think, for instance, *that spring has begun*, and that very same thing, *that spring has begun*, can be the case. That is truistic, and it cannot embody something metaphysically contentious, like slighting the independence of reality. (27)

As in the case of Putnam, this is a typical example of the reduction of philosophy to the level of natural or ordinary language, and to a triviality. The defense against putatively metaphysically problematic assertions proves in fact to be the renunciation of non-trivial philosophical explanations of what the authors' own assertions *actually* maintain or would have to maintain for them to be taken seriously as philosophical theses.

[5] There are other examples of false or otherwise unsatisfactory "solutions" to the cut-or-gap problem. One quite interesting example is the position of one of the best systematically thinking philosophers of the present, Nicholas Rescher. Rescher (1992–94: I:289) operates on the basis of the fundamental *epistemological* distinction "between aim and achievement, between what we set out to do and the extent to which we actually manage to do it." He introduces three forms of this fundamental distinction: (1) "'our putative reality'" and "'reality as such'" (296), (2) "our achieved putative scientific truth" and "the real truth of the matter" (297), and (3) "science in the *present* state of the art" and "*ideal* or perfected science" (299). He describes the interrelations of these three domains as follows:

> [O]ne must maintain a clear distinction between "our conception of reality" and "reality as it really is." Given the equation
> Our (conception of) reality = the condition of things as seen from the standpoint of "our *putative* truth" (= the truth as we see it = the science of the day),
> we realize full well that there is little justification for holding that our present-day science indeed describes reality and depicts the world as it really is. . . . In science as elsewhere, there is a decisive difference between achievement and aim—between what science *accomplishes* and what it *endeavors* to do. (283)

It is clear that this perspective is purely *epistemic*. The gap or cut between thinking, theory, etc., on the one hand, and world or reality on the other could be wholly overcome only in the *ideal case*, but the ideal case is, by definition, concretely or in fact unattainable. Charles S. Peirce maintains that perfect and complete science—and thus 491 the overcoming of the gap or cut—must be identified with the definitive condition of science at the end of the history of science. Rescher (298) remarks,

> Peircean convergentism is geared to the supposition that ultimate science—the science of the very distant future—will somehow prove to be an idea[l] or perfected science freed from the sorts of imperfections that afflict its predecessors. But the potential gap that arises here can be closed only by metaphysical assumptions of a most problematic sort.

[6] The further course of this chapter reveals that the epistemic perspective is a real one that must therefore be taken into consideration, but that it is not the decisive perspective. Other considerations also cast new light on the epistemic perspective, making clearer its status with respect to the problem of gap or cut. Above all, the putatively rigid distinction between "ideal science" (or "reality in itself" or "truth in itself") and "factical science" (or "reality for us" or "truth for us") proves to be inadequate and one-sided.

As far as Rescher's epistemic perspective is concerned, it is not an overcoming but instead (so to speak) a widening of the gap or cut. The following section develops a radically different conception. The central point can be introduced, in the current context, by means of this question: is it possible or indeed philosophically necessary to thematize the epistemic distinction "for us vs. in itself"? If this distinction is the expression (or at least *an* expression) of the gap, then the thematization of the distinction is a thematization of the gap. But the thematization itself cannot develop within any framework governed by the gap, because the distinction encompasses both sides of the putative gap.

This fundamental situation is ignored by all the authors who introduce distinctions of the form "for us vs. in itself," and then reduce or situate every assertion, every theory, etc., in terms of this distinction. What they fail to do is to reflect on the fact that if the distinction—understood as acknowledging an unbridgeable gap—were indeed absolutely universal, their assertions about it would be neither intelligible nor indeed possible, precisely because those assertions thematize the relation between the two dimensions between which, they claim, there can be no bridge. Worth noting is that this is a long-standing problem in philosophy that is exhibited in exemplary fashion by the case of transcendental-critical philosophy. If Kant's transcendental philosophy were universally valid, then the transcendental-philosophical assertions found in the *Critique of Pure Reason* would not be possible or would not be correct, because they would all hold not for the actual structure of the human intellect as it is "in itself," but only of an (its!) "appearance." It is, however, clear that the assertions made in the *Critique* claim to articulate the structure of the cognitive subject "in itself."

In confronting this problem, John McDowell takes a notable step in that in a specific respect he effectively overcomes the gap:

> [The] image of openness to reality is at our disposal because of how we place the reality that makes its impression on a subject in experience. Although reality is independent of our thinking, it is not to be pictured as outside an outer boundary that encloses the conceptual sphere. *That things are thus and so* is the conceptual content of an experience, but if the subject of experience is not misled, that very same thing, *that things are thus and so*, is also a perceptible fact, an aspect of the perceptible world. (1994/1996: 26)
>
> [T]alk of impingements on our senses is not an invitation to suppose that the whole dynamic system, the medium within which we think, is held in place by extra-conceptual links to something outside it. . . .
>
> We find ourselves always already engaging with the world in conceptual activity within such a dynamic system. Any understanding of this condition that it makes sense to hope for must be from within the system. It cannot be a matter of picturing the system's adjustments to the world from sideways on: that is, with the system circumscribed within a boundary, and the world outside it. (34)

This state of affairs—that the world does not lie outside the dimension of the conceptual—McDowell (24ff) names "the unboundedness of the conceptual." If the talk here is of "our dynamic system," then what is meant is "the system of concepts" or, in a phrase stemming from Wilfred Sellars, "the logical space (of reasons)." The gap is here in principle overcome or denied, but only from the perspective of the subjective pole of the distinction, which is here termed the dimension of the "conceptual." The basic thought that grounds this thesis is essentially identical with a fundamental aspect of the thought, developed above, of the *universal expressibility* of reality, of being, of the system (where "system" designates not the subjective or conceptual but the objective or genuinely ontological side or pole of the putative gap).

The thought of *expressibility* develops from the perspective of the pole of being and, as indicated above, it entails a means of expression; that means is designated above, globally, as the dimension of language. It must be specified, however, that this means of expression consists not only of symbols (signs) but includes concepts as well, so that of course the concept of the concept must be clarified, as it is in Chapter 3. The universal expressibility of being, it must now be said, includes the dimension of the linguistic and the conceptual. Language and conceptuality *do not lie outside of* the dimension of being. One can therefore appropriately speak of the unboundedness of being.

That McDowell does not speak of the unboundedness of being is a result of the fact that he lacks even the starting points for a theory of being, an ontology. For this reason his comprehensive conception, despite several accurate insights and formulations, is characterized by one-sidedness and limitation. Its limitation is to the single side of the subject that is said to have access to the other side, that of being; about this other side, nothing more is said. His view remains fundamentally a transcendentally determined view, despite essential corrections that he makes to Kant's conception. Because of its lack of an ontology, his position as presented cannot develop a comprehensive philosophical systematics.

5.1.3 Comprehensive Clarification of the Problem of the Putative Gap as Starting Point for a Theory of Comprehensive Systematics: Four Fundamental Theses

The systematic clarification of the basic problem of the gap or cut, and thus of the status and program of comprehensive systematics, is based essentially on four central theses, two of which (the first and second) are introduced above; the other two (the third and fourth) are intimated in one way or another. The following sections in part summarize them and in part present them more extensively within the systematic context.

5.1.3.1 Thesis One: The Appropriate Form of Presentation for the Structural-Systematic Philosophy Requires Sentences in the Purely Theoretical Form

Sentences in the purely theoretical form are those that meet the criterion presented in Section 2.2.3.1 as the general linguistic criterion for indicativity and thus theoreticity. These are sentences that can be preceded, implicitly or explicitly, by the operator, "It is the case that,"—hence, "It is the case that such-and-such" or, semi-formalized, "It is the case that φ." Instead of "criterion" one can say "appropriate form of theoretical presentation." Negatively, the decisive point is that this form is absolute in the sense of not being determined by any external points of reference; of particular importance is, again, that sentences of this form make no explicit reference to subjects, speakers, or conditions of any sort, including those of place, time, culture, language, etc. This absoluteness excludes positions that take any of these external points of reference as measures for or as essential ingredients of theories; yet more specifically, it excludes all positions for which transcendental philosophy serves in any way as a model.

One point made above bears repeating here, because of the disastrous consequences that result if it is ignored. Some might be tempted to understand the formulation "It is the case that such-and-such" as absolute or dogmatic in the sense of purporting to be somehow the last word on the relevant such-and-such. The source of the temptation would be the fact that, in ordinary language, such formulations often connote definitiveness of the sort just described. This connotation is, however, a result simply of linguistic habit; the connotation, which is a significant qualification, cannot be derived from the formulation itself. "It is the case that such-and-such" says nothing about the (definitive or absolute or probable or . . .) manner or form in which the such-and-such is the case.

In the systematic conception presented here, the formulation "It is the case that such-and-such" is understood in a *relative* sense, *not* in the sense of any relativity to any such non-theoretical factors as subjects, cultures, etc., but instead in the sense of relativity to some specific theoretical framework. To be sure, this brings with it a quite serious problem; this is the problem of relativism, which must be taken in all serious-ness particularly by systematically oriented philosophies (this problem is treated in Section 3.3.4.3).

5.1.3.2 Thesis Two: Semantics and Theories of Beings and of Being Are Fundamentally Interrelated

The second thesis concerns the relation between semantics and ontology. In various contexts, this book emphasizes that semantics and ontology are two sides of the same coin. How this metaphor is to be understood precisely is made clear by the presentations of the basic features of this book's semantics and ontology in Chapter 3. Here, an addi-tional and central distinction is to be emphasized. To this point in the book, "ontology" is used in a broad sense, such that the theory of being would fall within the scope of ontology. Viewed etymologically, however, "ontology" means "theory of beings." That is how the term is to be understood here. This chapter is the systematic locus for intro-ducing and explaining the distinction between theories of being and theories of beings. One could also simply say "worldly ontology" and "metaphysical ontology," or "specific ontology" and "universal ontology." The briefest and most linguistically appropriate way to make the distinction, however, uses the terms "ontology" and "theory of being."[8]

Chapter 1's quasi-definition of philosophy uses the phrase "unrestricted universe of discourse." The intended signification of this phrase can be specified further by means

---

[8] In certain contexts this book continues to use the term "ontology" in the broader sense that includes "theory of being." The reason for this is that it makes possible the introduction of succinct formulations. Thus, for example, Section 5.1.5.2.2.1 uses "ontologization of the theoreti-cal dimension." The contexts in which the term appears make clear whether the broader or the narrower sense is intended.

of the distinction between ontology and theory of being. For the distinction between being and beings, Heidegger introduces the term "ontological difference"; this distinction or difference is the central theme in his thought. One can of course acknowledge the importance of this issue and of what Heidegger accomplishes concerning it without endorsing the manner in which he deals with it.

In a first approximation, one can say that the theory of being has as its object the totality of beings. Is the totality of beings itself a being? If it were, then there would be no distinction between ontology and theory of being. Ultimately, it depends upon how one understands the totality of beings. In any case, the term "being" proves to be the contentually adequate one to designate the subject matter in question. To this topic, an entire section of this chapter (5.2.1) is devoted.

5.1.3.3 Thesis Three: Expressibility Is a Factor Fundamental to the Structurality of Beings and of Being

Thesis three articulates a fundamental structural feature of ontology and of semantics. Sections 2.2.6, 2.4.3.1, and 5.1.2.2 introduce and explain the term and concept *expressibility*. To be added here is that this term/concept actually articulates and makes manifest what is indicated by the phrases "unrestricted universe of discourse" and both "unboundedness of the conceptual" and "unboundedness of being." It would be difficult to exaggerate the systematic significance of this third thesis. It provides the genuine foundation for a comprehensive clarification of various central philosophical problems that are currently topics of intensive philosophical discussion, including the issue of realism and anti-realism. Decisive deepening of this thesis requires attainment of a clarification of the understanding of *language*. That clarification is the task of Sections 5.1.3.4 and 5.1.4.

496

5.1.3.4 Thesis Four: Philosophical Languages Are Languages of Presentation

This thesis recalls the most fundamental feature of the language appropriate for philosophy (as well as for science) introduced particularly in Section 2.2. Closer clarification of this essential feature and of its consequences requires the extensive account presented in the next section.

5.1.4 *The Adequate Concept of Theoretical-Philosophical Language*

Language is among the phenomena that are most familiar to human beings. As with everything that is immediately familiar and reliable, however, the very familiarity of language tends to make it more difficult to bring into view the *entirety* of the phenomenon, with all its dimensions and consequences. This holds in a quite particular manner for the language that is used in philosophy. Throughout the history of philosophy and especially at present, philosophical language is *fundamentally* identified with natural or ordinary language. "Fundamentally" here points to the fact that philosophers, particularly in the analytic tradition, view ordinary language as the indispensable measure for philosophy, even when they use logical means to make this language more precise and indeed to correct it; even then, the fundamental structures of ordinary language are not brought into question. As is made clear in Chapter 2, this book utterly rejects this conception; this section explains further how its rejection is to be understood. To be presented are the central aspects or structural features of the strictly theoretical concept of philosophical (scientific) language.

5.1.4.1 Language, Communication, and Presentation

There is no question that the philosopher can and indeed—under certain circumstances—must rely on ordinary language, but this in no way entails that the basic structures of ordinary language simply be accepted. As noted above in various contexts, the reason for this is that ordinary language, in the first instance, serves the aim of *communication*. It does make room, so to speak, for *presentation*, but only in a restricted respect, i.e., in the service of communication (exceptions include such segments of language as poetry, prayer, etc.). Communication, for its part, has various aspects, particularly practical ones. Ordinary languages, as serving communication, are therefore structured primarily to satisfy practical aims. Central for such languages are structures that regulate communicative exchanges. This is syntactically evident from the fact that the basic structures of communicative languages are situative, practical, and self-expressive ones, including indexical sentences (sentences including such terms as "I," "now," "here," etc.), imperative sentences, sentences expressing wishes and values, etc. In other words, situative, pragmatic, and self-expressive sentences comprise the most important linguistic structures of languages of communication.

Philosophical language is (or should be) a *language of presentation*, in the sense that presentation, not communication, is its primary function. Languages of presentation are theoretical; for this reason, their sentences are of the form "It is the case that such-and-such" (or "It is the case that φ"). Sentences of this form have, in the first instance, the function not of communication but of presentation. The distinction lies in the fact that in the case of communication what is central is the relation to other speakers or partners in conversation, whereas in the case of presentation the primary role is played by whatever it is that a given account presents. This distinction might appear to be artificial, abstract, or perhaps even nit-picking. Despite this possible appearance, however, it in fact articulates one of the most fundamental elements of the distinction—so often ignored, neglected, or even denied by philosophers—between ordinary and philosophical languages. The goal of ordinary language is communication; the goal of philosophical language is presentation, the expression or articulation of subject matters. To be sure, just as ordinary language does not exclude the aspect of presentation, philosophical language does not preclude the possibility of communication. But in each case, the aspect central to the one is marginal to the other: within ordinary language, presentation is marginal, and within philosophical language, communication is marginal.

One might object that presentation entails communication or indeed is a form of communication: when philosophers or scientists present what they present, what are they doing if not communicating with one another? Closer consideration is required. One can—and many do, with great frequency—use "communication" and "presentation" in quite broad and unprecise senses; when they are so used, it can appear that presentation either entails communication or indeed is a form of communication. That it is not is revealed by two essential points.

First, if the terms "present" and "communicate" are used not in their vague, ordinary-language senses, but in the senses specified here, the decisive difference is clearly visible.

The distinguishing moment is that of the unambiguously distinct *goals* or *determining factors*: for communication, the goal or determining factor is one or more interlocutors; for presentation, it is whatever is presented. This distinction is irreducible.

Second, presentation is the expression or articulation of what is presented within what Frege terms the "third realm,"[9] Karl Popper (1972: esp. Ch. III, IV) the "third world," Wilfrid Sellars (1956: 298–99) the "space of reasons,"[10] and John McDowell (1994/1996; Ch. II) "the unboundedness of the conceptual."[11] In the technical terms used in this book, a given simple or complex primary fact enters this arena explicitly when it is articulated by means of a true primary sentence (i.e., a primary sentence expressing a true primary proposition); more broadly, a given state of affairs enters the arena when it is articulated within the framework of a theory. By articulating the sentence or the theory, the theoretician presents its subject matter within this space of reasons; the reader or listener who, also within the space of reasons, accesses the sentence or theory thereby gives it the additional feature of being a presentation-to. If one termed this situation one of "communication," then one would have to say that presentation entails communication. It must be granted, however, that on the one hand this would be a contrived notion of "communication," and that on the other such "communication" would still fail to constitute the goal of the presentation of subject matters. In any case, granting this point would in no way threaten or weaken the thesis defended here. Philosophical language, as theoretical, is defined by its presentational character, no matter what use one might thereafter make of any of its presentations. If some form of "communication" enters the realm of presentation—by the back door, so to speak—then with its entrance, presentation does not cease to be what it is: the theoretical articulation of facts or states of affairs.

499

5.1.4.2 The Fundamental Criterion for the Determination of the Basic Structures of an Adequately Clarified Philosophical Language

One of the most important consequences resulting from the strict distinction between communication and presentation concerns the question of what structures a philosophical language ought to have. If philosophical language is fundamentally distinct from ordinary language, then it cannot be presupposed to have or involve the same structures. Instead, to clarify the question of what structures a philosophical language

---

[9] Frege writes (1918: 337/69) as follows:

> A third realm must be recognized. Anything belonging to this realm has it in common with ideas that it cannot be perceived by the senses, but has it in common with things that it does not need an owner so as to belong to the contents of his consciousness. Thus for example the thought we have expressed in the Pythagorean theorem is timelessly true, true independently of whether anyone takes it to be true. It needs no owner. It is not true only from the time when it is discovered; just as a planet, even before anyone saw it, was in interaction with other planets.

[10] "In characterizing an episode or a state as that of *knowing*, we are not giving an empirical description of that episode or state; we are placing it in the logical space of reasons, of justifying and being able to justify what one says" (Sellars 1956: 298–99).

[11] The terms "third realm," "third world," "space of reasons," and "unboundedness of the conceptual" are here introduced only as articulations of a certain fundamental insight that this book shares with the cited authors. This does not of course mean that the respective systematic positions of those authors are accepted here.

should have, a criterion must be recognized and applied that is wholly different from the criterion that is relevant to the structures of ordinary language. The central criterion for the latter is the simplest and most practicable possible intercourse among individuals within the human lifeworld. The intelligibility of ordinary language is thus oriented by a factor that is here termed "practicability." How mundane things are grasped *and* named, how they are articulated, etc., is oriented exclusively with respect to the communicative interactions of human beings. "The world" is then only the space within which communication takes place, because in such interactions, the primary concern is not with conceiving and articulating the world itself, but with getting around within the world in such a manner that exchanges among people are made both possible and as smooth as possible.

The situation is completely different with the linguistic structures for philosophical (and, more generally, scientific) language. Here, the criterion is not practicability, even in a quite broad sense, but instead solely the intelligibility of the subject matter. Ordinary language in no way dictates or determines how reality (the world) is constituted; for this reason, theoretical language, considered semantically, must be conceived as inseparable not from ordinary language but from the ontological dimension. This interrelation of semantics and ontology is determined solely and completely by the criterion of the intelligibility of the subject matter. If a specific linguistic structure has an ontological counterpart that is unintelligible or insufficiently intelligible, the linguistic structure must be abandoned. This matter is treated thoroughly in Chapter 3, which introduces an innovative semantics, with correspondingly innovative structures; the basis for its introduction is the demonstration that the semantics based on ordinary language is philosophically unacceptable because it in one respect presupposes and in another respect entails an unintelligible ontology (i.e., substance ontology).

5.1.4.3 Philosophical Language as a Semiotic System with Uncountably Many Expressions

As its title indicates, Section 5.1.4.3, including its various subsections, is devoted to the presentation of philosophical language as a semiotic system having uncountably many expressions. Its first subsection serves as an introduction to the topic and presents the motivation behind the acceptance of philosophical language as so understood. The motivation lies in the fact that one of the central philosophical problems under current discussion is insoluble unless philosophical language is recognized as having uncountably many expressions, but is soluble easily if it is so recognized.

5.1.4.3.1 *The Realism/Anti-Realism Debate as a Dead End: Reasons and Consequences*

In addition to having its own basic structures, philosophical language must be so conceived that it makes possible the solving of fundamental philosophical problems. Particularly important here is the structural moment that can be termed the *extent* or *expressive power* of philosophical language. Current debate concerning the now-famous realism/anti-realism problem can provide initial clarification of the relevant task. The debate itself revives the earlier opposition between metaphysical realisms and transcendental or absolute idealisms, but transforms it on the basis of two new factors that have relocated the basic data of the problem to a fundamentally new level. The first factor is the central position that is now accorded to language and thus to semantics; the second is the no less important acknowledgment of the central role played by the natural and social sciences with respect to formulating and solving philosophical problems. These changes are in part responsible for the fact that the problem is the topic of what is now generally termed "the realism/anti-realism debate." One of the central points of contention in this debate concerns whether one must assume that there is a world independent of language (and/or of mind) or whether such a world 501 exists, or not. Those who accept or assume such a world are termed "metaphysical realists"; those who reject or deny it are "anti-realists" (or, more recently, "internal," "pragmatic," "direct," or "natural" realists).[12]

It has been clear for some time that the debate has arrived at a dead end. It is not difficult to discover why. In what follows the two presumably most important reasons are analyzed in detail, along with one consequence that emerges from them. The two reasons are two errors that those who take part in the debate continue to make.

[1] The first reason or error is the failure to problematize the putatively clear notions of dependence on, and independence from, mind and/or language. Virtually all parties in the debate understand the relevant mind and language to be *our* language or *our* minds. If this is presupposed, a radical aporia is unavoidable: because *our* minds and *our* language are contingent, they are from the outset factors that are too negligible to serve as absolute with respect to the determination (i.e., here, the existence and the structuration) of the world or universe, which, after all, was around long before there were human minds or human languages. But if, from their inadequacy as such decisive factors, one concludes that the world or universe must be absolutely independent of language and thus of structuration, what results appears to be unintelligible: a non-structured, utterly unarticulated and inarticulable, hence un-understood and ununderstandable "world." This is the metaphysical realist's aporia.

Anti-realists are confronted by an aporia that is, in a certain respect, precisely the opposite of that of the metaphysical realists: if they make *our* minds and/or languages the unconditional factor and thus measure for any intelligible world, then they can no longer coherently maintain the existence of what both ordinary language (in an intuitive and imprecise manner) and particularly the language of physics (yet more specifically, that of cosmological physics) term the "real world" or "the real universe, in all of its dimensions." The only coherent (which is not to say acceptable) way out of this aporia, under anti-realistic presuppositions, is to *restrict* the dimension of the world, i.e., concretely, *to reduce* it to the world-as-accessible-to-us. This is the world known among philosophers for roughly the past hundred years as the lifeworld; it is the world that comes all too clearly into appearance in Putnam's most recent developmental phase. But this reduction is philosophically untenable, because it yields (among vari- 502 ous other things) the consequence that whatever is not compatible with the so-called lifeworld—and that would include much of the knowledge of the world that arises

---

[12] The debate is not of course unrelated to the problem of the gap, which is thematized in this chapter. The current section focuses on somewhat different aspects of the debate.

from the immense labor and the undeniable progress of the natural sciences—would have to be deemed empty and senseless, indeed fictitious.[13]

If the dependence or independence in question is on or from *our* language and/or *our* minds, the dilemma just introduced is inescapable. But why should this dogma be maintained? The following paragraphs argue that it should not.

[2] The second reason or error is, in the case of anti-realism, the *anti-ontological* orientation of all philosophical movements standing under the influence of Kant and, in the case of (metaphysical) realism, in its *insufficient and indeed inconsequent ontological perspective*. More specifically, it is a matter in the former case of a total misrecognition and in the latter of a short-sighted and superficial treatment of the thought, articulated above, of *ontological expressibility*. What, after all, does it mean to say that something is "conceived" or "explained" or "known" or, in a word, "expressed"?

[i] The anti-ontological position characteristic of anti-realism interprets everything involved with conceiving, understanding, explaining, expressing, etc., as accomplished exclusively by us as knowers or speakers. This position is articulated sharply in a historically accurate manner in such famous passages from Kant as the following:

> Understanding does not acquire its laws (a priori) from nature, but prescribes them to nature. (*Proleg* §36)
>
> [L]aws exist just as little in the appearances, but rather exist only relative to the subject in which the appearances inhere, insofar as it has understanding, as appearances do not exist in themselves, but only relative to the same being, insofar as it has senses. The lawfulness of things in themselves would necessarily pertain to them even without an understanding that cognizes them. But appearances are only representations of things that exist without cognition of what they might be in themselves. As mere representations, however, they stand under no law of connection at all except that which the connecting faculty prescribes. (*CPR*: B164)

503

Understanding as the legislator of nature—this expresses the heart of the anti-realist position. Here, no place is left for a sensible ontology; on the contrary, everything revolves exclusively around subjectivity, whose highest point is its understanding (see B134). This position has given rise, in the history of philosophy and quite particularly in recent years, to a wide variety of philosophical anti-realisms.

One can only be amazed that such conceptions have been championed at all, and for so many different reasons. One reason is gotten at by means of this question: what

---

[13] Worth noting again is that this anti-realist conception is a variant of the transcendental-philosophical turn inaugurated by Kant; the turn results from introduction of the fateful distinction between "things in themselves" and "appearances," thus between the "world-in-itself" and the "world-as-appearance." Kantians of all sorts then accept this distinction as central to philosophy, taking it then to be obvious that we can know only the world-as-appearance. Kant of course does not reduce the world as it is in itself to the world as it appears; current anti-realists retain this basic view in a certain respect, in that as a rule they do not directly deny the existence of a world independent from our minds and our language. They do, however, assert that it makes no sense to acknowledge such a world, because nothing is to be done with it. In this respect, the anti-realist position is deeply obscure and ambivalent—in opposition to Kant's position, which in this respect is at least clear (which, again, does not mean that it is tenable).

is gained if we do not grasp and articulate the world itself or the things themselves, but instead consider *our* conceptual and theoretical structures as making the laws to which nature is subjected? What is thereby attained? It is often said of late that this position opens the way for scientific assertions and theories that make possible and indeed at times achieve secure predictions. But how is the success of predictions to be understood if the things themselves are not grasped and articulated, if what occurs is merely that *our* subjective structural apparatus is forced onto nature? *Nature itself* does not allow itself to be forced; *our* theoretical game thus remains merely subjective, without any genuinely objective significance. It could have objective significance only if the "things themselves" entered the picture in a significant way, but for the anti-realists, they do not.

[ii] In opposition to the anti-ontological position of anti-realism, metaphysical realism fully acknowledges the importance of the thought of *ontological expressibility*. According to this position, our scientific concepts, assertions, theories, etc., articulate reality or the world as it actually is. Even if the realist position is not generally formulated so explicitly, it can be understood as taking expressibility to be an immanent— genuinely ontological—structural moment of the things themselves.

Nevertheless, however, this realism stops halfway. Recognition of the genuine ontological expressibility of things themselves or the world itself remains a grand intuition or thesis or assertion, lacking sufficient philosophical intelligibility. Above all, insistence that the "real" world, which we seek in philosophy and the sciences to grasp and to articulate, is a dimension that is *independent of us* (of *our* language, of *our* minds) is an assertion that is intelligible and plausible only at first glance and thus only superficially, but that is ultimately unintelligible and anything but robust. Like the anti-realists, the realists understand independence from *us*—from *our* languages and/ or *our* minds—as independence from language *in general* and from minds *in general*. But this makes the expressibility of the world something simply mysterious. One fails to note that expressibility is a relation that is explicable and thus intelligible only when its inverse is also acknowledged: X is expressible only if there is a Y that can express X. If one maintains that X is in this case fully independent of (every!) Y, then the concept of expressibility is contradicted. How one designates the means of expression is an insignificant matter of terminology. The obvious candidate is of course "language" (in a quite broad sense). And indeed—as shown in the following paragraphs—there are no good reasons not to select this candidate.

504

[3] Before the wide-reaching consequences of the preceding considerations can be drawn, thesis three (Section 5.1.3.3) must be recalled explicitly: the utter totality of the world (the universe, etc.), with all the things it contains, is expressible; expressibility is an immanent structural characteristic of the world (universe), with the consequence that it is *coextensive* with the world (the universe). What grounds this central assumption or thesis is that it is presupposed by even the slightest and least significant theoretical step. If one did not *assume in advance* that that to which a given theoretical statement related was expressible, then making the statement would be nonsensical. *Universal ontological expressibility* is an absolutely immovable basis for any and every theoretical undertaking. Any such undertaking consists precisely of the attempt to *make manifest* the expressibility of the world (i.e., to *express* what is *expressible*).

5.1.4.3.2 *An Essential Presupposition for the Universal Expressibility of the World (of Being): Theoretical Languages with Uncountably Many Expressions*

The preceding analysis makes clear the following fundamental fact: the universal expressibility of the world presupposes or (in another respect) entails a universal means of expression that is here called "language." It should be immediately clear that this presupposed or entailed language cannot—certainly not without extensive specification and basic corrections—be any natural or ordinary language, at least if such languages are understood, as they usually are, as purely contingent human products that are quite limited in scope. The universal language required here must be of a wholly different character. For reasons presented in the following sections, it must be understood as an abstract semiotic system consisting of uncountably many expressions.

5.1.4.3.2.1 **The Possibility in Principle of Semiotic Systems with Uncountably Many Signs/Expressions**

[1] With respect to any semiotic system, a distinction is to be drawn between signs as types and signs as tokens. Sign-types are abstract entities, whereas sign-tokens are instances (typically utterances or inscriptions), thus entities in space and time. In the following paragraphs, the signs in those semiotic systems we call "languages" are usually termed "expressions."

If one wanted to answer the question of how many expressions (for example) the English language consists of by identifying "linguistic expressions" with tokenings, the only possible answer would be that English has finitely many expressions. But how could the number of tokenings be determined? In principle, one could identify a specific period of time, such as the current day. But if one wanted in addition to determine the total number of tokenings, one would have to include *the entire past and future*—a fantastic and senseless undertaking. Even ignoring the practical problems, however, the number of expressions a language includes or consists of cannot be simply identified with the number n of tokenings, because this would be a purely extensional determination that paid no attention at all to the internal relations among the signs. It quickly becomes clear that a given sentence, e.g., "Snow is white," can have various tokenings, including "Snow is white," "*Snow is white*," "**Snow is white**," "SNOW IS WHITE," etc. The (complex) expression that is here named the sentence itself is an abstract entity, a type.

A given semiotic system consists of simple or basic signs and the chains or strings that can be formed from them.[14] The basic signs are the elements of the alphabet and the punctuation marks; in the case of ordinary languages, the number of these is unquestionably finite. Whether languages with countably infinitely many basic signs or in addition ones with uncountably many basic signs can be conceived of is not considered because (as becomes clear in the following paragraphs) that issue is not relevant here (or at least not decisively so). It is clear, however, that from the basic signs, countably infinitely many strings can be formed. The string that is of simply central linguistic importance is the sentence. Ordinary languages thus contain at least countably infinitely many sentence types.

---

[14] The basic signs considered here are those of written semiotic systems. That there are semiotic systems that are not written is not relevant to the issue under consideration, because for any such system, a written counterpart can be constructed.

[2] Could there be languages containing *uncountably many expressions*? Almost always in the past and still into the present, analytically oriented philosophers have accepted the following assumption or thesis, which has by now become a kind of dogma:

(LC)   The set of expressions constituting any language is countable.

As a rule, this thesis is simply presupposed, along with the semantics that goes with it. In some cases, supporting reasons are provided, although even then the support is generally indirect. Thus, for example, it is suggested that to be learnable, a given language must be countable. But this line of argument is questionable, because it presupposes a purely atomistic understanding of language—it presupposes the unrestricted validity of the semantic principle of compositionality. Chapter 3 shows, however, that this principle is to be rejected in the case of philosophical languages. An additional and more important problem with the argument from learnability is that it presupposes that to qualify as a *language*, a given semiotic system must be *learnable*. The following paragraphs demonstrate that semiotic systems with uncountably many expressions can qualify as languages even though they are not learnable in the usual sense; talk of the learnability of such systems can only mean that the basically mathematical procedures with whose aid such systems are constructed can be mastered and applied correctly.[15] Further clarification of this issue is provided by Section 5.1.4.3.2.3, "The Status Of Tokening Systems for Theoretical Languages." 507

As noted in various contexts above, the philosophical language envisaged here can be—and, in the expected normal case, is—connected to an ordinary language. From the basic signs of a language and the (at least) countably infinitely many strings of those signs that can unproblematically be formed from them, one can, with simple set-theoretical means, construct a semiotic system with uncountably many expressions. The construction works as follows (see Hugly and Sayward 1983, 1986). Strings are understood most clearly as sequences, in the mathematical sense. The sequence $<s_1, \ldots, s_n>$ is defined as a string or sequence of basic signs if, for every positive natural number i, s_i is either a basic sign or a sequence of basic signs. A sequence s is infinite if and only if for every positive natural number n, $s \neq <s_1, \ldots, s_n>$, or some term from s is an infinite sequence (1986: 46).

Set-theoretical considerations make it clear that, from a finite vocabulary and exclusively finite sequences, only a *countably* infinite semiotic system can be formed. But if from the finite vocabulary even countably infinite sequences can be formed, then it is set-theoretically possible to form a system with *uncountably* many sequences or expressions. The proof involves the application of Cantor's procedure of diagonalization: with the introduction of the axiom of the power set, the path is opened to uncountability. That countably infinite sequences must be accepted is obvious from the fact that (for example) there are infinite conjunctions of those sequences that are called "sentences."

---

[15] This use of the term "constructed" in no way indicates that the philosophical conception of mathematics termed "constructivism" is here relevant. "Constructed" here refers simply to the concrete procedure used to develop such systems.

An infinite conjunction might be characterized as an infinite sequence whose n*th* coordinate is "and" if n is even, and whose n*th* coordinate is a sentence of English if n is odd. Using trivial truths about English plus set theory, the existence of such an entity can be established. (Hugly and Sayward 1986: 59)

5.1.4.3.2.2 A Fundamental Problem: Language and "Tokening System" (the Position of Hugly and Sayward)

[1] How, more precisely, are such semiotic systems with uncountably many signs (expressions) to be understood? Are they languages? To this question, Hugly and Sayward answer in the negative. They base their answer on the following criterion: semiotic systems with uncountably many signs or expressions are *abstract* systems, resulting from set-theoretical constructions. Languages, on the other hand, are semiotic systems of *communication*. In principle, communication does not exclude the possibility that the corresponding system be abstract. Abstract semiotic systems that are to count as languages must however satisfy an additional criterion: all signs and expressions of any such system must be tokenable as "perceptible particulars" (1983: 75). The standard form for tokening is the inscription of the abstract sign or expression.

Hugly and Sayward attempt to prove that no semiotic system with uncountably many signs and expressions can have a tokening system, i.e., "a finitely formulable set of instructions for constructing perceptible particulars." The argument is as follows:

1. For every language L, the set of expressions of L is a formal system.

2. For every formal system A, A has a complete tokening system if and only if A is countable.

3. For every language L, the set of expressions of L has a complete tokening system.

4. Thus, for every language L, the set of expressions of L is countable. (1983: 75)

Premise (1) is explained easily. The expressions of a language comprise a set $A = B \cup C$, where B is the set of basic signs of the language A, and C "some set of non-unit strings from B." According to Hugly and Sayward, such expressions are *abstract objects* in that they satisfy the following two conditions: "they are not sensible and are available, and evidently so, in infinite supply" (75). That the expressions, as abstract objects, are not "sensible" follows from the fact that the negation of this contention would entail the non-existence of all non-sensible or untokened expressions—a conclusion that is untenable. That a language can contain (countably) infinitely many expressions is accepted by Hugly and Sayward as evident (see above).

Premise 2 is the decisive one. To understand it correctly, it is helpful to consider an example of a formal system with uncountably many expressions. Hugly and Sayward rely on Joseph Shoenfield's (1967) formal system $L(\mathfrak{a})$, which is a first-order predicate language. The first step is to define, for such a language, a structure that contains an interpretation of the language. The structure consists of (1) a non-empty set, the domain of $\mathfrak{a}$ (symbolized $|\mathfrak{a}|$), whose elements are called, in first-order predicate language, "individuals" or "objects," and (2) an interpretation of the function-signs and predicate-signs of L. As for the domain of $\mathfrak{a}$, the following is stipulated:

Let $\mathfrak{a}$ be a structure for L. For each individual a of $[|\mathfrak{a}|$, the domain of] $\mathfrak{a}$, we choose a new constant, called the *name* of a. It is understood that different names are chosen for different individuals. The first-order language obtained from L by adding all the names of individuals of $[|\mathfrak{A}|$, the domain of] $\mathfrak{A}$ is designated by $L(\mathfrak{A})$. (Shoenfield 1967: 18)

Hugly and Sayward next make a quite significant point. They note correctly that Shoenfield relies upon a substitutional definition of truth. It is to be understood as follows: a substitutionally understood existential quantification is true in $L(\mathfrak{A})$ if and only if it has a substitution instance that is true in $L(\mathfrak{A})$; a substitutionally interpreted universal quantification is true if and only if the formula preceded by the universal quantifier is true for all grammatically allowable substitutions for the variable governed by the quantifier.

[2] There arises here a problem identified particularly by Quine. Often—indeed usually—quantifiers are understood not substitutionally, but *objectually*. So understood, the universal quantifier "$\forall x$" in first-order predicate logic means "every x is such that," and yields a true sentence if and only if the formula that it precedes is satisfied by every object that, within the chosen or presupposed domain, is a value for the bound variable x. The existential quantifier, for its part, means "there is at least one x such that," and yields a true sentence if and only if the formula that follows it is satisfied by some object from the domain of quantification. Objectually interpreted, the quantifiers are thus understood in a genuinely *ontological* manner. This however leads to the question whether the two interpretations of the quantifiers correspond (or coincide or are equivalent). They do, *if* one makes or accepts the following assumption: there is a one-to-one correspondence between the names contained in $L(\mathfrak{A})$ and the items within the domain of $\mathfrak{A}$: to every individual in the domain of the universal quantifier, a name in $L(\mathfrak{A})$ is ascribed, such that different objects correspond to different names.

Assume now that the domain is an uncountable set, such as the set of real numbers. In this case, $L(\mathfrak{A})$ must also have uncountably many names and thus also uncountably many sentences. But how could we demonstrate that in this case there is a one-to-one correspondence between the names and the objects (or the sentences and the facts)? According to Hugly and Sayward, we cannot, because the demonstration would require a tokening system that is neither available nor possible.

The proof consists essentially of the clarification and development of such a   510
tokening system in a manner yielding as a consequence that no semiotic system with uncountably many signs (or characters) can have any such system. The central point is the assumption that the tokening of a sign is to be understood *in terms of communication*. The tokening system is defined as follows:

A tokening system T for a formal system $A = B \cup C$ is a finitely formulable set of instructions which associates with each $x \in A$ a rule for constructing a perceptible particular and further satisfies at least these three conditions:

(i)   for each $x \in A$ and possible world W and $y \in B$ with i $(i > 1)$ occurrences in x, if some a of W is a T-perceptualization of x, then a has i T-perceptualizations of y as parts;

(ii)   for each $x \in C$ there is a possible world W in which a T-perceptualization of x results from a finite assembling of T-perceptualizations, one for each occurrence of each $y \in B$ in x;

(iii)   for each $x \in A$ there is a possible world W in which there are distinct T-perceptualizations and some finite processing of at least two such T-perceptualizations which establishes that they each are T-perceptualizations of x. (Hugly and Sayward 1983: 80)

Given this definition, Hugly and Sayward reason as follows:

[A] formal system $A = B \cup C$ is uncountable only if its set B of basic characters is uncountable or its class C of non-unit strings from B has among its elements infinite strings from B. In the latter case items (i)-(ii) above preclude there being a tokening system for A. In the case where B is uncountable item (iii) above precludes there being a tokening system for A. (81)

[3] In their second essay, Hugly and Sayward make this thesis and argument more precise (and in part correct it) by presenting the concept *tokening system* more thoroughly and in more detail. They argue that on one interpretation of "tokenable," not even all the finite expressions of a given language are tokenable. For example, no human being could be in the position to token

(X)   the sentence of English beginning with 'John is' followed by 1000^{1000} occurrences of 'the father of' and ending with 'someone' (1986: 47; there "(1)," not "(X)").

Hugly and Sayward distinguish sharply the "series of marks which tokens a *description* of an expression" from "a token *of that expression*" (emphasis added). They emphasize that a string that tokens a description of an expression must not be mistaken for a token of this "expression itself"—and add that (X) is a series of marks that tokens such a description, but does not token the sentence that is described. From this correct observation the two draw a problematic conclusion, as shown below.

Hugly and Sayward next introduce the concept of the representation of a language: a representation (R) of a language L has three components: one syntactic, one semantic, and one communicative. The syntactic component is a set A containing the set B of basic characters and the set C of strings formed from elements of B—thus (as before) $A = B \cup C$. The semantic component is a model M for A. The communicative component is a function T whose definition includes an $A' \subset A$ (as argument domain) and, for every $x \in A'$, an associated instruction for the construction of a unique, perceptible individual object (as value); the resulting triple is R = <A, M, T>. It then holds that "x is an expression of the L relative to R only if $x \in A$ and T is defined for x" (49).

Hugly and Sayward then (60) distinguish between the abstract representation of a language and the language's "concrete social-material reality."[16] This also introduces a fundamental distinction with respect to the concept of tokenability. If one takes this procedure only abstractly, the result is the following:

To say that a string ψ is tokenable in L is to say that, for some representation <A, M, T> of L, T is defined on ψ. In this sense infinite sentences are tokenable in English. (60)

---

[16] Hugly and Sayward (1986: 60) write, "There is a fundamental distinction between an abstract representation of a language and its concrete social-material reality." In response to an e-mail query in August 2004, Sayward confirmed that the "its" is meant to have "language," not "representation," as its antecedent.

But Hugly and Sayward then restrict the meaning of "expression," writing,

> [T]okenability of an infinite sentence ψ is not sufficient for its being an infinite expression. A further condition is that no finite sentence expresses ψ's truth-conditions. So the question
>> Does English contain infinite expressions?
> comes to
>> Is there a representation of English which is reducible to no token-finitistic representation of English?
>> This question would have a fairly obvious positive solution if it were possible for a tokening system to have an uncountable domain. But this is not possible. (60)

The last assertion results from a more precise (and strongly restricted) determination   512
of "tokening system":

> Let S be any set of strings from set B of basic characters. Say that a basic character $b \in B$ occurs in a string $s \in S$ just in case, for some positive integer i, $b = s_i$ or b occurs in s_i (s_i being the ith element of s). Then
>
>> (***) T is a tokening system for S only if T associates with each string $s \in S$ an instruction for constructing a perceptible particular which (if constructed) would result from a finite assembling of tokens, one for each basic character b which occurs in s and with just as many tokens of b as there are occurrences of b in s. (54–55)

Hugly and Sayward consider thoroughly one basic objection that can be raised against (***). The objection is that the problem here is merely practical. Assume that a speaker U is familiar with the syntax of a fragment of English, and also the definition of a function g. On this basis the speaker can easily develop and apply a tokening system for infinite strings of English. To this, Hugly and Sayward respond by distinguishing between strings and expressions, insisting that not every string is an expression. They concede that even infinite strings are tokenable. But "tokenability of an infinite sentence $f(\varphi)$ does not suffice for calling it an infinite expression" (57). To be an expression, a given string must satisfy an additional condition: the truth-conditions of the infinite sentence must be expressible (i.e., tokenable). This is however possible only for "token-finitistic" representations of languages, which rule out the possibility of uncountably many linguistic characters. Two definitions are introduced to clarify and justify this claim:

Df. 1   A representation <A, M, T> is token-finitistic just in case, for no $x \in$ the domain of T, x is an infinite string.

Df. 2   If R = <A, M, T> is a representation of a language L and ψ is an infinite sentence of A, then ψ is an infinite expression of L, relative to R, just in case (i) ψ belongs to the domain of T and (ii) R is not reducible to a token-finitistic representation of L.

The truth conditions of any infinite expression (i.e., sentence) therefore (58) cannot be satisfied. The authors support this thesis as follows:

> For there to exist a complete tokening system for $B \cup C$ there must exist a finitely formulable set G of general rules from which a specific tokening instruction of each

x ∈ B ∪ C can be derived. (. . .) By observing G it must be possible to determine with respect to a pair of objects *a* and *b* that they token the same expression, since otherwise communication fails. This condition cannot be met if B ∪ C is uncountable. (58–59)

To clarify: T is a tokening system for a semiotic system B ∪ C only if for every string z from B ∪ C there is an instruction for the construction of a perceptible individual thing that *makes the structure of z perceptible (visible)*. If, for example, a name occurs five times in a sequence z, then a tokening of z must, according to the instructions for T, have as parts five tokens of this name. This is realizable only if B ∪ C is countable.

To understand this conception more precisely, it is helpful to consider more closely an example that Hugly and Sayward introduce (59):

> The essentials of the situation can be grasped by considering the case in which one rule serves for G [i.e., presumably, the finitely formulatable set G of universal rules consists of a single rule]. Such a rule *functionally* correlates some variable property differentiating the elements of B ∪ C with some variable perceptual property. For example, suppose B ∪ C is denumerable and is enumerated by e. Then a general rule for tokening the elements of B ∪ C might be
> To token x ∈ B ∪ C draw a line e(x) centimeters long.
> The variable property discriminating the elements of B ∪ C here is the property of being the n*th* B ∪ C relative to e. The variable perceptual property is the property of being a line j centimeters long. Suppose now that B ∪ C is uncountable. Then the variable property differentiating the elements of B ∪ C will be *continuous*. In that case the functionally correlated perceptual property will be a continuum property. And in that case it will be impossible to determine with respect to a pair of perceptible particulars *a* and *b* that they are tokens of the same expression.
> Consider an example. Suppose C is the set of all infinite decimals standing for the reals between 0 and 1. Let G consist of this rule:
>
> To token an infinite decimal <a_1, . . .> construct a line segment r centimeters long where r is the real number represented by <a_1, . . .>. (58–59)

According to this rule, with every infinite decimal is associated an instruction for the construction of a single, perceptible, individual thing. This, according to Hugly and Sayward, has the following consequence:

> The rule associates with each infinite decimal in C an instruction for constructing a unique perceptible particular. But in using the rule you can *never* tell that distinct line segments *a* and *b* token the same infinite decimal. No matter how fine your discrimination, line segment *a* could be indenumerably many distinct lengths other than *b*, but only indiscriminably so. (59)

Hugly and Sayward conclude that their tokenability conditions must be satisfied by every possible tokening system, for otherwise communication could not occur. They conclude further that no tokening system can be given for any semiotic system with uncountably many expressions, so no such system is a language.

5.1.4.3.2.3 The Status of Tokening Systems for Theoretical Languages
The theory presented by Hugly and Sayward is of great consequence, especially because it touches on central aspects of systematic philosophy and raises an extensive series of

fundamental questions. In the following, the most important aspects and questions are developed and treated in detail.

[1] As emphasized above, particularly in Section 5.1.3.1, philosophical and scientific languages are languages of presentation, not of communication (in the specified senses). This basic fact has radical and far-reaching consequences for the issues under current consideration. Because philosophical languages are a languages of presentation, linguistic aspects grounded in structures of communication play for them at most a derivative role—if they indeed play any role at all. This holds with respect both to the tokening systems Hugly and Sayward describe and to the distinction they introduce between the abstract representation of a language and its concrete social-material actuality. On the whole, abstract representation is characteristic of presentation, and concrete social-material actuality of communication. The significance of these facts must now be considered.

Hugly and Sayward's distinctions among the sentence itself, descriptions of the sentence, and perceptible tokenings of the sentence are quite problematic in more than one respect. The two appear to identify the sentence itself with the token of the sentence, but this cannot be correct. The sentence itself is an abstract entity that can have various modes of presentation and thus various tokenings. Among these are descriptions of the sentence. For this reason, *the social-material or perceptual tokening* of a sentence is an ambiguous concept, precisely because the description is also a mode of presentation and thus a tokening having a differently configured form. This form of tokening has the immense advantage that it enables us to present sentences that we could not present in tokenings of the sort envisaged by Hugly and Sayward as the only one. This is clear from their own example, introduced above: 515

(X)   the sentence of English beginning with 'John is' followed by 1000^{1000} occurrences of 'the father of' and ending with 'someone.'

If it is objected that the description of the sentence does not present the sentence itself because it leaves it unqualified, an adequate response is at hand. One must merely make the qualification explicit in one way or another. For example,

(X')   the sentence of English beginning with 'John is' followed by 1000^{1000} occurrences of 'the father of' and ending with 'someone' *is true* or *has been asserted* (etc.).

With respect to the status of the sentence, this description is better than a Hugly-Sayward tokening, which does not make the status explicit.

[2] Example (X) yields another extremely important insight, because it shows that even for the language from which it is drawn—ordinary language, even understood as a countable (finite or infinite) semiotic system—it is not the case that all sentences with all their components can be explicitly tokened in complete detail. This shows that language is an immense dimension that encompasses us, as speakers, in that we can access—realize, concretize, token, present, use (inscribe, utter), etc.—at most a determinate and relatively tiny segment from it. It would be simply arbitrary and unrealistic to identify language with whatever segment of language some one or

more language users happen actually to use. Given this, it is hard to see any obstacles to conceiving of philosophical and scientific languages as semiotic systems with uncountably many expressions. And indeed, as considerations now to be introduced show, there are good reasons for holding the theses that there are such systems and that they are indispensable.

What is shown above with respect to languages consisting of only finitely many or of countably infinitely many expressions—that only small segments of such languages can ever be tokened or used—holds yet more obviously for languages consisting of uncountably many expressions. At no time can more than a minuscule segment of these languages be effectively extracted or realized or used. A consequence is that *all three* sorts of languages, albeit in different ways, require specific conventions and stipulations that are necessary consequences of the segmental—extremely limited—real forms (really *used* forms) of these languages. Such a consequence is visible in ordinary languages in the fact that their segmental reality originates and develops historically, changes virtually daily, etc. The segmentality of these languages appears as what Hugly and Sayward call their "concrete social-material reality."

The segmental reality of philosophical and theoretical languages is to be conceived of somewhat differently. These languages, too, have historical origins, and they too develop; moreover, they are connected in various ways with ordinary languages. But what distinguishes them fundamentally from ordinary languages is that they do *not* develop naturally (i.e., through generally unreflective communicate use). On the contrary, they are precisely *regimented* languages, regimented with respect to the aim of the most adequate possible presentation of theoretical subject matters and interconnections. One result of this regimentation is that determinate theoretical frameworks are specified within which theories can then develop. Any such theoretical framework establishes (or should establish) how the theoretician is to deal with the framework's semiotic system consisting in principle of uncountably many expressions. For this purpose, many means, primarily formal ones, are available.

[3] The preceding considerations reveal that it is not requisite to provide, for the assumed language with uncountably many expressions, a tokening system of the sort that Hugly and Sayward demand. To insist that the concrete and effective use of a philosophical or scientific language consisting of uncountably many expressions requires, for each of those expressions, "an instruction for constructing a perceptible particular which would render the structure [of the expression] perceptible" (1986: 55) would be to insist that, as shown above, that language meet a demand that cannot be met even by finite languages. Moreover, there is no reason to require that scientific or philosophical accounts identify explicitly all the individual items covered by universally quantified sentences. For philosophy and science, it is fully appropriate to formulate universally quantified sentences; that is how principles, axioms, laws, etc., are ultimately formulated. That the detailed social-material structure of the expression is not thereby made "perceptible" is neither a disadvantage nor a deficiency. A procedure like quantification provides an elegant and highly efficient scientific form of presentation of states of affairs that obtain universally.

A second form of efficient presentation is, as has been noted, *describing* sentences whose structures are not materially tokenable in complete detail. Again, the sentence described in (X) is a good example.

A third form of presentation is the *definition*. It is scarcely contestable that the concepts *uncountable set* and *determinate* or *particular uncountable set* can be defined. Such definitions are available, and they play important roles in the formal sciences and thus in the sciences in general.

There are additional forms of theoretical presentation for which it is not necessary and not sensible to grant Hugly and Sayward's demand for the development and use of their sort of tokening system. In this respect, formal means provide virtually inexhaustible possibilities.

5.1.4.3.3 *The Segmental Character of Effective Theoretical Languages*

The segmental character of ordinary as well as philosophical languages, introduced above, broadens the perspective of what is to be considered in relation to the problematic of philosophical language as a semiotic system with uncountably many expressions. From within this broader perspective, several additional points are to be emphasized and explained; they clarify central aspects of the problematic in a manner that reveals the continuum argument used by Hugly and Sayward in support of their thesis to be untenable. Without that argument, the thesis is unsupported.

[1] The first point concerns the question of how to determine precisely the relationship between language and world (actuality, universe, being). If language and world were two fully separate dimensions, fully independent from each another, there would be the insoluble problem of how they could ever connect. How could they intersect in such a way that the world would be articulated by means of language? So fully externalist a conception cannot explain the interconnection.

In this book, the interconnection of language and world is articulated in various contexts, particularly in the formulation of the thesis that semantics and ontology are two sides of the same coin. In the current context, a simply fundamental aspect of this thesis is crucial. The thesis itself can be understood in various ways. This book relies on a strong version, according to which the fundamental unity of language (semantics) and ontology is conceived as entailing that access to the world (to the universe, to being) is absolutely language-mediated, occurring from inside language, so to speak, or within the linguistic dimension itself. If "language" is specified to be a fully developed theoretical language and then, abbreviating, one identifies language and theory, then Quine's (1981: 21–22) contention follows: 518

> Truth is immanent, and there is no higher. We must speak from within a theory, albeit any of various.

This thesis can also, however, be understood in a manner that is untenable; it is untenable if it makes the world (and thus also truth) simply reducible to the dimension of *our* language, where "our language" is taken in a quite ordinary or banal sense. As shown above, this reduction can take either of two forms, first that of a *complete* reduction of *the world* to *our world*, second, that of a dichotomous reduction that introduces a gap between "our world" and "the world in itself," whereby the latter is taken to be wholly inaccessible to us. As shown above, neither form is defensible.

Everything changes when language is identified not with our language, in any usual sense, but with a semiotic system with uncountably many expressions. There is then no reduction to anything like *our* language, nor is there any commitment to any dichotomy of for us vs. in itself. Instead, language is understood as the counterpart of the universal expressibility of the world.

In what follows, when—for sake of simplicity—there is talk of "linguistic expressions," the expressions in question are always *primary sentences* of the philosophical language that this book introduces and upon which it relies. "Language as semiotic system consisting of uncountably many expressions" is thus to be understood as "language as semiotic system consisting of uncountably many primary sentences." This is a consequence of the semantics developed in Chapter 3.

On the basis of this fundamental and central thesis, two additional specifications are to be made with respect to the "correspondence" between the set of linguistic expressions (i.e., primary sentences) and the set of entities in the world (i.e., primary facts). First, the cardinal number of the set of primary sentences in the philosophical language must be at least as great as the cardinal number of the set of the entities constituting the world. Second, the mapping of the set of primary sentences to the set of entities cannot be merely *injective*,[17] because that would conflict with the basic principle of the expressibility of being as a whole. Third, the mapping also cannot be only *surjective*,[18] for if it were, the possibility could not be excluded that two (or more) primary sentences could be mapped to one and the same entity (one and the same *true primary proposition* or *primary fact*). According to the semantics defended here, however, it is not even the case that two (or more) tokens of one and the same type of primary sentence express exactly the same primary proposition or primary fact; *a fortiori*, it cannot be the case that two different sentence types *and* their tokens express the same proposition. Strictly speaking, concepts like *synonymy* have in principle no application in this semantics (which does not exclude a pragmatic synonymy—one established for pragmatic purposes). This thesis is both extraordinarily strong and extraordinarily counterintuitive. But it is indispensable for the clarification of an extensive series of fundamental questions and the avoidance of various difficulties (see Section 3.2.2.4.1.3).

Whether the mapping of the philosophical language to the dimension of the world is *bijective*[19] is a question that brings with it a difficult problem. The considerations introduced thus far are grounded in three theses that form the basis for a theory of the correspondence of language and world. First, the possibility must be excluded that there be entities (elements of the world, of being as a whole) for which there are no primary sentences. Globally: no entity evades the dimension of language. The reason is that such an entity would not be expressible. Given the rejection of such entities, it

519

---

[17] A mapping from a set *A* to a set *B* is *injective* if the relation is unambiguously reversible. Formally, *f* is injective if and only if $\forall x, y \in A \; [f(x) = f(y) \to x = y]$. This mapping thus does not exclude the possibility that the range of values is not identical to the set *B*. If the relation between primary sentences and primary facts were injective, there could thus be primary facts to which no primary sentences would correspond.

[18] A function $f : A \to B$ is *onto B* or is *surjective* when every element of *B* is the value of *f* for some element of *A*: for all *b* in *B*, $b = f(a)$ for some *a* in *A*. Surjection does not exclude the possibility that some member of *B* is assigned to more than one member of *A*.

[19] A function is a *bijection* when it is both injective and surjective. A bijective mapping or function is thus always a one-to-one correspondence, but it is more than that in that it is also surjective (injective mappings are also one-to-one correspondences).

follows that the philosophical language must have at least the cardinality of the world (i.e., the number of primary sentences must at least equal the number of primary facts). Second, there must be assumed to be a one-to-one correspondence between individual primary sentences and individual entities in the world. Third, the answer to the question whether the cardinality of language is greater than the cardinality of the world, thus whether there are more primary sentences than there are entities in the world, must be subtly differentiated. That is, the concept *world* must be differentiated.

For the sake of simplicity, this book often uses the term "world" to designate   520
the most comprehensive of all concepts: the concept of the unrestricted universe of discourse or of being as a whole. In the contexts wherein it is so used (chiefly in Chapters 1–3) there is no danger of misunderstanding, because these contexts make clear how the use is intended. The difference between *the world* and *being as a whole* is not relevant in those passages; in them, reference is clearly to the *ontological dimension* in its entirety. In Section 4.1.1., however, the concept of the world is determined more precisely. It is there made clear that the world, understood in a more precise sense, is *not identical* to being as a whole or the unrestricted universe of discourse. The concept *world* is there subjected to a twofold restriction: first, it is restricted to the domain of actuality (the *actual* world), and second, *within the domain of actuality* it designates only a specific dimension: the dimension designated in Section 5.3 as the *contingent dimension of being*. Because Section 5.3 demonstrates that the *absolute dimension of being*—the *absolute*—is itself actual (that it exists), and because, terminologically, the absolute cannot adequately be designated a "part" of the world, the term/concept "world" is further restricted so as to designate only the *contingent* or *relative dimension of being*. "World" then designates the domain of actual, finite entities.

On this basis, the question whether the cardinality of language is greater than the cardinality of the world, thus whether there are more primary sentences than there are entities in the world, can be answered clearly. The cardinality of language (in the sense presupposed here, as a semiotic system with uncountably many expressions) is greater than the cardinality of the world in the sense just specified—thus, of the world as the contingent or non-absolute dimension of being. But there is a clearly *bijective* mapping from this language to being as a whole (in the strict sense). This consequence follows directly from the central theses explained and supported in this book.

This informally presented thesis concerning the world in the narrower sense can be specified further because it does not exclude a quite determinate—restricted—bijective mapping from language to the so-understood world. If A is a philosophical language, understood as a semiotic system consisting of uncountably many expressions (primary sentences), and B the set of entities in the world (as the contingent or non-absolute dimension of being), then there is a subset $A' \subseteq A$, such that there is a bijective mapping from the subset A' to the set B.

[2] An additional point that is decisive for clarification of the issues under consideration is the thesis, formulated above in various contexts, that every theoretically relevant   521
sentence presupposes some determinate theoretical framework or arises only within such a framework. One of the most essential constituents of any well-defined theoretical framework is the framework's language. If the relevant language is an abstract semiotic system consisting of uncountably many expressions (primary sentences), particular

problems arise. Such a language, as such and as a whole, cannot be universally and concretely used or applied in the manner that a finite or countably infinite language can. It is, however, the task of the theoretical framework to clarify and regiment how the theoretician is to deal with its uncountable language. The use of such a language must in any case be quite limited. It would be illusory to think that one could on any occasion draw unrestrictedly on the entire breadth or potential of such a language, for every theoretical assertion. Thus, in terms introduced in Section 5.1.3.3.3 and repeated in the title of this section, the use of such a language is always segmental: its resources are never effectively available in their entirety.[20]

[3] Given the preceding, the question arises why philosophical or scientific languages must be conceived of as uncountable semiotic systems. To this, the answer is twofold.

First, in a specific albeit quite limited arena, an uncountable language is thoroughly applicable. Section 5.1.3.3.2.3[3] introduces various forms of presentation that make this evident; they include quantifications, descriptions, and definitions. That such forms of presentation be relied upon is essential to any theory. Renouncing them would have fatal consequences for theories in various domains, including particularly those of philosophy and of mathematics.

It is to be kept in mind that the use of such a language makes possible the presentation only of quite general structures. As is emphasized above, as an example, it would be nonsensical to demand that, in the case of a universal quantification in the theoretical domain, each individual instance be explicitly tokened.

Second, the explicit assumption of an uncountable language has an indispensable *explanatory* function and vitally significant consequences. The central consideration is introduced and emphasized above in various contexts: without an uncountable language, the basic phenomenon of universal expressibility is inconceivable.

[4] The preceding considerations provide the basis for a critical analysis of the central thesis of Hugly and Sayward.

[i] The thesis that semiotic systems should be called or considered to be *languages* only if they have or can be provided with tokening systems of the sort described by Hugly and Sayward is one that cannot be grounded adequately. It holds only for semiotic systems that serve either primarily or exclusively for communication, and not for semiotic systems oriented toward presenting theoretical matters. Whether one wants to call systems of the latter sort "languages" is a question of terminology. Given,

522

---

[20] A comparison may be helpful. The fact that a speaker is conversant in English might be taken to entail that the speaker had somehow assimilated English in its entirety; that such an entailment is taken to hold is suggested by talk of "mastering" the language. Little reflection is required, however, to reveal that no human being can ever be in the position of entertaining all possible English sentence types, even if the vocabulary is (arbitrarily and artificially) restricted to that available at some specific temporal point. This reveals that instead of English being "contained," so to speak, in the minds or intellects of its users, the users are "contained" or embedded within it, in that they will never be in positions from which they can survey it exhaustively. In a slogan: language is not within us, we are within language. In philosophizing, we are within a philosophical language.

however, that formal systems are typically called languages, there are no good reasons for refusing to use that word for semiotic systems with uncountably many elements.

[ii] The decisive consideration for Hugly and Sayward (1986: 55) is the demand, "for T to be a tokening system for [the set of strings of basic characters] S, T must associate with each string s in S an instruction for constructing a perceptible particular which would render the structure of s perceptible." This demand is, however, unjustified in various respects. The demand is for the "making visible" of the fine-grained structure of a given expression. That Hugly and Sayward understand this in an extreme fashion is clear from one of their examples: "If s is a string with five occurrences of a name, a token which would result by following the instructions of T must have five tokens of that name as parts." This is not a theoretically sensible demand. There are various ways to do justice to the five occurrences of a name in a sentence without explicitly and "visibly" presenting five tokens of the name.

It also becomes clear that Hugly and Sayward's conception contains a fundamental unclarity: the two speak on the one hand of (A) "a string with five occurrences of a name" and assert on the other that (B) "a token which would result by following the instructions of T must have five tokens of that name as parts." What is the distinction between A and B? What, here, is a "token"? Is "a string with five occurrences of a name" not itself a token? This is the same unclarity, indeed confusion, pointed out above with respect to example (X).   523

What can, does, or should it mean to "render the structure of s perceptible"? Is the relevant structure syntactic or semantic? Because the talk is of strings, it would appear that what the authors have in mind is fine-grained *syntactic* structure. But the syntactic structure of a given sentence is in no way governing or decisive. As shown in Chapter 3, a language can retain a subject-predicate structure for its sentences *on the syntactic level* while eliminating subjects and predicates *on the semantic level*. Within the framework of structural semantics, the syntactic subject-predicate form is reinterpreted, with singular terms (syntactic subjects) understood as convenient and practical *abbreviations* of quite complex configurations of primary sentences, thus of sentences that lack subjects and predicates. Half-formalized, the universal form for primary sentences is, as is noted often above, "It's F'ing." It would be senseless, indeed absurd to demand for complex configurations of such sentences that their fine-grained *syntactic* structure be made "perceptible." With respect to the intended semantics, the syntactic structures of such sentences are in no way governing or decisive. To be sure, a philosophical or scientific language having a semantics like the one developed in Chapter 3 must also take care to develop a syntactics corresponding to its semantics, thus developing the most complete possible correspondence between its semantic and syntactic structures. This correspondence would be fully complete only in the case of an ideal language. As long as philosophy and science must or do, for pragmatic reasons, attach to ordinary languages, the disparity between syntactic and semantic structures will continue.

[iii] Something may now be said in response to the Hugly and Sayward argument illustrated above by the continuum example. The question is whether two perceptible "objects" a and b can be taken to be tokens of one and the same expression. In an uncountable language this is not possible because what in the example makes an expression determinate, and thus distinct from other expressions, is the distinguishing property

of being the n*th* element of B ∪ C. If however B ∪ C is uncountable, the relevant property would be a continuum property, and the perceptible property correlated to it (its token) likewise would be a continuum property. It is then not possible to "determine" *a* and *b* in such a way that they can be said to be tokens of one and the same expression.

Before the decisive point can be made, some aspects of this argument or example must be analyzed. Hugly and Sayward speak quite determinately of the "objects *a* and *b*" and of "the same expression." In what language is this determination accomplished? Obviously, in a presupposed background language. They speak further of the "set of all infinite decimals standing for the reals between 0 and 1"; they then speak of the decimals themselves being tokened by means of line segments of length r, such that "r is the real number represented by [the decimal] <a₁,...>." These formulations motivate the question, what, here, is a "token"? Hugly and Sayward speak also of a series of "things (entities)." Aren't their "expressions," thus their "concrete sentences," themselves "social-material realities" and thus "tokens"?

Basic clarifications are required before there can be talk of a compelling argument. But this issue is not decisive, indeed not even particularly important for the conception defended in this book, because within this conception the uncountable philosophical or scientific language is not conceived as more than segmentally available; as a whole, it is a type of *regulative idea* or metaphysical assumption whose chief function is to play a significant *explanatory role*.

5.1.4.4 Are There Uncountably Many Entities?

Within structural ontology the question raised by this section's title would be, more precisely, are there uncountably many primary facts? But present purposes do not require that this specification be explicit.

[1] A beginning may be made with the presentation and analysis of an argument presented by Nicholas Rescher purporting to establish an affirmative answer to the question. The argument itself is based on a series of problematic assumptions concerning various domains, above all those of semantics and ontology (of the problems, more below). Rescher's thesis (1987: 112–13) is that "[o]ur concept of the real world is such that there will always be non-denumerably many facts about a real thing." He supports the thesis by an argument using a formalization relying upon Cantor's procedure of diagonalization. The argument develops as follows:

[A]ny infinite list or inventory of distinct facts about something will inevitably be such that there will always be *further* facts, not yet included, that do not occur anywhere on it. Thus, let us suppose, for the sake of *reductio ad absurdum*, that we had a non-redundant complete enumeration of *all* of the distinct facts about something:

$$f_1, f_2, f_3, \ldots$$

Then, by the supposition of *factuality* we have $(\forall i)f_i$. And by the supposition of *completeness* we have it that, as regards claims about this item:

$$(\forall p)[p \rightarrow (\exists i)(p \leftrightarrow f_i)]$$

Moreover, by the supposition of *non-redundancy*, each member of the sequence adds something new to what has gone before.

$$(\forall i)(\forall j)(i < j \rightarrow \neg\,[(f_1 \,\&\, f_2 \,\&\, \ldots \,\&\, f_i\,) \rightarrow f_j]$$

Consider now the following course of reasoning:

(1) $(\forall i)\, f_i$ — by "factuality"

(2) $(\exists j)\,[(\forall i)\, f_i \leftrightarrow f_j]$ — from (1) by "completeness"

(3) $(\forall_i)\, f_i \leftrightarrow f_j$ — from (2) by existential instantiation

(4) $f_j \rightarrow f_{j+1}$ — from (3) by universal instantiation [and from the principle of non-redundancy]

(5) $\neg\,[(f_1 \,\&\, f_2 \,\&\, \ldots \,\&\, f_i\,) \rightarrow f_{j+1}]$ — from "non-redundancy" by universal instantiation

(6) (4) and (5) are contradictory

One of the argument's chief premises, the principle of non-redundancy, is of a clearly epistemic nature—and this introduces an enormous problem. Given a list of facts, Rescher assumes, we can in principle always discover additional facts. Rescher thus simply assumes (113) that reality is absolutely inexhaustible:

> The domain of fact inevitably transcends the limits of our capacity to *express* it, and *a fortiori* those of our capacity to canvass it in overt detail. There are always bound to be more facts than we are able to capture in our linguistic terminology.

But these epistemic claims do not suffice to establish the thesis that there are uncountably many facts, particularly about every individual "thing." From the fact that, on the epistemic level, only finitely many facts can be grasped, it cannot logically be concluded that there are uncountably many facts. Can one epistemically justify the assumption that in the sequence $f_1 \,\&\, f_2 \,\&\, \ldots \,\&\, f_i \,\&\, f_{i+1}$, after f_{i+1} there must be an f_{i+2}? In fact, Rescher first assumes that there are countably infinitely many facts, and then, on the basis of a powerful epistemic extrapolation, "proves" uncountability.

[2] It is not particularly difficult to prove, without recourse to questionable epistemic assumptions, that there are uncountably many facts. For structural ontology, according to which the world (the universe, being as a whole) consists only of primary facts, there are none of the problems that arise for ontologies based on objects or substances. As Chapter 3 shows, in such ontologies the status of facts remains unclear.

If one takes seriously the mathematics of uncountable sets, it is clear that there are uncountably many mathematical facts, because every element of any such set corresponds to a primary fact or, more precisely, every element of any such set *is* a primary fact. To be noted in addition is that these sets themselves and all of their subsets are likewise (complex) primary facts. This thesis is based on the thesis that mathematical structures or entities are to be interpreted as genuinely ontological. The latter is one of the theses fundamental to the structural-systematic philosophy.

With respect to the domain of non-formal entities it may also be shown that there are uncountably many such entities. The argument is similar to the one introduced above (see Section 5.1.3.3.2.1) in support of the thesis that philosophical language is to be grasped as a semiotic system with uncountably many expressions. One starts with a finite number of *basic* entities (primary facts), analogous to the alphabet and punctuation marks (i.e., the basic signs of a semiotic system). From these basic items, one can easily form countably infinite sequences (i.e., combinations or configurations of basic facts). This "forming" is not simply a construction, but is instead the *reconstruction* of *real* interconnections (i.e., ones that are ontological in the genuine sense). That is because every individual ontological entity (i.e., primary fact) has at least countably infinitely many relations to other facts, to other configurations of facts, etc. Every individual fact is, that is to say, determined systematically—embedded within the entire system of the world (in the sense of the universe, of being as a whole).

Given this, relatively simple set-theoretical considerations show that there must be uncountably many entities—the argument, again, is analogous to the one used for philosophical and scientific languages. To the objection that application of set-theoretical tools could lead only to abstract constructions, the response is that set-theoretical tools are, just like non-mathematical concepts, ways in which reality is *understood*, and thereby ways in which reality itself presents or manifests itself. The objection is based on an ultimately self-contradictory and thus senseless conception of the relation between concept and reality.

527

5.1.4.5   Is Philosophical or Scientific Language a Purely Human Production? Or What, Ultimately, Is (a) Language?

[1] The preceding considerations give rise to the question announced in this section's title. This general and comprehensively formulated question has many aspects; not all can be considered in this book. With respect to present purposes—the clarification of the concept and the program of comprehensive systematics—two aspects of the question are particularly important; they can be articulated as specific questions. First, is language (in the sense developed above) a human product, or does it have some other ontological status? Second, is there only a single philosophical or scientific language—or, modally formulated, is only a single philosophical or scientific language possible or conceivable—or is there instead a plurality of such languages? The relation between actuality and possibility with respect to philosophical or scientific language(s) is itself a topic internal to the second question.

This chapter treats both questions in appropriately cursory fashion, the first in this section, the second in the next; these two sections articulate two aspects of a single fundamental issue. For the sake of convenience, the treatment of the first question uses the singular, "philosophical or scientific language"; this usage should not be taken to prejudge the issue raised by the second question.

[2] Many philosophers—for example, Rescher (1987: 112, 161–62n2)—distinguish between actual and possible languages. When they say, as Rescher does, that the set of facts is larger than the set of linguistic signs, they are always concerned with our *actual*

language(s). But this distinction, which at first glance appears to be clear and illuminating, proves on closer analysis to be highly problematic and indeed untenable. What precisely is our *actual* language (e.g., the English language?) What precisely belongs to it, and what precisely does not? No doubt, the set of past and current tokens so belongs, as do, incontestably, sentence types. But how many sentence types are there?

As noted above, it is easy to show that the English language, *considered as a whole in the manner described above*, has *uncountably many sentence types*. For this reason, the distinction between actual and possible languages becomes blurred. Indeed, it must   528 be either reinterpreted fundamentally or abandoned altogether. Ultimately, the phenomenon of language(s) is not clarified adequately by the introduction of this putative distinction. One does better to speak of languages that in fact are spoken or used and ones that in fact are not. Any language that is in fact spoken is only a segment of a far more comprehensive language.

[3] If languages are such grand semiotic totalities (systems), however, how are they to be understood more precisely? According to one current notion, indeed the one that is dominant at present, languages are purely human productions that arise in the course of history; they are thus comparable with other human accomplishments like the developments of civilizations, of systems of justice, of the metric system, etc. But this view is a simplification.

To do justice to the *comprehensive phenomenon* of language, one must consider several basic distinctions and specifications.

[i] The *universal expressibility* introduced above in various contexts is a factor fundamental to the structuration (or an immanent structural characteristic—see Section 5.2.4[2]) of the world (in the sense of the universe, of being as a whole). With the converse of the relation that is co-posited with it, the phenomenon (of) *language* is *given*. Gadamer (1960/2003: xxxiv/*GW*2:444) expresses this insight drastically, thus too baldly, as follows: "Being that can be understood is language." This sentence in no doubt too vague to be adequately intelligible, but it does voice an absolutely fundamental comprehensive-systematic insight. The vagueness is avoided if one formulates the insight as "being that can be understood *is the unrestricted universe of discourse*." Being is thus immanently related to language or contains an essentially immanent relation to language. One could speak of the *linguisticality* of being, just as one at times speaks of the conceivability, explicability, etc., of the world or of being. Because the world or being is also the object or subject matter of theories, it must be said that the world (being) conforms to theory, is theorizable, etc.

Language in this comprehensive sense of the linguisticality of being could be termed "maximal language" or "the absolutely universal language": the language that expresses the world or being, the language that is the necessary counterpart of the expressibility of the world or of being as a whole. This language is simply coextensive or coextensional with the world, with being as a whole. It is the language that is identical with the discourse of the formulation "unrestricted universe of discourse."

But what more could be said of this maximal or universal language? It does not "exist" anywhere (no one speaks it), although in a certain fundamental respect it exists   529 everywhere, in that there is nothing that is not "bespoken"—structured or articulated— by it. But its "non-existence" says little about the reality of such a language, because something similar holds for any natural language, unless one identifies such a language with some arbitrarily delimited totality of tokens. At least the following is now clear: if this maximal language is actually understood as the expressing factor that is the necessary and unavoidable counterpart to universal expressibility, then it must be able to fulfill this function; this requires that it be a semiotic system that is structured not only syntactically but semantically and ontologically as well. To say more about this absolutely universal language, other linguistic forms and additional distinctions must be introduced and considered.

[ii] "Language" can also—especially in the formal sciences, sometimes exclusively and sometimes in the sense of a first stage in the construction of a logic or a mathematical language—be taken to be a semiotic system that is configured only syntactically. Such a language is to be called a pure semiotic (syntactic) language. There are no fundamental or insoluble problems with respect to the countability or uncountability of any such language. Different categories of such languages are conceivable, depending upon how one considers or applies the following two factors: the distinction between basic characters and strings of characters, and the kinds of characters that are introduced. With respect to the former factor, one can in principle conceive of a pure semiotic system, which could be uncountable, consisting only of basic characters, with no strings at all; the distinction between basic and non-basic characters would then disappear. But a pure semiotic language could include strings as well. As far as the second factor is concerned, the most important difference is between arbitrary characters and characters that, in human history, are understood as linguistic and have been or perhaps continue to be so used.

[iii] The form of language fundamental for philosophical and scientific purposes is the form that is structured not only syntactically but also—most importantly— semantically and ontologically. Only languages so structured can with full justification be termed languages in the genuine *and fully philosophical* sense. Chapter 3 shows how semantic-ontological structuration is to be understood more precisely. Particularly important here is that—as indicated above in several contexts—there need not be a strict one-to-one correspondence between syntactic structuration on the one hand and semantic-ontological structuration on the other. A given language that is syntactically structured in a determinate manner (e.g., any Western language, all of which rely fundamentally upon the subject-predicate sentence form) can non-problematically be semantically-ontologically *reinterpreted*. In any case, however, that there must be a strict one-to-one correspondence between semantic structures and ontological structures follows from what is established in Chapter 3.

[4] It is now possible to respond to the question, is philosophical or scientific language a purely human product? Or: what, ultimately, is (a) language? Language in the maximal sense (i.e., absolutely universal language) is not a human product, because it is intrinsic to the world in the broad sense—reality, the universe, being as a whole. It is thus also the case that language as a pure semiotic system with uncountably many characters, *if* these characters are not specified further and thus cannot be understood as linguistic characters in the sense explained above, is also not a human product. But *linguistic* characters in the familiar sense are in play (in any form whatsoever) only when human beings—with their own history, their own products and contingencies—enter the scene. Without human beings, *these specific, determinate linguistic characters* do not exist.

What is most important but also most difficult is answering the question concerning *(syntactically-)semantically-ontologically structured language(s)*. This book answers that question as follows: in the case of those languages, the purely semiotic part, insofar as it consists of *linguistic* characters, is a human product. This accounts for the historical contingencies that cannot be denied. But the situation is different with the semantic-ontological structurations of such languages. This structuration is not a human product or invention; it emerges from the expressibility of reality itself.

At this point, the thesis just introduced is not immediately illuminating; it requires scrupulous analysis and development. The semantic-ontological structuration of a given language would be relatively easy to understand and to present if the following presupposition were satisfied: there is a *single* semantic-ontological structuration. If this were the case, it would be immediately clear that this structuration was the structuration of reality itself and thus not a human product. This result would follow from the additional thesis (decisively defended in this book) that semantic-ontological structuration must unavoidably be understood as being the structuration of reality itself.

This presupposition is, however, not satisfied, because there is a *plurality* of (syntactically-)semantically-ontologically structured languages. Given the basic theses   531
of this book, it follows that there is a plurality of structurations of reality. How is that to be understood?

5.1.5 The Plurality of Languages: Its Ontological Interpretation and Several Consequences

5.1.5.1 In What Sense and on What Basis Is There a Plurality of (Theoretical) Languages?

That there is a plurality of natural or ordinary languages is a fact. It is also a fact that there exists a plurality of scientific languages. Finally, it is a fact that various different scientific or philosophical languages can be conceived of and in fact have been and are conceived of.[21] What must be clarified is how this plurality is to be understood more precisely. What is of decisive importance is the discovery of where lie the genuinely relevant distinctions among different philosophical and scientific languages.

Distinctions with respect to basic linguistic characters are not decisive here. Whether and in what way an alphabet is used, whether quite different graphic or phonic characters are used, is ultimately of little significance. How the characters are configured syntactically has a certain albeit not decisive relevance, as previous considerations show. What in the end is exclusively decisive, however, is the *ontological (and with it the inseparably semantic) structuration* of the language(s). The basic characters and their syntactic configurations are in principle arbitrarily interpretable semantically and ontologically. If however one speaks of a plurality of scientific or philosophical languages that is relevant to science or philosophy, then the languages must differ markedly with respect to their semantic-ontological structurations.

It is relatively easy to show, by means of two examples, where such a distinction would lie. The first example could be any language that has the subject-predicate form as its essential syntactic element *and* that is interpreted in such a way that a semantics   532

---

[21] It suffices here to note Carnap's numerous attempts to construct different scientific languages (see, e.g., Carnap 1958).

and its associated ontology are directly derivable from this syntactical structuration; the form of articulation of the semantics and ontology is then provided by a specific formalization of the language. Such a philosophical or scientific language is developed from ordinary English or German (for example) with the aid of first-order predicate logic; the result is the standard ontology of substances. Language so understood and used is, as is well known, *the* language of analytic philosophy; it should be noted, however, that in the course of its history, and especially over the last two decades, these internal connections have been noticed and articulated, slowly and piecemeal, by increasingly many analytic philosophers.

It should be clear that a philosophical or scientific language that did *not* recognize such ontological and semantic structures as substances, properties, subjects, predicates, etc., instead introducing wholly different semantic and ontological structures, would be *a fundamentally different language*. This remains the case even if the basic signs, the vocabulary items, and the only syntactically configured signs are partly or wholly identical—which in practice is usually the case, as is emphasized in various places in this book.

The second example is the plurality of *conceptual schemes*, a plurality accepted by many and perhaps indeed by most analytic philosophers.[22] This notion "conceptual scheme" is usually taken in a quite broad sense, such that it encompasses a broad spectrum of concepts or categories. This notion is relevant for present purposes only if it involves an ontological as well as a semantic aspect. There are doubtless various conceptual schemes that indeed have both aspects. Thus, for example, the semantic-ontological *compositional* theoretical framework often discussed in this book is fundamentally different from the contextual-structural theoretical framework based on the context principle (see Chapter 3).

5.1.5.2 The Ontological Ramifications of the Plurality of Theoretical Languages

5.1.5.2.1 *On Various Approaches to the Problem*

[1] Nicholas Rescher (1982: 40) formulates the current issue and task, with respect to the genuinely ontological ramifications of this distinction, as follows:

533

> The difference between conceptual schemes is *not* a matter of treating the *same issues* discordantly, distributing the truth values T and F differently over otherwise invariant propositions. Different conceptual schemes embody different theories, and not just different theories about 'the same things' (so that divergence inevitably reflects disagreement as to the truth or falsity of propositions), but different theories about different things. To move from one conceptual scheme to another is not a matter of disagreement about the same old issues, it is *in some way to change the subject*. (Emphasis in final sentence added)

This passage articulates the fundamental issue that arises from the acknowledgment of a plurality of languages (in the currently relevant sense), but it neither makes the decisive point clear nor treats it correctly. What does it mean to say that to change languages or

---

[22] Donald Davidson (see esp. 1974/1984) notoriously denies such a plurality. Nicholas Rescher (1982: Ch. 2) provides an on-target critique. See also Puntel (1999a: 118ff).

schemes is "in some way to change the subject"? To say simply that the change from one language to another—and thus from the theory corresponding to one conceptual scheme to a different theory corresponding to a different conceptual scheme—is such that the different languages and theories "speak" not of the "same things" but instead of "different things" misses the decisive ontological point. In accordance with the semantics and ontology developed in Chapter 3, what should be said instead is that the change within the theoretical domain from one language to another has as its immediate consequence that there is also a change from one theoretical framework to another theoretical framework. With this change necessarily comes a change from one type of ontological structure to another type of ontological structure.

With respect to the structuration of the world (of the universe, of being as such and as a whole), what follows is this: if one were to understand these different ontological structures such that they would not be different structures "of the same things" or, more precisely, that they would not constitute the same things (or, yet more precisely, the same thing), that they instead were or constituted structures of *different things* (or constituted different things), then not only the *unity* of the world (of the universe, of being) but also the relationships among the putatively so-constituted "different things" would remain fully unthematized and unconceptualized. One would instead have fully disparate, unrelated, incommensurable "things," whatever "thing" might mean.

Nevertheless, Rescher does articulate one essential point: different conceptual schemes (and thus different theories and languages) indeed articulate different ontological features—but what must then be determined is what the features are of, or with respect to what they are features. Concerning them, Rescher provides an explicitly negative determination and in the second part of the passage a minimal positive one in saying that the different conceptual schemes (hence languages and theories) articulate (or bring to expression) *not* something different about the same things, but simply different things (i.e., presumably, ontological structures that are wholly different, that have nothing whatsoever to do with one another). This thesis remains insufficient and indeed, if taken as definitive, is false; this is shown more clearly in what follows.

[2] How is the plurality under current consideration to be interpreted? A beginning may be made with the explication of three approaches to or ways of dealing with the problem.

[i] The first way of dealing with the problem is inspired by Charles Peirce and can be termed the *endpoint approach*. Its central thesis is that all languages/theories/schemes, however different they may be, present within the framework of the process of investigation an infinite but converging series that approaches an ultimate limiting value, an endpoint. In Peirce's (1878/1965: 268) words:

> Different minds may set out with the most antagonistic views, but the progress of investigation carries them by a force outside of themselves to one and the same conclusion. This activity of thought by which we are carried, not where we wish, but to a fore-ordained goal, is like the operation of destiny. No modification of the point of view taken, no selection of other facts for study, no natural bent of mind even, can enable a man to escape the pre-destined opinion. This great hope [in the first draft "law"] is embodied in the conception of truth and reality. *The opinion which is fated to be ultimately agreed to by all who investigate is what we mean by the truth, and the object represented in this opinion is the real.*

[ii] A second approach relies on the thought of the plurality of worlds. Every language, every theory, every conceptual scheme corresponds to its own world. On the question of how these worlds are to be understood precisely, opinions diverge widely. The variant developed by Nelson Goodman (1978) rejects "the world" and recognizes only worlds, but ultimately leaves unclear just how these "worlds" are to be determined more precisely. The most plausible interpretation appears to be that "worlds" are "versions of the world."

Another variant ascribes to every comprehensive theoretical language or theory its own world, but considers these worlds to be subworlds of an all-encompassing, highest world, such that this all-encompassing world corresponds to a likewise all-encompassing language and theory (see Puntel 1990: 250–94).

[iii] A third approach relies upon the idea of approximation of the truth. Karl Popper introduces this idea (usually designated as that of *verisimilitude*; see Popper 1972: 47–48, n18); since then, it has been discussed widely, but thus far nothing resembling consensus has been attained.[23] Popper himself (1972: 47) defines what he calls the "notion" of verisimilitude as the combination of two notions taken from Tarski: "(a) the notion of truth, and (b) the notion of the (logical) *content* of a statement; that is, the class of all statements logically entailed by it." This content has two parts: (i) the truth-content, the class of all true statements that follow from a given statement (or from a given deductive system), and (ii) the falsity-content, the class of all false statements that follow from the statement (or deductive system) (this falsity-content does not have the properties typical of Tarskian sets of consequences). On this basis, Popper characterizes the concept of verisimilitude as follows:

> [A] theory T_1 has less verisimilitude than a theory T_2 if and only if (a) their truth contents and falsity contents (or their measures) are comparable, and either (b) the truth content, but not the falsity content, of T_1 is smaller than that of T_2, or else (c) the truth content of T_1 is not greater than that of T_2, but its falsity content is greater. In brief, we say that T_2 is nearer to the truth, or more similar to the truth, than T_1, if and only if more true statements follow from it, but not more false statements, or at least equally many true statements but fewer false statements. (52)

This definition has been subjected to harsh critiques and at present has virtually no defenders.[24] Alternatives have appeared that are similar in that they exclusively thematize purely formal aspects and use corresponding concepts as *definientia*.

The deficiencies of approaches [i]–[iii] are so severe as to make them untenable. Peirce's initiative postpones until the end of the research process the arrival of a universal theory using an appropriate language, the theory that articulates what reality actually is. But that is a fantastic notion that has no possible philosophical use, because this endpoint is wholly unthinkable. Given all the contingencies characterizing human

---

[23] Niiniluoto (1987) is something of a *magnum opus* on this topic.
[24] The first radical critiques of Popper's definition are Miller (1964) and Tichý (1964). A determination of verisimilitude that is "structuralist" (in the sense of Sneed, Stegmüller, and Moulines) is found in Kuipers (2000, 2001).

life and human beings—and thus also the process of research—Peirce's notion is not one that philosophers can or should take seriously.[25]

Goodman's initiative is likewise untenable. Distinguishing different worlds to which different languages correspond without ever considering their interrelations or their reality avoids the task of philosophical reflection and theorization. The other variant, which takes the various worlds to be subworlds of an encompassing world, is an interesting philosophical suggestion, but it remains too global and is insufficiently developed. The conception sketched below can be understood as one that in part clarifies and in part corrects this suggestion.

A conception based on a concept of approximation or verisimilitude, no matter how the concept is defined, is scarcely helpful ontologically. If one understands by truth a concept that has an essential ontological import, then the concept of approximation of truth would have to include that of an approximation of the world or of reality. Given the intuitive content of the concept of approximation, however, the concept of approximating reality or the world appears quite problematic, particularly because of the basic presupposition that reality or the world is a dimension that we can, so to speak, approach. This posits an absolute and fundamental distance between reality or the world, on the one hand, and us (our languages, theories, concepts) on the other. Intuitively plausible and attractive though the concept of verisimilitude may be, it is philosophically insufficient precisely because it involves the assumption of a gap that is in principle unbridgeable, whereas what philosophy and science are concerned with conceiving or conceptualizing is nothing other than reality or the world *itself*. Nevertheless, as becomes clear below, the concept of verisimilitude does contain one important aspect that must be taken into account if an adequate response to the current question is to be provided.

5.1.5.2.2 *A Suggested Three-Step Solution to the Problem*

As a comprehensive result of the various critical considerations developed in this book on the relation of language/theory and world (universe, being as a whole), the following thesis emerges: *every transparent and comprehensive theoretical language grasps or articulates the world or reality itself; there is no separation or gap, no matter how conceived, between these two domains.* In addition, however, a plurality of relevant theoretical languages must be accepted as in part actually at hand and in part arbitrarily conceivable. From this arises a problem: how is the plurality of such languages to be brought into harmony with the thesis that one indispensable and ineliminable characteristic of language is the ontological import introduced above?[26]

The requisite approach must both avoid the deficiencies identified above and take into account the various considerations that are there presented as essential. This novel and comprehensive approach is guided by the thought of an *increasing adequacy of the articulation of the world or reality as such.* How this increasing adequacy is to be thought

536

537

---

[25] See Rescher's critique of Peirce (esp. 1992: 47–52).

[26] This section thematizes only some ontological aspects of this problem. The central aspect of relativism (with respect to truth) is treated in Section 3.3.4.3. Chapter 6 includes a concluding analysis of some additional facets of this complex issue.

is indicated, in a coarse outline, in the following subsections, which formulate the core of the suggestion in three steps.

5.1.5.2.2.1 First Step: The Ontologization of the Theoretical Sphere

The first step is the thesis that knowledge, theories, science, and thus also philosophy can be adequately grasped and undertaken only if one *ontologizes* the entire theoretical dimension. This ontologization[27] is to be understood as follows: not only the knower—the theorizing scientist and philosopher—but also the entire theoretical dimension in the objective sense (the dimension of concepts, theories, etc.) must be seen as *parts of nature* or *of the world* or of *being as a whole*. The ontologization of the entire theoretical dimension in this sense then involves, among other things, the radical overcoming of the fateful gap that continues to present such formidable problems within modern philosophy: the putative gap between the subject (knower, theoretician) and nature or the world or being as a whole.[28]

Is this thesis strange or even astonishing? It will appear so to many philosophers, particularly analytic philosophers. But what is truly astonishing is that the thesis is not seen by analytic philosophers as obvious: what is it, after all, other than the expression of one of the implications of the basic truism presented to us by modern science? The truism—the thesis that has become basic and that is accepted as true—is that humanity as a whole—and thus the thinking, theorizing subject, using all formal, conceptual, etc., means of thinking—is a part and indeed a product of the evolution of nature, something that comes to be at a specific temporal point in the course of cosmic evolution. Given this, what—other than something ontological, something that *is*—could the theoretical or cognitive dimension itself possibly *be*? Even the question cannot be formulated coherently, because what, save a *being*, could whatever is named "the theoretical dimension" possibly *be*?

5.1.5.2.2.2 Second Step: Changing the Focus of the (Philosophical/Scientific) Perspective from Subjectivity to Being (Nature, the World)

The second step is a significant consequence that may be drawn from the just-formulated first thesis.

[1] If human beings belong to the world (to nature or being), then everything that human beings do, accomplish, etc., likewise belongs to the world. With respect to the theoretical dimension in the objective sense, this means that it is itself, to put it one way,

---

[27] "Ontologization" is to be understood as wholly analogous to "naturalization," but in a strictly ontological sense (and not in the methodological sense in which this expression is used, for example, by Quine in such formulations as "naturalized epistemology"). But the term that is more adequate philosophically is without doubt "ontologization," because this term relates not only to nature or the world, but instead, and directly, to being as a whole.

[28] Worth noting in passing is the striking phenomenon that most analytic philosophers, who with respect to epistemology tend to be naturalists and to ontology strict materialists, and who thus acknowledge only an exclusively materially/physically constituted world, proceed in a fully incoherent manner when they then construe the theoretical dimension as one that is wholly separated from the dimension of nature or the world or reality—as, so to speak, an absolute, Platonic realm. Quine is a characteristic example.

an ontological arena, an ontological showcase. Seeing the theoretical dimension as an ontological arena involves a change of perspective so radical that it can be understood in a fundamental respect as a *reversal* of Kant's "Copernican turn" to the subject, but one that, despite its rejection of Kant's thesis of the centrality of the subject, retains certain Kantian insights concerning subjectivity. For this reason the reversal could be characterized more accurately as a quasi-Hegelian *Aufhebung* or suspension of the (transcendental) subject.

The best way to characterize this post- and indeed anti-Kantian turn is perhaps as follows: knowledge (conceptualization, ultimately theorization) is *primarily* or *fundamentally* not something *accomplished by a subject*, but something *occurring within nature or the world or being*. This is not to deny that knowledge is the accomplishment of a subject, but instead to deny that this is what it is *primarily* or *fundamentally*. Primarily and fundamentally, knowledge is a way in which nature or the world or being *manifests* or *articulates* or *expresses* itself. The most accurate sentence form for this self-manifestation is therefore one in which the subject does not appear as a factor—or at least not as a determining factor.

This sentence form is identified in this book in various places as "It is the case that such-and-such" (or "It is the case that φ"). As Chapter 2 demonstrates thoroughly, this sentence form is the genuine form for theoretical sentences. With respect to the topic of this section, the significance of this theoretical sentence form must be explained in more detail. For this purpose, it is helpful to compare it with the form of sentence characteristic of Kant's transcendental philosophy: the transcendental sentence form. The latter involves—almost always only implicitly—a *transcendental operator* that precedes every sentence presented in this philosophy (or that can be presupposed by this philosophy to precede every sentence it presents). This operator is the following: "From the perspective of transcendental subjectivity (it is the case that φ)." An example: "From the perspective of transcendental subjectivity it is the case that the sun warms the stone." The transcendental operator is the expression of a wholly determinate form of subjectivity (i.e., transcendental subjectivity). This subjectivity is not particularistic, but it is also not unrestrictedly universal subjectivity.

[2] The operator "It is the case that. . ." is the absolutely universal theoretical operator; it is restricted in no manner whatsoever. It introduces no talk of a subject or of subjectivity. Does this mean that the subject, or subjectivity, does not at all come into play, that it has no role whatsoever? It is said above that the subject plays no *determining* role. This can now be made more precise. One must distinguish between the universal subject (or universal subjectivity) and every other level of subject/subjectivity, *beneath* the level of universality. Transcendental subjectivity is perhaps the best-known and most important form of non-universal subjectivity. How universal subjectivity is to be understood is explained thoroughly in Chapter 2 (see Section 2.3.2.4). There it is also shown that universal subjectivity, which coincides with the simply universal standpoint, for just that reason makes itself superfluous in the following sense: it need not be explicitly named as such. With respect to the universal theoretical operator "It is the case that. . .," this means that subjectivity is always given along with this operator and thus always accompanies it. It thus makes no difference whether one says "It is the case that. . . ." or "From the perspective of universal subjectivity, it is the case that. . . ."

539

An additional step may however be taken by means of a comprehensively systematic interpretation (traditionally, a metaphysical interpretation) of the universal operator and of universal subjectivity. That is to say, one can somewhat daringly ascribe to the particle "it" a particular function and significance.[29] As is well known, the precise interpretation of this particle, when it is used in conjunction with impersonal verbs, is a hotly contested topic within philology. Whereas many accord to this word, so used, no contentual value at all, taking it to be purely formal or a pseudo-subject,[30] others see in it the linguistic expression for the presence or working of impersonal forces. One can, however, instead interpret this particle as the linguistic indicator of a dimension that is not designated more precisely. In the theoretical context of systematic philosophy, such an interpretation can be well defended, presupposing that one does not interpret the dimension in any manner that is at all mythic (e.g., as designation for obscure forces of any sort whatsoever) or substantialistic (as designation for some underlying substance or subject of whatever sort) or anything similar. One can understand the "it" in formulations of the form "It is the case that such-and-such" as a general and unspecified indication of *the world* (*reality, the universe, being*) as a whole, although in the case of theories that articulate restricted domains, the precise domain or object of the theory in question must be named explicitly.[a] The result is formulated as follows: in every theoretical sentence, even the most minor, that is (or must be) preceded by the operator "It is the case that...," the world (as a whole or in part) is presented in that it reveals or discloses itself in its structuration.

The interpretation of the particle "it" described in the preceding paragraph is not directly applicable within theoretical frameworks that rely on sentences having a different form (specifically, the subject-predicate form), but the fundamental ontological state of affairs can nevertheless be generalized.

Here, one encounters a fundamental problem. If in sentences with the operator "It is the case that...," the world itself (nature, being as a whole) expresses itself or accomplishes

---

[29] Here, explicit reliance is made upon the interpretation of the formulation "It is the case that such-and-such (or that φ)" presented in earlier chapters (esp. Chapters 2 and 3). The interpretation presented here does not contradict the earlier passages; instead, it expands upon them.

[30] There are of course languages, such as Italian, Spanish, and Portuguese, that do not use subject pronouns in the comparable impersonal constructions.

[a] This is the point at which syntactic differences between "*es verhält sich so dass*" and "it is the case that," alluded to above in note a to Chapter 2 (p.90), become relevant to the structural-systematic philosophy. One relevant difference is that whereas "is the case" can be preceded only by the neuter third-person-singular pronoun "it," "*verhält sich*" can, with changes in the verb form required for agreement, be preceded by any personal pronoun, thus by the German counterparts to "I," "you," "she," "they," etc., and by various nouns (e.g., "*Die Sache verhält sich so*," "This is how it is"). "*Er verhielt sich musterhaft*" could be rendered, "He behaved (himself) (or comported himself) in exemplary fashion." Interpreted as articulating being, the phrase takes its semantic sense from the non-impersonal formulations, but also, like the impersonal formulation, is completed by indicative sentences rather than by adverbs. Anticipating section 5.2.4: the "behaving" or "comporting" of being is its self-explication, self-unfolding, or self-disclosing. "It is the case that," interpreted by way of "*Es verhält sich so dass*" as indicating being, may therefore be reformulated as "It's self-disclosing such that ..." or "It's being such that it's self-disclosing such that...."

a self-articulation, how is the plurality of theoretical frameworks within which such sentences emerge to be understood? This question is addressed in step three, presented below in Section 5.1.5.2.2.3.

[3] The preceding reflections clearly reveal two points: (i) the location or situatedness of human beings within nature or the world (and within being as a whole), their relation to nature or the world, is not one-dimensional; instead, there is an extensive scale with many different levels and degrees of situatedness of the human being within nature or the world and therefore also of how the human being relates to nature or the world. To each of these levels and to each of these degrees there corresponds a specific mode of articulation that is here termed a specific theoretical framework (or conceptual scheme). (ii) Theoretical frameworks (or conceptual schemes) can be compared with one another by means of their identification as different forms of the relation of human beings to nature or the world. The scale itself provides the measure for the comparison: the more particularistic is the determinate mode of the subject's relating to nature or the world, the more restricted and inadequate, and the less objective, is the corresponding theoretical framework.

5.1.5.2.2.3 Third Step: Three Pairs of Concepts as Criteria for Judging the Strength or Weakness of the Ontological Adequacy of Theoretical Frameworks

The third step treats the point that is decisive in clarifying the *ontological consequences* of the first two steps. It is shown above that every theoretical framework is an ontological arena. What must now be shown are the ways in which these ontological arenas and the "versions" of the world or being as a whole that are manifested within them relate to one another.

[1] First, a fundamental aspect of the universal theoretical operator "It is the case that" must be clarified in conjunction with the concept of the subject or of subjectivity. Sentences with this operator appear within every theoretical framework, including the simplest, that of the everyday lifeworld. Within that framework, sentences such as the following are used: "It is the case that the sun rises and sets every day." With respect to their form, such sentences are universal theoretical sentences. It is shown above, however, that in such sentences the subject is also involved, even if it is not named explicitly. But which subject is involved, or the subject in which of its forms? How is this complex of issues to be understood and clarified?

There are three possible ways to answer this question. The first is supported by the thesis that in the case of such everyday sentences one either cannot or should not speak of a theoretical framework. This is the easiest and most comfortable option, but it is philosophically short-sighted and ultimately inconsequent. Even in the lifeworld, the language that is used has an *indicative* segment and thus—in the terminology of this book—a *theoretical* segment. According to one of the most fundamental of this book's theses, no theoretical sentence can be understood and evaluated in, so to speak, splendid isolation; every such sentence presupposes one or another theoretical framework. No doubt, the theoretical framework of the lifeworld is extremely incomplete, and it is rife with unclarities, incorrectnesses, confusions, etc. Yet it is nonetheless a theoretical framework, because sentences that appear within it *speak of the world*. The first option is therefore rejected.

The second option involves recognizing indicative/theoretical sentences of ordinary or lifeworldly language as indicative/theoretical in terms of their form, but as ones that are always *implicitly* determined by a *particularistic* operator that articulates the perspective of a *particularistic subject*. The form of the relevant particularity of the subject is *lifeworldly* particularity. In accordance with this option, the precise paraphrasing of the example introduced above would be, "From the perspective of the particularistic-lifeworldly subject, it is the case that the sun rises and sets every day." Using the notation introduced in Chapters 2 (Sections 2.2.3.1 and 2.3.2.4) and 4 (Section 4.3.2.1.2), this would be formalized as follows: $\text{S}_{\text{PL}}(\text{T}(\phi))$ (with S = "from the perspective of the subject...," "$_{\text{PL}}$" = "particularistic-lifeworldly," T = "it is the case that...," and "ϕ" = the sentence "the sun rises and sets every day"). To be emphasized is that the theoretical dimension (represented by the theoretical operator T) here lies within the scope of the particularistic-lifeworldly subject.

This option appears to have a decisive deficiency: it must interpret the indicative/theoretical sentences of the lifeworldly theoretical framework as sentences that articulate only a view that is *subjective* in the strong sense, but this subjective character then appears to have as a consequence that such sentences are deprived of any genuinely *ontological meanings*. Yet against this overly quick and abbreviated interpretation and critique two points can be made. First, such sentences do speak about the world and in such a way that their users get along quite well in the world about which they speak. It is implausible that the sentences that are expressed in this manner fail to articulate anything that belongs to the world in any way or to any degree. Second, the option criticized here fails to recognize the fundamental fact that every subject is a part of the world (of nature, of being) and thus with it *everything* that constitutes the subject and everything that the subject achieves. It is paradoxical only at first glance to say that there is no subject and nothing in any subject that is purely subjective (in the usual sense and in the sense of this second option or interpretation). If the subject is an entity that is an element of the world (of being), then everything within it and everything it does is likewise *ontological*. This includes the subject's view of the world. The question is how this is to be understood precisely. The third option attempts to answer this question.

The third option attempts to do justice both to the subjective and to the ontological status of the example sentence. It does so by thematizing and thus conceptualizing the second option on a *higher* level. This higher level is what this book terms the genuinely structural-systematic standpoint. Formally and methodologically, the formulation of this higher, systematic view is quite simple: one precedes the sentence with a higher operator that formalizes the second optional interpretation. This operator is the *theoretical* operator (at the higher level). What results is the following: "It is the case that from the perspective of the particularistic, life-worldly subject it is the case that the sun rises and sets every day." Formalized: $\text{T}[\text{S}_{\text{PL}}(\text{T}(\phi))]$. The entire formula that articulates the second option—the second interpretation—now lies within the scope of the all-determining theoretical "superoperator" "T." But how is this paraphrasing or this formula to be interpreted philosophically?

Within the framework of the systematic conception defended here, the so reconstructed (paraphrased and formalized) sentence from the lifeworldly theoretical framework is to be interpreted such that it is the case of the world itself that the sun rises and sets every day. Thereby, a determinate facet of the world itself is presented. The sentence says that the world, in a specific situation—as immediately perceived by a subject—is such that the sun *appears* as rising and setting every day. This in no way means that "the world in itself" (whatever that might be) is thereby presented; instead, the world is here articulated in complete relativity to a location determined by an immediately perceiving subject. Such a location is, however, a location within the world itself, so the appearing of the world at or to this location is also a genuine appearing of the world. If the location changes, that is, if the subject does not simply perceive immediately, but is instead rational, reflective, and investigative—in a word, if it develops the capacities of its theoretical attitude—then the appearing of the world itself automatically changes. To be emphasized is what actually occurs here: the immediate perceiving of the world and the reflective-theoretical attitude with respect to the world are not occurrences that are somehow outside of the world; instead, they are occurrences within the world. One could say that they are forms or shapes of the world. The example sentence expresses the fact that there is a form or shape of the world such that the sun appears as rising and setting (and is perceived and articulated as so doing).

This state of affairs can also be articulated as follows: the sentence, "The sun rises and sets every day," is a *true* sentence, but only in the sense that it is true *within* or *relative to the lifeworldly theoretical framework*. That it is correct to designate it as true (in this sense) becomes clear as soon as one is confronted with the negation of the sentence, "It is not the case that the sun rises and sets every day." This sentence, if asserted within the lifeworldly theoretical framework, is false. Within or relative to a different theoretical framework, such as a scientific one, the example sentence is of course not true, whereas its negation is. This makes it clear that the qualification of a sentence as true within one (here, the lifeworldly) theoretical framework has an *ontological* implication: one must show which facet of the world is articulated by means of the true sentence.

[2] There is a plurality of theoretical frameworks, and the differences among frameworks can be vast. The only ones relied upon by the sciences and philosophy are not ones like the lifeworldly one, but instead ones whose components (language, conceptuality, logic, etc.) have been subjected to thorough analyses. In the first two steps, the thesis is maintained that theoretical frameworks are ontological arenas within which the world presents or articulates itself. Because for the most part or at least often the various theoretical frameworks differ from one another markedly, the ontological consequences of this fact must be determined. How is "the world" to be understood, if every theoretical framework has an ontological import? Are there theoretical frameworks that are *more adequate* than others? What would greater adequacy mean with respect to ontology? If in this context one thematizes the example sentence about the rising and setting of the sun, it becomes clear that the facet of the world that this sentence articulates is an extremely limited one: it is a *contingent* and *punctual* appearing of the world that results from a quite determinate (contingent and punctual) constellation of factors (*immediately* perceiving subject, *contingently* perceived phenomena...). Other theoretical frameworks, those that are developed on the basis of explicit, reflective, theoretical undertakings, are far deeper ontologically. The following account focuses on frameworks of this latter sort.

Because theoretical frameworks are ways in which the world manifests or articulates itself in its structuration, one must assume that there are various levels of ontological structuration. Such structural levels must, however, be understood correctly. It is crucial to note that it is extremely difficult to explain this state of affairs. The following account is merely a first approach to this topic.

The different manifestations or (self-)presentations of the world that articulate different structural levels of the world do not simply *coexist* disparately alongside one another; instead, they form an immanently coherent comprehensive presentation and thus, consequently, a comprehensive structuration of the world. Thereby, the world itself does not become some kind of comprehensive construct consisting of individual, fully self-sufficient real levels or spheres that stand in some mysterious relation to one another. Instead, the different (self-)presentations and thus structural levels, as structures or structurations of the world, must be interpreted as standing in a *wholly exceptional relation* to one another: in the relation that holds between coarse-grained and fine-grained structures or structurations of one and the same entity. The plurality of structures and structurations, interpreted in this manner, is a wholly exceptional type of plurality: one that does not only not exclude unity but indeed includes it; one could speak of a "plural unity of one and the same fundamental structure or structuration."

The relation of coarse- and fine-grained structures and structurations is a wholly exceptional relation that is not simply to be equated with the precisely mathematically definable relation between structures and substructures, as shown in what follows.

[3] One can attempt to explain how this relation is to be understood more precisely by introducing three conceptual pairs that are criteria for judging the ontological strengths and weaknesses of the ontological adequacies of theoretical frameworks. These conceptual pairs can also be termed *structural-conceptual continua*. The first articulates degrees of *depth* of structuration by distinguishing between surface structures and deep structures. The second conceptual pair thematizes a different aspect, that of grainedness; this leads to the distinction between fine- and coarse-grained structures and structurations. The third conceptual pair, finally, articulates different degrees of *coherence*.

[i] *Surface* structures (or surface levels or aspects of structures) are those that are articulated on the basis of *particularistic* perspectives or of *particularistic* theoretical frameworks; they are therefore just as partial and limited as are the perspectives or theoretical frameworks within which they develop. An example is the one introduced above concerning the rising and setting of the sun. As is shown there, the world manifests itself in that example in and from a particularistic perspective, whereby, however, it is again to be emphasized that such a perspective *belongs to the world itself*. It is a constitutive aspect of the *surface structure or structuration* of the world.

Deep structures are those that develop within increasingly universal theoretical frameworks. Within such frameworks, the world manifests or articulates itself increasingly in its own structuration, one that is ever less particularistic and ever more universalistic. Because, as noted above, the conceptual pair surface structures/deep structures is to be understood as a kind of continuum, no sharp borders can be drawn. On the level of deep structures, thus of *universal* structures, there is a more or less, there are degrees, there are deeper and less deep structures. The attempt to articulate the absolutely deepest or absolutely universal structure, the simply deepest/most universal structure, is a vain undertaking. The simply deepest/most universal structures would be those thematized within the absolutely highest and most adequate theoretical framework, but the thought that one could ever develop such an absolute theoretical framework is untenable in every respect, as is clear from the central theses of the structural-systematic philosophy defended in this book.

[ii] Can surface and deep structures be determined in a way different from and more precise than by drawing sharp borders? This is an immensely difficult and laborious task, but a beginning may be made by reintroducing the conceptual pair *fine-grained* and *coarse-grained*. Surface structures are still immediately global structures that bring to manifestation external forms or initially approximative forms (of segments) of the world, whereby the external form would not be or remain an absolutely unchanged magnitude. Specifically, this means that surface structures lack what can be termed differentiation, detail, and specificity. Structures that do not lack these, that thus exhibit or are characterized by the just-named factors, are fine-grained (or more finely grained) structures.

It is to be emphasized that coarse-grained and fine-grained structures form a continuum without precisely determinable boundary lines. This relationship is therefore not to be confused with other kinds of relationships. For example, it cannot simply be equated with the precisely mathematically definable relationship between structures and substructures; only in a specific respect can one compare the relationship between fine- and coarse-grained structures, on the one hand, with the relationship between structures and substructures, on the other. In the mathematical sense or domain, structures that are (further) differentiated or determined by means of the embedding of substructures remain absolutely what they are or were; they do not change. In this sense, one can designate their differentiation or determination as something external. That, however, is not the case with the relation between coarse-grained and fine-grained structures, as that relation is understood here. In a way that is difficult to understand and to articulate, coarse-grained structures change in a certain respect in that the process of their articulation proceeds in the direction of the working out of fine-grained structures (i.e., in the direction of ever more differentiated determination).

Two examples can at least approximatively illustrate this state of affairs.   547

The first example is taken from the domain of the lifeworld, and thus presupposes a lifeworldly theoretical framework. An observer who sees a house from relatively close range can utter an initial and accurate theoretical sentence about it by saying, "At the next corner there is a red house." This sentence articulates a purely global, coarse-grained fact—and thus structure—about or of the world. For various pragmatic and even theoretical purposes, this utterance might well be not only acceptable but even optimal in that, in the concrete case, it fully suffices to identify the house in question. But from the same point of observation, the observer can communicate much more information about the house (e.g., concerning the colors of doors and of window trim, the number and location of windows, the roofing material, etc.). What is interesting about this example is the fact that the house as a whole and with respect to its specific aspects does not remain absolutely the same, even if in everyday life one speaks relatively thoughtlessly of "the same house." In that the house is observed and articulated in ever-greater detail, in a certain respect it constantly changes. What takes place here could be termed the process of progressing from coarse-grained to fine-grained structures. That thereby a structural change of a quite particular sort takes place is clear from details, such as the following: the quality "red," which is initially articulated as a global and thus coarse-grained structure, *changes* in that it is concretized by means of the introduction of differentiations (nuances) that differ from one another; such a concretization obviously contains a modification, an alteration. Thereby, the sentence "At the next corner there is a red house" does not become untrue, but the truth that it articulates is revealed to be, so to speak, only a coarse-grained truth.

To be sure, such an illustration does not serve in every respect to show how the relation between coarse- and fine-grainedness in the lifeworldly example compares to the relations among the different ontological structures and structurations of the world that can be classified by theoretical languages. Nevertheless, it does serve to illustrate some fundamental aspects of this relationship. A second, particularly informative philosophical example may be more illuminating; this is the example, often used in this book and considered particularly closely in Chapter 3, of the relation between the compositional or substance ontology this book rejects and the contextual or configuration ontology upon which it relies. Formulated with maximal brevity, it is as follows: according to the compositional ontology, the things in the world are substances (in analytic philosophy, usually termed "objects") that have properties and stand in relations to one another; according to the contextual or configuration ontology, the things are configurations of primary facts. The two ontologies ascribe to a given entity (a thing) that qualifies in ordinary language as "one and the same thing" two fully different and mutually incompatible structurations. But what is the status of this incompatibility? If the two ontologies are mutually exclusive, how can their structures and structurations be understood as two "forms" of one and the same fundamental structure or of the world?

To be noted at the outset is that one has good reasons to say that the configuration ontology preserves a fundamental intuition basic for the substance ontology; indeed, it brings it to expression in a significantly more intelligible manner. This is the intuition that a plurality of aspects in no way excludes a unity, indeed a center. The configuration ontology articulates this intuition better than does substance ontology, which situates this unity in an entity interpreted as a substratum. This shows that one can interpret the relationship between the two ontologies as follows: if one takes them as such (i.e., in mutual isolation), they exclude each other, in that it is clear that the configuration ontology does not recognize anything like substrata. But if one compares the two ontologies with each other, it becomes clear, as has just been shown, that the configuration ontology conceives of their shared fundamental intuition in a more finely grained fashion.

[iii] The third conceptual pair, which articulates a third form of structural continuum, is that of higher (greater) or lower (lesser) coherence. "Coherence" is not here understood as simply synonymous with "consistency" in the strictly logical sense; coherence in the sense intended here is not determined purely negatively, as lack of contradiction, but instead positively. A higher or greater coherence is then distinguished by two factors: first, by a greater number of aspects or structural moments (this is the "material" aspect), and second, by the totality of the relations by means of which what may be termed the "material elements" are interconnected (see Rescher 1973). The more precisely, the more tightly the configuration of the relations is articulated, the greater or higher is the coherence (in the sense relevant here). To articulate and demonstrate this in detail is an extensive and fascinating task. Here, an abstract indication must suffice.

To clarify the question of how the *ontological import* of different theoretical frameworks is to be evaluated, the entities or domains of entities in question must be analyzed with respect to the spectra of the three just-sketched conceptual pairs (or structural continua). Only on this basis can one—at least in principle—provide a   549 clarification of the central concept of the greater or lesser *ontological adequacies* of the many theoretical frameworks.

[4] These last reflections could suggest the objection that the relation between different ontological structures and structurations does not concern *genuinely ontological* structures or structurations, but instead *our views that are projected onto the ontological level*. Such an objection is wholly characteristic of all schools of thought that develop under the influence of Kant's transcendental turn. As this book shows in various places, this turn should not be taken. If one proceeds on the basis of the thesis that every consistent and comprehensive theoretical language determines its own wholly determinate ontology, then it will no longer be possible somehow subsequently to construct a genuine ontology. We do not "project" our concepts or structures onto the world or reality; instead, the world or reality presents itself in that it shows that it is the case that such and such.

To be added, however, is that above all the thought of increasing ontological adequacy, characterized by the three conceptual continua, as explication for the factual (and justifiable) plurality of theoretical languages and ontologies and for their integration within the unity of the world, can be presented here only globally and intuitively. The more precise development and presentation of this thought is one of the greatest and most difficult tasks for the philosophy of the present and of the future. The difficulty this task brings with it rests in a quite peculiar manner on the fact that, in the course of the philosophical tradition since Kant, the notion has become dominant that the knowing subject and the world are dimensions that are separated by the fundamentally unbridgeable gap that is thoroughly considered at the beginning of this chapter. If one recognizes the untenability of the thesis of the gap, then it appears not only as not strange but indeed as wholly consequent and appropriate to say that to different theoretical languages there correspond distinct ontological structures that, however, present more and less adequate forms of self-presentation of a unified, comprehensive structure.

5.1.6 *Summary: Comprehensive Systematics as Universal Theory*

It is shown in the preceding sections that the thesis that there is a gap or cut of the sort described and analyzed at the beginning of this chapter and that it serves as the basic obstacle to the development of comprehensive systematics is untenable, and thus must be one that can be rejected. The gap or cut in question is between the observer—more precisely, indeed comprehensively, the theoretician—and the "system" (i.e., the world,   550 the universe, being). This problem has many quite different facets, as is revealed clearly in the extensive preceding sections.

[1] The first step toward overcoming the problem of the gap is accomplished above; it is the demonstration of the fundamental untenability of the thesis. Its accomplishment is based chiefly on four fundamental considerations. The first is that it is an arbitrary and unfounded assumption. The Kantian and post-Kantian retreat from the dimension of reality to a putative dimension of isolated subjectivity cannot be

grounded adequately. All attempts to rely upon it share a fundamental error: they presuppose that it is not only possible but indeed unconditionally necessary to recognize the dimension of subjectivity (in whatever form whatsoever, whether the traditional mentalistic one or one of subjects characterized as linguistic or even as logical) as the absolute and all-determinative point of theoretical reference. This is tantamount to removing subjectivity from its location in the world (in the universe, in being) and ripping a gap between it and the world.

A second way in which the preceding account shows the assumption of the gap to be untenable is that it is incoherent and self-contradictory. Not only our ordinary, life-worldly dealings with all sorts of real-worldly things, but also—and above all—science and philosophy would be senseless endeavors if the subject were indeed entrapped ultimately, inescapably, or unavoidably in the presumed manner within a dimension of its own. What sense would it make—in concrete life, in science, or in philosophy—to talk about the world if the world consisted only of subjective projections?

Third, the preceding account reveals that the assumption of the gap contradicts both the fundamental intention and the fundamental structure of both ordinary and scientific-philosophical languages, because in all these languages we speak about the world in ways that make no references to subjects. The basic linguistic form for talk about the world is "It is the case that such-and-such." This linguistic form, however, is understood most adequately as bringing to expression the world (reality, being) itself, such that, in a non-metaphorical sense, the world (reality, being) *presents itself* in or by means of such formulations.

Fourth, preceding sections show that the *effective* overcoming of the gap—an overcoming that is accomplished rather than merely asserted—presupposes clarification of an extensive series of fundamental issues. The most important clarification is that of theoretical language. This topic is approached via the idea of the *universal expressibility* of the world (the universe, being). Doing justice to this idea involves establishing the following point as fundamental: to universal expressibility there must correspond a universal language that itself must consequently be understood as a semiotic system with uncountably many expressions. The fact that a plurality of such languages is actual and others are conceivable as possible introduces a fundamental problem. As shown above, solving this problem without retreating into a sphere of isolated subjectivity is a fascinating but difficult task.

[2] What the preceding sections of this chapter present makes possible the specification of the philosophical status of comprehensive systematics as *universal philosophical theory*. But this concept itself requires further specification. Here, two central considerations are introduced.

The first concerns the relation between the fundamental formal, semantic, and ontological structures introduced in this book and the dimensions or domains of the world or of being. It is shown above that all these structures have *immediately ontological* statuses and that everything ontologically valid (no matter how designated) is to be grasped only as an ontological structure or structurality.

There is thus no gap or cut between the two dimensions or domains. The distinction between them rests on a more primordial unity; talk of distinct dimensions make sense, or is accurate, only if these dimensions are understood as belonging to a comprehensive dimension that contains them both. Within this comprehensive dimension, all individual domains of being and all individual entities can be articulated in accordance with their own specific ontological characters.

In force throughout is the basic assumption that can be designated the fundamental axiom of comprehensive systematics: comprehensive systematics is concerned throughout with the self-presentation of the primordial and all-encompassing dimension—no matter what it is called—announced by the "it" in all "It is the case that such and such" formulations.

At this point the *second* central consideration making possible the specification of comprehensive systematics as universal philosophical theory can be presented and characterized: this primordial and all-encompassing dimension must *itself* be clarified. Its clarification is the task of the next section.

5.2 Basic Features of a Theory of Being as Such and as a Whole

To what extent, why, and how can or indeed must the primordial and comprehensive dimension identified in the preceding section be a topic for a systematic philosophy? One might think that the philosophical task was sufficiently and even completely accomplished when the problem of the cut or gap is solved, as it is in Section 5.1. The solution lies in understanding the difference between thinking and being, subject and object, concept and actuality, structure and the ontological dimension, or whatever terms are used to designate the duality that leads to the problem of the gap or cut, not as a dichotomy, but instead as always already contained within a more fundamental unity, one that reveals itself in every instance of knowledge, even the simplest and most modest, and in every sentence that qualifies as true. Can philosophy go further than this? What would going further involve?

5.2.1 *What Is Being as Such and as a Whole?*

[1] Throughout the history of philosophy, whenever the point described in the preceding paragraph has in one way or another been reached, there is always the impetus to go further. The impetus of philosophical questioning and of the striving for intelligibility presses further. With the indication of the need to overcome the putative dichotomy and even with the recognition of a comprehensive and more fundamental unity, the philosophical task has not been accomplished fully. This is clear from the simple fact that human beings in general, and especially philosophers, have questions that push them further. As long as such questions in the theoretical domain continue to arise and remain unanswered, the human potential for intelligibility is not yet exhausted. This of course presupposes that such questions are clear, coherent, and thus sensible. If there are questions of this sort, what sense would it make, at some specific point, to forbid additional questioning, to enact something like a prohibition of thinking? The human mind strives for universal questioning and maximal intelligibility.

The question that presses here is the following: can this primordial, comprehensive unity or dimension be made the explicit theme of philosophical theorization? Given the fundamental characteristics of the project pursued in this book, the only answer that can be given here is the following: its thematization is not only possible; it is indispensable.

553   [2] In the history of philosophy, the dimension identified above as primordial and all-encompassing has been given various designations. To be sure, this dimension is not understood there in exactly the way that it is understood here. For the most part, the dimension is taken there to be "objective" in the sense of having no relation, or at best an unspecified relation, to the theoretical dimension. The approach taken here is far more radical and comprehensive in that it recognizes and *explicitly* thematizes the entire "perspectival" (hence linguistic-logical-conceptual-semantic or—most succinctly—theoretical) dimension as one of the subdimensions that co-constitute the primordial and most comprehensive unity. If this is not kept in mind, the following discussions of the primordial, all-encompassing dimension will be misunderstood.

The most famous and probably also the most philosophically neutral (in the positive sense of being the most open to interpretations of all sorts), as well as the most comprehensive designation for the dimension in question, is "being." This book also uses that term, above as well as below. Because of the term's neutrality, however, further specification is required.

[i] The terms "being" and "existence" have lengthy histories and continue today to be used in different ways and to be accorded various meanings. In the history of metaphysics, "being" and "existence" are sometimes clearly distinguished and sometimes simply identified; that history is not considered here (see, e.g., Gilson 1948 and Keller 1968). Heidegger draws a strict distinction between them, whereas Quine (1969: 100) identifies the two this way:

> It has been fairly common in philosophy early and late to distinguish between being, as the broadest concept, and existence, as narrower. This is no distinction of mine; I mean "exists" to cover all there is, and such of course is the force of the quantifier.

Quine's well-known position is highly instructive with respect to what happens to the traditional question concerning being in extensive parts of contemporary analytic philosophy. Of "existence" (and thus, for him, also of "being"), he writes (97),

> Existence is what existential quantification expresses. There are things of kind *F* if and only if $(\exists x)Fx$. This is as unhelpful as it is undebatable, since it is how one explains the symbolic notation of quantification to begin with. *The fact is that it is unreasonable to ask for an explication of existence in simpler terms.* We found an explication of singular existence, "*a* exists," as "$(\exists x)(x = a)$"; but explication in turn of the existential quantifier itself, "there is," "there are," explication of general existence, *is a forlorn cause.* (Emphasis added)

554   Quine's "explication" is clearly circular: "existence" is explained by means of the existential quantifier, but the quantifier is itself understood or interpreted by means of "existence." Moreover, Quine simply maintains that it is a *fact* (!) that it would be "unreasonable to ask for an explication of existence in simpler terms." This may be the case, but even if it is, it is also the case that explications need not involve simpler terms; they can instead involve situating terms or concepts to be explained within one or more of the broader semantic-ontological fields within which they belong. Quine fails even to consider such fields. The claim that asking about "general existence" is a "forlorn cause" is thus arbitrary and dogmatic.

A second, less comprehensive and less open designation for the primordial, all-encompassing dimension is "nature." Since the beginning of philosophy, both "being" and "nature" have been fundamental terminological and conceptual resources for philosophy, but their meanings and uses of course have changed over time. Whereas "being" has, for the most part—unaffected by historical developments—retained its maximally open and maximally comprehensive signification, indeed at times becoming more radical (e.g., with Heidegger's attempt to pose "the question of being" anew and more radically, *Being and Time* §1, of which more below), the conceptual content of "nature" has changed significantly and fundamentally. Currently, "nature" is used comprehensively only in conjunction with a specific metaphysical position (i.e., materialist or physicalist metaphysics).

In certain movements and traditions other designations and concepts are used, including "spirit," "idea," "God," "the absolute," etc. But these more often articulate some contentually quite determinate aspect of actuality as a whole, generally on the basis of central contentual presuppositions.

It is thus clear that all of these terms save "being" fail to designate the primordial and comprehensive dimension with appropriate openness. Here, in terminological harmony with the tradition of philosophies of being but within a radically different theoretical framework, it is designated "being," or "the dimension of being."

[ii] In this book "being" and "existence" are distinguished in contexts that present the structural conception (when other positions are discussed, ones in which the terms are identified, the terms are also identified here). The reason for the distinction is that in ordinary language, in most of the languages used in the history of philosophy, and indeed in contemporary philosophy, the signification and connotations of "existence" are for the most part significantly narrower than are those of "being." Because this is simply a terminological clarification, this issue need not be considered further here.

555

"Being" must, however, be specified more precisely on the basis of what is done in the preceding sections of this chapter. The two most important specifications are now to be presented. The first concerns a strong distinction that plays an important role in many passages in this book (see esp. Section 1.3): the distinction is between "being" in a purely objective sense, thus as designating only the "objective" pole in distinction from the dimension of theory, and "being" as designating the primordial dimension encompassing both poles. This point is clarified by a passage from a letter from Heidegger to Husserl dated October 22, 1927. Heidegger opposes Husserl's procedure of *epoché* and thereby his absolute privileging of transcendental subjectivity, arguing as follows:

> What constitutes is not nothing, thus something that is—although not in the sense of the positive. Universally, the problem of being thus relates both to what constitutes and to what is constituted. (Heidegger 1927b/1967: 602)

Here, "being" is clearly not used to designate the objective counterpole to subjectivity, to the theoretical dimension, etc., but instead to designate the dimension that is primordial and all-encompassing.

The second specification appears in the title of this chapter: "Theory of Being as Such and as a Whole." This specification introduces a differentiation *within* being *as* the primordial and comprehensive dimension, as well as designating being as the primordial and comprehensive dimension, and not as the objective counterpole to subjectivity.

So that the two significations of "being" not be confused, this book relies upon two terminological conventions. First, the term "dimension of being" is generally used to designate being as the primordial, all-encompassing dimension, thus for being as such *and* being as a whole. Second, "being" is often used, for sake of simplicity, as an abbreviation for "dimension of being." For the counterpole to the dimension of subjectivity and/or the theoretical dimension *within* the distinction "subject(ivity)/theoretical dimension—being," the qualifier "in the purely objective sense" is added to "being."

556   The distinction between *being as such* and *being as a whole*—to be explained at this point only programmatically—announces the two grand perspectives that are determinative for the theory of being sketched here: being must be conceived *both* with respect to what constitutes it itself, *and* in its totality, as encompassing anything and everything, thus all entities of any sort whatsoever. The latter specification, understood wholly generally, should be adequately clear immediately and intuitively; not so with the former, despite the fact that it recurs, in various forms, in the history of metaphysics. The task of this section is to make precise both of these two qualifications or perspectives, and to raise and address the questions that thereby arise.

[3] How then should the dimension of being be understood? The following sections answer this question step by step. To begin, four traditional ways of conceiving this dimension—ways to be found in the history of philosophy and in recent and current works—are introduced briefly.

The first form can be termed the "non-reductive objectivistic" conception. It is characteristic of classical metaphysics, particularly in its Christian forms. Perhaps the most important of its advocates is Thomas Aquinas. This conception attempts to grasp the dimension of being *itself*, but in such a way that the conceiving subject (with all that belongs to it) is understood as *remaining external to* the dimension of being. In this sense, the dimension of being is understood objectivistically. Only subsequently is the conceiving subject related to the dimension of being, and in such a way that it is not reduced to any other of the entities constituting this dimension; this is the basis for the qualifier, "non-reductive." On the explicit level, the relation between the dimension of being and the dimension of the subject is conceived purely externally, but implicitly or in intention it is envisaged as an internal relation. Aquinas's assumption of an absolute (God) as being subsisting of itself [*esse per se subsistens*] gives this form of conceptualizing the dimension of being an imposing culmination, at least to a degree, but one that must be characterized only as a subsequently contrived global coherence.

557   A second option can be termed the "transcendental-dualistic" view. Here, the dimension of being itself (in the sense explained above) is missed altogether thanks to the distinction between the world "in itself" (or "things in themselves") and the world as the totality of appearances. The dimension of the in-itself is supposed to be unattainable by the cognitive subject; this is not a negation of the dimension of being, but an exiling of it into inaccessibility. This position is the typical manner of affirming the gap or cut; it is thoroughly criticized in Section 5.1.

The third option can be designated the "reductivist-objectivist" view: the dimension of the cognitive subject (with everything belonging to it) is utterly reduced to entities and domains of nature that constitute the dimension of being understood purely objectively. This is radically materialist metaphysics. It claims completely to grasp and to explain the so-understood dimension of being. The dimension of being is then simply matter, in an all-encompassing sense.

The fourth option is the one taken by this book. For the sake of simplicity, it can be termed the "comprehensively systematic" view of the dimension of being.

[4] Closer explication of the dimension of being is introduced through reference to Heidegger, who on the one hand saw as no other philosopher had the questions and tasks posed to philosophical thinking by thematization of the primordial and comprehensive dimension of being, but who, on the other hand, did not succeed with his own thematizations of this dimension. His failure results chiefly from his attempts, particularly in his later years, to follow paths irreconcilable with rationality, clarity, theorization, etc. (see Puntel 1997). Here, two particularly relevant aspects of Heidegger's later thought are introduced.

The first aspect directly concerns the clarification of the question raised above: what is being as such, and what is being as a whole? In the programmatic essay "Time and Being," Heidegger attempts to surpass all prior attempts to "think being" as such (being as being, being itself). His attempt is instructive in that it shows both what the sense of this question is and what it cannot be. He writes (1969, 1972: 2/2) as follows:

> We want to say something about the attempt to think Being without regard to its being grounded in terms of beings. The attempt to think Being without beings becomes necessary because otherwise, it seems to me, there is no longer any possibility of explicitly bringing into view the Being of what *is* today all over the earth, let alone of adequately determining the relation of man to what has been called "Being" up to now.

558

At the end of the lecture Heidegger summarizes, in a manner typical of him, by bringing into play metaphysics *as he understands it* (24/25):

> The task of our thinking has been to trace Being to its own from Appropriation—by way of looking through true time without regard to the relation of Being to beings.
> To think Being without beings means: to think Being without regard to metaphysics.

The summary published with the lecture characteristically stresses that such phrases, taken together, are "the abbreviated formulation of 'to think being without regard to grounding Being in terms of beings'" (33/35–36). In further clarification, he writes,

> "To think Being without beings" thus does not mean that the relation to beings is inessential to Being, that we should disregard this relation. Rather, it means that Being is not to be thought in the manner of metaphysics.

As is well known, Heidegger has a quite idiosyncratic and demonstrably wholly inaccurate interpretation of the philosophical tradition he sweepingly designates as "metaphysics"; that is not considered further here. But the following may be said of the passages cited above: if Heidegger is concerned with the feature of classical metaphysics referred to above as the "purely external relation," he is correct, although it must be added that the problematic externality involves not only the relation between the dimension of being and that of humans (as cognitive subjects) but also that of being itself to every (particular) being and to beings generally. But Heidegger thinks along quite different lines, as shown shortly below.

To be emphasized at the outset is the significant fact that Heidegger, correcting himself or his own misleading formulations, at least in part explicitly recognizes the two distinct aspects of the being-question introduced above: adequately thematizing the dimension of being involves thematizing both being as such and being as a whole. Being as a whole is to be understood as being-in-its-relation-to-beings or being-together-with-beings, or, yet more explicitly, being-as-the-interconnection-of-(or: among-)beings. Nevertheless, Heidegger fails virtually completely to attend explicitly to the second aspect; he appears even to acknowledge this topic only extremely unwillingly. The reason for this is presumably that he is inclined to view all talk of any relation of beings to being or of being to beings as "metaphysics," in a pejorative sense. This however places in question his entire approach to the thematization of the primordial dimension of being. It suffices here to cite a highly symptomatic passage from his *Contributions to Philosophy (From Enowning* [Vom Ereignis]), written from 1936–38:

> §135. The Essencing [*Die Wesung*][b] of Being as Occurring [*Ereignis*]
> (The Relation of Dasein and Being)
> [This relation] includes the occurring of Dasein. Accordingly, and strictly speaking, talk of a relation of Dasein *to* being is misleading, insofar as this suggests that being essences "for itself" and that Dasein takes up the relating to being.
> The relation of Dasein *to* being belongs in the essencing of being itself. This can also be said as follows: being needs Dasein and does not essence at all without this occurring.
> Occurring is so strange that it seems to be complemented primarily *by* this relation to the other, whereas from the ground up occurring does not essence in any other way.
> Talk of a relation of Dasein to being obscures being and turns being into something over-against [*ein Gegenüber*]—which being is not, because being itself is what first occurs [*er-eignet*] *that to which* it is supposed to essence as over-against. For this reason also this relation is entirely incomparable to the subject-object relation. (1936–38/1999: 179-80/*GA* 65: 254; translation altered)

This text shows, first, that Heidegger himself claims to do precisely what he does not allow Western metaphysics to do: to use formulations that are often inappropriate and often misleading. It is moreover clear that Heidegger—at least on the basis of this text and others like it—views the "relation of being and beings" as a central

---

[b] Heidegger's peculiar use of *Ereignis* is briefly considered in Section 2.2.2(6). *Dasein*, as is widely known, is an ordinary-language German term usually translated by "existence," but one that Heidegger uses to designate the mode of being specifically of *human* beings. A brief explanation of *Wesung*, etc.: the German past participle for *sein* ("being"), *gewesen*, is irregular, but it would be the regular past participle of the verb *wesen*, which did not occur in the ordinary (colloquial) German of Heidegger's day, although it was used frequently in German poetry (especially by Goethe). As a noun, *Wesen* is also the German counterpart to "essence." Words that appear to be forms of the verb *wesen* thus suggest, to those conversant in German, both *being* and *essence*. In the English version of Heidegger (1936–38/1999), Emad and Maly render *Wesung* "essential swaying" (and *Ereignis*, "enowning").

aspect of his philosophy of the primordial dimension of being. Unfortunately, his contribution is minimal because he focuses almost exclusively on being as such. Being as such, in this text, is called "occurring" (*Ereignis*; in other translations, "appropriation" or "enowning"). Heidegger speaks of this "occurring," without exception, in general and often strange, indeed cryptic, formulations. Here is an additional example:

> Occurring occurs [*Das Ereignis ereignet*, "appropriating appropriates," "enowning enowns"]. Saying this, we say the Same in terms of the Same about the Same. . . .
> If overcoming [metaphysics] remains necessary, it concerns that thinking that explicitly [*eigens*] enters occurring [*Ereignis*] in order to say It in terms of It about It. (1969/1972: 24/24–25; translation altered)

This mode of proceeding and of speaking is the result of a problem that concerns the second aspect of Heidegger's thinking to be considered here: the form of thought and thus the means of presentation he uses to thematize the dimension of being. Here, too, it is relevant that, particularly in his later works, these means are irreconcilable with rationality, clarity, theoreticity, etc. In the summary published with the lecture "Time and Being," this is fully clear. The following expresses one issue decisive for him:

> A few grammatical discussions about the It in "It gives,"[c] about the kind of sentences characterized by grammar as impersonal or subjectless sentences, and also a short reminder about the Greek metaphysical foundations of the interpretation of the sentence as a relation of subject and predicate, today a matter of course, hinted at the possibility of understanding the saying of "It gives Being," "It gives time" other than as sentences [*Aussagen*]. (1969/1972: 40/43; translation altered)

560

Apparently, Heidegger here understands "sentences" as "subject-predicate sentences," and he apparently would like to express himself without using such sentences.[31] Heidegger thus simply identifies (indicative) "sentence" with "subject-predicate sentence." It is presumably quite rare that a thinker's error is so obvious; to show that Heidegger's identification *is* an error, it suffices to refer to the philosophy of language—and particularly to the semantics—whose basic features are presented in this book, along with the corresponding ontological consequences. Moreover, Heidegger's radical rejection of what he terms "logic," "formal thinking," "theory," etc., testifies to an astonishing ignorance of the essence of logic, of the immensely broad possibilities for philosophical language, etc. To develop a conception that is a radical alternative to a purely externally conceived philosophy of being and to a substantialist ontology, one need by no means get rid of the dimensions of logic, of semantics, etc.

[5] The preceding considerations make possible first versions of comprehensive and programmatically more precise characterizations of *being as such* and *being as a whole*. Negatively, being as a whole is not the totality of all beings in the sense either of

---

[c] "It gives" literally translates "*es gibt*," which is however the idiomatic German counterpart to the English "there is" or "there are," the French "*il y a*," etc.

[31] See the detailed analysis of the relevant Heideggerian texts in Puntel (1996, esp. pp. 319ff).

the set of all beings or the sum of all parts or any other *purely extensional* determination.[32]
561 Still negatively, this means that it is not to be conceived of by means of the relations of
universal and particular, set and elements, sum and parts, or anything similar; instead, the
relation of being as a whole to beings is absolutely *sui generis*. The decisive point is the fact
that any and every specific being is *a determinate or specific configuration of all and only of
all of what is termed "being."* To avoid any confusion with the contemporary purely extensional (mis)understanding of the term "totality" in such formulations as "the totality of
(or of all) beings," in this book the formulation "being as a whole"[d] is preferred, and is used
throughout. This formulation encompasses and articulates precisely the just-articulated
relationship between *being* and *beings*. As a synonym for "being as a whole," one can also
use "totality of being" (in strict distinction from "totality of (or of all) beings").

A formulation of Wittgenstein's, from a journal entry from 1916, helps make this
basic thought more intuitive:

Whatever is the case is God.
God is whatever is the case (1914–16/1961: 79e; translation altered)

If one replaces "God" with "being," one gets the following:

Whatever is the case is being. Being is whatever is the case.

---

[32] The formulation, "totality of beings," is not to be rejected in every respect. Whether or not it is
acceptable within the framework of the structural-systematic philosophy defended here depends
upon how "totality" is understood. If it is not understood as anything like a set or sum or fusion,
etc., but instead (for example) as the interconnection of all beings, then that already indicates
that "totality as interconnection" has a higher or more comprehensive status than does totality
as determined *purely extensionally*. The term "whole (totality)" is at least often understood, traditionally, on the basis of the axiom, dating to Euclid and Aristotle, "The whole is more than the
sum of its parts (see Euclid, *Elements* 1 and Aristotle, *Politica* I, 2, 1253a19ff). The question then
is that of how this "more" is to be determined in the case of being. Currently, however, the term
"totality" (especially in the literature of logic and semantics) is almost always understood purely
extensionally, which amounts to a reduction of the whole to the extension of its parts. Given this
understanding, the question is, what after all *is* the "extension itself" of the parts in relation to
the parts themselves? This question is rarely raised at present. For this reason, the formulations
"totality of all beings" and "totality of beings" are better avoided. The quite extensive following
section, in its entirety, is devoted to the problematic of talk about "totality" and "totalities."

A different problem concerns the designations "being(s)" and "entity/entities." The German
language makes the fundamental distinction between *Sein* (being) and *Seiendem/n* (being(s)).
Some languages, such as English, make no comparable distinction. The term "entity" has become
an utterly general designation in contemporary philosophy.

[d] "Being as a whole" is less satisfactory, with respect to resisting purely extensionalist interpretations, than is its German counterpart, *Sein im Ganzen* (with its "*im*," literally "in the," rather
than "*als*," "as"). "*Im Ganzen*" could be rendered, "taking all aspects into consideration" and,
although that phrase is too bulky to be used regularly, it is introduced here to supplement the
clarification in the passage to which this note is appended. The central point is, however, the
one emphasized in the main text: the "as a whole" must be understood *as it is explained and
used in this book*, and *not* as it may be used or understood anywhere else.

Section 5.3 shows that the two formulations, using "God" and "being," are not in contradiction; on the contrary, the formulation with "being" is the initial and fundamental characterization of the primordial and all-encompassing dimension, whereas the formulation with "God" is, in a specific fundamental respect, the *ultimate* (fully determined, fully explicated) characterization of the primordial dimension.

Decisive is the following: the distinction between being as a whole and being as such is, from a perspective that is both contentual and heuristic, not only useful but also indispensable. Talk of being as such serves as a programmatic indication of the absolute   562
uniqueness of this topic. If one is not attentive to this uniqueness from the outset, one develops superficial "theories of being" that miss what is decisive about being. But it must likewise be emphasized that being itself or as such would also be missed if one were to understand the "as such" in an abstract, hypostatizing manner. That is precisely the position that Heidegger, in passages cited above, characterizes as "thinking being without beings" or "without the relation of being to beings." This position or attempt fails because it makes of being some sort of abstract Platonic entity that one then attempts in vain to apprehend "in and as itself." This leads to peculiar formulations like the following:

> But being—what is being? It "is" Itself. The thinking of the future must learn to experience this and to say it. (Heidegger 1946/1998: 252/*GA*9:331; translation altered)

In the course of analysis of the two lines of questioning traced above, it soon becomes clear that a point comes at which they are no longer to be separated: they condition each other reciprocally, indeed they ultimately fuse together. To name these two perspectives on what is thus ultimately one subject matter, the account that follows uses—as noted above—the unusual expression "dimension of being"; it is meant to indicate that within both perspectives, being is understood not in the purely objective sense, but instead in the primordial sense.

5.2.2 Talk of "the Whole (the Totality)": Semantics, Logic/ Mathematics, and Philosophy

From the beginning, philosophers have always spoken of "the whole (of the world, of being, of reality, of nature, etc.)." Their talk about it has been questioned regularly, but fundamentally only in a specific respect, i.e., with respect to the possibility of such talk about the whole, insofar as such talk claims to be meaningful or contentual. Section 5.1 treats in detail the form of this fundamental questioning most characteristic of modernity (i.e., the problem of the cut or gap). The preceding section provides initial explanations of how being as a whole and being as such are to be understood. Before however a more extensive treatment of this comprehensive topic can be undertaken, consideration must be taken of a problem that—in part—is an utterly different form of questioning the possibility of talk about "the whole." The problem arises on the one hand from a specific analysis, in semantics, philosophy of science, and critiques of metaphysics, of the term/concept "world" (and other, similar terms and concepts), and on the other hand from specific formal-logical and mathematical states of affairs.

[1] From the first (semantic) perspective, many objections can be raised against the   563
thesis that there is a world as the totality of all that is. As a sort of collection of many and presumably the most important objections from this perspective, one can consider the

thesis formulated as the title of the well-known essay by Bas van Fraassen, "'World' Is Not a Count Noun" (1995). Van Fraassen analyzes the uses of the term "world" and the meanings associated with them in ordinary language, as they are provided, for example, by the *Oxford English Dictionary*. It is clear that the result of such an analysis yields no unitary meaning. Van Fraassen next devotes himself to philosophical uses of "world" in connection with the sciences, particularly physical cosmology. This leads him to the following:

> To conclude then: whether or not the world exists is not settled by the success or acceptance of physical cosmology, *except* relative to certain philosophical points of view. The disturbing corollary for analytic ontology is then that it is never a simple bringing to light of existential commitments in our theories. At best it does so relative to some more basic philosophical stance which is taken for granted. (145)

Van Fraassen's "conclusion" is no surprise, given that it is the result of his own highly controversial theory of science, constructive empiricism, which he characterizes briefly as follows: "According to *constructive empiricism*, the aim is only to construct models in which the observable phenomena can be isomorphically embedded (empirical adequacy)" (143). If no ontology in the genuine sense is possible or can be accepted, then it indeed makes no sense to speak of an "existing world." Such an interpretation of science is, however, highly problematic, as current discussion shows.

Positively, van Fraassen suggests making a "schematic use" of "world." He summarizes what he thereby means as follows:

> [H]ere is my suggested alternative to the idea that world is a count noun. It is instead a context-dependent term which indicates the domain of discourse of the sentence in which it occurs, on the occasion of utterance. It plays this role sometimes by denoting the domain (a set), and sometimes by purporting to denote an entity of which the members of the domain are parts. In the latter case we need not take that very seriously (it may be metaphor, colorful language, rhetorical extravagance); important is only the indicated domain of discourse. (153)

This "clarification" of the world "concept" is vaguer and indeed more confused than the putatively vague and confused ordinary concept of the world. The distinction between "domain of discourse" and "an entity of which the members of the domain are parts" is scarcely meaningful or clarificatory, because in the first case it is not said exactly how "domain" is to be understood, and in the second case the concept *entity* is characterized purely mereologically, but without making explicit the mereological determination of the world itself (i.e., the world as "sum"). Here, a fundamental factor of every theoretical undertaking is ignored: the question concerning the *interconnection* of the "elements" that constitute a domain (however these elements may be designated: as things, beings, entities, elements of a region, etc.). In the first case, the "domain of discourse" is identified by van Fraassen, in passing, with a set, but this provides a merely extensional determination of the domain, not a characterization of the interconnections of the elements of the domain *qua* set. As far as the second case is concerned, the simple identification of the "world" with an implicitly mereologically characterized "entity" comes close to a wholly unreflective and superficial notion of a "concept of the world": the "world" would be a "sum" (others would say "a fusion"). But what exactly is that; what sort of interconnections would it involve? The normal mereological determination of "sum/fusion" is philosophically unsatisfactory. In addition, it must be emphasized that an explicitly mereological determination of "world" is only *one* way of understanding the world as a complex whole—more precisely, as a *grand, complex interconnection* of elements that are, as a rule, extremely heterogeneous. Van Fraassen implicitly rejects *any and every* way of understanding the world as a grand, complex nexus.

Clearly, van Fraassen simply shoves aside the task here posed for philosophy. The task is to bring to clarity the intuitive idea of "world as complex interconnection or configuration." This task is a theoretical undertaking that arises from the fact that human beings can fully utilize their irrepressible capacities for intelligibility. Van Fraassen fundamentally misrepresents this undertaking when, at the end of his essay, he writes,

> We can sit in our closets and in a perfectly meaningful way, kneading and manipulating language, create new theories of everything and, thereby, important contributions to ontology. In other words, to put it a little more bluntly, this "world play" we engaged in here is but idle word play, though shown to be meaningful, just idle word play nevertheless. . . . [O]ntology is not what it purports to be. (156)

Such contentions are basically nothing more than purely rhetorical formulations that illuminate nothing and justify nothing.

[2] Much more important are objections from the logical and mathematical perspective raised against the possibility of talk of totalities. The *specifically logical* objections are not objections in the genuine sense, but instead are to be understood as problems and difficulties that arise from logically formulated talk about totalities. They concern the precise interpretation of quantified sentences in first-order predicate languages. If a quantification is presented as concerning absolutely everything possible, the question presses of how the "domain" is to be understood within which the values of the bound variables of the quantified sentences are supposed to be situated. That such a domain is to be taken *in some sense or other* appears scarcely contestable, given that we formulate sentences that relate in one way or another to everything. But the question that then presses is the following: in what sense is the relevant domain *absolutely all-encompassing*? Formulated yet more precisely, how or as what is such a domain to be understood? As an entity that would be in some sense the only entity, at least on its—the highest—level (such as a set, a class, a collection, etc.)?

Here lies the genuine problem under current discussion.[33] Philosophically, the problem is extraordinarily interesting and important, because a *positive answer* that would not be won from decisively philosophical considerations appears scarcely possible. The perhaps most neutral interpretation available is the so-called plural interpretation of quantification or of the presupposed domain. According to this interpretation, formulations of the form "there is a domain D_i such that. . . (such and such)," are to be interpreted as "there are things—the D_is—such that. . . (such and such)," and formulations of the form "x is a member of the domain D_i," as "x is one of the D_is." Thereby, nothing specific is said about the domain D_i; the logical quantification leaves everything open, and yet such a quantification succeeds in speaking "of everything" (although only if one ignores the presupposed or fundamental ontology). Philosophically, one can understand this problem such that it is not a task for logic

565

---

[33] See, e.g., Cartwright (1994), McGee (2000), and Rayo (2003).

somehow to determine domain D_i further. That is a task for philosophy. And this philosophical task is unavoidable, because it arises from a thesis that is philosophically compelling: if there are things that are D_is, then it is presupposed that they are somehow interconnected, because otherwise it would be senseless and arbitrary to speak of things in such a way that they are designated as D_is. If, as usually happens, this interconnection of D_is is designated a "domain," this term appears misleading. It is therefore understandable that at present the "plural" interpretation of "domain" is preferred, because it is neutral in a strong sense. But that does nothing at all to clarify the philosophical question.

566

[3] The entire problem changes if mathematical (set-theoretical) means are applied, when the "domain" is understood as a set, class, collection, etc., because there then arises, for meaningful talk about totalities, a wholly new problem. In his book *The Incomplete Universe: Totality, Knowledge, and Truth*, Patrick Grim attempts to show that talk about totalities is inconsistent and therefore impossible. He draws this conclusion from a treatment of the truth-paradox (the so-called liar paradox), from the paradox of the knower, and from results of Gödel's incompleteness theorem for the concept of universal knowledge or of a totality of all truths. He understands all these problems as different forms of one and the same fundamental problem. He writes, at the beginning of his Chapter 4,

> [W]hat follows may be the cleanest and most concise form in which we have yet seen [the problem]. By a simple Cantorian argument, it appears, there can be no set of all truths. (1991: 91)

The example of the totality of truths is only one case; the same conclusion arises in all other cases of totalities: the totality of all propositions, the totality of all facts, and—of particular importance for this chapter—the totality of all beings. In the following, the basic idea of the proof is first sketched briefly for the example of the totality of all truths (see 91ff). Details and more exact clarifications of the mathematical terms and structures cannot be provided here.

Grim understands totalities as sets; his thesis with respect to the totality of all truths is that "there can be no set of all truths" (91). The following brief sketch of the proof follows Grim. Let $T = \{t_1, t_2, t_3 \ldots\}$ be the set of all truths. Then consider the subsets of T, thus the elements of the power set $\wp(T)$:

$$\varnothing$$
$$\{t_1\}$$
$$\{t_2\}$$
$$\{t_3\}$$
$$\vdots$$
$$\{t_1, t_2\}$$
$$\{t_1, t_3\}$$
$$\vdots$$

Note that to every element in this power set there corresponds its own truth. With respect to every set within the power set, it then holds (for example) that t_1 either does or does not belong to the set:

567

$$t_1 \notin \varnothing$$

$$t_1 \in \{t_1\}$$

$$t_1 \notin \{t_2\}$$

$$t_1 \notin \{t_3\}$$

$$\vdots$$

$$t_1 \in \{t_1, t_2\}$$

$$t_1 \in \{t_1, t_3\}$$

$$\vdots$$

It results that there are at least as many truths as there are elements in the power set $\wp$. Cantor's famous procedure of diagonalization, however, shows that the power set of every set is *greater* than the set itself. From this it follows that there are *more* truths than there are elements of T: T does not encompass all truths.

[4] Grim's thesis is of great significance for philosophy. A thorough response to it is a significant task that cannot be undertaken in this book.[34] The following account introduces only four brief remarks on the subject.

[i] Grim considers alternative set theories in order to support the thesis that none of them makes possible the avoidance of the Cantorian diagonalization argument, but in this matter he appears to have overlooked something important. Although he does consider the set theory of Kelley-Morse (KM)—albeit quite briefly—he appears not to have noticed that, in this set theory, the Cantorian argument is ineffective. In all brevity: KM[35] distinguishes between sets and classes. The universal class (not set) is defined as: $\mathfrak{U} = \{x: x = x\}$. Theorem 37 of KM is the following: $\mathfrak{U} = \wp\mathfrak{U}$: the universal class $\mathfrak{U}$ is not smaller than its own power class, but identical to it. This is not so with the power set of any set; there, the power set is larger than the set itself. Applied to the example of the totality of all truths, this means that the power class of the universal class does not contain more truths than does the universal class itself. To be sure, this state of affairs must be clarified precisely with respect to all of the aspects belonging to the formation of a consistent *and efficient* set theory (or class theory). Here it suffices to indicate the fundamental and presumably decisive factor that makes possible a solution of the problem.[36]

[ii] A second remark concerns a peculiar incoherence or, if one will, paradox in 568 Grim's argumentation. His argument presupposes that we grasp and articulate both

---

[34] The extensive exchange between Grim and Alvin Plantinga leads to interesting philosophical clarifications. See Plantinga and Grim (1993).

[35] For details on KM, see the appendix to Kelley (1955), titled "Elementary Set Theory."

[36] Schneider (2006b) contains a detailed critique of Grim that, in essence, thoroughly presents and grounds the thesis presented in the main text.

truths that are contained within the set of all truths, and truths that are not contained in this set. We thus indeed grasp and articulate a totality that has a higher character (i.e., a genuinely comprehensive character), such that no truths remain outside this higher totality. This is a consequence of an informal philosophical analysis. But this conclusion results as well from a detailed analysis of the more precise form or formulation of the argument itself: Grim can formulate his thesis that there is no totality of truths only by using premises *that involve quantification over just such a (comprehensive!) totality.* This shows that Grim relies (and must rely) precisely upon what he wants to prove to be impossible.[37]

[iii] The third remark is a purely philosophical one. It concerns a presupposition that Grim simply makes, apparently assuming it to be non-problematic—that recourse to set-theoretical means for philosophical theorization is not only possible and meaningful but also indispensable. The assumption underlying this presupposition is presumably the following: set-theoretical (and logical) means are the only theoretical means that make possible precise formulations; only they bring clarity to scientific and/or philosophical theories.[38] Both the presupposition and the assumption on which it is based are, however, quite problematic, and not only philosophically. Already within the frameworks of the formal sciences, the long-privileged position of set theory is increasingly being questioned, specifically by the mathematical theory of categories.[39] With respect to philosophy, it is quite questionable whether set-theoretical means, used either alone or in a manner that is decisive (rather than merely helpful), are means that are adequate for the articulation of philosophical issues. Section 4.3.1.2.1 examines this problem as it relates to the configuration that this book understands the human individual to be. Ultimately, all depends upon the concept of the set that is "defined" by means of the concept of the relation of being an element ($\in$). Different items (of whatsoever sort) can be collected together into wholes on the basis of their having a common property. To say that the items, with respect to this common factor, "belong together" means that they form a set or belong to a set. What this common factor, this *interconnection*, is to be understood as remains unexplained. It is considered only, so to speak, as a purely abstract, contentless belonging-to or belonging-together. How such a common factor binds the items together and of what, precisely, this

[margin: 569]

---

[37] For details see Plantinga and Grim (1993, especially sections 9 (291–97) and 11 (301–05), both "Plantinga to Grim").

[38] Another author who accepts Grim's basic thesis, John Bigelow, explicitly formulates this presupposition and this assumption as follows in his essay "God and the New Math":

> If we need to abandon minimal pantheism [in a quite unusual and misleading terminology, Bigelow understands by "minimal pantheism," "the doctrine that there is such a thing as the totality of all that there is" (130–31)] in order to make space for set theory—then we should abandon minimal pantheism. That is a small price to pay for the peace of mind which can flow, for any scientific realist, from the freedom to take set theory at face value as literally true. (1996: 145–46)

Such a contention may be a self-description of Bigelow's own unquestioned theoretical preferences, but it is not an argument.

[39] On this topic, see Section 3.2.1.2, section 5 of Corry (1992: 332ff), Goldblatt (1979/1984), and MacLane and Moerdijk (1992).

binding consists are not explained. It is obvious that such a conceptuality or structure is not suitable for explaining the complex contentual interconnections with which philosophy is concerned.

It is revealing that the authors who accept Grim's thesis always understand "totality" as "one big thing" (see Bigelow 1996: 127 and passim). If totality is thought of in this manner, then it must indeed be rejected: it is not an "additional thing," so to speak, on the same level as all the other "things" that it is supposed to contain. Tom Richards provides an accurate formulation of this state of affairs:

> I do not see how the Universe can be said to be a thing of any *sort* at all, or even a *thing*, since it encompasses everything. If the Universe were a thing, then in contradiction to the Axiom of Regularity it would be self-membered, since everything is a member of the Universe. (1975: 107; quoted in Bigelow 1996: 127–28)

Richards, however, appears to assume that there is no alternative to the determination or characterization of the universe as a set. In addition, he appears to hold that only one ontological category, "thing," comes into question or is thinkable for articulating the totality, such that it is immediately obvious that totality is impossible as an all-encompassing great thing. He thus says: "The way things are is not a thing of any sort at all," and "By using the phrase, 'the way things are,' one does not refer to anything" (106; cited by Bigelow 1996: 127). Such contentions raise the question of what kind of ontology Richards, Grim, and others maintain. If the basic category of ontology is the category "thing" or "object" (and thus, ultimately, "substance as substratum"), then one can understand the totality only as a thing/object. Chapter 3 not only presents an alternative ontology that is defensible but also reveals substance ontology to be untenable.

[iv] The fourth remark concerns a partial aspect of a central thesis of the structural-systematic philosophy presented in this book. This is the thesis that every theoretical sentence and every theory always already presuppose some determinate theoretical framework and are situated within such a framework. This thesis has two specific aspects: that there is in fact a plurality of theoretical frameworks and that the structural-systematic framework presented here is by no means the highest or the absolute theoretical framework. This latter aspect is decisively relevant to the question under consideration concerning the possibility of talk of totality (or totalities).

This book presents and defends the position that the structural-systematic theoretical framework is not the highest, but is only superior to or more adequate than other currently available theoretical frameworks. As Chapter 6 shows, this is a *metasystematic* thesis, a thesis that is located methodologically and scientifically within a more comprehensive theoretical framework. What does this mean with respect to the question concerning totalities? The theoretical situation here is to a certain point analogous to the situation that results for formal systems from Gödel's incompleteness theorem, i.e., that no complete deductive system is possible even for so small a fragment of mathematics as elementary number theory. Differently stated: for every axiomatizable, non-contradictory system for arithmetic, it is the case that there are true arithmetic sentences that are not provable as true within the system. Gödel's proof is based on so-called Gödelization—a procedure of encoding by means of which linguistic entities can be assigned to numbers. In a broader, more comprehensive system, or with the means of such a system, it is possible to prove as true sentences not provable as true within the first system.

In a certain analogy to this situation, the following holds with respect to the structural-systematic philosophy: within a broader, more comprehensive theoretical framework, it is possible to overcome the limitations of the structural-systematic theoretical framework, but not in an absolute manner, because it cannot be assumed that the broader, more comprehensive theoretical framework is the one that is simply ultimate or absolute (at least, this assumption is not made in this book). Thus, one must also assume an *incompleteness* of the structural-systematic philosophy. With respect to the topic of truth, one can characterize this incompleteness as follows: the higher or indeed ultimate truth *about* the system of the structural-systematic philosophy cannot be established *within this system*. Such a truth is a *metasystematic* (more precisely, meta-structural-systematic) truth; in the terminology of Chapter 6, it is an *external* meta-structural-systematic truth (see Section 6.3). The structural-systematic philosophy is *incomplete* in the sense that it is an *open system*: it is open to broader, higher, more adequate theoretical frameworks. Such a characterization must, however, be understood correctly. This is shown in what follows to counter certain philosophical lines of argument that are drawn from Grim's theses.

571

[5] The strongest conclusion drawn thus far is the following: there is no universe; there is no totality of all that there is (see Bigelow 1996, which cites various authors). This is a gravely imprecise thesis that depends upon a point of confusion and results in a misunderstanding. What can be meant is only the following: the universe *as a set* does not exist. As shown above, however, those who defend this thesis tacitly assume that if there were a universe, it could be conceived of only as a set (the comprehensive set). As also shown above, such an assumption is neither established nor in any way whatsoever even plausible.[40] If the understanding of the universe as a set always leaves something *outside the universe understood as a set* (because the power set of this set has a larger cardinality), what follows is only that the *so understood universe* does not also contain whatever remains outside the universe-as-set. In the casual formulation, "a or the universe does not exist," the term "universe" is used *equivocally*, first in the sense of "totality of everything that there is, understood as a set" (=universe$_1$), and then in the sense of "totality of everything that there is *simpliciter*, i.e., as including *both* everything that is in contained in the totality understood as a set *and* everything that remains outside the totality understood as a set" (=universe$_2$). With respect to the equivocal formulation, "a or the universe does not exist," what results is the following: what is proven or provable is only the sentence, "a or the universe$_1$ does not exist," not however, "a or the universe$_2$ does not exist." The criticized formulation, however, itself establishes that the non-existence of the universe *simpliciter*, of universe$_2$, is not in question.

---

[40] In this respect, Grim is on the whole more cautious (and also more precise) than other authors. The title of his book speaks only of "the *incomplete* universe."

Various authors attempt to save talk about totality, but they pay a high price. They assume only what may be termed a "halved totality"; that is, they understand the assumed totality in an extremely restricted sense, in that they speak of a *submaximal* world or a *submaximal* universe. These submaximal "totalities" are characterized in various ways (see Bigelow 1996: 144ff). It is said, for example, that "although there is no aggregate of all the things there are, there may yet be an aggregate of all things which are contingent in the sense that it is logically possible that they might not have existed (in the widest-open sense of 'existed')" (152). Such attempts fail because of the argument that is presented above under [4][iii]. It is astonishing how, without reflection, these authors speak about *absolutely everything*—and thus, in their arguments, quantify over absolutely everything—precisely when they attempt to prove the impossibility of talk about the unrestricted totality. The proper conclusion to be drawn is that the genuine question is not and should not be whether such a totality is possible or exists, but *how it is to be understood*.

[6] For the structural-systematic philosophy presented in this book, the preceding considerations lead to the following important question: is it not the case that what is presented above in [4][iii] as solving the problem posed by Grim—that the structural-systematic philosophy is based on a *non-absolute theoretical framework*—actually poses a serious problem for this philosophy itself? The reason would be that within a non-absolute framework, it would appear that only a non-absolute and thus only a limited, a "submaximal" totality, could be thematized. This is an issue that indeed arises and must therefore be addressed.

Resolving this issue is not difficult. Its resolution emerges from a distinction that is strictly to be attended to here and elsewhere: it is one thing to speak of a non-absolute, (i.e., not absolutely adequate) *articulation* of a whole (in this book: of being as a whole), and it is something wholly different to acknowledge or take into consideration only a "submaximal totality" (a "submaximal being as a whole" or a "submaximal universe"). The thesis that the theoretical framework introduced, developed, and used in this book is not the absolute, the highest theoretical framework, in no way entails that the totality articulated on its basis is only a limited or submaximal totality *in the sense in which there is talk of "submaximality"* in the conception referred to above. On the contrary, a non-absolute articulation of a maximal totality always remains an articulation of the maximal totality (i.e., a so-and-so determined articulation in distinction to other (more or less adequate) articulations that arise within other (more or less adequate) theoretical frameworks). There is a distinction between non-absolutely adequate articulations of the *maximal totality* and the absolutely adequate articulation of one and the same totality; this distinction is by no means to be equated with that between the maximal totality and submaximal (forms of) the totality or of totalities, nor does it entail any such distinction. If, purely terminologically, one wanted to *call* a not absolutely adequately articulated totality a "submaximal totality," that would be a matter only of terminology, not affecting the subject matter. But such a terminological convention would be the result of the confusion, noted above, of articulation with what is articulated—of the distinction between, on the one hand, an absolutely adequate articulation and a not absolutely adequate articulation of totality with, on the other hand, the distinction between the maximal totality and one or more submaximal totalities. A maximal totality that is not an absolutely adequately articulated totality remains a maximal totality: this (and not any sort of subtotality) is in such a case what is not absolutely adequately articulated.

To be sure, it is correct to say that by means of a more adequate articulation *more* structural moments of being as a whole and thereby *more* truths concerning being as a whole are brought to light, but, for many reasons, that does not introduce any new problem that is relevant here. First, it is here assumed and maintained that a totality absolutely adequately articulated *in every respect* would be attainable *only* on the level of the *absolute* theoretical framework, and this book makes no claim to rely on that framework. This is a *meta-structural-systematic* thesis that acknowledges explicitly that within the absolutely adequate theoretical framework *or already in any superior theoretical framework, more* structural moments and *more* truths are explainable. This is, however, wholly consistent with the structural-systematic philosophy, which explicitly accepts so-understood superior levels, but designates them, too, as *non-absolute*.

Second, it is to be noted that the talk of *more* structural moments or of *more* truths can be misleading. Strictly speaking, it is not a matter of quantitatively more structural moments or truths, but instead of a more finely grained presentation of structural moments or a more precise form of truths.

Third, if, in a purely formal respect, one wanted to consider the unrestricted or absolute totality (being as a whole) as a *universal class*, one would, in the light of what is noted in [4][i] above concerning the universal class, have no problem arising from the fact that a distinction is drawn between the structural-systematically non-absolutely adequately articulated totality—being as a whole, non-absolutely adequately articulated—and the (here neither attempted nor attained) absolutely adequately articulated totality—being as a whole, absolutely adequately articulated. There would be no problem because one would understand the non-absolutely adequately articulated totalities as *partial classes* of the absolutely adequately articulated totality, the latter understood as a universal class. And the result of this would be the following: the universal class would be *identical* to its power class, in accordance with the abovementioned theorem of Kelley-Morse: $\mathbb{U} = \wp\mathbb{U}$.

[7] As a consequence of the refutation of the logical and mathematical objections to the possibility of talk about totalities, the intelligibility of the concept *being as a whole*, explained in Section 5.2.1, is solidly grounded. Being as a whole, as here understood, is not the purely extensionally understood totality of all beings. If "totality (of being)" is understood differently, for example in the sense of "the being (of all beings)," then its intelligibility depends upon how, precisely, one determines the relationship between being and beings. In this case and in this sense, as indicated above, one can and should speak not of the "totality of beings" (or "of all beings"), but instead, avoiding misunderstanding, of the "totality of being" (on this point, see Section 5.2.1, particularly [5], pp. 419–21). The ontological difference between being and beings thus reveals itself to be absolutely decisive for the development of the structural-systematic philosophy.[41]

---

[41] In a specific stage of his development, Heidegger puts the term "ontological difference" between being (*Sein*, often translated as "Being") and beings (*Seienden*) at the center of his philosophizing. As the considerations in previous sections of this chapter show, and the later sections continue to show, Heidegger's specific understanding of this difference is not here presupposed or indeed accepted.

5.2.3 *The Primordial Dimension of Being, the Actual World, and the Plurality of Possible Worlds*

Scarcely any issue is likely to be of more current importance for the program of a comprehensive systematics than the issue of *possible worlds*. Clarification of this issue and of some basic aspects of current discussions of it can contribute significantly to the clarification and sharpening of this program. There can of course be no question of here doing anything approaching full justice to the complete array of current treatments of the topic; the following account aims exclusively at clarifying the goal pursued in this book.

[1] The central concept in the theory of possible worlds, the one around which all reflections orbit, is the concept *actuality* in such formulations as "the actual world." Despite possible appearances to the contrary, this concept is anything other than clear and unambiguous. For the classification of different conceptions of possible worlds, of central significance is the distinction between two opposed and mutually exclusive perspectives that open paths leading in different directions. The two perspectives are generally termed "actualism" and "possibilism."

According to actualism, the fundamental and indeed exclusive starting point for all considerations of possible worlds is the so-called *actual world*: anything that can be designated a possible world is understood and determined on the basis of the actual world (i.e., as a *possible variant* of it). Possibilism, on the other hand, starts with a plurality of possible worlds as the primary dimension and on its basis determines the so-called actual world as one possible world among the many possible worlds. There are various species of both actualism and possibilism, but the species are not centrally relevant to present purposes.

With respect to the comprehensive systematics under current consideration, an important question is how these two positions relate to the primordial dimension of being. More generally formulated: in theories of possible worlds, in either or both of the variants introduced above, are the questions concerning being as a whole and being as such posed and treated? Does either variant or do both variants thematize a dimension of being in the sense introduced above? As can be shown in exemplary fashion, the answer to these questions is no, particularly in the case of possibilism. For this reason, the following account does not treat this issue only in general, but instead examines in some detail the specific position that is presumably the most important of those that are available; this is the theory of David Lewis (1986).

[2] Lewis defends a *possibilistic* conception, but one sufficiently peculiar that he terms it "modal realism." His reason for assuming a plurality of possible worlds is that he deems this hypothesis "serviceable" (3) with respect to clarifying various central questions in the realms of logic, language, philosophy of mind, philosophy of science, and metaphysics. Lewis's starting point is the analysis of modalities on the basis of the concept of possible worlds. Formulations like "It could have been that there were blue swans" are analyzed by means of quantification over possible worlds and thereby translated into the language of the theory of possible worlds: "It could have been that there were blue swans if and only if there is a world *W* such that in *W* there are blue swans." From this is easily derived, "There is a (non-actual) possible world in which there are blue swans."

575

The first question that presses is whether this is anything more than a convenient but *purely intuitive illustration* of the modality "It could have been." Lewis thinks that it is, insisting that his conception of possible worlds is genuinely ontological. The chief points, in the formulation of Gideon Rosen (1990: 333; numbering altered), are the following:

1. Reality consists in a plurality of *universes* or 'worlds.'

2. One of these is what we ordinarily call *the* universe: the largest connected spatiotemporal system of which we are parts.

576

3. The others are things of roughly the same kind: systems of objects, many of them concrete, connected by a network of external relations like the spatiotemporal distances that connect objects in our universe (Lewis 1986: 2, 74–76).

4. Each universe is isolated from the others; that is, particulars in distinct universes are not spatiotemporally related. (It follows that universes do not overlap; no particular inhabits two universes; 78.)

5. The totality of universes is closed under a principle of recombination. Roughly: for any collection of objects from any number of universes, there is a single universe containing any number of duplicates of each, provided there is a spacetime large enough to hold them (87–90).

6. There are no arbitrary limits on the plenitude of universes (103).

7. Our universe is not special. That is, there is nothing remarkable about it from the point of view of the system of universes.

A question that is decisive, given present purposes, is how Lewis deals with the concepts *existence, being, actuality, possibility*, etc. That question is posed with a view to the additional question concerning what conditions his talk about a plurality and indeed a system of possible worlds must satisfy in order to be meaningful and intelligible. Is it not the case that a universal dimension must be presupposed? Is it not the case that the designation "world," used for all the worlds, presupposes that they all have something in common?

[3] To assume a plurality of possible worlds is to speak of them. That involves the use of such concepts as *world, possible, plural*, etc. Are these concepts to be understood as universals having the individual possible worlds as instantiations? Strangely, this is a question that is not only not answered but indeed not even raised explicitly in current theories of possible worlds. Nevertheless, it is unavoidable, because unless it is answered the talk of "worlds," "possible," "plural," etc., remains ultimately unintelligible. The problem becomes sharper when one considers more carefully the concept *plural*. This concept directly designates entities considered in mutual isolation. But the concept itself relates the entities to each other, thus either presupposing or establishing (in usual terms: analytically) an interrelation, a unity. In Kantian terms: talk of the plurality of possible worlds entails as a condition of its possibility the assumption of a comprehensive collection containing the distinct items and thus also of a unity that makes the plurality of items possible. But how is the collection or unity to be thought?

577

Lewis knows and uses a specific concept to which he accords an unambiguous ontological priority and for whose designation he uses two terms that he appears to consider, albeit not always and in all contexts, generally interchangeable and thus synonymous: the two are "actuality" and "existence." In his view, every possible world is actual and existent:

> [E]very world is *actual at* itself, and thereby all worlds are on a par. This is *not* to say that all worlds are actual—there's no world at which that is true, any more than there's ever a time when all times are present. The "actual at" relation between worlds is simply identity. (93)

This passage can be understood only if one notes that Lewis understands "actual" and "existent" quite idiosyncratically: Lewis understands these words on the basis of an "indexical analysis" yielding the simple meaning, "this-worldly" (92). This shows how his easily misunderstood formulations are to be interpreted. He often says, for example, that *only our world* is a or the *actual world*, whereas the other worlds are to be conceived of as *unactualized worlds*. This means that we, as inhabitants of *this* (i.e., *our*) world, consider this world to be *the actual* world; inhabitants of other worlds would proceed in the same manner with respect to their worlds, viewing those worlds as actual.

Lewis stresses that *actuality* (and thus also *existence*) is, for him, a fundamentally *relative* concept. In part to support this thesis and in part to elaborate its far-reaching consequences, he argues as follows:

> Given my acceptance of the plurality of worlds, the relativity is unavoidable. I have no tenable alternative. For suppose instead that one world alone is *absolutely* actual. There is some special distinction which that one world alone possesses, not relative to its inhabitants or to anything else but *simpliciter*. I have no idea how this supposed absolute distinction might be understood. (93)

This formulation, in conjunction with the passage previously cited—"[T]his is *not* to say that all worlds are actual—there's no world at which that is true"—articulates Lewis's strongest and indeed only argument against the assumption of a primordial and all-encompassing dimension of being in the sense explained in this chapter. It therefore requires careful analysis.

In his denial that "all worlds are actual," Lewis appears not to use "actual" as an indexical term. The "actuality" he is denying must therefore be something putatively absolute, something taken to hold *simpliciter*, something that would characterize not a specific world but instead *all worlds*, and thus something common to all worlds. Against such an actuality Lewis raises two objections: first, he reports having "no idea" how the relevant absoluteness could be understood; second, he asserts that "there's no world at which that is true." The first objection is countered by the adequate explication of the relevant absoluteness. The second, however, is obviously circular in that it presupposes the conclusion it purports to establish—the contention "there's no world at which that is true" presupposes that "actual" is understood purely indexically. To make explicit the argument that Lewis formulates only elliptically: if "actual" were an absolute characteristic (i.e., one that would hold non-indexically for every possible world), then there would have to be some specific world having this characteristic, but there is no such world, so there is no such characteristic.

This argument can be countered in at least two ways. The first notes the circularity identified in the preceding paragraph. This circularity can also be exposed in a different way: Lewis simply presupposes a plurality of worlds that are "on a par." Only on the basis of *this* presupposition can he consequently argue that if one wanted to understand actuality (existence, being) *absolutely* or *simpliciter* (i.e., for Lewis, nonindexically), one would have to say, "one world alone is *absolutely* actual." This, however, is a *petitio principii* because it presupposes that *absolute actuality* can be conceived of only as a characteristic of *one* (single) world, a world selected from the plurality of worlds presupposed somehow to be "on a par." This presupposition is, however, exactly what was to have been proved. In other words, what Lewis would have to prove, but does not even argue, is that "absolute" actuality or existence *cannot be thought and conceived of other* than as a characteristic of a single one of the plural possible worlds.

The second counterargument is based on the fact that Lewis overlooks, indeed utterly ignores, a simply fundamental point: the one raised by the question, to what world (even in some highly restricted sense) does the theoretician (Lewis, or any theoretician) *who talks about* all these infinitely many worlds and makes significant assertions about them, etc., belong? If the theoretician were an inhabitant of only a single, wholly specific one of the worlds, *in Lewis's sense*, how could the theoretical dimension of this theoretician extend to all the other worlds; how could the theoretician have access to them? Such access or such an extension *must be assumed*, because sentences taken to be true are made about all the worlds, and true sentences are ontologically revelatory.

A condition of possibility for a theoretician's presenting a theory that, like Lewis's, is about *all* the worlds reveals Lewis's contention that there is no world at which the thesis that all worlds are actual is true to be vulnerable to a *reductio ad absurdum*. Not only must it be the case that there is at least one world at which this thesis is true, and with it the thesis that the theoretical dimension is all-encompassing; it must also be the case that these theses are true in *all* worlds (no matter how these worlds are understood more precisely, given only that a plurality of them is assumed). The reason for this is that the theoretical sentences are true for *all worlds*. Lewis simply fails to raise the question of the location or status of the theoretician or of the theoretical dimension, and thus also that of the status of theoretical theses.

This failure has, in principle, two fully opposed consequences; which of the two emerges depends upon which of the following two opposed assumptions one makes. Either one considers all the theoretician's activity—more precisely, the theoretical dimension, with everything it contains—as something that is fully *external* to *all* worlds, and that thus in no way belongs either to any specific world or to the totality of worlds—or one considers the theoretician's activity—the theoretical dimension—to belong to the totality that is the subject matter of the theory itself. In the first case, one accepts an absolutely unbridgeable gap between theory and actuality or "system"; in the second, one assumes that the theoretical dimension is integrated into the whole to which it relates. If, on the basis of the reasons presented in detail in Section 5.1, one rejects the former alternative, then one must accept the latter. But this latter alternative entails the thesis that the so-called plurality of worlds is to be conceived of *on the basis of a primordial and all-encompassing dimension of being*. This is now to be shown briefly.

Within the theoretical dimension, the worlds about which the theory speaks are bound to one another in such a way that they form a grand nexus, no matter how this is designated and explained more precisely. In Lewis's language, the theoretical dimension "is" or "exists" in or at all worlds because it relates to or encompasses all worlds; it must, because true statements about all worlds can be made. The theoretical dimension, with respect to its objective content, thus belongs unrestrictedly and in the genuine sense to the inner constitution of the totality of worlds. With this, one of the major theses 580 expressed in this book in various contexts attains full force: the mind, particularly with respect to the theoretical dimension that characterizes it, is intentionally *coextensive* with anything and everything, with the universe, with being as a whole. The fact—obviously not to be ignored when there is theoretical activity—that there is a theoretical dimension and that this dimension (in the language of theories of possible worlds) exists in or belongs to all worlds reveals a fundamentally primordial and all-encompassing *unity of all worlds*. But this unity is not only conceptual, it is in the strictest sense *ontological*, because the theoretical dimension is the integrating part of the unity of that totality that Lewis calls the "plurality of worlds."

[4] How can—indeed, how must—this unity be conceived? If one wants to continue to speak of a plurality of "worlds," then this plurality will clearly be a *secondary phenomenon* in comparison with the primordial and all-encompassing unity, secondary in the sense of presupposing it and being conceivable only in terms of it. In principle, this unity can be designated in various ways but, for reasons given in Section 5.2.1, the most appropriate designations are "being" and "dimension of being."

The thesis that the plurality of possible worlds is the fundamental point of reference with respect to which all statements about actuality/existence/being must be situated, with the consequence that our world is "actual" only "for us," is untenable. It results from ignoring fundamental questions and interrelations and is itself incoherent. One cannot avoid acknowledging a primordial and comprehensive dimension of being that is in every respect a dimension one cannot get around and cannot go beyond. It provides the basis for addressing questions concerning the metaphysical or comprehensively systematic concept of the possibility of a plurality of (possible) worlds, etc. The following paragraphs provide, on the basis of conclusions already drawn, some indications of the manner in which the structural-systematic philosophy responds to these questions.

In the first place, the concept *world* must be subjected to critical analysis. Is the dimension of being, as explained above, a "world"? Here, two questions must be addressed separately; one is terminological, the other is substantive. Terminologically, one could certainly choose to use the term "world" (and/or "universe") to designate the dimension of being. But that opens the substantive question: are there *in addition to* or *outside of* the world-as-dimension-of-being *other* (possible or actual) worlds-as-dimensions-of-being? There cannot be, because it is unthinkable that in addition to the dimension of being there could be a "dimension-of-being$_1$," a "dimension-of-being$_2$," . . . a "dimension-of-being$_n$," etc. Even this brief reflection reveals that it is not 581 illuminating to designate this primordial, comprehensive dimension as "world." This further supports this book's practice of ascribing to the term "world" (when it is used strictly rather than quite generally and even loosely) a restricted signification (i.e., the one specified in Chapter 4).

As shown above, the terms "actuality," "actual," etc., are far from clear. If this terminology is retained, then it follows from what is said above that there is *only one* actual dimension of being = actual world. Better, however, is to distinguish between being *in the primary sense* and being in a *secondary* or *derivative* sense. Being in this secondary sense would be the entire dimension of the possible. Possible or derivative being can be conceived of only on the basis of actual being, of being in the primary sense.

If one wants to continue to use the formulation "possible world," preceding considerations establish that a *plurality of possible worlds* is indeed to be assumed. *These* possible worlds, however, unlike those introduced by Lewis, are in no sense "on a par" with the dimension of being (= the only actual "world"); they "are" or "exist" *only* in the secondary and derivative sense.

This conception, at whose center is the thesis of the absolute uniqueness of the dimension of being, can be made fully clear only by closer investigation of the dimension of being. This is an extensive and difficult topic that is barely considered, much less adequately treated, by theories of possible worlds. To it, Section 5.2.4 and all of Section 5.3 are devoted.

5.2.4 *The Inner Structurality of the Primordial Dimension of Being: The Most Universal Immanent Characteristics*

[1] Philosophical treatment of the dimension of being presupposes a positive answer to the question whether it makes any sense whatsoever to raise such questions as, "What is being itself or as such?" If philosophy is to achieve anything at all in this domain, then that will require further thought about the status of some questions that at first glance appear illuminating and obvious. Among these belong particularly questions beginning and "What is/are. . .?"[42] No doubt, many questions that so begin are reasonable, such as "What are atoms?" "What is knowledge?" But questions that so begin are generally reasonable only when they seek answers articulating something or other that is distinguished from other "things" that can be explicitly identified. But from what is *being* to be distinguished? The obvious answer is *nothing*. But *nothing*, or *the nothing*, is a limiting concept, a defective—because totally negative—"concept." The question, "What is (the dimension of) being," taken literally, is therefore not a reasonable one.

What questions, then, can or should be raised about the dimension of being? Only ones that concern perspectives from or within which the dimension of being can be thematized. But the identification of such perspectives is a part of the thematization itself and thus of its problematic. What are the relevant perspectives? Here it is particularly difficult to find and use appropriate language. If one spoke of "properties" of the dimension of being, then one would introduce an immense and unquestioned ontological ballast, in that properties require substrata within which they are supposed to subsist or reside. Even the word "perspective" is fundamentally problematic in that it suggests the particular viewpoint of a somehow external observer. The thematization of the dimension of being needs, however, to allow the dimension of being to—so to speak—bespeak itself, to achieve a self-presentation.

582

---

[42] See the idiosyncratic and obstinate but nonetheless noteworthy reflections in Heidegger (1956/1958).

The word "thematize," in its various forms, is of course itself problematic, because it has the explicit connotation of being the act of a subject (i.e., the theoretician). Of course, one cannot contest that there is such an act, but a strong or exclusive emphasis on the thematizing act suggests that it is somehow outside the dimension of being: the theoretician makes the dimension of being a topic of theorization. This formulation, however, distorts the relevant state of affairs. For this reason, "explication" or even "self-explication" of the dimension of being is preferable. It is no accident that "*ex-plicatio*" has long been used in the Western metaphysical tradition. To be sure, "explication" is also generally understood as connoting the act of a subject, but this connotation can at least be kept in the background, particularly when the passive voice is used along with the operator, "it is the case that"—hence "it is the case that (it is explicated that...)." In what follows, this is the preferred formulation.

What does an explication or self-explication of the dimension of being bring to   583
expression? Given the state of affairs just described, the designation chosen here as less inappropriate (because less heavily burdened) is "immanent structural features or characteristics of being as such or of the dimension of being." The explication or self-explication of being thus articulates immanent structural characteristics of being. But can immanent characteristics of being be identified? This question is to be answered in the positive, as shown by the following account. To begin, it presents the most universal immanent characteristics of being. They are designated appropriately as immanent characteristics *of being as such* (and indeed as *the most universal or most fundamental* immanent characteristics of the dimension of being), in distinction from the characteristics of being as a whole further determined in one way or another (thus, characteristics of regions of being).

The following account aims only to formulate a first analysis and explanation of the most universal immanent structural moments or characteristics. Developing a genuine theory of the dimension of being would require, among other things, presenting the most universal immanent structural characteristics in the form of the *most universal comprehensively systematic sentences* having strictly theoretical status. Such presentation is, however, relatively straightforward; it would result from the accomplishment of a diligent but routine task.

[2] The most universal immanent characteristics of being emerge from or as a (self-)explication of the dimension of being. This dimension itself emerges above as the constellation including the interconnection of thinking/mind/language on the one hand and world/universe/being-in-the-objective-sense on the other, such that the constellation itself is the primordial dimension that encompasses the two.

More detailed explication of this constellation makes explicit the characteristics that are presented in what follows: the (better: some of the) most universal immanent characteristics of the dimension of being. Always to be kept in mind is that "dimension of being" is not here understood as anything like a Platonic entity from which—in some incomprehensible manner—certain "immanent characteristics" could be derived. Instead, the dimension of being appears as the constellation that includes within itself the nexus of thinking/mind/language and world/universe/being-in-the-objective-sense. It is the constellation-character of the dimension of being that makes the explication of the dimension of being both possible and requisite.

584   [i] The first most universal, immanent structural feature or characteristic is the absolutely universal *intelligibility* of the dimension of being, the intelligibility coextensive with the dimension of being. Because the dimension of being *appears as* a complex network or a constellation in the sense explained above, it is simply unthinkable that the dimension of being could be outside the sphere of thinking/mind/language. Only because the dimension of being appears essentially *as* this network or constellation is it accessible to thinking/mind and to language. Such accessibility is precisely what its intelligibility involves: it is conceivable, understandable, articulable, knowable, etc. It should be clear that the universal intelligibility of the dimension of being does not mean that we, as finite knowers, are in a position to articulate it fully; to the contrary, we can grasp, in a determinate fashion, only segments of the vast and total intelligibility of the dimension of being.

    This fundamental, most universal immanent structural characteristic of the dimension of being is envisaged in various ways in the history of philosophy, and often explicitly articulated in a variety of terminologies. Beginning from the famous sentence of Parmenides, "τὸ γὰρ αὐτὸ νοεῖν ἐστίν τε καὶ εἶναι" ["for thinking and being are the same"],[43] through the central tenet of the grand metaphysical tradition, "*Ens et verum convertuntur*" ["being and truth are interchangeable"], through Hegel's equation of the idea (in his sense) and actuality, to Heidegger's equation of being and truth (in his sense): throughout, there is a central intuition that is, however, often expressed by means of exaggerated and/or cryptic formulations. This history is particularly instructive because it makes clear how differently the same terms are used by different philosophers. This holds particularly for "thinking/idea" in Hegel and "truth" in Heidegger. Highly revealing is the way Heidegger conceives of "truth" as an explication of "being." He interprets "truth" on the basis of what he understands to be the ancient or original sense of the Greek word ἀλήθεια as "unconcealment," or the manifestness of being.[44] There is a philosophical tradition that uses the term "ontological truth," but this tradition differs from Heidegger in not simply equating "truth" with "ontological truth"; it relies instead on the thesis that truth in the genuine sense

585 is in the judgment (*veritas est in iudicio*).[45] Heidegger misses the phenomenon of truth precisely because he does not directly consider the relation of language and being.

    The theory of truth sketched in Chapter 3 shows in all clarity that there can be talk of truth only when the dimension of language is introduced explicitly and radically into

---

[43] Gallop (1984: 56–57; translation altered). Heidegger, characteristically, provides an idiosyncratic translation: "For the same perceiving (thinking) as well as being" (1969/1972: 27/18).

[44] As is indicated in Section 2.5.1.1[2][iii] (p. 143), Heidegger himself comes to revise this interpretation.

[45] In exemplary fashion, Thomas Aquinas introduces a threefold distinction concerning the word and concept " truth":

> [T]ruth or the true has been defined in three ways. First of all, it is defined according to that which precedes truth and is the basis of truth. This is why Augustine writes: "The truth is that which is. . . .
>
> Truth is also defined in another way—according to that in which its intelligible determination is formally completed. Thus, Isaac writes: "Truth is the conformity of thing and intellect [*veritas est adaequatio rei et intellectus*]. . . .
>
> The third way of defining truth is according to the effect following upon it. Thus, Hilary says that the true is that which manifests and proclaims existence. (*Truth*: 6 [I Q 1 A 1])

the determination of the truth-concept. According to that theory, a true sentence is one that expresses a true proposition; the true proposition, as fully determined proposition, is *identical* to a fact in the world, and the fact itself is thus an entity identical to the true proposition. This theory clearly accords to truth a radically ontological import. It would, however, be inappropriate, indeed strictly speaking incorrect, to speak of "true facts." "True" is an operator that can take as its arguments only sentences, propositions, and utterances—not facts.

[ii] From *intelligibility* as the fundamental immanent structural characteristic of the dimension of being, two other immanent characteristics can be derived. One can be termed the *universal coherence* of the dimension of being. This designation is here given a specific and comprehensive signification. Coherence in this sense is not simply identical with consistency; in addition to the absence of contradictions, it involves *positive interconnections*. The genitive, "universal coherence *of* the dimension of being" is understood as the subjective genitive: the dimension of being *as* universal coherence, thus *as* universal network or nexus.[46] This characteristic derives from that of intelligibility because to conceive of, understand, explain, etc., anything involves, among other things, grasping or articulating the context within which it exists. Coherence is, in short, systematicity. Determined more precisely and more explicitly, universal coherence is *universal structuration*. Being itself is thus conceived of as *the structure of all structures*, the absolutely primordial and comprehensive structure.

With this point, the systematic location is finally reached from which one of the absolutely fundamental theses from which the conception of systematic philosophy developed here emerges can be brought to complete (i.e., comprehensively systematic) clarity; this is the thesis that everything centers upon or revolves around the fundamental structures. Three major types of such structures are distinguished: the fundamental formal, semantic, and ontological structures. It is now clear that these structures simply exhibit the intelligibility and the coherence (in the sense explained above) of anything and everything, of all domains of being, and indeed of the dimension of being itself. To conceive of anything at all, no matter how trivial, is to relate it to or situate it within the comprehensive structuration of the dimension of being. This makes fully clear what the fundamental structures actually are: they are structures *of being as such and as a whole*.

[iii] An additional fundamental and immanent structural characteristic derivable from the characteristic of intelligibility is the *universal expressibility* of the dimension of being. This book considers this characteristic in detail, particularly in Section 5.1.

[iv] A fourth most universal and immanent structural characteristic differs significantly from the first three. Those three emerge from the explication of the dimension of being *with respect to the intellect*; in the terminology common since Aristotle (although problematic in the context of this book), they would be termed "theoretical characteristics." This designation can be used if it is correctly understood. Given this reservation, one

586

---

[46] The genitive or possessive in "Hegel's book," or "the book of Hegel," is *objective* if the book in question *belongs* to Hegel, and *subjective* if the book was *written by* Hegel, thus, if Hegel (as subject) is responsible for its existence. The universal coherence is, so to speak, accomplished by the dimension of being, rather than somehow happening to be true of it, or something that comes to be true of it because of something somehow outside of it.

could say that the fourth immanent characteristic is "practical," in that it characterizes the dimension of being *with respect to another human capacity equiprimordial with the intellect*: i.e., *the will*. Just as the dimension of being must be understood as being the "object" of the intellect, the will has the dimension of being as its definitively absolute point of reference (i.e., it is that by which willing must take its bearings), but *not in the same respect*. The dimension of being is the absolutely unrestricted and complete point of reference for the intellect with respect to its first three immanent characteristics, termed above the "theoretical" ones. What is its counterpart with respect to the will?

Within the metaphysical tradition, this fourth characteristic is termed *the good*. For this reason, an axiom of this metaphysics is *omne ens est bonum* [every entity is good], or *ens et bonum convertuntur* [being and good are interchangeable]. The "idea of the good" standing at the center of Plato's philosophy has been understood, in the course of the history of philosophy, in various ways, usually without the connection to "ideas" as Plato understands them. In traditional metaphysics, attempts to determine the good are made in two ways. The first begins with the will and treats the good as its "formal object," as that which guides the will in its relating to any "material object" ("*sub ratione boni*," "with respect to the conceptual content of 'good'"). The second traditional metaphysical determination of the good is based on being: the good is the immanent characteristic of being that the will appeals or corresponds to: whatever the will does in any specific case, it always does with a view to: the good, because that is the defining characteristic of the will. The fourth (the "practical") most universal immanent characteristic of being itself is therefore the characteristic of *goodness*.

The perspective of being relevant to a characterization of the good or of goodness has a long history within the framework of the tradition of Christian metaphysics, particularly in the works of Thomas Aquinas, who (following Aristotle) articulates it as the thought of perfection (*perfectio*).[47]

[v] Sometimes, in the metaphysical tradition, *beauty* is also presented as an immanent structural characteristic of being. It is determined by means of the thought of the harmony or consonance (*consonantia*) of the previously introduced immanent structural moments of being. It is indeed a determination that emerges consequently from the question of how the unity or interconnection of these immanent structural moments is to be understood. The essential clarifications are provided in Chapter 4, in conjunction with the presentation of the *aesthetic world* (see Section 4.4.2[2][i], pp. 315–16).

[3] These five most universal immanent characteristics of the dimension of being essentially (but only in part) correspond to the thesis expressed in the traditional metaphysical axiom as *omne ens est unum, verum, bonum* (*et pulchrum*). It is important to note, however, that the axiom speaks only of individual entities, of every individual entity, whereas the concern here is with the dimension of being. The characteristic of the *unum* is, in the explication sketched here, contained in the characteristic of universal *coherence*: the *unum* articulates the character of the concrete coherences or configurations that form so-called individual beings.

---

[47] Consider formulations like the following: "the goodness of a thing is its perfection" (*ScG*: 86 [I 40]); "the goodness of a thing consists in its being well disposed according to the mode of its nature" (*STh* FS Q 71 A 1 Obj 3).

5.3 Starting Points for a Theory of Absolute Being

5.3.1 *Preliminary Clarifications*

The articulation of the immanent characteristics of being as such is the first stage of the explication of the primordial dimension of being and thus of the development of a structural-systematic theory of being. It fundamentally considers being as such. The second stage focuses on being as a whole. With the explication of its immanent structurality, the dimension of being is only determined preliminarily. Is further determination possible or indispensable? One cannot say that one can *prove* that further determination is possible or indispensable, etc. Instead, one must see that the situation is as follows: the human intellect has an enormous and irrepressible potential for intelligibility; that is, it is both capable of and pressed to raise and to address questions with limitless scopes, to investigate and articulate interconnections of all sorts—in short, it is so situated within being as a whole that anything and everything, and therewith being as a whole, appears as fully intelligible. For the human intellect, there are no prohibitions of any sort whatsoever. One cannot simply maintain, *a priori*, that something is not possible for it, that it is striving to surpass some unsurpassable boundary, that it is doing something senseless in investigating something or other, etc. Whether or not it is can be determined only on the basis of the relevant experiences and the results that either are or are not attained.

Given this, it is both sensible and rational to ask additional questions concerning the possibility of further determination of being as a whole. But how should one proceed? There are various possible paths. Here, the following path is taken: the intellectual instruments available to human beings are examined with the aim of determining whether there are any that can accomplish the task at hand. Concretely: one seeks concepts (in accordance with the ordinary and non-problematic understanding of "concept") that can and/or should be used to further determine being as a whole. If one seeks such concepts, one quickly finds that among the concepts indispensable for the adequate understanding of interconnections of all sorts are those designated as the modalities: *necessity, possibility, contingency.* No major philosophical theory is thinkable without these concepts. For this reason, it is easy to understand that and why they have played a simply central role throughout the history of philosophy. At present, we have a modal logic that makes these concepts significantly more precise.

In accordance with the basic theses of this book, this means that in this section, a more determinate and richer theoretical framework is concretized: it contains, in addition to the concepts of the primordial dimension of being, of being as such, and of being as a whole, also the modalities. Thereby, the horizon for questioning and the possibilities of articulation are broadened significantly.

To be sure, modal logic is not as free of problems as is elementary logic (propositional logic and predicate logic). For decades, there has been a significant problem with respect to the *interpretation* of modal logic,[48] and it cannot be said that a consensus

---

[48] The problem is raised by Quine in his 1947 essay "The Problem of Interpreting Modal Logic."

concerning the solution to this problem has been attained.[49] One interpretation is the metaphysical (or ontological) one. In this book and particularly in this section, the modalities are understood as ontological/metaphysical structures. In a specific (and widespread) terminology, they are taken as *de re* modalities, in an explicitly ontological sense. This means, for the case of *de re* necessity, that to a non-linguistic item (entity) is ascribed a necessary factor (depending upon semantics and ontology: a necessary property, a necessary structure, etc.). A sentence of the form "$\Box Fa$" says of item *a* that it necessarily has the property *F*.

In modal logic, the modalities are understood as operators that have as their arguments sentences or the propositions expressed by sentences. If one wants to treat a topic modally, and thereby to proceed strictly in accordance with the demands of modal logic, one must apply all these determinations quite precisely. The following presentation of the starting points for a theory of absolutely necessary being does not proceed in so strict a fashion. Instead, a train of thought is presented programmatically and only informally. This means, among other things, that here the modal terms are not used strictly or even principally as operators, but instead, for the most part—as is common both in ordinary language and throughout most of the history of philosophy—in predicate and/or attributive positions; they are used as ontological factors characterizing specific entities (e.g., "there is both a necessary and a contingent dimension of being").

The following presentation has an extremely simplified and abbreviated character also in another respect. It does not develop in strict accordance with the fundamental determinations of the contextual semantics whose basic features are presented in Chapter 3. At the same time, it does not contradict that semantics. To recall the central point, which concerns the subject-predicate form of theoretical-philosophical sentences not recognized by such a semantics: Chapter 3 shows and emphasizes that the elimination of singular terms and predicates in favor of primary sentences does not have the consequence that syntactically subject-predicate sentences cannot or should not be used; the elimination concerns only the semantic interpretation. Subject-predicate sentences (and thus singular terms and predicates) are understood as abbreviations of complexes of primary sentences. The subject-predicate sentences used in what follows are to be understood in this manner.

The modalities "it is necessary that," "it is possible that," and "it is contingent that" are to be understood in accordance with the definitions that are usual in modal logic

---

[49] Roberta Ballarin (2004) accurately investigates the discussions generated by Quine (1947). She shows that much in the literature, particularly in textbooks, is inaccurate and/or unfounded. She concludes that Quine and Kripke, despite their divergences with respect to details, share a remarkable conviction that she designates the "Quine-Kripke conjecture" (QK) and formulates as follows:

(QK)   An interpretation of 'necessarily' (or '$\Box$') has to be grounded in actual essential predications: whereby some but not all truths are metaphysically necessary. (637)

On this topic, see also Shalkowski (2004).

(here, the following symbols are used: "□" for necessity, "◊" for possibility, and the (unusual) symbol "∇" for contingency):

Necessity: $\Box\, P = \neg\, \Diamond\, \neg\, P$

Possibility: $\Diamond\, P = \neg\, \Box\, \neg\, P$

Contingency[50]: $\nabla\, P_{df} = \Diamond\, P \wedge \Diamond\, \neg\, P$ or $\neg\, \Box\, P \wedge \neg\, \Box\, \neg\, P$

The determinations of the meanings of necessity and possibility are not here understood as definitions—hence the absence of the symbol "df." The reason is that those two meanings cannot be simultaneously understood as definitions, because in that case they would define each other, which is not allowable. One must accept one of   591
the two modalities as a primitive (undefined) concept and then use it in the *definiens* of the other operators. It does not matter which of the two operators one accepts as primitive/undefined. For the understanding of the primitive/undefined operator, one usually relies on the meaning attached to the term in ordinary language (see Hughes and Cresswell 1996: 14). As they are written, the formulas provided above give only a general understanding of the modalities, so they are left as they are.

Yet to be clarified is the concept *absolutely necessary*. This concept is not simply synonymous with the concept *necessary*. *Absolutely necessary* involves *more* than *necessary*; it also involves independence from anything else and not being conditioned by anything at all. This independence and complete non-conditionedness cannot be explained by or derived from the concept of necessity. In this sense and for this reason, the following account generally relies on the concept *the absolutely necessary*, and understands *the absolute* to be interchangeable with it.[51]

5.3.2 *The Decisive Step: The Primordial Difference with Respect to Being as the Difference Between the Absolutely Necessary and the Contingent Dimensions of Being*

[1] The following account attempts further to explain the primordial dimension of being, in the sense developed above. This is accomplished by the application of the

---

[50] This definition corresponds precisely to the understanding of contingency in the Christian-metaphysical tradition. See, e.g., Thomas Aquinas: "contingency is a potentiality to be or not to be [*contingens est quod potest esse et non esse*]" (*STh* FP Q 83 A 3). It also articulates the understanding of contingency currently common in philosophical-logical literature. If for example "contingent (submaximal) worlds" are assumed, the usual clarification is "in the sense that it is logically possible that they might not have existed (in the widest-open sense of 'existed')" (Bigelow 1996: 152; see Section 5.2.2[6], pp.429–30).

[51] In the Christian-oriented metaphysics of the Middle Ages (Aquinas), one does not find the term or concept "absolute." Kant uses, in addition to the concept *necessary*, above all the concept *unconditioned*. It is the German Idealists who first extensively use the concept *absolute* (in the form, "the absolute") in a comprehensive manner to characterize the necessary and unconditioned being (God). Thereafter, it is common to use this concept in connection with talk about God.

modalities to the dimension of being. One can ask in advance: what grounds the attempt (further) to understand the dimension of being with the aid of the modalities? Differently stated: why should we use the modalities to comprehend the dimension of being? To this question, which is characteristic of modern philosophy and one that deters many philosophers from the attempt to develop metaphysical theories, a succinct answer is this: what grounds the attempt is the fact that these concepts have enormous potential for explanation or, more generally, for intelligibility. They are aspects of the human capacity to ask about anything and everything and thus also about being as a whole.

592   The modalities can be utilized in various ways. Here, the following way is chosen: the starting point is the question whether anything and everything, thus being as a whole, has a *contingent* character or status. Differently stated: the "all-is-contingent" thesis is to be refuted; this is the thesis that maintains that everything that is, is contingent. From the falsity of this thesis it follows immediately that being as a whole has an absolutely necessary dimension. Because however it is a fact that there are contingent beings—beings that are but might not have been, because they come to be and pass away—it is also the case that there is a contingent dimension of being. The primordial dimension of being, being as a whole, is therefore understood as two-dimensional. The proof is thus an indirect proof, a *reductio ad absurdum*.

The proof of this thesis is, as is indicated above, presented informally, but it would be easy to formalize it, presupposing that one had solved a number of problems concerning modal logic. That cannot be undertaken in this book; the concern here is only to develop the starting points for a theory of absolutely necessary being.

[2] The individual steps of the proof are as follows:

i. The starting point is the acceptance of the primordial dimension of being and of the task of further determining it. Although that dimension is explained preliminarily in this chapter, it appears appropriate at this point to point out a possible unclarity or a possible fundamental misunderstanding. This unclarity or misunderstanding arises when questions like the following are raised: does the primordial dimension of being exist? The answer to this question cannot be direct, because the question is nonsensical. One can say neither that the dimension of being exists nor that it does not exist. The concept *existence* presupposed by the question makes sense only on the presupposition that there is a radical dichotomy between thinking (mind, language, theories . . .) and reality (or being or the world) in the sense of a domain utterly distinct from thinking. Given this presupposition, existence would involve being situated in this domain. But the primordial dimension of being is precisely the dimension that encompasses both thinking and the "objective" domain that is understood as the other to thinking. The primordial dimension of being is the dimension that is the basis of and that articulates the avoidance of the putatively unbridgeable gap or cut, criticized in Section 5.1, between thinking (theories, etc.) and "system" (in the sense of "objective world" or "purely objective being"). The sensible question is therefore whether the being of the primordial dimension of

593   being is to be or could be understood as being-*possible*. This too is excluded, because being-possible is already a derivative form of primordial being. If one seeks and wants to have a qualifying term, one could say that "being" here means "actual being," if one understands this designation such that being-possible is to be understood only on the basis of actual being.

ii. If absolutely everything and thus the primordial dimension of being itself were contingent, then it *could have been the case* that neither the dimension of being nor any kind of item belonging to it (no entity) had ever come to be.[e] This, however, would mean that one would have to assume the *possibility of absolute nothingness* (of the *nihilum absolutum*). To be emphasized is that the talk here is of the necessity of the assumption of the *possibility* of absolute nothingness as a consequence of the thesis that anything and everything, and thus the dimension of being itself, is contingent. This consequence results directly from the concepts that are used: the all-is-contingent thesis ascribes contingency to absolutely everything and thus to the dimension of being itself: absolutely everything and thus the dimension of being itself could, possibly, not have been. This is the assumption of *the possibility of absolute nothingness.*

iii. This implication or assumption is, however, utterly unacceptable for two reasons. (a) The "concept" of absolute nothingness is a non-concept, an unthinkable, totally self-contradictory pseudo-concept. If, that is, one wants in any way to think it, one must ascribe to it precisely what it is supposed to exclude: one "determines" it if one names it any*thing* at all, but any "thing," of no matter what sort, is a determinate mode of being. One can speak of the "concept" of absolute nothingness only in a paradoxical manner, and that only to articulate its absurdity. If one says, for example, that absolute nothingness is the total negation of being, total non-being, these designations are meaningful or sensible only if they designate something or other—and thus some mode of being. But the putative concept *the-possibility-of-absolute-nothingness* is absolutely paradoxical—better, self-contradictory—because possibility is always possible *being*; it would therefore be contradictory to accept a *possibility-of-being* for absolute nothingness.

(b) The thesis that everything is contingent entails not only the assumption of the possibility of absolute nothingness but also an additional assumption: that beings could somehow emerge from absolute nothingness into the dimension of being (or, because there are things that are, that they in fact emerged from absolute nothingness). But the thought of even a possible emergence of beings from the unthinkable "dimension" of absolute nothingness, or of a transition from the "dimension" of absolute nothingness to the dimension of being, is a simply senseless, impossible pseudo-thought: no *being* of any sort whatsoever could come from absolute nothingness. To formulate the point

---

[e] Here and below, formulations using "exist" would be, at least at first glance, less awkard—"it could have been the case that absolutely everything did not exist." But the use of "exist" in this context would be highly problematic, particularly given the sense in which that term is used in the structural-systematic philosophy. If "exist" were used here, that would make of *existence* a superordinate, more comprehensive concept that would encompass even *being* and *nothingness*. There can, however, be no such "concept," because it could not explain itself and could not be explained. The attempt to explain it would lead back to what this book terms "the primordial dimension of being." For just this reason, reliance on *the dimension of* being and, correspondingly, on forms of "to be" are requisite here.

Worth noting in addition is that the structural-systematic philosophy understands *existence* to mean *belonging to the "objective" pole of being* (i.e., to the world in the sense of "world" used in Chapter 4—to what Aquinas terms "the nature of things [*de rerum natura*]," in distinction from "in the mind alone" [*solum in intellectu*]).

594   by means of an image: there could not *be* any emerging from the dimension of absolute nothingness into the dimension of being. The dimension of being and absolute nothingness are absolutely incompatible with each other (i.e., incompatible in every respect). One could only try to say that the dimension of being *is* the absolute negation, the absolute exclusion of anything like absolute nothingness, and in the case of an absolute negation, there is no possibility of any kind of transition from one "pole" of the negation to the other, because when contradictories rather than contraries are opposed, there are no continua between the poles.

iv. Because the thesis that everything is contingent entails an absurd consequence, it follows that not everything is contingent, i.e., that there is absolutely necessary being; absolutely necessary being is here designated quite neutrally and generally as an *absolutely necessary dimension of being.*[f]

v. Because it is a fact that there are contingent entities, it follows from the preceding that the dimension of being is to be understood as *two-dimensional*—as consisting of an absolutely necessary dimension and a contingent dimension. Q.E.D.

The core of the proof can also be summarily presented as follows:

— If absolutely everything and therewith the dimension of being itself were contingent, one would have to accept the *possibility* of *absolute nothingness.*

— Absolute nothingness is not possible.

— Therefore, not everything (i.e., not the entire dimension of being itself) is contingent.

5.3.3 *Additional Remarks and Clarifications*

[1] The proof articulates one of the most fundamental systematic-metaphysical theses. Presumably for this reason, this thesis appears to many philosophers as extremely abstract, even as an empty play with concepts or even only with words. In opposition to this view, it is to be emphasized that the proof ultimately only explicitly thematizes the immense superficiality and at the same time the immensely negative consequences of certain putatively comprehensive ideas and contentions that are either formulated explicitly or—as is more usual—assumed tacitly. These are ideas that are articulated in formulations like the following: "Everything is contingent—all things are contingent," "Everything comes to be and passes away," etc. Such contentions are, whether so intended or not, absolutely comprehensive; indeed, they are comprehensive *metaphysical* contentions. In philosophy, one then encounters the peculiar phenomenon that entire philosophies allow themselves to be based decisively on such generally unanalyzed, much less justified or grounded, grand contentions. But even aside from this aspect, characteristic of the contemporary philosophical situation, it is to be noted that the proof presented above provides the basis for

595   the possibility of developing a coherent, comprehensive metaphysical view. That it does is shown in the remainder of this chapter.

[2] Correct evaluation of the result of the proof presented above is aided by its comparison with the traditional so-called proofs of God's existence. The proof is not a proof,

---

[f] The consequence formulated in iv is articulable in a primary sentence as "It follows that it is not only the case that it is contingent that it's being; it is also the case that it is absolutely necessary that it's being."

in the usual sense, that God exists. A first reason that it is not is that this book strictly distinguishes *being* from *existence*. One of the consequences of its distinction is that formulations like "the existence of the absolutely necessary dimension of being" here make no sense and are therefore unacceptable. *Existence*, in this book, pertains *not* to the absolutely necessary dimension of being, but only to a secondary dimension (see note e, p. 445 above). The second reason that the proof is not of God's existence is that it leads not to God, but to the absolutely necessary dimension of being; it can therefore be said only that it provides the *metaphysical basis* for a *theory of God*. This must be clarified in two ways.

First, "God" is originally not a philosophical concept, but a term arising in religions, and one with which many in part quite heterogeneous ideas have been and continue to be connected. Like many others, the topic of "God," arising in religion, finds its way into philosophy at the very beginning. In the philosophical schools of thought that in one way or another assume something necessary or absolute (thus something like an absolutely necessary dimension of being, under whatever designation whatsoever), one can find chiefly three forms or types of theories about God.

The first form of theory identifies God—without differentiation—with the totality of being, of the universe, and quite particularly with nature; this theory is named "pantheism." The second form is the traditional metaphysical form: God is understood as the highest entity, the one that, as one contemporary author puts it, is "a person without a body (i.e. a spirit) who necessarily is eternal, perfectly free, omnipotent, omniscient, perfectly good, and the creator of all things" (Swinburne 1979/2004: 7).[52] Third, there is the explicitly religiously oriented theory about God, which treats God as the highest, free, personal entity, who acts in the history of humanity, in world history on the whole (in Christianity particularly, in that God *reveals* or *manifests* Himself), and is therefore an object of worship or prayer, cultic activities, etc. To be emphasized with respect to this third form is that different religions do not all lie at the same level or even on similar levels; only Christianity recognizes, develops, and defends a theory about God, a theology in the genuine (i.e., theoretical) sense. If "theologies" are also ascribed to other religions, then they are not theologies with theoretical status, but instead theologies in the sense of authoritative clarifications of writings viewed as holy, of religious traditions, etc. For this reason, when in the following account there is mention of this third form of theory about God, the reference is exclusively to *Christian* theology.   596

The proof given above provides the basis for all three forms of theories about God; more precisely, it *must* provide the basis for the corresponding theory about God to be minimally coherent, if "God," as word and concept, is to have a genuinely theoretical meaning.

Second, the proof sketched above is *fundamentally different* from every previously known proof[g] that God is, particularly in two respects. The first is that the previous

---

[52] One must in any case attend to the fact that a philosopher and theologian such as Thomas Aquinas identifies "God" not only with "highest being" (*ens summum*) but also—and above all—with "*esse per se subsistens*," being subsisting of itself. Also to be noted is that Aquinas does not develop this thought in a sufficiently systematic manner.

[g] Although the main text argues that available procedures that purport to prove that God exists fail to do so, these other procedures are, in accordance with common practice and for sake of convenience, referred to simply as "proofs," not as "putative proofs," "attempted proofs," etc.

proofs—with the exception, in one specific respect, of the ontological proof—consider only single aspects of the universe (no matter how it is understood), emphasizing that aspect in order to move from it to a different point, a first or highest point. Thus, from the phenomenon of motion the move is made to a first mover, from that of innerworldly causality to a first cause, from the fact that there are more and less complete things to a first or perfect completeness, from innerworldly order to a first intelligence, etc. To be sure, one would have to distinguish scrupulously among the traditional proofs that God exists, but the fundamental point remains: one moves from some specific point (i.e., phenomenon) within the universe to different highest points.

This makes it clear that in these previous proofs, the universe as *a whole*, thus what is above termed the primordial dimension of being or being as a whole, is lost from view or, more precisely, that it is never taken into view. This distinction is radical, because it conditions the entire contentual conception and the entire methodological procedure. If one begins with the primordial dimension of being, as does the proof sketched above, then the task to be accomplished is to explain this dimension in the sense of further determining it, adding specifications such that, at the end, being as a whole comes into view as wholly structured, at least in its basic features. In this respect, it no longer makes sense to speak of *proofs*, in the plural, that God exists. One way or another, one can, as traditional theology does, add these specific highest points after the fact, in that one shows that the first mover is *also* the first cause and that what is both is *also* the highest completeness or perfection, etc. Philosophers and theologians like Thomas Aquinas do this in one way or another. But such a procedure is external to the matter at hand and therefore inappropriate.

The second respect in which the proof sketched above differs fundamentally from the traditional or classical and from the modern proofs that God exists concerns the conclusion of the proof. Strictly speaking, the previous proofs always get to a highest point in the sense explained above (first mover, first cause, etc.); if successful, they would indeed lead to these various highest points. But then, by means of some kind of commentary or interpretation, the highest point is immediately identified with God or simply named "God." Aquinas's formulations of this point have become classic; his proofs that God exists (his "five ways," *quinquae viae*) conclude, respectively, as follows:

597

> And this [first mover] everyone understands to be God. . . . [I]t is necessary to admit a first efficient cause, to which everyone gives the name of God. . . . [W]e cannot but accept that there is some being having of itself its own necessity. . . . This all men speak of as God. . . . [T]here must also be something which is to all beings the cause of their being, goodness, and every other perfection; and this we call God. . . . There *is* some intelligent being by whom all natural things are directed to their end; and this being we call God (*STh* FP Q 2 A 3; translation altered).

Here, the usually extraordinarily careful thinker overlooks that fact that he—and with him the entire tradition of the classical proofs that God exists—commits a significant methodological error: between such highest points and what "we all call God" there is an enormous distinction. In fact, Aquinas takes this distinction fully into consideration in his other works, but the fact remains that the entire undertaking of proving that

God exists does not go far enough, in that it fails adequately to develop the appropriate metaphysical basis either methodically or contentually.

[3] To clarify these two respects, Aquinas's third proof (*tertia via*) is considered briefly here. This proof—if one ignores its interpreted conclusion—has a certain similarity with the structural-systematic proof sketched above, but it differs from it in two fundamental points. To begin, here is the text:

> The third way is taken from possibility and necessity, and runs thus: we find in nature things that have the possibility of being and of not being, because they are found to be generated and to pass away, and consequently have the possibility of being and of not being. It is however impossible for these always to be [textual variant: . . . that all that is is of this sort], for there is a time when that which has the possibility of not being is not. Therefore, if everything has the possibility of not being, then at one time there was nothing. Now if this is true, even now there would be nothing, because that which is not begins to be only by something that already is. Therefore, if (at one time) there was nothing, then it was impossible for anything to have begun to be; and thus even now nothing would be—which is manifestly false. Therefore, not all beings are merely possible, but there must be something the being of which is necessary.
>
> [But everything necessary either has its necessity caused from elsewhere, or not. It is however impossible to go on to infinity in necessary (things) that have causes of their necessity, just as this is not possible in the case of efficient causes (as is shown above, in the second proof). It is therefore necessary to assume something that is necessary by itself and that does not have the cause of its necessity elsewhere, but that instead is the cause of the necessity of the other (things). This all call God.] (*STh* FP Q 2 A 3; translation altered to accord more closely with *Struktur und Sein*)[53]

598

As Franz von Kutschera notes correctly, the argument in the text enclosed in square brackets is, strictly speaking, superfluous (see Kutschera 1991: 26n17). As far as the structural-systematic proof is concerned, above all two fundamental distinctions must be drawn between it and the proof provided by Aquinas.

[i] The first is that Aquinas begins with a phenomenon—"we find in nature things that have the possibility of being and of not being"—thus from the phenomenon that some things are contingent. Step ii of the structural-systematic train of thought

---

[53] The Latin original reads:

> *Tertia via est sumpta ex possibili et necessario: quae talis est. Invenimus enim in rebus quaedam quae sunt possibilia esse et non esse; cum quaedam inveniantur generari et corrumpi, et per consequens possibilia esse et non esse. Impossibile est autem omnia quae sunt talia, semper esse [variant text: . . . omnia quae sunt talia esse]: quia quod possibile est non esse, quandoque non est. Si igitur omnia sunt possibilia non esse, aliquando nihil fuit in rebus. Sed si hoc est verum, etiam nunc nihil esset: quia quod non est, non incipit esse nisi per aliquid quod est; si igitur nihil fuit ens, impossibile fuit quod aliquid inciperet esse, et sic modo nihil esset: quod patet esse falsum. Non ergo omnia entia sunt possibilia: sed oportet aliquid esse necessarium in rebus.*
>
> *[Omne autem necessarium vel habet causam suae necessitatis aliunde, vel non habet. Non est autem possibile quod procedatur in infinitum in ncessariis, quae habent causam suae necessitatis sicut nec in causis efficientibus . . . Ergo necesse est ponere aliquid quod sit per se necessarium, non habens causam suae necessitatis aliunde, sed quod est causa necessitatis aliis: quod omnes dicunt Deum].*
> (*STh* I q. 2 a. 3)

presented above develops instead as an explication of the dimension of being, of being as a whole; it thus has a holistic status, in opposition to the particularistic starting point of Aquinas's proof. Aquinas's conclusion is that the necessary being that all name God exists. As is shown in Section 5.3.3[2] (pp. 448–49), this conclusion is the result of a deep methodological error.

[ii] The second distinction is immediately clear if one pays attention to the fact that Aquinas accords a simply central significance to *time*. This introduces a number of fundamental problems that clearly show that his argument is inconclusive. This is shown below within the framework of a commentary on Kutschera's interpretation and critique. Aside from these considerations, however, the basic intuition that grounds Aquinas's line of argumentation corresponds fundamentally to the holistic-systematic thesis that grounds the proof introduced above.

Franz von Kutschera reconstructs Aquinas's argument as follows:

a) Something exists.

b) Everything that exists exists either contingently or necessarily.

c) Everything that exists contingently came to be at some time.

d) Everything that comes to be does so by means of or because of something that already exists.

e) If the existence of everything that exists were contingent, then there would have to have been a time at which there would not yet have been anything in existence. (This does not correspond exactly to the formulation of Aquinas, who puts his second sentence not in the subjunctive but in the indicative: "at one time there was not yet anything that existed. . . .")

f) But then, according to d), nothing could have come to be.

g) There is therefore something that exists necessarily, and we call what does "God." (Kutschera 1991: 26)

This reconstruction does not correspond precisely to Aquinas's text. Premises b) and c) are not—at least not explicitly—contained in the text. But that is not of central importance here. Important is Kutschera's contention that e) does not follow from d), and that this is the decisive error in the argument. He explains it as follows: from the sentence

(S1)   For all objects that have come to be, there is a point in time at which they did not yet exist.

the sentence (S2) does not follow:

(S2)   There is a point in time at which all objects that have come to be did not yet exist.

Kutschera adds that e) would hold only if the set of all objects that up to now have come to be were finite. He then notes correctly that Aquinas commits an unallowable quantifier shift fallacy. This is easily shown if one formalizes S1 and S2. (Symbols/abbreviations: "CB" = have come to be; "TP" = point in time; "E!" = existence predicate.)

S1: $\forall x \, (\mathrm{CB}x \rightarrow \exists t \, (\mathrm{TP}t \wedge \neg \mathrm{E}!(x, t)))$

S2: $\exists t \, (\mathrm{TP}t \wedge \forall x \, (\mathrm{CB}x \wedge \neg \mathrm{E}!(x, t)))$

It is easy to see that S2 does not follow from S1, because the two have shifted quantifiers. To be emphasized again is that the proof sketched in Section 5.3.2 contains nothing of the sort: it neither involves the factor of time nor in any way exchanges or confuses quantifiers.

5.3.4 Additional Steps in the Explication of the Absolutely Necessary Dimension of Being

600

The proof introduced above establishes that an absolutely necessary dimension of being must be recognized along with a contingent dimension of being. The primordial dimension of being is thus understood, more determinately, as a two-dimensionality of being. How the contingent dimension of being is to be understood more precisely does not pose a difficult task because all the things familiar to us in the world are contingent things. How the absolutely necessary dimension is to be understood remains, however, wholly open. The task this poses is undertaken in the following account only incipiently. The thought that one could somehow derive the absolutely necessary dimension's more precise determination from the pure or abstract concept of it is a fantastic, indeed nearly absurd thought. The only possible way one can proceed is to attempt to clarify the question of how the relation between the two dimensions of being is to be determined. The concept of the absolutely necessary dimension of being is thus contentually determined at least to the extent that this relation can be investigated. Clarification of the relation provides additional determination of the absolutely necessary dimension of being.

[1] What belongs to the non-contingent dimension of being? The logical/mathematical structures are often reckoned to it. For example, authors who assume only the existence of a "submaximal world" (see Section 5.2.2[5], pp. 428–29) designate that world as "contingent." According to these authors, "contingency" is a consequence of the fact that these worlds are "submathematical realms":

> There might be several ways of making more precise the notion of the submathematical realm. Take something to be submathematical just in case its existence is contingent. Then consider this proposal: that "the world" could be taken to be the aggregate of all contingent beings. Suppose, that is, that there is such a thing as the aggregate of all contingent beings, and define "the world" to be that thing. All the mathematical entities like sets—the things which are too numerous to aggregate into any single entity—will then lie outside "the world," in this more restricted sense of the term. (Bigelow 1996: 148)

What does it mean to say that the mathematical entities "lie outside of" the submaximal, the contingent, world? Does it mean that they are non-contingent, thus necessary? But then there *is*—in whatever more precise sense—a necessary dimension of being. These authors do not pursue this question. The question to be clarified would be that of how the "necessity" in question is to be understood *ontologically*. If one said for example that (ontologically understood) necessity meant that something *existed in all possible worlds*, then a problem would arise: how would these worlds themselves be understood? Would they be contingent, in the usual sense? If so, then one would have a peculiar concept of (*ontological*) *necessity*: the necessary would be whatever existed in all possible *contingent* worlds.

For the question treated in this section, that of how the absolutely necessary dimension of being is to be explicated more precisely, it suffices to say that the logical/mathematical entities have only a *derivative* non-contingent status *within* the absolutely necessary dimension of being.[54]

[2] The following account shows that the absolutely necessary dimension of being is to be understood as *absolutely necessary minded being*, whereby "minded being" means being possessing intelligence, will, and freedom. (In conjunction with a terminology that is currently widespread, one can also say *absolutely necessary personal being*, although the qualifier "personal" can easily give rise to misunderstandings.) In favor of this thesis, two arguments are introduced, each only briefly. A thorough treatment of this immense topic is beyond the scope of this book.

[i] The first argument focuses on the relation between the contingent and the absolutely necessary dimensions of being. It should be clear that, within a *theory*, it would be not only fully unsatisfactory but also not acceptable to determine the relation purely negatively (i.e., as the distinction of the two dimensions of being), because that would say only that neither of the dimensions is the other. If one wanted to deny that there is a positive relation between the two, that would not do justice to the human intellect, which has the unconditional need *and the capacity* to ask and to understand how a *positive* relation is to be conceived of. If it were ignored, the genuinely philosophical task would be left untackled. Because however the human mind is often thematized above in this book, in various contexts, some theoretical results concerning it are available and can serve as resources for the current task. The question can therefore be formulated as follows: how is the absolutely necessary dimension of being more closely to be conceived of so that its relation to the human mind, which is at the center of the contingent dimension of being, can be determined appropriately and coherently?

Chapter 4 articulates the specific character of the human mind particularly with this brief formula: *the human mind is intentionally coextensive with being as a whole*. It is thus intentionally coextensive with the absolutely necessary dimension of being and the

---

[54] How one can or must understand this derivative status depends upon how one understands the absolutely necessary dimension of being itself. The conception sketched here likewise yields the fundamental consequence that logical/mathematical entities are to be understood as pure or abstract (precisely, as formal) *structural constellations* that co-structure the dimension of being itself. *In that* they *co-structure* the dimension of being, they are not abstract or pure, but are always, to put it one way, concretized in the sense that they always already have *their primordial and full ontological status*. They are, however, to be understood *derivatively as pure or abstract* (or again, as *formal*) structural constellations, in that they abstract (or are abstracted) from this *primordial and full ontological status*.

If the absolutely necessary dimension of being is determined more precisely as intelligent absolute being, the possibility arises of further determining these formal entities (e.g., in that one understands them as ideas in the mind of absolute intelligent being, as has often happened in the metaphysical tradition—for example, with Leibniz). In this case, the meaning of "idea" would have to be analyzed precisely. That task cannot be undertaken in this book.

601

602

contingent dimension of being. Assuming that the absolutely necessary dimension were not mentally constituted, but instead something other, perhaps a purely abstract principle (however one might conceive of that) or something like primordial non-minded nature or something similar, one would have a peculiar juxtaposition: on the one hand a mentally constituted contingent entity that is intentionally coextensive with being as a whole and thus also with the absolutely necessary dimension of being, and on the other, a purely abstract or purely naturally constituted absolutely necessary dimension of being. By definition, the so-understood absolutely necessary dimension of being could not be intentionally coextensive either with itself or—much less—with being as a whole. Is that conceivable or coherent?

The answer to the question just raised depends upon clarification of the question of what criteria or measures can be applied to the evaluation of a so-conceived relationship. It would be possible to present a principle or axiom concerning the relation between the two dimensions of being and from it then to derive a specific conclusion. Then the entire discussion would center on the groundedness or acceptability of the principle or axiom. As a rule, the discussion would be inconclusive, and the question would then not have been answered by the introduction of the principle or axiom. Be that as it may, the following account proceeds differently. To be sure, it introduces principles or axioms, but not in such a way that they can be simply presented explicitly; instead, the argument is that principles and/or axioms must be presented *along with* their groundings. The argumentative considerations rely upon the criteria of (greater) intelligibility and of the (greater) coherence that results from it.

Intelligibility is an *ontological-epistemic* concept: on the one hand, it articulates the ontological structuration of the subject matter in question, and on the other hand, it thematizes the stance or attitude of the intellect with respect to the so-structured subject matter. If a specific ontological structuration enlightens the intellect, then *intelligibility* (with respect to this structuration) in the genuine sense is attained, although the 603 intelligibility can be of varying degrees (see Section 6.1.2[3], pp. 464–67). A conception or theory based on a higher intelligibility in this sense would then be the result of a specific form of *inference to the best explanation.*[55]

The thesis that the absolutely necessary dimension of being must be conceived of as mentally constituted exhibits *incomparably higher intelligibility* than does the opposed thesis (i.e., that this dimension is not mentally constituted). Two lines of argument support this thesis. First, the contingent human mind is *totally* dependent upon the absolutely necessary dimension of being. This total dependence must be thought of strictly ontologically, which means that the human mind, *with respect to its entire being*, is wholly and in every respect reliant upon the absolutely necessary dimension of being. This reliance can be explained in various ways, but the inalterable core remains always the same: *without* the absolutely necessary dimension of being, the human mind could not *actually be.* The reason for this lies in the central thought of the argument presented above in the refutation of the all-is-contingent thesis and as a ground for the thesis of the two-dimensionality of being. The question in the current context is the following: how can it be made intelligible that a contingent minded entity can be totally dependent, *in every respect*, upon a non-mentally-constituted absolutely necessary dimension of being? It cannot. Non-mentally-constituted absolutely necessary being would have

---

[55] See the illuminating considerations in Forrest (1996, esp. 26–35, 41–42, 117–21).

to be so constituted that the total dependence of the contingent mentally-constituted entity would be intelligible or explicable. As a non-mentally-constituted dimension of being, however, it would contain nothing that could be the basis for making this intelligible or explicable. The contingent human mind contains within itself capacities that the non-mentally-constituted absolutely necessary dimension of being would not have and, by definition, could not have. Because, as shown above, the relation between the two dimensions of being, the absolutely necessary and the contingent, must be determined *positively*, the denial that the absolutely necessary dimension of being is mentally constituted would be tantamount to the assumption of a simply inexplicable and hence unintelligible metaphysical gap. Every attempt to explain the relationship positively on this basis would utterly fail to do justice to the mentally constituted human being.

The second line of argument depends on the thesis, treated in detail in this chapter (see esp. Section 5.1.3.3, p. 371), of the *universal expressibility* of being as a whole. Even an absolutely necessary dimension of being conceived of as non-mentally constituted would be expressible—it would be structured in one way or another. But it would be so structured, so to speak, only in itself; it would be unable to grasp and articulate itself *as expressible*. In distinction thereto, the contingent human mind has the incomparably higher status of having a mode of being that is expressible *and that expresses (articulates)*. This opens to the contingent human mind capacities with respect to action and to configurations of every form that would be absolutely beyond the non-mentally constituted absolutely necessary dimension of being. It then could in no way be made conceivable that the contingent human mind is, in its being, totally dependent upon a non-mentally-constituted absolutely necessary dimension of being.

[ii] The second argument differs from the first only in form; it articulates *in a generalized form* the essential aspect of the human mind that is introduced in the first argument. The generalized form is that of a principle or axiom.

The starting point is the contention that the following ontological principle, the principle of ontological rank[56] (POR), must count as a fundamental principle for any ontology that can do justice to the specific characters of different entities and of different domains of being:

(POR)   Something of a higher ontological rank cannot arise exclusively from or be explained exclusively by anything of a lower ontological rank.

The application of the POR to the problem of the determination of the relation between the absolutely necessary and the contingent dimensions of being leads to an unambiguous conclusion. The ontological rank of a non-mentally-constituted dimension of being would be lower than that of the contingent human mind; the latter can therefore not be explained by the so-conceived absolutely necessary dimension of being. Because however the absolutely necessary dimension of being is absolutely necessary, it must be understood as constituted in such a manner that the contingent dimension of being is

---

[56] For an explanation and grounding of this principle, the most important case of such an ontological rank—the unique status of the human being as mentally constituted being within the universe—is of decisive importance. What is most important about this status is presented in Chapter 4, particularly Sections 4.3.1.2.2.3 (pp. 275–78) and 4.3.1.2.3.2 (pp. 287–89).

explicable on its basis. This requires that it have an ontological rank that is, at the least, not lower than the ontological rank of any contingent beings, including contingent minded beings. Therefore, the absolutely necessary dimension of being must be understood as a mentally constituted dimension of being.

[iii] A few explanations and specifications are appropriate here. The POR in no way entails a rejection of any specific phenomena or of natural-scientific theories, particularly the phenomenon of evolution or theories of evolution—presupposing that this phenomenon is explained adequately and that the theories contain no ungrounded extrapolations. The (physical) cosmos indeed evolves in the direction of life and in the direction of mind, species originate and diversify through evolution, etc. This is to be accepted in part as fact and in part as well-supported theory. But how is this evolution to be understood precisely? Is not the arising of beings with higher ontological ranks from those with lower ontological ranks maintained or at least implied, and indeed *without any specification or differentiation*? Is the higher not here simply "explained" by means of the lower? This is in fact a contention or an implication that is contained within the most widespread understanding of evolution and/or of evolutionary theory. Such an understanding, however, must be termed a simplistic understanding of evolution and evolutionary theory. This understanding is rejected on the basis of the POR, but that does not mean, to stress the point once more, that evolution is rejected or that evolutionary theory, correctly understood, is false. What are required are scrupulous differentiations.

The decisive point is presented relatively easily. If it is said that (for example) living beings—which, in accordance with the usual understanding and on the basis of the philosophical considerations presented in Chapter 4, are taken to be elements within a higher domain of nature—develop from beings that are located within a lower domain of nature, then this can say either of two things or be understood in either of two ways. The first way is the following: the lower or more basic beings (in the most radical case, the purely physical beings[57]) from which something higher develops are considered *purely as such and only as such*, thus without any further qualification; such qualification would be required if, for example, the process of development from these entities were to be *explained adequately*. The second way also understands the lower or more basic (physical) entities in terms of their physical structures, but in such a way that the additional qualifications are supplied that are required *to explain adequately* various other specific phenomena in conjunction with these beings.

[a] If one interprets evolution strictly on the basis of the first way, what results is a theory of evolution that, oversimplifying, is one of two types of theories, whereby the second type has two completely different variants. In accordance with the first type, the theory of evolution is understood *exclusively* as a natural-scientific theory in the strict

605

606

---

[57] For the sake of simplicity, the following account explicitly considers only this *most radical case* of an evolution: the process of development of higher entities (life, mind) from purely physical entities (in the narrow sense of "physical"). It is clear that evolution also has many more concrete forms. Particularly in biology, "evolution" designates the course of the history of species from the lowest stages of the organization of life up to the highly organized forms that exist at present. In these more concrete cases, the lower or more basic entities are no longer purely physical. To be made fully adequate, the course of argument developed in the main text would have to be concretized for such cases, and modified appropriately.

sense. In that such a theory considers the purely physical basic entities *as such and only as such*, it articulates and/or explains the specific interconnections *exclusively on this basis and in this sense*. If a theory of evolution is explicitly understood in this strictly natural-scientific sense, there can be no philosophical objections. The most important consequence that results for philosophy is that the *explanation* of the process of evolution provided by such a theory is only a restricted one; this explanation cannot claim to be *complete and adequate*.

Matters are utterly different with theories of evolution of the second sort. These claim to provide, with the natural-scientific theory of evolution, complete and adequate explanations of the process of evolution. This type has two variants. The first is radically reductionistic in the sense that it holds that all evolved phases or beings or phenomena can be explained as *purely physical* transformations of the physical basis. One says, for example, that there arises an additional, higher complexity of physical elements, etc. But the entire evolutionary process remains, according to this mode of explanation, a purely physical process. The second variant of this type of theory of evolution is oriented utterly differently; here, the development of higher stages is "explained" as due to *chance*. Any sort of structural plan that would explain the course of evolution is rejected. Questions concerning the conditions of possibility of a development of higher stages are not posed; indeed in most cases they are rejected explicitly. There remains only the reference to the "chance" that then serves as a sort of magical "explanation."

That the first, the radically reductionistic variant of the second type of theories of evolution, is not acceptable is established in Chapter 4, at least as far as the human mind is concerned. But the second variant is likewise not a conception that can be seriously defended philosophically. The reason is that this type of theory is the result of a rejection of serious philosophical effort. Such a theory contents itself with accepting the series of the successive phases of an evolutionary process as proof of the complete and theoretically incontestable explicability of every phase simply on the basis of the preceding phase. But thereby, it simply ignores the eminently philosophical question that asks *under what conditions* the series of phases in an evolutionary process can be considered as explaining the entire process and its individual phases.

[b] How this critique is to be understood more precisely is made clear in the following account within the framework of a characterization of the second way that the development of higher-staged beings or domains can be understood on the basis of lower-staged beings or domains. The basic (physical) beings from which something ontologically higher develops are not considered in isolation as purely physical beings; instead, they are understood as co-constituents of the entire universe, of being as a whole. So understood, they can be adequately determined ontologically only in conjunction with everything that happens or can happen to or with them *within* this whole. This means that to them must be ascribed those ontological factors that would for the first time explain how an evolution to something higher can take place on the basis of those beings *as physical*. What these ontological factors are is clear from the following ultimately incontestable and thus, philosophically considered, simple insight: *when something happens, it was possible for it to happen.* But this means that one must ask about the *conditions of this possibility*.

The conditions of the possibility of the development from lower-staged to higher-staged beings can be designated with the general term "ontological potentialities"; these are the ontological factors belonging to the physical beings that make possible the development from them of something higher. The genuine question is then, how are such ontological potentialities to be understood? Potentialities are, as the word says, factors that are not (yet) actualized but that can become actualized. To be emphasized is that such factors are not some sort of abstract considerations or anything of the sort, but instead genuinely ontological *determinations* of the physical beings. More precise explanation of these potentialities is to be sought in the *systematic contexts* within which these beings are found; these beings are, as is indicated above, embedded within the entirety of actuality, no matter how this entirety may be understood more precisely. This embeddedness is such that these physical beings have *immanent tendencies* to *transcend* themselves insofar as they are purely physically structured, tendencies to point or lead beyond themselves, etc. These tendencies are manifest in the process of realization or actualization of the   608 potentialities of these physical beings. What these potentialities are, specifically, cannot be determined a priori or somehow derived from axioms; one can discover them only concretely, by studying the evolution of the physical cosmos. These are the potentialities that are the bases of the development of the domains of life and of mind.[58]

[iv] From these considerations emerges the following: if one understands evolution as an ontological development not solely from purely physical entities, but from the physical entities *together with the potentialities that lie within them* (in the sense explained above) in relation to ontologically higher entities, then it is obvious that any theory of evolution that articulates this understanding of evolution in no way contradicts the principle of ontological rank (POR) introduced above. It is then clear that it is not the case that life and mind develop simply out of the physical (the latter not further differentiated). Life and mind are ontological factors that, to be sure, arise in the course of the process of evolution from the physical, but in such a way that they are actualizations or realizations of *potentialities* that are *non-physical* but that nonetheless lie within the physical domain. (A consequence of these considerations that makes this matter yet more precise is drawn below, in connection with additional considerations concerning absolute being.)

The preceding consideration of theories of evolution has the goal only of showing that the principle of ontological rank (POR) cannot be put into question, much less refuted, by any such theory. The second argument formulated above in support of the thesis that the absolutely necessary dimension of being is to be understood as minded, the argument based on this principle, can now be briefly summarized as follows: the contingent dimension of being is completely dependent upon the absolutely necessary dimension of being, which entails that the contingent dimension of being owes

---

[58] See the in part similar and quite interesting considerations in Forrest, who writes, for example (1996: 48):

> [A]lthough a scientific theory might explain the suitability of the universe for life by assuming the occurrence of an appropriate set of laws, it cannot explain why there are life-friendly rather than life-hostile laws. Here by life-friendly laws I mean ones that make it probable that a universe in which they hold will be suited to life. That there are life-friendly laws can, however, be explained in terms of the divine purpose in creating.

its own being to the absolutely necessary dimension of being. Because the contingent dimension of being also includes *minded* beings (i.e., human beings), the possibility is excluded that the absolutely necessary dimension of being could have an ontological rank lower than that of the human being, as minded. Because however mindedness is the highest of all possible ontological ranks—because, as is shown above in various contexts, mindedness entails intentional coextensivity with being as a whole, and no more comprehensive coextensivity is possible—the absolutely necessary dimension of being must be understood as minded.

609

[3] Establishing that the absolutely necessary dimension of being is mentally constituted means that it is to be determined more precisely as absolutely necessary minded personal being. It is "absolute being," not "an absolute being" or "the absolute being" because the status of absolutely necessary minded personal being is absolutely unique: it is not an additional being somehow conjoined with contingent beings, and thus not even a or the first or highest being. The position presented here is not an "onto-theology" in Heidegger's sense, so Heidegger's (justified or unjustified) objections to such positions are irrelevant to it.[59] Instead, the theory presented here stands in a certain respect in continuity with a conception whose starting points are formulated—although not consequently—especially by Thomas Aquinas. Particularly relevant are his contentions concerning the absolute (God) as being subsisting in itself [*ipsum esse per se subsistens*].[60] To develop this in detail is an enormous task that lies beyond the scope of the present work. In conclusion, only a few indications concerning additional steps toward the determination of the absolutely necessary dimension of being are added.

It is shown above that the contingent dimension of being is *totally* dependent upon the absolutely necessary dimension of being or, as it may now be put more precisely, on absolute, personal being. The mode of this dependence can now at least incipiently be determined more precisely. As absolutely minded, absolute being possesses absolute intelligence and absolute will (i.e., absolutely free will). From this it follows that the total dependence of the contingent dimension of being on absolute being rests on the free decision of absolute being to constitute the contingent dimension of being from non-being. Total dependency then is the status of having been produced. In classical Christian metaphysics, this free act of absolute being is termed "creation," and more precisely determined as production out of nothingness, although it must be emphasized immediately that "nothingness" here in no way means "absolute nothingness." The Christian metaphysical tradition characterizes the concept of creation more precisely as follows: production (of a given being) out of the non-being (out of the non-existence) of this being and without reliance upon any sort of fundamental or presupposed "stuff" or "matter."[61] This work cannot examine more closely the explanation and grounding of this thesis.

610

---

[59] See his essay "The Onto-Theo-Logical Constitution of Metaphysics," in Heidegger (1957/1969: 35–73/42–74).

[60] See, e.g., *STh* FP Q 4 A 2 Rp 3; *Power of God* Q 7 A 2 Rp 5. On Aquinas's position as a whole, see Puntel (2007: Chapter II).

[61] See, e.g., a formulation characteristic of Aquinas: "creation is the production of a thing in its entire substance, nothing of this thing being presupposed either uncreated or created" (*STh* FP Q 65 A 3; translation altered). In the tradition of Christian metaphysics, the following terse and fitting formulation develops: "Creation is the production of a being out of the nonbeing of the being itself and of any substratum [*Creatio est producto entis ex nihilo sui et subiecti*]."

From the thesis just introduced arises a problem relating to theories of evolution. Sometimes, the thesis that the contingent dimension of being is created is interpreted as requiring that the creator (often termed "intelligent designer") must be involved *directly and immediately* not only at the starting point but also at every progressive phase throughout the process; the consequence is then that an evolution, strictly speaking, simply does not take place. Such an understanding of the idea of creation is, however, an immense misunderstanding that has disastrous consequences. The idea of creation in no way requires that the creator (no matter how designated) either does or must involve itself in created, self-developing actuality in the manner just indicated. The created world *as a whole* is indeed totally dependent upon absolute being, and thus has the status of having been created (in the traditional terminology, the world as a whole is "maintained in being" by absolute being). But the created world, *as* created, is a thoroughly structured whole, with *its own* laws (including laws of development), potentialities, etc. And it develops itself *in accordance with these laws and by means of these potentialities* (see pp. 456–58).

[4] In conclusion, with respect to a comprehensively systematic view of the problem of the (further) determination of absolute being, a final consideration may be introduced. When systematic philosophy has gotten to the point of explaining the absolutely necessary dimension of being as *creative absolute*, the question arises whether additional determinations of creative, absolute being are possible or even indeed indispensable. At this point, the systematic development takes an immense turn that requires a far-reaching methodological break. Additional determinations of creative, absolute being, beyond the determinations of absolute intelligence, absolute will or absolutely free will, and personality are no longer in any way derivable, but depend instead upon a decisive factor: the freedom of creative, absolute being. This factor is clarified by consideration of the human being as free. If one wants to provide additional determinations of any free human being (i.e., of that human being as free), there is only one way to discover them: from investigation of the history of the human being's free decisions and acts. The situation is wholly analogous in the case of further determinations of creative, absolute being: such determinations could emerge only from investigation of the free history of absolute being. To determine whether or not there is such a history, one must investigate world history, and particularly the history of the major religions. With this, the point is reached where the word "God" can and perhaps must, methodologically, be introduced. "God" is the word that monotheistic religions use to designate personal, absolute being as manifesting itself and acting within world history.

To be noted in this context is that, at least as it is understood by Christianity in the wake of extensive and controversial discussion, the "acting" of God in human history is determined, philosophically, in a way that does not at all conflict with evolution. According to this understanding, God's "acting" is not an alien intervention, from the outside, countering the laws, potentialities, etc., proper to the created world as a structured whole; on the contrary, this acting is the actualization of potentialities native to the structure of the human being. Because however the human being is itself a part of the (natural) world and thus an element within the evolution of this (natural) world, *all* human potentialities, thus also those potentialities that relate to human participation in what the Christian religion calls the "history of revelation" (or "history of salvation"), are themselves wholly within the natural world. What this shows or presupposes, among other things, is a conception of the (natural) world that understands it not as absolutely encapsulated or enclosed within itself, but as an open, contingent whole, also and particularly with respect to the relation of this whole to the absolutely necessary dimension of being.

With the detailed and precise investigation and interpretation of this subject matter, a wholly new chapter of systematic philosophy begins, but that chapter requires its own investigation.

Metasystematics: Theory of the Relatively Maximal Self-Determination of Systematic Philosophy

6.1 The Status of Metasystematics

6.1.1 *Metasystematics and Metaphilosophy*

It is scarcely contestable that correct determination of the status of any theoretical discipline can result only from consideration of it as a constitutive or integral part of the comprehensive corpus of theoretical knowledge. One consequence of this is that every theoretical discipline brings with it (in a sense to be characterized more precisely) one or more "metalevels" that situate the discipline systematically within the body of theoretical knowledge as a whole. Emphasizing this structural moment of all theoretical activities and disciplines is made the more important by the fact that it is not always made explicit. One of the most famous *explicit* cases is that of *metamathematics*, which arose in the 1870s,[1] but which, as currently understood, traces back to David Hilbert (1922/1935, 174); this discipline has had a significant history and is today among those that are central in the philosophy of mathematics. By now, there is an extensive array of metadisciplines, in all theoretical domains: in addition to metamathematics and to the longstanding metaphysics, there are now also metalogic, metasemantics, metalinguistics, etc., and finally metaphilosophy. Stated most broadly, a metadiscipline is the study or theorization of a discipline, as well as the theory that emerges from such study or theorization.

Most broadly, metaphilosophy is thus the theorization of philosophy, but the distinction between a given discipline and its theorization is not always sharp. The unclarity of the distinction in the concrete case of presentation arises from and is visible in the fact that it cannot be established *a priori* just where to draw the line between the self-explication of a discipline and the theorization of the discipline. That disciplines explain themselves occurs constantly within the theoretical domain. Thus, for example, mathematics books can define mathematics, indicate how mathematics understands

614

---

[1] See *"Metamathematik"* in Ritter et al. 1977ff, Vol. 5, Columns 1175–77.

itself, what its method is, etc. On the one hand, all of that can be considered to be talk about mathematics; on the other hand, it can instead be taken to be the self-explication of the discipline.

If one considers a definition of a discipline—initially purely syntactically—to belong to a metalevel, then one must also acknowledge an additional, higher metalevel, thus a second-order metalevel at which the discipline is not defined, but instead situated in relation to other disciplines. Talk about a discipline that is already defined in this sense articulates a consideration and thus a theory of the discipline that itself includes a first-order metalevel; the additional talk thus develops on a second-order metalevel, and is thus a second-order metatheory.

Given this situation, it is helpful to introduce an important distinction, that between metaphilosophy in a *broader* and a *narrower* sense. Taken in the broader sense, metaphilosophy involves discourse *about* philosophy in some manner that is not further specified; an inquiry is metaphilosophical in the narrower sense when the manner of talk is understood and clarified quite precisely. Comparable distinctions hold for all theoretical metadisciplines. Many passages in this book are metaphilosophical in the broader sense. That is not inconsequent, nor does it lead to confusion; on the contrary, it serves to situate and to clarify particular issues with respect to the conception as a whole. In all theoretical domains of philosophy, comments that are metaphilosophical *in the broader sense* are typical and appropriate. It is also clear that passages that are metaphilosophical in this sense should be considered as belonging to philosophy itself.

The situation is quite different with accounts that are metaphilosophical *in the narrower sense*. These must be explained and characterized precisely. One central question is, are such accounts still *philosophical?* The answer must be that they are, due to a characteristic exclusive to philosophy as such and as a whole: the distinction between object level and metalevel has, in the *sole* case of philosophy, only a quite relative and indeed intraphilosophical significance. In the strict sense, metaphilosophy remains philosophy. How the relations between metalevel and object level (in the narrower sense) are or should be understood with respect to other theoretical disciplines is not considered here, but in no other science can there be so close a connection between levels as there is in philosophy. The reason for this is the universality of philosophy. This must of course be correctly understood; the following sections articulate its correct understanding step by step.

These sections also further determine the concept of metaphilosophy *in the narrower sense* on the basis of the distinction between the two basic forms of such metaphilosophy that are relevant to this book: immanent and external metasystematics. Immanent metasystematics is theory that remains entirely within the framework of the philosophical theory developed here. Its topics are the individual systematics presented, respectively, in the five preceding chapters of this book: global systematics, the systematics of theoreticity, the systematics of structure, world-systematics, and comprehensive systematics. These distinct systematics constitute the object level for what is here termed "immanent metasystematics." In the complex expression "immanent metasystematics," the component "metasystematics" designates every single specific systematics that constitutes a part of systematic philosophy, i.e., of the philosophical system.

External metasystematics has as its object level the structural-systematic philosophy as a whole, including both its five distinct systematics and its immanent meta-systematics. Yet to be shown is how to understand the status of such an external metasystematics.

6.1.2 *The Metasystematic Self-Determination of the Structural-Systematic Philosophy and the Criterion of Relatively Maximal Intelligibility and Coherence*

[1] Usually, the concept *self-determination* is used in the domain of practical philosophy to characterize free agency. This practical signification is not intended here; the current concern is instead with self-determination that is exclusively theoretical.

A theory or discipline is determined by means of the specification of its status with respect to adequacy. In the case of self-determination, this specification develops within the discipline or theory itself, although what it means within the discipline or theory itself must be clarified. As the following account makes clear, this point constitutes an essential aspect of the distinction, developed in greater detail below, between immanent and external metasystematics.

The theoretical status of a given theory or discipline has many quite different aspects, depending upon which level of the theory or discipline is in question: among the possibilities are the levels of linguistic clarity, of conceptuality, of argumentation, of the interconnections among components, etc. In a stricter sense, the theoretical status involves all the levels or factors that constitute the theoretical framework and that are, for that reason, essential to the concept of the theory. Theoretical (self-)determinations can also have various forms. The two relevant to this book can be designated by means of the distinctions between total and partial (self-)determination or absolute and relative (self-)determination. This distinction is best suited to the aims of this book, so it is the primary focus of the following account.   616

More precise clarification of the concept of theoretical or systematic (self-)determination requires a criterion. This criterion is that of *relatively maximal coherence and intelligibility.*

[2] To be noted at the outset is that this is unquestionably a strictly theoretical criterion. That it is harmonizes perfectly with the conception of philosophy defended in this book: that of a theoretical activity and undertaking. If philosophy is understood differently (e.g. as any sort of practical activity), then the criterion used here would scarcely serve philosophy's (self-)determination. In that case, a quite different criterion would be required (e.g., orientation with respect to life or utility in solving human problems).

Why this specific theoretical criterion is accepted rather than some other likewise theoretical criterion deserves a thorough explanation, but providing such an explanation is beyond the scope of this book. Instead, the answer given here is a general one that emerges from the basic theses of the structural-systematic philosophy. The criterion exhibits a positive coherence with this philosophy as a whole, as it is presented in this book. To the objection that using this criterion therefore introduces a fundamental or (so to speak) systematic circularity, the response is that there is no circularity because the concern is not with a definition or a concept, but instead with the grounding of the choice of a criterion. That the criterion emerges from the basic theses is not a disadvantage, but instead an explicitly positive and quite significant advantage, because it grants the criterion, *considered in itself or as such*, a high degree of plausibility, as can easily be shown by brief consideration of other possible criteria.

One could, for example, object that the truth itself—the truth of this philosophy—should be, if not the only criterion, nevertheless the best of all criteria for the (self-)determination of systematic philosophy. If this criterion were actually and effectively available, it would indeed be the best, indeed the sole and definitive criterion—for what could be better than to establish the truth of a specific philosophy? But truth is not suitable as a criterion for the (self-)determination of this philosophy, because a criterion must be available; truth itself is the goal of the endeavor, not the criterion.[2] Evidence and self-evidence are also unsuitable as criteria, because these phenomena are vague and insufficiently determined.

To be sure, *relatively maximal intelligibility and coherence* are not determined absolutely or in every respect, but they are sufficiently unambiguous and sufficiently available to serve as criteria. Making these criteria sufficiently clear is the task of the following subsection.

[3] As is noted at various points above, *coherence*, as it is understood here, must not be identified or confused with (logical) *consistency* (*freedom from contradiction*). Consistency is a purely *negative concept* indicating only that the simultaneous acceptance or derivation of a sentence and its negation is excluded. The concept of coherence is a *positive* one that presupposes consistency, but involves beyond that a certain interrelatedness among concepts, sentences, theories, etc. Coherence is thus an eminently *contentual* concept, and involves interrelations in all contentual contexts.

The concept of coherence includes an extensive scale of forms and degrees, from weak coherence to the strongest coherence. The strongest form is attained when the interconnections are purely analytic or logically inferential. A sentence with the status of being a theorem in the strict sense exhibits the strongest possible form of coherence within the theory to which it belongs. At the other end of the graduated coherence scale are loose interconnections, such as intuitively plausible agreements among concepts, sentences, and theories. For a systematic philosophy like the one presented here, of fundamental importance is the form of coherence that can be termed "systematic coherence." This is achieved by locating or situating every item (concept, sentence, theory, etc.) within the system as a whole; this can occur in various ways. Finding the adequate place for every item within the system is of decisive importance. A given philosophical system is improved in this respect as its internal systematic coherence is developed with increasing adequacy and transparency.

The concept of *intelligibility* is not identical with the concept of coherence; instead, in one respect it is presupposed by it and in another is a consequence of it. It is presupposed in that it is more primordial and fundamental than are analytic interconnection, derivability, and indeed coherence—in a word, only what is intelligible can possibly qualify as coherent. The status of a given subject matter is intelligible, in the sense intended here, if it fully satisfies theoretical criteria; "subject matter" here covers anything whatsoever that is or can be the object or topic of theoretical activity

---

[2] See Rescher (1973), which defends the thesis that the concept of correspondence defines truth, whereas the concept of coherence is the criterion for truth.

(thus, "reality," "phenomenon," etc.). To be sure, as so characterized *intelligibility* is quite general and therefore quite vague. Despite this vagueness, indisputable at first glance, intelligibility as thus characterized can nevertheless be the source—better, the source of inspiration—for the adequate theoretical comprehension and articulation of the subject matter. This is in any case so if the intellectual activity of theorization is qualified more precisely as follows: it is the actualization of intellectual capacities. This actualization is the full development of the intellect in the sense of the utilization, at least in principle, of all the structural moments of the intellect. Conscious reliance upon all the theoretical capacities of the intellect requires an unlimited *openness* of the intellect, a basic stance such that no prohibitions of thinking or questioning, etc., are accepted, such that all possibilities for thinking that appear sensible are investigated and attempted, etc. Only from the utilization of all potentialities and the conscious maintenance of an unrestrictedly open attitude does there first arise the possibility of developing the *inventory* of items and issues requisite for the attainment of a deeper and more adequate understanding of the *intelligibility* of a given subject matter.

How then is intelligibility in one respect not a presupposition but instead a consequence of (explicitly developed) coherence? The explanation is simple. The intelligibility in question is of a *second order*, in the sense that it is the result of effectively developed coherence: a given item becomes *more* intelligible as it is related, coherently, to other items. To be sure, such increases in intelligibility can themselves be the starting points for new investigations, so this second-order intelligibility can itself be the basis for more extensive and more adequate coherence. The question of how far this process does or can go is clarified by what follows.

The concept *maximal* must also be clarified. Particularly in this respect, the concepts of intelligibility and coherence are distinguished quite clearly from the concept of consistency. The concept *maximally consistent subset* is used in set theory and metalogic, where it is defined as follows: if S is the set $S = \{p_1, p_2, \ldots\}$ of sentences of language L that are mutually consistent or inconsistent, then the subset S_i of S is *maximally consistent* if (1) S_i is not empty, (2) S_i is consistent, and (3) no element of S that is not an element of S_i can be added to S_i without producing an inconsistency (it thus holds that for every sentence p in S that is not in S_i, the set $S_i \cup \{p\}$ is inconsistent). Because consistency does not allow for degrees (there is no more or less, or maximum or minimum, of consistency), in the phrase "maximally consistent subset," "maximal" has a quite peculiar meaning (i.e., the one stipulated in the definition just introduced).[3]

619

---

[3] This peculiar meaning can be explained more precisely. Strictly speaking, *consistency* is not determined as "maximal" (as though it could have a maximal degree). Instead, the "maximal" of the definition indicates a specific factor that must be either given or excluded for there to be talk of a "maximally consistent" subset; this is the factor *determinate size* or, more precisely, the number of members (*cardinality*); if the cardinality of the set in question is increased by even one element, the resulting set is inconsistent. It must still be added that sets themselves do not have "degrees" in the strict sense, because sets are determined by the principle of extensionality: $\forall X \forall Y (\forall z (z \in X \leftrightarrow z \in Y) \rightarrow X = Y)$ (i.e., two sets that contain the same elements are themselves the same (are identical)). From this it follows that a given set cannot have a "degree" in the sense of being able to contain more or fewer elements; if a set has even one element more than the "same" set without the element, then there are two different sets.

The situation is quite different with intelligibility and coherence, both of which *do* allow for degrees. Purely intuitively and informally, they can be considered to lie on scales whose extremes are "minimal" and "maximal." But even these extremes, considered in themselves, can have (sub)degrees, as is shown by the additional qualifications (hence subdegrees) of *relatively maximal* and *absolutely maximal* intelligibility and coherence. The more precise determination of these concepts is presented below and in the sections on immanent and external metasystematics.

Yet to be clarified is the concept—quite important for this book—of *relatively* maximal intelligibility and coherence. There are two relevant forms of relativity: a basal or fundamental or theoretical-framework-dependent relativity, and a relativity immanent to a given theoretical framework. The first form is entailed by the thesis that every theory, even the most comprehensive theory (i.e., systematic philosophy), involves an intelligibility and coherence *exclusively relative to the theoretical framework* within which it develops. This form of relativity is a consequence of two theses that are formulated and analyzed in this book in various places: first, that there is a plurality of possible and available theoretical frameworks, and second, that *only one* theoretical framework can *ever* be in current application. From these two theses follows the necessity that precisely one theoretical framework, and one that is wholly determined, be chosen and used. All of a given theory's sentences are thus *relative to the theory's theoretical framework. Absolute intelligibility and coherence*—intelligibility and coherence having no relation to any theoretical framework—is an empty concept.

620

The other form of relativity, the form that is immanent to every theoretical framework, presupposes as its basis an acknowledged theoretical framework *and its relativity*; it is thus a relativity *within* the basal or fundamental or theoretical-framework-dependent relativity. It involves neither comparisons between different theoretical frameworks nor the necessity of selecting a single theoretical framework; it involves instead the realization of a specific level (or a specific degree) of intelligibility and coherence *within* the theoretical framework being used.

As is noted above, with respect to these two concepts theoreticians must distinguish and accept various levels and degrees. Structural-systematic theories (component theories and the comprehensive theory) strive to attain the level or degree that is *relatively maximal* in the following sense: they strive to attain the highest levels or degrees of intelligibility and coherence that, given all the *concrete* factors determining the theoretical undertaking, are realistically attainable. Among these concrete factors are included above all epistemic-pragmatic factors. The maximal intelligibility and coherence spoken of here are thus *relative* to the particular concrete situation, especially the epistemic-pragmatic situation. In or relative to this situation, the maximally achievable degree of intelligibility and coherence is envisaged. This degree is *never the absolutely maximal* degree.

The second form of relativity has two quite different variants. The first can be termed *immanent relativity in principle*. The second has the character of being contingent rather than a matter of principle; it can be termed *contingent-pragmatic relativity*. "Pragmatic" here designates the factors that determine the domain of our human possibilities.

The forms and variants of relativity that are introduced and explained in the preceding paragraphs quite generally and abstractly are explicated more fully in the more detailed presentation, in what follows, of immanent and external metasystematics.

According to this book, philosophically immanent metasystematics can be understood only as relatively maximal theoretical self-determination. This self-determination has various aspects that must be investigated and presented in detail.

6.2 Immanent Metasystematics

6.2.1 *What Is Immanent Metasystematics?*

Immanent systematics is the theory of the interconnection of all the structures and dimensions of being of the structural-systematic philosophy. One can also adopt a term from Kant and call immanent metasystematics the architectonic of this philosophy. Kant defines this concept as follows:

> By **architectonic** I understand the art of systems. Since systematic unity is that which first makes ordinary cognition into science, i.e., makes a system out of a mere aggregate of it, architectonic is the doctrine of that which is scientific in our cognition in general, and therefore necessarily belongs to the doctrine of method. (B860)

Kant's definition almost exactly fits the concept of immanent systematics in use here, although one must of course note that Kant understands some of the terms he uses in the definition quite differently from the way they are understood in this book (this holds above all for "knowledge/cognition"). Given Kant's division of the *Critique of Pure Reason* into the Transcendental Doctrines of Elements and of Method, it makes sense to place the treatment of architectonic in the latter; the systematic philosophy developed here is of course structured quite differently.

6.2.2 *Three Aspects of Immanent Metasystematics*

The description of three of its aspects—[1] a comprehensive overview of the components of the structural-systematic philosophy; [2] the explicit presentation of the method applied in the development of the comprehensive corpus of the theory; and [3] the articulation of the inner coherence of the individual systematic components of the theory—serves as a brief introduction to immanent metasystematics.

[1] The Table of Contents provides a comprehensive overview of the components of the structural-systematic philosophy. This overview is (so to speak) the global-holistic aspect: the entirety of the theory comes into view. To be sure, in a specific respect the articulation of the theory provided by the Table of Contents is quite external, but it is a genuine articulation whose significance should not be underestimated.

[2] As is explained at the beginning of Chapter 1, this book presents the comprehensive theoretical framework of the structural-systematic philosophy. Within the framework of immanent metasystematics arises the pressing question concerning the method that is thereby applied. Chapter 1 also presents the complete or idealized structural-systematic method as four-staged. The presentation of the structural system would be complete in every respect only if the method were fully applied to every theory or topic in every domain. As often noted above, however, such completeness is an ideal that can be attained, if at all, only in the cases of a few specific theories or domains. Nevertheless, the four-staged method has, as a *regulative idea*, invaluable importance in that it contributes decisively to the clarification of the methodological status of this book's conception of philosophy and of its presentation of its own philosophy.

Does the philosophical-systematic method apply also to the presentation of the comprehensive theoretical framework of the structural-systematic philosophy? Within the framework of immanent metasystematics, this question is to be answered in the affirmative, because *immanent* metasystematics is precisely this theory's self-explication of its comprehensive theoretical framework. Because metasystematics *is* self-explication, there is a self-application of the philosophical-systematic method. The self-application is, however, only partial, in that within the framework of *immanent* metasystematics only the first three of the stages can, in principle, be applied. The final stage of the method goes beyond the system understood immanently in that it relates it to other (philosophical-systematic) conceptions, comparing it with them.

This book does not apply the second stage of the method, the stage at which rigorously structured holistic networks or axiomatic theories would be constituted. A completely developed immanent metasystematics would be the correct (strictly systematic) location for a comprehensive, systematic, rigorously structured holistic network explicitly interrelating all of the components of this philosophy, or an axiomatic theory integrating all parts of this systematic philosophy. This task is not undertaken here because this book is limited to the presentation of the comprehensive systematic theoretical *framework*. In addition, undertaking this task would be sensible only if, in advance, all the individual parts of the structural-systematic philosophy had themselves been developed completely clearly. This latter task is without doubt the one that is currently the greatest desideratum for a complete systematic philosophy.

623   This book applies only the first and third stages of the method of systematic philosophy. In that Sections [3] and [4] show how these stages are at work, the book illuminates an important aspect of the immanently metasystematic theory of the structural-systematic philosophy. The methodic stages are taken as a unity, but in such a way that the third is clearly accorded the greater weight.

[3] At various places in this book, there are immanently systematic reflections. The reason for this is not that they would be required by a systematic presentation that was developed strictly methodically, but instead that they make the overall presentation more easily understandable. Above all, the basic idea of the systematic philosophy presented here is explicitly articulated in various places. Although such reflections belong, in the strictly methodic-systematic respect, in the domain of immanent metasystematics and thus in this section, they are here presupposed rather than repeated.

The entire presentation—and thus the entire self-explication—(of the comprehensive theoretical framework) of the structural-systematic philosophy develops on the basis of one overarching thesis or, in epistemic terms, intuition: the task of systematic philosophy is, in an initial description, the theoretical articulation of the universal datum, whereby *universal datum* serves as a term of art designating what in the course of the history of philosophy has been named "being (as a whole)," "(comprehensive) actuality," "the universe," "the world," etc.

A somewhat more precise initial formulation of this thesis/intuition is provided by the quasi-definition of the structural-systematic philosophy presented in Chapter 1. The rest of Chapter 1 and Chapters 2–5 present the self-explication of this quasi-definition. The self-explication initially reveals the status of the quasi-defined concept: it has the form of theoreticity, and this is the articulation of the dimension of structures. All the major questions concerning language, cognition, theory-formation, etc., are here treated and brought to systematic clarity. Thereafter, the three grand dimensions of structures are developed: the fundamental formal structures, semantic structures, and ontological structures. Here, the first and third methodic stages are applied in a particular manner. The comprehensive undertaking has both reconstructive and constructive aspects, as well as, in a decisive manner, aspects of coherence. The undertaking is not a linear-axiomatic one, but instead one of developing a network: all the relevant items or elements are brought into a comprehensive, theoretically articulated, coherent whole. What is not of decisive importance therefore is derivation (of theorems or "truths") from "truths" (true sentences) axiomatically established or introduced at the outset; the resulting theoretical whole is not a deductively determined nexus. In other words, the relations among the individual items or elements within the whole are not of a strictly deductive nature; instead, although they involve inferential aspects, they are fundamentally and decisively determined by another consideration or by another measure or criterion: the consideration or criterion of (*greater or relatively maximal*) *intelligibility and coherence*. The sought-for comprehensive, theoretically articulated whole is developed with a view to intelligibility and coherence as understood in this way.   624

This is particularly clear in Chapter 3, in which the dimension of theoreticity is presented as consisting essentially of the three types of fundamental structures. In Chapter 3 there is particular concern with the coherence of these elements with respect to relatively maximal intelligibility. This explains particularly the wholly novel semantics and ontology whose basic features are presented in that chapter, as well as the innovative theory of truth.

Chapters 4 and 5 are contentual in the sense that they directly investigate relations between the fundamental structures and the universal datum. Chapter 4 examines the universal datum in the form of the domains of the world, Chapter 5, in the form of the unrestricted totality (being as such and as a whole). Each applies the fundamental structures presented in Chapter 3 to these domains, although (as is emphasized in Chapter 4) "applies" has a quite external and potentially misleading sense, in that it has the connotation of suggesting that the two dimensions (the fundamental structures and the universal datum, i.e., the various domains of being as well as being itself) are initially fully separate and indeed autonomous domains that are later brought into relation to each other. This is by no means the case. To be sure, the fundamental structures are initially presented as such, i.e., without reference to or reliance upon the universal datum and its elements. The fundamental structures are thus understood and presented *purely abstractly*, in the genuine sense of the term; at the same time, however, they have from the beginning *immanent ontological imports*. Considerable theoretical efforts are of course required to articulate, in all their complexity, the immanent ontological imports of the fundamental structures. Chapters 4 and 5 present an outline of this articulation.

6.3   External Metasystematics   625

6.3.1   *What Is External Metasystematics?*

External metasystematics brings into play dimensions that are neither totally nor partially identical with the philosophy developed in this book; thus, these dimensions are clearly "meta" in relation to this philosophy. There are two wholly different types of such dimensions: other theoretical dimensions and non-theoretical dimensions.

External intratheoretical metasystematics considers the former dimensions; *external extratheoretical* metasystematics considers the latter. External intratheoretical meta-systematics itself has two wholly different forms: external philosophical and external non-philosophical metasystematics. All these forms are purely theoretical.

How is the metacharacter of external metasystematics to be understood precisely? One must note at the outset that *in one specific respect* there is and can be no metadimension for systematic philosophy. The reason for this is that every "thing," every domain, theoretical or non-theoretical (every discipline, every activity, every phenomenon of whatever sort) can and indeed must be considered to be a topic that in principle is a possible topic *for* systematic philosophy. In this *thematic* sense there is nothing that could be a "meta" in relation to systematic philosophy in the sense of *not* being the object of a theory that would be a component of systematic philosophy. The comprehensive theme or topic or subject matter of systematic philosophy is, after all, the unrestricted universe of discourse.

If not as something thematic, some subject matter, then how is the "meta" of external metasystematics to be understood? This meta is a *perspective of theoretical consideration* that is not simply identical with the proper perspective of the theoretical consideration of the structural-systematic philosophy; the metaperspective is more than and different from the structural-systematic perspective. What "perspective of theoretical consideration" means can be explained with the aid of the concept of the theoretical framework, which is used throughout this book. An external metasystematics is one that has the presented structural-systematic philosophy as its topic or object or subject matter; its thematization requires another, broader theoretical framework—to be precise, an *externally metasystematic theoretical framework*. *How* this broader metasystematic theoretical framework is constituted depends upon the external dimension that is thematized. In the case of external intratheoretical *philosophical* metasystematics, the external metasystematic theoretical framework must include all the elements (language, concepts, principles, logical instruments, etc.) that are required for the thematization of the structural-systematic philosophy *and one or more other* (systematic but also non-systematic) philosophies. This framework must make possible a comparison of the structural-systematic philosophy with these other (systematic or non-systematic) philosophies. Corresponding theses hold with respect to the other external dimensions. If such a dimension is extratheoretical (e.g., the life-world with its non-theoretical structures and activities), the externally metasystematic theoretical framework that thematizes this extratheoretical dimension must include all the elements that make possible the thematization of the relationship between the structural-systematic philosophy and this extra-theoretical dimension, even though some of the elements of the latter are not themselves theoretical.

6.3.2 *External Intratheoretical Metasystematics*

6.3.2.1 External Intratheoretical Interphilosophical Metasystematics

[1] The task here is to put the structural-systematic philosophy into relation to other (systematically and non-systematically oriented) philosophies. This relationship is thus *interphilosophical* (and thereby also *intraphilosophical*: it is both *between* different philosophies and *within* the domain of philosophy). The determination and explication

of relationships of this sort constitute a considerable part of philosophical activity, particularly when a central significance is accorded to the entire history of philosophy.

In distinction from a *general* consideration of the history of philosophy, however understood, interphilosophical metasystematics is a *meta*systematics; it thus takes its bearings by the structural-systematic philosophy, relating it to other philosophies. The thematization of this relationship can take different forms: historical, comparative, critical-polemical, systematically constructive, etc. All these forms are metasystematic in character, and all presuppose, explicitly or implicitly, some metasystematic theoretical framework that has in particular one central component: a criterion that makes possible a comprehensive comparative analysis of the relation by answering the question of which of the philosophies has the superior or more adequate theoretical framework.

In this book's presentation of the general theoretical framework of the structural-systematic philosophy, external intratheoretical interphilosophical metasystematics is *anticipatorily* applied in many specific contexts, in that the individually developed (partial) theories, those articulating specific domains of systematic philosophy, are often compared with competing theories. This is done for the sake of the contemporary reader, who, with respect to *every specific domain* of philosophy, is *immediately* confronted with a flood of often radically divergent positions. Because of this situation, it appears appropriate to introduce discussions of competing conceptions in the course of the exposition rather than presenting them only at the end of the comprehensive presentation. These anticipatory metasystematic accounts develop within external or non-systematic perspectives. The accounts would be strictly systematic, in the sense used in this book, only if the comparisons with other positions came at the *end* of the comprehensive presentation, and within the perspective that is then available and grounded.

[2] When the structural-systematic philosophy is put into relation to some other philosophy, the comparison—most importantly, that between the relevant theoretical frameworks—can yield one of *four* results: [i] the structural-systematic theoretical framework proves to be superior to the alternative; [ii] the other theoretical framework proves to be superior to the structural-systematic framework; [iii] the two theoretical frameworks prove to be equivalent; [iv] both theoretical frameworks prove to be deficient and therefore in need of correction.

It should be clear what conclusions would follow from these four results. In case [i] the other philosophy and in case [ii] the structural-systematic philosophy would be refuted, whereas in case [iii] the two philosophies would prove to be different presentations based on one and the same theoretical framework, and in case [iv] neither of the two theoretical frameworks would emerge as acceptable. There not only can be, but have been and continue to be, many instances of cases like [i] and [ii].[4] Whether

627

---

[4] As of the initial presentation of the structural-systematic philosophy, there cannot of course have been comparisons of other philosophies with it. But there have unquestionably been comparisons of various sorts between other philosophies and/or their theoretical frameworks. Such comparisons, unsatisfactory though they may be, have always been, and continue to be—perhaps more than ever—central to philosophy. The following account treats the *methodological* or *theoretical* problematic of such comparisons quite generally (i.e., not as it relates specifically to the structural-systematic philosophy).

case [iii] represents anything other than an abstract possibility that has no identifiable instances and perhaps cannot have instances is a question to be taken seriously, but one whose extensive investigation is not undertaken here because answering the question would not contribute significantly to the metasystematics required for this book.[5] Far more important (and difficult) is case [iv], likely the type most common in normal, actually sensible philosophical discussions. If both of the two theoretical frameworks being compared are deficient, the deficiencies must be remedied. But how? Of decisive importance is the fact that the deficiencies must be identified *from within a metaframework*. The metaframework provides the criteria determining the ways in which the deficiencies of the theoretical frameworks under consideration can and must be remedied, how the requisite corrections can be made, etc. What then results is an instance of one of the cases [i]–[iii].[6]

[3] The decisive question here is the following: on the basis of what *criteria* can or should this comparison be made? To appreciate the precise meaning and the entire scope of this question, it is helpful to begin by considering the actual practice of philosophical discussion. Discussions are everyday philosophical events, but there are quite different types of discussions: discussions about particular topics, discussions about ways of posing questions, discussions within a given philosophical school of thought, discussions between different schools of thought, etc. But rarely if ever is the question posed, under what conditions—on the basis of what presuppositions—do such discussions in fact take place? Also rarely if ever raised is the question about the possibility *in principle* of such discussions being sensible.

As a matter of fact, philosophers from various schools of thought often have conversations; far less often do they have genuine (i.e., sensible) philosophical discussions. "Having conversations," in a general and unspecified sense, is something that is in principle possible among philosophers of all sorts, but to enter into a genuine, sensible discussion is something that happens far more rarely and that is in many cases impossible. Without question, analytic philosophers have discussions of many topics with other analytic philosophers, Hegelian philosophers can and do have discussions

---

[5] A particular variant of this third possibility is Quine's (1975) famous underdetermination thesis, the thesis that it is possible to formulate empirically equivalent but logically incompatible theories. Strictly speaking, the equivalence in question is not one of theoretical frameworks, but of the empirical bases of the theories in question.

[6] With respect to philosophical practice, an extremely important question arises: are corrections of a given, specific philosophical conception possible *only* within the framework of the comparison of that conception's theoretical framework with one or more others? Answering this question is easy if one first draws a distinction between *two kinds* of deficiencies and required corrections: (1) those that are immanent to the relevant theoretical framework and (2) those that concern the theoretical framework itself. Those of the first kind, as their designation makes clear, are accessible and can be accomplished *within* the relevant theoretical framework. These are deficiencies and corrections that come into play because the theoretical framework itself has not been correctly developed or concretized. It is clear that such *immanent* deficiencies can and indeed must be remedied, and *immanent* corrections made, *independently* from any comparison with any other theoretical framework. With deficiencies and corrections of the second type, things are of course quite different: these indeed enter the scene only when one theoretical framework is compared with another.

with other philosophers of specific sorts who take positions regarding Hegel, etc.; it is also a fact that Kantian philosophers can and do have fruitful discussions with analytic philosophers. But can there be genuine, sensible discussions between Heideggerians or Hegelians, on the one hand, and analytic philosophers, on the other? It is far from clear that an affirmative response can be given to this question. To be sure, it must be emphasized that authors such as Hegel and Heidegger are often "interpreted" quite loosely or superficially, and when they are, they can be put into conversations with analytic philosophers. But the situation is quite different if such authors are taken at their words (i.e., if one takes seriously the radicality of their respective modes of thinking) and interprets them accordingly.

If one analyzes the current philosophical situation as a whole, one sees that genuine, sensible discussions often take place, but also that genuine, sensible, fruitful discussions between representatives of some distinct schools of thought neither take place nor appear to be possible. What is the reason for this disparity? The answer is obvious: genuine and sensible discussion presupposes at least minimal common ground. This minimal common ground is not provided by the mere fact that the interlocutors identify themselves as "philosophers" or say they are doing "philosophy," because all that some "philosophers" and "philosophies" have in common are such self-descriptions. Even minimal common ground must include the factors that first make possible the argumentatively oriented treatment of questions, topics, etc. These include fundamental logical rules or structures, methods for clarifying concepts, clarity of formulations, etc. When authors like Heidegger and Hegel, in their quite different ways, reject formal logic, this makes it impossible for analytic philosophers to have genuine, sensible discussions with them. This does not of course exclude the possibility of raising objections   630
to these authors by presenting reasons for rejecting their ways of thinking (see Puntel 1996, 1997); neither does it exclude the possibility of learning by reflecting on their works, nor that of attempting to develop theoretical frameworks within which insights taken so to have been learned could be presented with adequate theoretical rigor.

[4] To determine more precisely the status of the discussions in question one must attend to the central concept of the theoretical framework. All such discussions presuppose implicit or explicit reliance upon a specific theoretical framework that has a *metastatus* with respect to the schools of thought or conceptions of the participants in the discussion; this is thus a *metaframework*. In the case of the structural-systematic philosophy, such a theoretical metaframework has the status of being *external* to this philosophy. This theoretical metaframework includes the criteria on the basis of which is clarified how the comparison with other theoretical frameworks can develop. How is this systematic theoretical metaframework itself precisely to be conceived? This question is answered in two steps, taken in [5] and [6].

[5] To begin with, one must distinguish between two quite different categories of discussions: discussions within one and the same school of thought and discussions between one (or more) fundamentally different schools of thought.

[i] Within the former category, one must distinguish between two variants. In the first, a fundamentally *complete* theoretical framework is presupposed or used, one that *completely* characterizes the school of thought. "School of thought" is here taken in a restrictive sense, designating a closed conception that allows room only for discussions that do not fundamentally affect the theoretical framework. Such discussions can

concern either differences in modes of presentation or differences in the consequent concretization of the theoretical framework with respect to some specific topic within it. The term "discussion" is thus used here in an almost inauthentic sense. Many works that are designated "discussions" are instances of this variant. In historical work, this variant takes the form of disagreements about the correct or more adequate interpretation of texts of the philosopher who has founded a given school of thought or of that philosopher's successors who have retained the founder's framework. One of the most familiar designations for works of this sort is "reconstruction" (of the conception presented by a given philosopher).

631

The second variant arises when the theoretical framework characterizing a given school is not completely given, but instead allows some room for free play among possible ways of completing it. This happens for example when different conceptualities or methodologies are introduced and used. Contemporary analytic philosophy offers perhaps the best example of this variant. One can certainly speak of an "analytic theoretical framework," but this framework is limited to relatively few components, such as formal logic, argumentative methods, etc. If then the conceptual, semantic, ontological, etc., components required by any philosophical theoretical framework claiming completeness are added, distinctions become visible, some of them quite radical ones. In the philosophy of mind, for example, one variant relies on a purely physicalist conceptual scheme, another on a dualistic one; the consequences are wide ranging. Almost all of the intense discussions within contemporary analytic philosophy clearly exemplify this variant of the first category of discussions.

As the example of analytic philosophy clearly shows, one can in general speak of common ground only in a quite restricted sense. This is shown, for example, by the component *logic*. Perhaps the only non-controversial area of logic is propositional logic. Predicate logic is, as this book shows, not without problems. If one considers such extensions of elementary logic as modal logic, opinions diverge widely. These differences cannot be pursued further here.

[ii] The second category involves discussions that are between two (or more) fundamentally different schools of thought. These discussions differ from those just considered (the second variant within the first category) in that the common ground is absolutely minimal: it involves only those factors that are indispensable for any discussion, such as argumentation, clarity, etc. This minimal common ground includes certain logical rules, semantic capabilities, etc., but nothing more. It must also be noted that the distinction between this second category and the second variant of the first category is not fixed; in many and indeed perhaps most cases one must speak of flexible borders. (It is this circumstance that has led extensive parts of the philosophical community to hold that the concept *analytic philosophy* is not definable and indeed not even generally characterizable).

632

[6] The second step is the clarification of the question of how the metasystematic theoretical framework itself is to be understood. This question is here posed not somehow generally and abstractly, but as a question within the framework of the external metasystematics of the structural-systematic philosophy. A purely general and abstract questioning would not be senseless, but it would have to operate with empty variables that would play the role of indeterminate indications of a concretely developed philosophical account. The answer to such questioning could itself only be

general and abstract. But here a variable assignment is assumed: that provided by the structural-systematic philosophy.

It would be an illusion to assume or indeed to demand that the metasystematic theoretical framework should (or could) be absolutely neutral in the sense of being some kind of highest or even infallible authority for the clarification and decisive answering of external metasystematic questions. There is and can be no such authority, at least not on the basis of the conception developed in this book. In attempting to develop such a neutral theoretical framework or to treat an extant framework as neutral in this fashion, one would overlook the fact that any such theoretical framework would of necessity result from a *philosophical* conception or theory and thus could not be neutral in the sense in question. If a specific philosophical conception or theory claimed to be neutral in this sense, it would claim to be beyond discussion and would thus reveal itself to be dogmatic. In any case, it could not count as rational.

The metasystematic framework for the conception characterizing the structural-systematic philosophy has, in addition to such basic components as logical and semantic factors, a quite specific conceptual component that is its fundamental characteristic; this is the concept *greater intelligibility and coherence.* The central conceptual component of the theoretical framework thematized by the *immanent* metasystematics presented above is *relatively maximal intelligibility and coherence.* Immanent metasystematics is not concerned with a *comprehensive* comparison between competing conceptions, but with the articulation of the metasystematics of a single philosophical theory, the systematic philosophy presented in this book. To be sure, the problem of comparing diverging conceptions or theories with respect to specific topics, domains, etc., is also posed *within* this book, but this is also a consequence of reliance on the criterion of greater intelligibility and coherence. In the case of external interphilosophical metasystematics, however, this criterion is applied when comparing the structural-systematic philosophy with other philosophical theories.

633

Can the concept of *greater* intelligibility and coherence be made more precise? One should scarcely expect that a strict definition could be provided, but in accordance with the characteristic style of this book's presentation of the structural-systematic philosophy, two basic characteristics of higher grades of intelligibility and coherence can be provided.

[i] The first characteristic of a greater intelligibility and coherence is, in relation to other positions, a *more extensive consideration and comprehension of the data.* The term "data" is used here in the technical sense frequently used elsewhere in this book (see esp. Section 4.1.2). The data include all the items or elements that can and therefore must be topics of philosophical consideration. One could say, briefly, that the data include all philosophical *explananda,* although "*explanandum*" (or "*explicandum*") should not be taken in the narrow sense relevant to natural-scientific "explanation." Among the data are all the topics that are thematized in this book.

The data, in the philosophical sense presupposed here, are to be considered in both a *quantitative* and a *qualitative* respect. It is clear that one philosophical conception is superior to another if it considers (and thematizes) more data. This factor has great significance for systematically oriented philosophical theories, above all where the most important (the most central) data are concerned. For example, if a philosophical theory does not consider the datum this book calls *the semantic dimension,* it has a clearly lower degree of intelligibility and coherence than does an otherwise comparable theory that also thematizes this dimension. Similarly, a semantics that does not explicitly thematize ontology is less intelligible and coherent than a semantics that explicates the immanent reciprocity between semantics and ontology.

Equally important are the *qualitative* aspects of the data. These aspects concern the specific features of data of distinct qualitative sorts; this factor is not adequately taken into consideration or comprehended when the data are considered purely extensionally or quantitatively. Again, an example: as shown in part in Section 4.3.1, the data that are or must be treated by the philosophy of mind present a unique and revealing problem. The central data are *conscious phenomena*, no matter how they are described more precisely. A given subject "experiences" something, "undergoes" something, is convinced that something is the case, is conscious of its freedom, etc. These subject-related data cannot adequately be conceived of as objective in the scientific sense; they cannot be subsumed under a model that articulates (explains, etc.) purely "objective" data. One way or another, *all* data can be subsumed under *any* model. If a model is called "scientific," this classification suggests that the "explanation" it makes possible will be a scientific explanation. But this leaves unanswered the question whether the model in question actually considers or comprehends the *peculiarity* of the mental data. The latter requires the introduction of an appropriate conceptuality and the carrying through of a likewise appropriate analysis of these data. Precisely this point, the question concerning the appropriate conceptuality, is, as shown in Chapter 4, the central bone of contention in the contemporary philosophy of mind.

[ii] The second basic characteristic of greater intelligibility and coherence is the development of more determinate or more finely structured interconnections among the data. "Interconnection" is here taken in a broad sense, as designating any kind of relation, combination, etc. Data and interconnections are not entities or dimensions, etc., that exist somehow separately, each on their own; instead, interconnections are determinate ways in which the data themselves are constituted and therefore ways in which they are to be conceptualized. This "being conceptualized" is itself not something external to the data, but instead designates the inner constitution or, in traditional terms, the "essence" or the "nature" of a given datum itself. Greater intelligibility and coherence are attained when the thus-understood interconnections are conceptualized and presented in a more detailed, more finely grained, more precise manner. Thereby, the data are automatically comprehended or conceptualized with greater adequacy.

Throughout this book, this consideration is the guideline for the entire conception; it therefore serves as the measure for the book's success.

[7] The brief presentation of intratheoretical interphilosophical metasystematics provided above is quite abstract. It can be concretized only by means of the comparison of the philosophical theory presented here with one or more other philosophical theories. Such a comparison would show, *in each specific case*, what greater intelligibility and coherence would involve. This book thematizes such confrontations in various contexts, albeit usually with respect to specific issues. One task of metasystematics is to make explicit the theoretical framework within which a given such confrontation takes place.

6.3.2.2 External Intratheoretical Philosophical-Nonphilosophical Metasystematics

The metasystematics presented in this section thematizes the relationship of the structural-systematic philosophy to the non-philosophical sciences. This is a vast topic, aspects of which are treated at various places in this book. This section treats exclusively this metasystematic aspect *as such* or *abstractly*.

The quasi-definition of the structural-systematic philosophy indicates how this relationship is fundamentally to be understood: articulating the most general or universal structures of the unrestricted universe of discourse is clearly not the task of any natural or empirical science, nor that of any formal science; for this reason, the natural or empirical sciences can all be termed *particular* sciences. Their particularity becomes clear if one compares their theoretical frameworks with that of the system presented here; such comparison is in every respect the decisive means of determining the distinction between the two kinds of theoretical frameworks. To be sure, all the theoretical frameworks contain some of the same basic elements, such as a common (at least elementary) logic, semantics, etc., but the differences are far more important: they have different conceptualities, methodologies, etc. The distinction is most clearly visible if one considers particularly the constellation of topics and issues treated in Chapter 5, which presents a theory of being as such and as a whole. The distinction would disappear only if being as such and as a whole were simply identified with nature, as understood by the natural sciences. But if that identification is made, there arises the question of what philosophy is or could be. It is quite noteworthy that most analytic philosophers who presuppose this identification—usually implicitly—do not raise this compelling question.

A non-natural-scientific—thus genuinely philosophically oriented—theory of being as such and as a whole can develop only within a theoretical framework that rejects the identification of being with nature. This is clear from the fact that different fundamental concepts are used (e.g., the modalities, the distinction between the absolute and the contingent dimensions of being, the more radical concept of the beginning, etc.).

To be sure, there is another philosophical dimension that does not reveal the distinction as clearly. This is the dimension traditionally termed *special metaphysics* (*metaphysica specialis*), as significantly distinguished from the *general metaphysics* (*metaphysica generalis*) just considered. To the *special* structural-philosophical domain (in the sense just explained) belong disciplines like philosophy of mind, philosophy of world history, etc., as they are partially developed in Chapter 4. Here, the boundaries are not fully clear. Because the structures that articulate the theory of being as such and as a whole are clearly *universal* structures, they are designated as *absolutely universal structures*, in distinction from the structures whose investigation is the task of *special* structural-systematic philosophy. The latter are therefore designated *relatively universal* structures.

How are these structures clearly to be distinguished from those examined by the particular sciences? This book holds that there is no absolute criterion, none that clearly applies in every case. There are only criteria that are *relative* on the one hand to specific philosophical conceptions of the world, of mind, etc., and on the other to the situations of the particular sciences as they have developed at a given time.

636

As far as this book's conception of the world is concerned, the following should be illuminating: if a philosophy defends the conception that anything and everything is or can be explained *physicalistically*—as noted above, this is the view of most analytic philosophers—then one cannot see what more philosophy has to say. On the basis of what these philosophers themselves maintain (i.e., on the basis of the physicalist philosophies they champion), they must leave every explanation to the particular natural sciences, or they must identify everything that they present as putatively philosophical with the natural-scientific explanation. These philosophers thereby ascribe to the sciences this book designates as "particular" the characteristic of scientific universality. "Universal science" would then *not* mean a science existing alongside and in addition to the particular sciences; instead, it would be nothing other than what the particular sciences are, do, etc.

The situation is wholly different for any philosophical theory that rejects the thesis that phenomena such as those that concern the philosophy of mind can be adequately explained physicalistically. Excluding this thesis is possible (i.e, here: consequent and coherent) only if that theory accepts a broader, more comprehensive science—indeed, a universal science—that can itself only be identified with or understood as philosophy. The methodological starting point for all discussions in the domain of philosophy of mind is extremely revealing: the holistic position (in the terminology of the metaphysical tradition: the general metaphysical conception) that a given philosopher holds is decisive with respect both to how that philosopher sees the issues and to what the outcome of the conflict will be. The physical holist, for example, can proceed in no other manner than to use all means possible to include absolutely all phenomena in the physicalist model, despite the fact that this does violence to the phenomena.[7]

These reflections make clear how fundamentally important it is to have a clear view of these matters. That the philosopher whose orientation is not that of monistic physicalism recognizes the fundamental distinction between philosophy as universal science and the other sciences as particular sciences leads him or her directly to situate the articulation of those phenomena that cannot be explained within the theoretical frameworks of particular sciences in some other theoretical framework. But the only other available or indeed imaginable theoretical frameworks are *philosophical* frameworks. Philosophical frameworks are distinguished from those of particular sciences precisely in that they are universal, although with respect to certain topics, such as those in the philosophy of mind, philosophy's theoretical framework is *relatively* rather than *absolutely* universal.

This theoretical framework is relative in two respects.

[i] It makes possible the conceiving or explaining of specific phenomena such as mental phenomena *insofar as* these phenomena relate to the entire universe (however this may be designated: being, reality, world, nature, etc.), or more precisely, it makes possible the understanding of specific phenomena in their relationality to the entire universe. Differently stated: the articulation of such phenomena *in their specificity* requires the explicit identification of what distinguishes these phenomena or beings from other phenomena or beings. This identification can be accomplished only by

---

[7] There are also physicalistic "holists" who attempt to treat mental phenomena in putatively non-physicalistic manners; see the treatment of Searle in Section 4.3.1.2.3[3] (pp. 285–87).

situating these phenomena or beings within the entirety of the universe, the entirety of being as a whole.

Such a *relativity* in the sense of a *relationality to all other phenomena or beings* and thus to being as a whole is not something that can be thematized within the theoretical framework of any particular science. Yet there is a problem here, one that emerges directly from the preceding reflections. The formulation "the *specificity* of phenomena or beings *with respect to their relationality to all other phenomena or entities*" is somewhat underdetermined, indeed ambiguous. How is "the specificity" of these phenomena or entities itself to be seen? Can this specificity be understood without the explicit articulation of the relationality of the phenomena or entities to being as a whole? To answer this question in the affirmative would introduce an inconsistency into this book's account, because if the specificity of these phenomena or beings could be adequately thematized without the articulation of their situatedness within the universe (being as a whole), then their inclusion within the totality would be something fully external to them, something from which one could non-problematically abstract. Such a view would deprive the idea of a system of its genuine sense. One would be left instead with a radical atomism. Within the framework of the comprehensive conception presented in this book, such a view can have no place.

This thesis must be made more precise. It does not follow from it that without the explicit thematization of their relationality to being as a whole nothing important can be said about the phenomena or entities in question in their specificity; what follows is only that, in principle, no *complete or adequate* comprehension of them in their specificity is possible without the thematization of their holistic relationality. And that means that the always particular treatments of phenomena or beings in their specificity is precisely the area in which the particular sciences can and must make their fundamental contributions, contributions that, in their full scope, are to be acknowledged and appreciated by philosophy. The exact limits of these contributions cannot be fixed *a priori*; their locations depend decisively upon how the particular sciences—above all, physics—understand themselves, how they develop, and how philosophers use them.   639

[ii] The particular sciences do develop. This phenomenon cannot of course be considered here as such and as a whole. Only one aspect that concerns especially physics, the fundamental natural science, requires thematization at this point. Most analytic philosophers accept the natural sciences, and make this acceptance a dogma. A great many are of the opinion that all phenomena are physically explainable; they therefore advocate a wide-ranging reductionism. This opinion is based in a fundamental unclarity, indeed confusion, as the passage concerning John Searle (1995) in Chapter 4 (pp. 285–87) reveals.

Two aspects require brief explanation. First, one must be extremely careful in speaking of the (natural) sciences, and particularly of physics. Who can say how physics will develop? At present, this fundamental science can be characterized relatively precisely. The possibility can safely be excluded that the physics of the future could introduce and apply concepts and methods that would be adequate to phenomena of the mental, of thinking, etc. It can be excluded because *this* science cannot *fundamentally* change the concepts and methods it currently uses.

It appears, however, that the possibility of the following development cannot be excluded *a priori*: the term "physics" could in the future be used in such a way that it would have a far more extensive, indeed comprehensive, signification, such that it would be in a position to, for example, do full explicative justice to mental phenomena. This science (a or the *new* "physics")—in any case at first only with respect to the understanding and explanation of mental and similar phenomena—would no longer be distinct from philosophy.

Second, a clearly negative thesis must be introduced, one that articulates a clear 640 negative border. Every particular-scientific theoretical framework, *no matter how it is designated,* can have only a quite restricted significance for the philosophical theory of mind as long as it relies upon a conceptuality and methodology that are not appropriate for the dimension of the mental in its specificity.

6.3.3 *Extratheoretical Metasystematics*

The term "extratheoretical metasystematics" can be misleading; it does not indicate that this metasystematics itself has an extratheoretical status. To the contrary, extratheoretical metasystematics *theoretically* thematizes the relationship between the structural-systematic philosophy and the various non-theoretical activities, situations, and phenomena of the human world; among these belong the lifeworld in general, individual and social activities of all sorts, the "worlds" of politics, of economics, etc., and particularly that of culture. How is philosophy viewed in relation to these extratheoretical domains—and how should it be viewed? It is first necessary to clarify these questions, which appear only at first glance to be clear.

The current account is metasystematic and thus theoretical. Because no other theoretical activity has a more comprehensive character than philosophy, and because philosophy itself is the topic of this metasystematics, this is a philosophical metasystematics, albeit one that must develop within a theoretical framework that is not simply identical with the theoretical framework of the structural-systematic philosophy.

There are of course extraphilosophical and extratheoretical views concerning philosophy; these include the views of politicians, employers, workers, journalists, pedagogues, artists, therapists, indeed human beings of all sorts. Simplifying, one can speak of public opinions concerning philosophy. These various opinions are here considered because they are among the topics of extratheoretical metasystematics, but they do not provide the perspective from or within which metasystematics develops. That perspective is a philosophical one.

Extratheoretical metasystematics must consider two thematic aspects. The first is the question of how to evaluate the various extratheoretical opinions. It is a fact that philosophy (i.e., that which is designated by the word "philosophy") has changed 641 radically over the course of history; it is also a fact that public estimations of philosophy vary greatly from country to country. But no matter what public notions and evaluations of philosophy may in fact be held, it must be emphasized that in this book there is talk of philosophy only in a quite narrow and specific sense. Philosophy is here understood as a *specific theoretical or scientific activity or discipline* and what that activity or discipline produces: theories with the status of universality.

No public opinions concerning philosophy should be accepted in any way as definitive, much less as binding. Public opinion is quite volatile and problematic, and certainly not competent to produce a well-grounded determination or evaluation of a theoretical activity or discipline. To be sure, public opinion is often of importance, indeed of great importance, for philosophy; without the facilities and resources provided by the public (by non-philosophers), philosophy could not exist or continue as an academic discipline. This has, however, nothing to do with the intrinsic self-understanding of this discipline.

The second aspect to be considered here is the question of how philosophy, on the level of philosophical metasystematics, does or should or must view its relation to the extratheoretical world (in the sense described above). Philosophy is universal theorization; so conceived, it includes anything and everything within its subject matter—and thus the extratheoretical world. For this reason, it must include a theorization of its own relationship to that world: the relationship of the theoretical to the human world. This involves both a negative and a positive aspect.

Negatively, philosophy as theory must in no way be confused or commingled with practice in any form. This is particularly important in light of the peculiar demand that arises in the time following Hegel, particularly characterizing Marxism but continuing into the present, notably in the so-called Frankfurt School ("Critical Theory"): the demand for a or the *unity of theory and practice* (see Habermas [1963/1974, 1968/1972]). The demand is first made by the young Karl Marx, who writes (1845/2000: 173), in his *Theses on Feuerbach*, "Philosophers have only interpreted the world, in various ways; the point is to change it." The formulation is not unambiguous, but it is always understood as presenting, as the task of philosophy itself, the changing of the world. This also corresponds to the demand Marx makes elsewhere that philosophy, in its purely theoretical form, be overcome. As this book shows in detail, this demand is simply nonsensical. Theory is not practice. If one confuses the two, as has happened so often particularly (but not exclusively) in the history of the Marxist movement, one has neither theory nor practice, but pure confusion.

642

One can bring wishes to philosophy, make demands of it, and subject it to extratheoretical critique. But all of that makes sense only when one turns to philosophy as *theory*, recognizes it as theory. If one recognizes philosophy as theory, then its *positive* aspect is also visible, as is its relation to the extratheoretical dimension as explained above. As universal theory, philosophy can serve non-philosophers by clarifying the contexts within which human lives unfold. It can do this not only with respect to human life as a whole but also with respect to specific domains of human life, including the social, the ethical, the cultural, etc. Its "practical" function—if one wants to speak this way—consists in understanding and presenting itself *as theory*: it can decisively aid human beings and human societies in attaining *clarity* with respect to their involvements, great and small. Of late, this function is more than ever not only recognized but also used, particularly in relation to ethical and political issues.

6.4 Self-Determination, Metasystematics, and the Self-Grounding of the Structural-Systematic Philosophy

[1] This final section reconsiders the self-grounding of the structural-systematic philosophy. Three topics considered in this book are, in this respect, of decisive significance. The first concerns the thesis that the grounding—and thus also, in a sense clarified below, the self-grounding—of this book's theory is sensible and possible not at the outset, but only at the end, so to speak after the fact. The second concerns the fourth (and last)

stage of the method—that of confirmation or of determination of truth-status. The third concerns the concept of grounding and quite particularly the *systematic* concept of grounding. These three points provide the basis for clarifying the question of the self-grounding of the structural-systematic philosophical theory.

Because this theory is a universal science, it cannot be grounded in or by anything outside itself. For this reason, the grounding of this theory can only be a self-grounding. To avoid a possibly serious misinterpretation, however, it must be emphasized once again that this self-grounding is not anything like an *ultimate* grounding in the sense of a definitive or absolute self-establishment. This book rejects any and every such grounding as being out of the question; the notion that there could be such a grounding is based on a series of fundamental misunderstandings. The centrally relevant point is the fundamental fact that any and every philosophical theory (like any and every other scientific theory) always presupposes the specific theoretical framework within which alone it can develop. Because there is a plurality of theoretical frameworks that is in principle inexhaustible, it would be utter presumption to attempt in any way to take into account *all possible theoretical frameworks*, in order to present one's own framework as absolutely the best. And if the absolute, definitive theoretical framework cannot be identified, then ultimate grounding is not possible.

Chapter 1 develops the systematic grounding-process and its three forms or stages: the incipiently systematic, the innersystematic, and the metasystematic. These three stages are accomplished in part in Chapters 1–5. These chapters rely primarily on the innersystematic stage of grounding, as is fully consequent given that the chief aim there is to unfold the entire architectonic of the theory. It is also the case, however, that genuinely explicit sentences concerning the first two stages of systematic grounding have a metasystematic status and are therefore presented in this chapter (Section 6.2), along with the concept of the *self-determination* of this theory and in general under the title "immanent metasystematics."

The third stage, *metasystematic grounding*, is essentially identical with what this chapter presents as "external intratheoretical metasystematics" (Section 6.3). Here, the method reaches its fourth stage. As indicated already in Chapter 1, the test of truth-status has, for this theory, a quite specific form. It does not compare the theory with being (the universe, the world, reality, etc.); instead, it compares different theoretical frameworks and their concretizations in order to determine which is superior with respect to intelligibility and coherence. This is a test of *truth* only in a quite determinate sense: it is a matter of the truth that is given relative to the theoretical frameworks in question. There can be no talk of confirmation of the truth of the theory by anything like experience or experimentation. Instead, the relevant criterion is, as this chapter shows, greater intelligibility and coherence.

By means of the realization of the threefold systematic process of grounding, *the structural-systematic philosophy grounds itself*; this process of *philosophical self-grounding* is utterly distinct from any *foundationalist* form of the process of self-grounding.

[2] How then should this self-grounding be accomplished? What has already been said should make the answer obvious: the self-grounding cannot be accomplished at once or definitively; it is instead a *constant process* that could continue, in principle, as long as there are philosophical discussions, because it must involve the confrontation of the structural-systematic philosophy with other philosophical conceptions. The final

stage of self-grounding is this checking for confirmation in successive confrontations with other positions. There is no outside or higher authority, of any sort whatsoever, that could definitively confirm the truth of this or any other philosophical theory. Such confirmation must always be contingent.

Confrontations of this sort can in principle have three forms because the structural-systematic philosophy is strictly *holistically* oriented. One can speak of its periphery, of its core or center, and of domains between the two.[8] A confrontation at the periphery would be one that concerned only the *concretization* (*realization*) of the structural-systematic theoretical framework. Any changes that resulted would leave the theoretical framework and the conception as a whole, strictly speaking, *intact*. Confrontations of this sort are frequent in philosophical discussions.

Weightier confrontations occur in the intermediate domain; they concern parts or, in the extreme case, the whole of the inner structuration of the theory. A particularly important example in this book involves one of the decisive components of the internal domain: its non-compositional semantics (and the corresponding non-compositional ontology). Discussions about such intermediate issues constitute by far the most important parts of those philosophical discussions that are the most frequent and the most intense. Concerning this matter, the conception presented in this book shows at least the following: discussions about such topics can proceed sensibly only if careful attention is paid both to the presuppositions and to the comprehensive implications of the positions taken. Stated differently and more briefly: only *strictly systematic* considerations and ways of proceeding are sensible and appropriate.

If confrontations led to significant changes in this intermediate domain, then an important constituent of the conception presented here would be considered to have been refuted. A significant change would be one affecting an entire subdiscipline (e.g., the entirety of the semantics defended here). Such a change would not refute the basic structural-systematic idea—described shortly below—but it would reveal the untenability of the concretization of that basic idea presented in this book.

Finally, there can be confrontations concerning the absolutely central domain. These would concern, to use normal philosophical jargon, the basic idea of the structural-systematic conception. This idea involves two basic theses, each of which both conditions and entails the other; because they are presented and explained in many passages in this book, here they need only be mentioned. The first basic thesis concerns the absolutely central and unavoidable role accorded to the concept of the theoretical framework, which brings with it the assumption of a plurality of such frameworks, some actually existing and others possible in principle. The second basic thesis articulates both the distinction and the interconnection between the two central coordinates on which the theory's basic conception rests and that attain concrete form within its theoretical framework; these are the two basal dimensions, the dimensions of fundamental structures (formal, semantic, and ontological structures) and of the comprehensive datum ("world," "universe," "being" etc.).

If this distinction and this interconnection between the two dimensions should prove to be untenable, then the basic idea behind the structural-systematic philosophy

---

[8] Quine introduces the imagery of periphery and center or core. See esp. his essay "Two Dogmas of Empiricism" (1953/1979).

would also be untenable. Demonstrating this untenability could of course be achieved only by a philosophy that, with respect to what it achieved with its explanations, was not inferior to the structural-systematic philosophy. Its explicative achievements would concern all the methodological, logical, semantic, ontological, and comprehensively systematic concepts and lines of thinking that are guiding and determinative for the structural-systematic philosophy. The demonstration would have to show how these concepts and lines of thinking could be treated differently and better on the basis of a different and better theoretical framework, and then would have to succeed in so treating them. An undertaking that failed to do justice to this task could not be considered to overcome the structural-systematic philosophy, even if it were designated as "philosophical."

If such a demonstration could meet all the demands just posed, it would establish that the structural-systematic philosophy, even with respect to its basic idea, would not be the most adequate theoretical framework for a systematic philosophy. It would then have been a stage in the process of the development of ever more adequate theoretical frameworks within which are accomplished the understanding and articulation of the subject matter of thinking, and thus of that of philosophy.

Works Cited

Aaron, Richard I. 1971. *Knowing and the Function of Reason*. London: Oxford UP.

Albert, Hans. 1968. *Traktat über kritische Vernunft*. Tübingen: Mohr Siebeck.

Almeder, Robert. 1994. Defining Justification and Naturalizing Epistemology. *Philosophy and Phenomenological Research* 54: 669–681.

Apel, Karl-Otto. 1973/1980. *Transformation der Philosophie*. Vol. II: *Das Apriori der Kommunikationsgemeinschaft*. Frankfurt am Main: Suhrkamp. English: *Transformation of Philosophy*. Translated by Glyn Adey and David Frisby. London, Boston: Routledge & Kegan Paul, 1980.

————. 1990. *Diskurs und Verantwortung*. Frankfurt am Main: Suhrkamp.

Aristotle. *Categ. Categoriae*. In *Topica et Sophistici elenchi*. Edited by W. D. Ross. Oxford: Clarendon, 1956.

————. *De Anima*. Edited by W. D. Ross. Oxford: Clarendon, 1956.

————. *Metaph. Metaphysica*. Edited by W. D. Ross. Oxford: Clarendon, 1957.

————. *Politica*. Edited by W. D. Ross. Oxford: Clarendon, 1957.

Bacon, John. 1995. *Universals and Property Instances: The Alphabet of Being*. Blackwell: Oxford.

Ballarin, Roberta. 2004. The Interpretation of Necessity and the Necessity of Interpretation. *The Journal of Philosophy* 101: 609–638.

Baldwin, Thomas. 1991. The Identity Theory of Truth. *Mind* 100: 35–52.

Balzer, Wolfgang. 1986. Theoretical Terms: A New Perspective. *Journal of Philosophy* 83: 71–90.

————, Moulines, C. Ulises, and Sneed, Joseph D. 1987. *An Architectonic for Science. The Structuralist Program*. Second edition. Dordrecht: Reidel.

————, and Moulines, C. U. (eds.). 1996. *Structuralist Theory of Science. Focal Issues, New Results*. Berlin, New York: de Gruyter.

Barwise, Jon, and Perry, John. 1981. Semantic Innocence and Uncompromising Situations. In Peter A. French, Theodore E. Uehling, Jr., and Howard K. Wettstein (eds.), *Foundations of Analytic Philosophy. Midwest Studies in Philosophy*. Vol. VI. Minneapolis: U of Minnesota Press, 387–404.

Baumgartner, Hans M. 1972. *Kontinuität und Geschichte. Zur Kritik und Metakritik der historischen Vernunft*. Frankfurt am Main: Suhrkamp.

————, and Anacker, Ulrich. 1973. Geschichte. In Hermann Krings, Hans M. Baumgartner and Christoph Wild (eds.), *Handbuch philosophischer Grundbegriffe.* Munich: Kösel. I: 547–557.

Beall, J. C. 2000. On Mixed Inferences and Pluralism About Truth Predicates. *The Philosophical Quarterly* 50: 380–382.

Beaney, Michael (ed.). 1997. *The Frege Reader.* Oxford: Blackwell.

Beaufret, Jean. 1968. *L'endurance de la pensée. Pour saluer Jean Beaufret.* Paris: Plon.

Bell, John, and Machover, Moshe. 1977. *A Course in Mathematical Logic.* Amsterdam, New York, Oxford: North-Holland.

Belnap, Nuel. 1993. On Rigorous Definitions. *Philosophical Studies* 72: 115–146.

————, and Gupta, Anil. 1993. *The Revision Theory of Truth.* Cambridge (MA), London: The MIT Press.

Bien, Günther. 1971. Philosophy, I.C. Aristotle. In Ritter et al., Volume 7: columns 583–590.

Bigelow, John. 1996. God and the New Math. *Philosophical Studies* 82: 127–154.

Blau, Ulrich. 1978. *Die dreiwertige Logik der Sprache.* Berlin, New York: de Gruyter.

Bourbaki, Nicolas. 1950. The Architecture of Mathematics. Translated by Arnold Dresden. *American Mathematical Monthly* 67: 221–232.

Brandom, Robert. 1994. *Making It Explicit. Reasoning, Representing, and Discursive Commitment.* Cambridge (MA), London: Harvard UP.

Brink, David O. 1989. *Moral Realism and the Foundations of Ethics.* Cambridge: Cambridge UP.

Bunnin, N, and Tsui-James, E. P. (eds.). 1996. *The Blackwell Companion to Philosophy.* Oxford: Blackwell.

Campbell, Keith. 1990. *Abstract Particulars.* Oxford: Blackwell.

Candlish, Stewart. 1989. The Truth About F. H. Bradley. *Mind* 98: 331–348.

Cantor, Georg. 1859–1897. *Beiträge zur Begründung der transfiniten Mengenlehre. Mathematische Annalen* 46: 481–512; 49: 207–246.

Carnap, Rudolf. 1950/1956. Empiricism, Semantics, and Ontology. Included as Supplement A in Rudolf Carnap, *Meaning and Necessity. A Study in Semantics and Modal Logic.* Enlarged Edition. Chicago, London: Chicago UP. The essay originally appeared in *Revue Internationale de Philosophie* 4, 1950: 20–40.

————. 1958. *Introduction to Symbolic Logic and Its Applications.* New York: Dover.

————. 1962. On Explication. In Rudolf Carnap, *Logical Foundations of Probability.* Chicago: Chicago UP: 1–18.

Carter, Brandon. 1974/1990. Large Number Coincidences and the Anthropic Principle in Cosmology. In John Leslie (ed.), *Physical Cosmology and Philosophy.* New York: Macmillan, 1990: 125–133. The essay originally appeared in M. S. Longair (ed.) *Confrontation of Cosmological Theories with Observational Data.* International Astronomical Union. Dordrecht: Reidel, 291–298.

Cartwright, Richard. 1994. Speaking of Everything. *Noûs* 28: 1–20.

Chalmers, David. 1996. *The Conscious Mind.* Oxford, New York: Oxford UP.

Chisholm, Roderick. 1996. *A Realistic Theory of Categories. An Essay on Ontology.* Cambridge (UK): Cambridge UP.

Corry, Leo. 1992. Nicolas Bourbaki and the Concept of Mathematical Structure. *Synthese* 92: 315–348.

Daly, Chris. 1994. Tropes. *Proceedings of the Aristotelian Society.* New Series 94: 253–261.

Danto, Arthur. 1986. *The Philosophical Disenfranchisement of Art.* New York: Columbia UP.

Davidson, Donald. 1965/1984. Theories of Meaning and Learnable Languages. In Davidson 1984, 3–16.

———. 1974/1984. On the Very Idea of a Conceptual Scheme. In Davidson 1984, 183–198.

———. 1977/1984. Reality Without Reference. In Davidson 1984, 215–226.

———. 1984. *Inquiries into Truth and Interpretation.* Oxford: Clarendon.

———. 1985. Reply to Quine on Events. In Lepore and McLaughlin 1985, 172–176.

Denkel, Arda. 1996. *Object and Property.* Cambridge (UK): Cambridge UP.

Diederich, Werner. 1996. Structuralism Within the Model-Theoretical Approach. In Balzer and Moulines 1996, 15–21.

Dipert, Randall R. 1997. The Mathematical Structure of the World: The World as Graph. *The Journal of Philosophy* 94: 329–359.

Dodd, Julian. 1995. McDowell and Identity Theories of Truth. *Analysis* 55: 160–165.

———. 1999. Farewell to States of Affairs. *Australasian Journal of Philosophy* 77: 146–160.

———, and Hornsby, Jennifer. 1992. The Identity Theory of Truth: A Reply to Baldwin. *Mind* 101: 319–322.

Dorschel, Andreas; Kettner, Matthias; Kuhlmann, Wolfgang; and Niquet, Marcel (eds.). 1993. *Transzendentalpragmatik. Ein Symposion für K.-O. Apel.* Frankfurt am Main: Suhrkamp.

Dowty, David R; Wall, Robert E.; and Peters, Stanley. 1981. *Introduction to Montague Semantics.* Dordrecht, Boston, London: Reidel.

Dummett, Michael. 1977/1978. Can Analytic Philosophy Be Systematic, and Ought It to Be? *Hegel-Studien Beiheft* 17, 1977, 305–326; cited as reprinted in Dummett 1978, 437–456.

———. 1978. *Truth and Other Enigmas.* Cambridge (MA): Harvard UP.

———. 1984. An Unsuccessful Dig. In Crispin Wright (ed.), *Frege. Tradition and Influence.* Oxford: Blackwell, 195–220.

Ebbinghaus, Heinz-Dieter. 1977/2003. *Einführung in die Mengenlehre.* 4th ed. Heidelberg, Berlin: Spektrum Akademischer Verlag.

Engels, Friedrich. 1873/1940. *Dialectics of Nature.* Translated and edited by Clemens Dutt. London: Lawrence and Wishart.

Findlay, John. N. 1933. *Meinong's Theory of Objects and Values.* Oxford: Oxford UP.

Forge, John. 2002. Reflections on Structuralism and Scientific Explanation. *Synthese* 130: 109–121.

Forrest, Peter. 1996. *God Without the Supernatural. A Defense of Scientific Theism.* Ithaca, London: Cornell UP.

Frank, Manfred. 1991a. *Selbstbewusstsein und Selbsterkenntnis. Essays zur analytischen Philosophie der Subjektivität.* Stuttgart: Reclam.

———. 1991b. *Selbstbewusstseinstheorien von Fichte bis Sartre*. Edited and with an Afterword by Manfred Frank. Frankfurt am Main: Suhrkamp.

Frege, Gottlob. All cited English page number/German page number (German sources noted in Beaney (1997), German page numbers provided in margins by Beaney).

———. 1884. *The Foundations of Arithmetic*. Selections in Beaney 1997, 84–129.

———. 1891a. Function and Concept. In Beaney 1997, 130–148.

———. 1891b. Letter to Husserl, 24.5.1891. In Beaney 1997, 149–150.

———. 1892a. On *Sinn* and *Bedeutung*. In Beaney 1997, 151–171.

———. 1892b. [Comments on *Sinn* and *Bedeutung*]. In Beaney 1997, 172–180.

———. 1892c. On Concept and Object. In Beaney 1997, 181–193.

———. 1897. Logic. In Beaney 1997, 227–250.

———. 1918. Thought. In Beaney 1997, 325–345.

———. 1979. Logic in Mathematics. In Gottlob Frege, *Posthumous Writings*. Chicago: U of Chicago Press.

Fukuyama, Francis. 1992. *The End of History and the Last Man*. New York: The Free Press.

Gabbay, Dov, and Guenthner, Franz (eds.). 1983. *Handbook of Philosophical Logic*. Vol. I: *Elements of Classical Logic*. Dordrecht, Boston, London: Kluwer.

———. 1986. *Handbook of Philosophical Logic*. Vol. III: *Alternatives to Classical Logic*. Dordrecht, Boston, London: Kluwer.

Gadamer, Hans-Georg. 1960/2003. *Truth and Method*. Translation revised by J. Weinsheimer and D. G. Marshall. New York: Continuum. 2003. German: *Gesammelte Werke*. Vol. 1. Tübingen: Mohr. 1986. The Preface to the second German edition is in Vol. 2. Cited English page/German page.

Gallop, David. 1984. *Parmenides of Elea*. Toronto, Buffalo, London: U of Toronto Press.

Gardner, Sebastian. 1995. Aesthetics. In Grayling 1995, 583–627.

Gettier, Edmund. 1963. Is Justified True Belief Knowledge? *Analysis* 23: 121–123.

Gilson, Étienne. 1948. *L'être et l'essence*. Paris: Vrin.

Goldblatt, Robert. 1979/1984. *TOPOI. The Categorial Analysis of Logic*. Amsterdam, New York, Oxford: North Holland. Revised edition.

Goodman, Nelson. 1978. *Ways of Worldmaking*. Hassocks: Harvester.

Grayling, A. C. (ed.). 1995. *Philosophy. A Guide Through the Subject*. Oxford: Oxford UP.

Greimann, Dirk. 2000. Die ontologischen Verpflichtungen der normativen Ethik. In Peres and Greimann 2000, 123–143.

Grim, Patrick. 1991. *The Incomplete Universe: Totality, Knowledge, and Truth*. Cambridge (MA), London: The MIT Press.

Grover, Dorothy; Camp, Joseph L. Jr.; and Belnap, Nuel. 1975. A Prosentential Theory of Truth. *Philosophical Studies* 27: 73–125.

Grünbaum, Adolf. 2000. A New Critique of Theological Interpretations of Physical Cosmology. *British Journal for the Philosophy of Science* 51: 1–43.

Habermas, Jürgen. 1963/1974. *Theory and Practice*. Translated by John Viertel. Boston: Beacon (1974). German: *Theorie und Praxis*. 4th ed. Frankfurt am Main: Suhrkamp (1971). Cited English page/German page.

———. 1968/1972. *Knowledge and Human Interests*. Translated by Jeremy Shapiro. Boston: Beacon (1972).

———. 1973. *Wahrheitstheorien*. In Helmut Fahrenbach (ed.), *Wirklichkeit und Reflexion. Festschrift für W. Schulz*. Pfullingen: Neske, 211–265.

————. 1999. *Wahrheit und Rechtfertigung. Philosophische Aufsätze.* Frankfurt am Main: Suhrkamp.

Hahn, Lewis E., and Schilpp, Paul A. (eds.). 1986. *The Philosophy of W. V. Quine.* The Library of Living Philosophers, Vol. XVIII. La Salle (IL): Open Court.

Hales, Steven. 1997. A Consistent Relativism. *Mind* 106: 33–52.

Hanna, Joseph F. 1968. An Explication of "Explication." *Philosophy of Science* 35: 28–44.

Hawking, Stephen. 1988. *A Brief History of Time. From the Big Bang to Black Holes.* New York: Bantam.

————. 2001. *The Universe in a Nutshell.* New York: Bantam.

Hegel, Georg Wilhelm Friedrich. *Aesthetics.* Hegel, *Aesthetics: Lectures on Fine Art.* Translated by T. M. Knox. Two volumes. Oxford: Clarendon, 1975. German edition cited: Hegel, *TW*: 13–15. Cited English page/German page.

————. *Enc. Logic.* Hegel, *The Encyclopedia Logic.* Translated by Theodore Geraets, Wal Suchting, and Henry S. Harris. Indianapolis: Hackett, 1991. German edition: Hegel, *TW*: 8. Cited by section number.

————. *History.* Hegel, *Lectures on the Philosophy of World History. Introduction: Reason in History.* Translated by H. B. Nisbet. Cambridge (UK), London, New York, Melbourne: Cambridge UP, 1975. German: *Die Vernunft in der Geschichte.* Edited by Johannes Hoffmeister. 5th edition. Hamburg: Felix Meiner, 1955. Cited English page/German page.

————. *Logic.* Hegel, *Hegel's Science of Logic.* Translated by Arnold V. Miller. London: Allen & Unwin, New York: Humanities, 1969. German edition cited: Hegel, *TW*: 5–6. Cited English page/German page.

————. *TW.* Hegel, *Theoriewerkausgabe.* Twenty volumes. Frankfurt am Main: Suhrkamp, 1970.

Heidegger, Martin. *GA. Gesamtausgabe.* Frankfurt am Main: Klostermann. 1976ff.

————. 1927/1962. *Being and Time.* Translated by John Macquarrie and Edward Robinson. New York: Harper. German: *GA*: 2. Cited by section number.

————. 1927b/1962. Letter to Husserl 22 October 1927. In *Husserliana* IX (1962): 602.

————. 1936–38/1999. *Contributions to Philosophy (From Enowning).* Translated by Parvis Emad and Kenneth Maly. Bloomington and Indianapolis: Indiana UP. German: *GA*: 65. Cited English page/German page.

————. 1946/1998. Letter on Humanism. In *Pathmarks.* Edited by William McNeill. Cambridge (UK): Cambridge UP, 239–276. German: *GA*: 9, 313/364. Cited English page/German page.

————. 1956/1958. English: *What Is Philosophy?* Translated by William Kluback and Jean T. Wilde. New York: Twayne. German: *Was ist das–die Philosophie?* Pfullingen: Neske.

————. 1957/1969. *Identity and Difference.* Translated by Joan Stambaugh. New York: Harper and Row. German: M. Heidegger, *Identität und Differenz.* Pfullingen: Neske. Cited English page/German page.

————. 1969/1972. *Of Time and Being.* Translated by Joan Stambaugh. New York: Harper and Row. German: *Zur Sache des Denkens.* Tübingen: Niemeyer. Cited English page/German page.

Henrich, Dieter. 1970. *Selbstbewusstsein–Kritische Einleitung in eine Theorie*. In Rüdiger Bubner et al. (eds.), *Hermeneutik und Dialektik*. Tübingen: Mohr Siebeck. I:257–284.

———. 1986. *Selbstbewusstsein–ein Problemfeld mit offenen Grenzen*. In *Berichte aus der Forschung*, Nr. 68, Ludwig-Maximilians-Universität Munich, April: 2–8.

Hilbert, David. 1899/1902. *The Foundations of Geometry*. Translated by E. J. Townsend. Chicago: Open Court.

———. 1922/1935. *Neubegründung der Mathematik. Erste Mitteilung*. In *Abhandlungen aus dem Mathematischen Seminar der Hamburger Universität*. Vol. 1, 157–177. In *Gesammelte Abhandlungen*. Vol. 3. Berlin: Springer, 157–177.

Hodges, Wilfrid. 1983. *Elementary Predicate Logic*. In Gabbay and Guenthner 1983: 1–131.

———. 1993. *Model Theory*. Cambridge (UK): Cambridge UP.

Hoffman, Joshua, and Rosenkrantz, Gary S. 1994. *Substance Among Other Categories*. Cambridge: Cambridge UP.

Hooker, Brad (ed.). 1996. *Truth in Ethics*. Oxford: Blackwell.

Hornsby, Jennifer. 2001. Truth: The Identity Theory. In Michael. P. Lynch (ed.), *The Nature of Truth: Classic and Contemporary Perspectives*. Cambridge (MA), London: The MIT Press, 663–681.

Horwich, Paul. 1998. *Truth*. 2nd ed. Oxford: Oxford UP.

Hughes, George. E., and Cresswell, M. J. 1996. *A New Introduction to Modal Logic*. London, New York: Routledge.

Hugly, Philip, and Sayward, Charles. 1983. Can a Language Have Indenumerably Many Expressions? *History and Philosophy of Logic* 4: 73–82.

———. 1986. What Is an Infinite Expression? *Philosophia* 16(1): 45–60.

Jackson, Frank. 1982. Epiphenomenal Qualia. *Philosophical Quarterly* 32: 127–136.

———. 1986. What Mary Didn't Know. *Journal of Philosophy* 83: 291–295.

James, William. 1907. *Pragmatism. A New Name for Some Old Ways of Thinking*. New York, London, Bombay, and Calcutta: Longman Green.

Kant, Immanuel. *CPR. Critique of Pure Reason*. Translated by Paul Guyer and Allen Wood. Cambridge (UK), New York: Cambridge UP, 1998. Following long-standing practice, "A" indicates the first (1781) edition, "B" the second (1787). German: *Werke*, Vol. III.

———. *CJ. Critique of the Power of Judgment*. Translated by Paul Guyer and Eric Matthews. Cambridge (UK), New York: Cambridge UP, 2000. German: *Werke*, Vol. V. Cited by German page numbers, which are indicated in the margins of the English edition.

———. *History*. Idea for a Universal History from a Cosmopolitan Point of View. Translated by Lewis White Beck. In Immanuel Kant, *On History*. Edited, with an introduction, by Lewis White Beck. Indianapolis, New York: Bobbs Merrill, 11–26. German: in Kant, *Werke*, Vol. VIII, 15–31. Cited English page/German page.

———. *Proleg. Prolegomena to Any Future Metaphysics That Will Be Able to Come Forward as Science*. Translated and edited by Gary Hatfield. Revised edition. Cambridge (UK), New York: Cambridge UP, 2004. German: Kant *Werke*, Vol. IV. Cited by section number.

———. *Werke. Akademie-Textausgabe*. Berlin: de Gruyter, 1968ff.

Keller, Albert. 1968. *Sein oder Existenz? Die Auslegung des Seins bei Thomas von Aquin in der heutigen Scholastik*. Munich: Huebner.

Kelley, John L. 1955. *General Topology*. Toronto, New York, London: Van Nostrand. Appendix: Elementary Set Theory.

Ketchum, Richard J. 1991. The Paradox of Epistemology: A Defense of Naturalism. *Philosophical Studies* 62: 45–66.

Kirkham, Richard L. 1992. *Theories of Truth. A Critical Introduction*. Cambridge (MA): The MIT Press.

Kleene, Stephen C. 1952/1974. *Introduction to Metamathematics*. 7th ed. Groningen: Wolters-Nordhoff; Amsterdam, Oxford: North Holland; New York: American Elsevier Publishing Company.

Kneale, William, and Kneale, Martha. 1962/1991. *The Development of Logic*. 11th ed. Oxford: Clarendon.

König, Gert, and Pulte, Helmut. 1998. Theorie. In Ritter et al. 10: columns 1128–1154.

Koslow, Arnold. 1992. *A Structuralist Theory of Logic*. Cambridge (UK), New York: Cambridge UP.

Kuipers, Theo. 2000. *From Instrumentalism to Constructive Realism. On Some Relations between Confirmation, Empirical Progress, and Truth Approximation*. Dordrecht, Boston, London: Kluwer.

———. 2001. *Structure in Science. Heuristic Patterns Based on Cognitive Structures*. Dordrecht, Boston, London: Kluwer.

Kutschera, Franz v. 1982a. *Erkenntnistheorie*. Berlin, New York: de Gruyter.

———. 1982b. *Grundlagen der Ethik*. Berlin, New York: de Gruyter.

———. 1991. *Vernunft und Glaube*. Berlin, New York: de Gruyter.

———. 1998. *Die Teile der Philosophie und das Ganze der Wirklichkeit*. Berlin, New York: de Gruyter.

Landmesser, Cristof. 1999. *Wahrheit als Grundbegriff neutestamentlicher Wissenschaft*. Tübingen: Mohr Siebeck.

Legenhausen, Gary. 1985. New Semantics for the Lower Predicate Calculus. *Logique et Analyse* 28: 317–339.

Lepore, Ernest, and McLaughlin, Brian P. (eds.). 1985. *Actions and Events. Perspectives on the Philosophy of Donald Davidson*. Oxford: Blackwell.

Lewis, David. 1986. *On the Plurality of Worlds*. Oxford, New York: Blackwell.

———. 1991. *Parts of Classes*. Oxford: Blackwell.

Link, Godehard. 1979. *Montague-Grammatik. Die logischen Grundlagen*. Munich: Fink.

Loux, Michael J. 1998. *Metaphysics. A Contemporary Introduction*. London, New York: Routledge.

Lowe. E. J. 1998. *The Possibility of Metaphysics. Substance, Identity, and Time*. Oxford: Clarendon.

Ludwig, Günther. 1978. *Die Grundstrukturen einer physikalischen Theorie*. Berlin, Heidelberg, New York: Springer.

Mackie, John L. 1977. *Ethics. Inventing Right and Wrong*. Harmondsworth: Penguin.

MacLane, Saunders, and Moerdijk, Ieke. 1992. *Sheaves in Geometry and Logic. A First Introduction to Topos Theory*. Berlin, New York: Springer.

Marchal, J. H. 1975. On the Concept of a System. *Philosophy of Science* 42: 448–468.

Margolis, Eric, and Laurence, Stephen (eds.). 1999. *Concepts: Core Readings*. Cambridge (MA): The MIT Press.

Marx, Karl. 1845/2000. Theses on Feuerbach. In *Karl Marx. Selected Writings*. Edited by David McLellan. Oxford: Oxford UP.

McCarthy, Thomas. 1978. *The Critical Theory of Jürgen Habermas*. Cambridge (MA): The MIT Press.

McDowell, John. 1994/1996. *Mind and World*. Third printing with a New Introduction. Cambridge (MA), London: Harvard UP.

McGee, Vann. 2000. Everything. In Gila Sher and Richard Tieszen (eds.), *Between Logic and Intuition*. New York, Cambridge (UK): Cambridge UP, 54–78.

McGinn, Colin. 1997. *Ethics, Evil, and Fiction*. Oxford: Oxford UP.

———. 2003. Isn't It The Truth? *The New York Review of Books*, April 10, 70–73.

Mellor, Dov H. 1995. *The Facts of Causation*. London, New York: Routledge.

Michel, D. 1968. 'ÄMÄT. *Untersuchungen über, Wahrheit' im Hebräischen. Archiv für Begriffsgeschichte* 12: 30–57.

Miller, David. 1964. Popper's Qualitative Theory of Verisimilitude. *The British Journal for the Philosophy of Science* 25: 166–177.

Montague, Richard. 1974/1979. *Formal Philosophy. Selected Papers of Richard Montague*. Edited and with an Introduction by Richmond. H. Thomason. 3rd ed. New Haven, London: Yale UP.

Morgan, August de. 1847. *Formal Logic: or, The Calculus of Inference, Necessary and Probable*. London.

Mormann, Thomas. 1995. Trope Sheaves. A Topological Ontology of Tropes. *Logic and Logical Philosophy* 3: 129–150.

———. 1996. Topology for Philosophers. *The Monist* 79: 76–88.

Moser, P. K. 1984. Types, Tokens, and Propositions. *Philosophy and Phenomenological Research* 44: 361–375.

Moulines, C. Ulises. 1996. Structuralism: The Basic Ideas. In Balzer and Moulines 1996, 5–9.

———. 2002. Introduction: Structuralism as a Program for Modeling Theoretical Science. *Synthese* 130: 1–11.

Nagel, Thomas. 1974. What Is It Like to Be a Bat? *Philosophical Review* 83: 435–450.

———. 1987. *What Does It All Mean? A Very Short Introduction to Philosophy*. Oxford, New York: Oxford UP.

———. 1997. *The Last Word*. New York: Oxford UP.

Narlikar, Jayant V. 1977. *The Structure of the Universe*. Oxford: Oxford UP.

Neale, Stephen. 2001. *Facing Facts*. Oxford: Clarendon.

Nelson, Leonard. 1908. *Über das sogenannte Erkenntnisproblem*. Special edition of *Abhandlungen der Fries'schen Schule, Neue Folge*. Vol. 2, No. 3: 413–818. Göttingen.

Nietzsche, Friedrich. BGE. *Beyond Good and Evil*. In F. Nietzsche. *Basic Writings*. Translated and edited, with commentaries, by Walter Kaufmann. New York: The Modern Library, 1968. Cited by section number.

Niiniluoto, I. 1987. *Truthlikeness*. Dordrecht, Boston: Riedel.

O'Leary-Hawthorne, John, and Cover, Jan A. 1998. A World of Universals. *Philosophical Studies* 91: 205–219.

Otto, Rudolf. 1917/1958. *The Idea of the Holy: An Inquiry into the Non-rational Factor in the Idea of the Divine and its Relation to the Rational.* Translated by John W. Harvey. London, New York: Oxford UP.

Oxford English Dictionary. OED. Oxford: Clarendon, 1989.

Papineau, David. 1996. Philosophy of Science. In Bunnin and Tusi-James 1996, 290–324.

Parker, Sybil P., editor in chief. 1993. *McGraw-Hill Encyclopedia of Physics.* 2nd ed. New York: McGraw-Hill.

Parsons, Charles. 1986. Quine on the Philosophy of Mathematics. In Hahn and Schillpp 1986: 369–395.

Peacocke, Christopher. 1995. *A Study of Concepts.* Cambridge (MA), London: The MIT Press.

Peirce, Charles S. 1877/1965. The Fixation of Belief. In Peirce (1965), V:232–247.

———. 1878/1965. How To Make Our Ideas Clear. In Peirce (1965), V:248–271.

———. 1965. *Collected Papers of Charles Sanders Peirce.* Edited by Charles Hartshorne and Paul Weiss. Volumes V and VI. Two volumes in one. Cambridge (MA): The Belknap Press of Harvard UP.

Penrose, Roger. 1994. *Shadows of the Mind. A Search for the Missing Science of Consciousness.* Oxford, New York: Oxford UP.

Peres, Constanze. 2000. *Schönheit als ontosemantische Konstellation.* In Peres and Greimann (2000), 144–173.

———, and Greimann, Dirk (eds.). 2000. *Wahrheit–Sein–Struktur. Auseinandersetzungen mit Metaphysik. Festschrift für L. B. Puntel.* Hildesheim, Zürich: Olms Verlag, 144–173.

Plantinga, Alvin. 1993a. *Warrant: The Current Debate.* New York, Oxford: Oxford UP.

———. 1993b. *Warrant and Proper Function.* New York, Oxford: Oxford UP.

———, and Grim, P. 1993. Truth, Omniscience, and Cantorian Arguments: An Exchange. *Philosophical Studies* 71: 267–306.

Pompa, Leon. 1996. Philosophy of History. In Bunnin and Tsui-James 1996, 414–442.

Popper, Karl. 1972. *Objective Knowledge.* Oxford: Clarendon.

Puntel, Lorenz. B. 1983. *Transzendentaler und absoluter Idealismus.* In Dieter Henrich (ed.). *Kant oder Hegel? Über Formen der Begründung in der Philosophie.* Stuttgarter Hegel-Kongress 1981. Stuttgart: Klett-Cotta, 198–229.

——— (ed.). 1987. *Der Wahrheitsbegriff. Neue Erklärungsversuche.* Darmstadt: Wissenschaftliche Buchgesellschaft.

———. 1990. *Grundlagen einer Theorie der Wahrheit.* Berlin, New York: de Gruyter.

———. 1994. *Zur Situation der deutschen Philosophie der Gegenwart. Eine kritische Betrachtung. Information Philosophie* 22: 20–30.

———. 1995. *Der Wahrheitsbegriff in Philosophie und Theologie. Zeitschrift für Theologie und Kirche, Beiheft* 9, 16–45.

———. 1996. *Läßt sich der Begriff der Dialektik klären? Journal for General Philosophy of Science* 27: 131–165.

———. 1997. *Metaphysik bei Carnap und Heidegger: Analyse, Vergleich, Kritik. LOGOS,* Neue Folge 4(4): 294–332.

———. 1999a. On the Logical Positivists' Theory of Truth: The Fundamental Problem and a New Perspective. *Journal for General Philosophy of Science* 30: 101–130.

————. 1999b. "The Identity Theory of Truth": Semantic and Ontological Aspects. In Julian Nida-Rümelin (ed.), *Rationality, Realism, Revision. Perspectives in Analytical Philosophy*. Berlin, New York: de Gruyter. Vol. 23: 351–358.

————. 2001. Truth, Non-Compositionality, and Ontology. *Synthese* 176: 221–259.

————. 2002. The Concept of Ontological Category: A New Approach. In Richard Gale (ed.), *The Blackwell Guide to Metaphysics*. Oxford: Blackwell, 110–130.

————. 2004. *Der Wahrheitsbegriff in der Ethik*. In Franz-Joseph Bormann and Christian Schröer (eds.), *Abwägende Vernunft. Praktische Rationalität in historischer, systematischer und religionsphilosophischer Perspektive*. Berlin, New York: de Gruyter, 299–328.

————. 2007. *Auf der Suche nach dem Gegenstand und dem Theoriestatus der Philosophie. Philosophiegeschichtlich-kritische Studien*. Tübingen: Mohr Siebeck.

Putnam, Hilary. 1962. What Theories Are Not. In Ernst Nagel, Patrick Suppes, and Alfred Tarski (eds.). *Logic, Methodology, and Philosophy of Science: Proceedings of the 1960 International Congress*. Stanford: Stanford UP, 240–251.

————. 1980/1983. Models and Reality. In Paul Benacerraf and Hilary Putnam (eds.), *Philosophy of Mathematics. Selected Readings*. 2nd ed. Cambridge (UK): Cambridge UP, 421–444.

————. 1983. *Realism and Reason. Philosophical Papers*, Vol. 3. Cambridge (UK): Cambridge UP.

————. 1990. *Realism with a Human Face*. Cambridge (MA): Harvard UP.

————. 1993. Realism Without Absolutes. *International Journal of Philosophical Studies* 1: 179–192.

————. 1994. The Dewey Lectures 1994: Sense, Nonsense, and the Senses: An Inquiry into the Powers of the Human Mind. *The Journal of Philosophy* 91: 445–518.

Quine, Willard Van Orman. 1934/1979. Ontological Remarks on the Propositional Calculus. In Quine, *The Ways of Paradox and Other Essays*. Revised and enlarged edition. Cambridge: Harvard UP, 265–271.

————. 1947. The Problem of Interpreting Modal Logic. *Journal of Symbolic Logic* 12: 43–48.

————. 1953/1980. *From a Logical Point of View. Nine Logico-Philosophical Essays*. 4th ed. Cambridge (MA): Harvard UP.

————. 1960. *Word and Object*. Cambridge (MA): The MIT Press.

————. 1963. *Set Theory and Its Logic*. Cambridge (MA): Harvard UP.

————. 1969. *Ontological Relativity and Other Essays*. New York: Columbia UP.

————. 1970a. On the Reasons for Indeterminacy of Translation. *The Journal of Philosophy* 67: 178–183.

————. 1970b. *Philosophy of Logic*. Englewood Cliffs (NJ): Prentice-Hall.

————. 1974. *The Roots of Reference*. La Salle (IL): Open Court.

————. 1975. On Empirically Equivalent Systems of the World. *Erkenntnis* 9: 313–328.

————. 1981. *Theories and Things*. Cambridge (MA), London: Harvard UP.

————. 1985. Events and Reification. In LePore and McLaughlin 1985, 162–171.

————. 1986. Reply to Charles Parsons. In Hahn and Schilpp 1986, 396–403.

————. 1987. *Quiddities*. Cambridge (MA), London: Harvard UP.

————. 1992a. *Pursuit of Truth*. Revised edition. Cambridge: Harvard UP.

————. 1992b. Structure and Nature. *The Journal of Philosophy* 89: 5–9.

Rayo, Agustin. 2003. When Does "Everything" Mean *Everything*? *Analysis* 63: 100–106.

Rescher, Nicholas. 1973. *The Coherence Theory of Truth*. Oxford: Clarendon.

———. 1979. *Cognitive Systematization. A Systems-Theoretic Approach to a Coherentist Theory of Knowledge*. Oxford: Blackwell.

———. 1982. *Empirical Inquiry*. Totowa (NJ): Rowman and Littlefield.

———. 1987. *Scientific Realism. A Critical Reappraisal*. Dodrecht, Boston: Reidel.

———. 1991. How Wide Is the Gap Between Facts and Values? In Nicholas Rescher, *Baffling Phenomena and Other Studies in the Philosophy of Knowledge and Volition*. Savage (MD): Rowman and Littlefield, 29–57.

———. 1992–94. *A System of Pragmatic Idealism*. Three volumes. Princeton: Princeton UP.

Richards, Tom. 1975. The Worlds of David Lewis. *Australasian Journal of Philosophy* 53: 105–118.

Ricken, Friedo. 1993/1998. *Allgemeine Ethik*. 3rd ed. Stuttgart: Kohlmanner.

Ritter, Joachim, et. al. 1971ff. *Historisches Wörterbuch der Philosophie*. 12 vols. Darmstadt: Wissenschaftliches Buchgesellschaft.

Romanos, George D. 1983. *Quine and Analytic Philosophy*. Cambridge (MA), London: The MIT Press.

Rosen, Gideon. 1990. Modal Fictionalism. *Mind* 99: 327–354.

———. 1994. What Is Constructive Empiricism? *Philosophical Studies* 74: 143–178.

Rosenberg, Gregg. 2002. *A Place for Consciousness. Probing the Deep Structure of the Natural World*. Oxford: Oxford UP.

Rosenkranz, Kark. 1853/1996. *Ästhetik des Haßlichen*. Reprinted. Darmstadt: Wissenschaftliche Buchgesellschaft.

Russell, Bertrand. 1905/1956. On Denoting. In Russell, *Logic and Knowledge: Essays, 1901–1950*. Edited by Robert Charles Marsch. London: Allen and Unwin, 41–56.

———. 1928/1996. What Is the Soul? In Russell, *The Collected Papers of Bertrand Russell*. Vol. 10: *A Fresh Look at Empiricism, 1927–42*. Edited by John G. Slater, with the assistance of Peter Kollner. London, New York: Routledge, 203–205.

———. 1948. *Human Knowledge. Its Scope and Limits*. New York: Simon and Schuster.

Sartwell, Crispin. 1991. Knowledge as Merely True Belief. *American Philosophical Quarterly* 28: 157–165.

———. 1992. Why Knowledge Is Merely True Belief. *The Journal of Philosophy* 89: 167–180.

Schlapkohl, Corinna. 1999. *Persona est naturae rationabilis individua substantia. Boethius und die Debatte über den Personbegriff*. Marburg: Elwert.

Schneider, Christina. 1999. Two Proposals for Formalizing a Bundle Theory Based on Universals. In José Falguera (ed.), *Proceedings of the International Congress: Analytical Philosophy at the Turn of the Millennium–La Filosofia Analitica en el Cambio del Milenio*. Santiago de Compostela: Universidad de Santiago Publicacións, 229–237.

———. 2001. *Leibniz' Metaphysik. Ein Formaler Zugang*. Munich: Philosophia Verlag.

———. 2002. Relational Tropes–A Holistic Definition. *Metaphysica* 3: 97–112.

———. 2006a. Towards a Field Ontology. *Dialectica* 60: 5–27.

———. 2006b. Totalitäten–ein metaphysisch-logisches Problem. In Guido Imaguire and Christina Schneider (eds.), *Untersuchungen zur Ontologie*. Munich: Philosophia Verlag.

Searle, John. 1992. *The Rediscovery of the Mind*. Cambridge (MA): The MIT Press.

————. 1995. *The Construction of Social Reality*. New York: The Free Press.

Seibt, Johanna. 2004. *General Processes. A Study in Ontological Category Construction.* Habilitationsschrift, Universität Konstanz.

Sellars, Wilfrid. 1956. Empiricism and the Philosophy of Mind. In Herbert Feigl and Michael Scriven (eds.), *Minnesota Studies in the Philosophy of Science.* Minneapolis: U of Minnesota Press, 1: 253–329.

Shalkowski, Scott A. 2004. Logical and Absolute Necessity. *The Journal of Philosophy* 101, 55–82.

Shapiro, Stewart. 1997. *Philosophy of Mathematics. Structure and Ontology.* New York, Oxford: Oxford UP.

Sher, Gila. 1999. On the Possibility of a Substantive Theory of Truth. *Synthese* 117: 133–172.

Shoenfield, J. R. 1983. *Mathematical Logic.* Reading (MA), London: Addison-Wesley.

Shope, Robert K. 1983. *The Analysis of Knowing. A Decade of Research.* Princeton: Princeton UP.

Simons, Peter. 1987. *Parts. A Study in Ontology.* Oxford: Clarendon.

————. 1994. Particulars in Particular Clothing: Three Trope Theories of Substance. *Philosophy and Phenomenological Research* 54: 553–575.

Smith, Quentin. 1988. The Uncaused Beginning of the Universe. *Philosophy of Science* 55:39–57.

————. 1991. Atheism, Theism and Big Bang Cosmology. *Australasian Journal of Philosophy* 69: 48–66.

Sneed, Joseph. 1979. *The Logical Structure of Mathematical Physics.* Dordrecht: Reidel.

Spinoza, Benedictus de. *Ethics.* Edited and translated by G. H. R. Parkinson. Oxford, New York: Oxford UP, 2000.

Stegmüller, Wolfgang. *Probleme I/1969. Probleme und Resultate der Wissenschaftstheorie und Analytischen Philosophie.* Vol. I, 1st ed. Berlin, Heidelberg, New York: Springer.

————. *Probleme I/1983. Probleme und Resultate der Wissenschaftstheorie und Analytischen Philosophie.* Vol. I, 2nd ed. Berlin, Heidelberg, New York: Springer.

————. *Probleme II-1/1970. Probleme und Resultate der Wissenschaftstheorie und Analytischen Philosophie.* Vol. II, No. 1. Berlin, Heidelberg, New York: Springer.

————. *Probleme II-2/1973. Probleme und Resultate der Wissenschaftstheorie und Analytischen Philosophie.* Vol. II, No. 2. Berlin, Heidelberg, New York: Springer.

————. *Probleme II-3/1986. Probleme und Resultate der Wissenschaftstheorie und Analytischen Philosophie.* Vol. II, No. 3. Berlin, Heidelberg, New York: Springer.

————. 1980. *Neue Wege der Wissenschaftsphilosophie.* Berlin: Springer.

————, and Varga von Kibéd, Matthias. 1984. *Probleme und Resultate der Wissenschaftstheorie und Analytischen Philosophie.* Vol. III: *Strukturtypen der Logik.* Berlin, Heidelberg, New York: Springer.

————. 1989. *Hauptströmungen der Gegenwartsphilosophie. Eine kritische Einführung.* Bd. IV. Stuttgart: Kröner.

Stern, Robert. 1993. Did Hegel Hold an Identity Theory of Truth? *Mind* 102: 645–647.

Stoljar, Daniel. Physicalism. *Stanford Encyclopedia of Philosophy.* http://plato. stanford.edu/entries/physicalism (accessed March 29, 2008).

Suppe, Frederick. 1977. *The Structure of Scientific Theories.* 2nd ed. Chicago: U of Illinois Press.

————. 1989. *The Semantic Conception of Theories and Scientific Realism*. Chicago: U of Illinois Press.

Suppes, Patrick. 1957. *Introduction to Logic*. New York: van Nostrand.

Swinburne, Richard. 1979/2004. *The Existence of God*. 2nd edition. Oxford: Clarendon.

Tappolet, Christine. 1997. Mixed Inferences: A Problem for Pluralism About Truth Predicates. *Analysis* 57: 209–10.

————. 2000. Truth Pluralism and Many-valued Logics: A Reply to Beall. *The Philosophical Quarterly* 50: 382–385.

Tarski, Alfred. 1933/1956. The Concept of Truth in Formalized Languages. In Tarski, *Logic, Semantics, Metamathematics. Papers from 1923 to 1938*. Oxford: Clarendon, 152–278.

————. 1944. The Semantic Concept of Truth and the Foundations of Semantics. *Philosophy and Phenomenological Research* 4:341–376.

Thomas Aquinas. *Eternity. On the Eternity of the World*. Translated by Cyril Vollert. Milwaukee: Marquette University Press, 1964.

————. *Power of God. On the Power of God*. Translated by the English Dominican Fathers. London: Burns, Oates, and Washbourne, 1932–34.

————. *ScG. Summa Contra Gentiles*. Translated by the English Dominicans. London: Burns, Oates, and Washbourne, 1934.

————. *STh. Summa Theologiae*. Translated by the English Dominicans. London: Burns, Oates, and Washbourne, 1912–1936.

————. *Truth. The Disputed Questions on Truth*. Vol. 1 translated by Robert William Mulligan, S.J. Chicago: Regnery, 1952. Vol. 2 translated by James V. McGlynn, S.J. Chicago: Regnery, 1953. Vol. 3 translated by Robert W. Scmidt, S.J. Chicago: Regnery, 1954.

Tichý, Pavel. 1964. On Popper's Definition of Verisimilitude. *The British Journal for the Philosophy of Science* 25: 155–160.

Tugendhat, Ernst. 1967. *Der Wahrheitsbegriff bei Husserl und Heidegger*. Berlin, New York: de Gruyter.

————. 1969. Heideggers Idee von der Wahrheit. In Otto Pöggeler (ed.). *Heidegger. Perspektiven zur Deutung seines Werks*. Köln, Berlin: Kiepenheuer & Witsch, 286–297.

————. 1979/1997. *Selbstbewusstsein und Selbstbestimmung*. 7th ed. Frankfurt am Main: Surhkamp.

————. 1992. *Philosophische Aufsätze*, Frankfurt am Main: Surhkamp.

Vallicella, William F. 2000. Three Conceptions of States of Affairs. *Noûs* 34: 237–257.

van Cleve, James. 1985. Three Versions of the Bundle Theory. *Philosophical Studies* 47: 95–107.

van Fraassen, Bas. 1980. *The Scientific Image*. Oxford: Clarendon.

————. 1989. *Laws and Symmetry*. Oxford: Clarendon.

————. 1991. *Quantum Mechanics. An Empiricist View*. Oxford: Clarendon.

————. 1995. "World" Is Not a Count Noun. *Noûs* 29: 139–157.

Weingartner, Paul. 1976. *Wissenschaftstheorie* II, 1. *Grundlagenprobleme der Logik und Mathematik*. Stuttgart: Frommann-Holzboog.

Whitehead, Alfred North. 1929/1978. *Process and Reality. An Essay in Cosmology*. Corrected edition. New York: The Free Press.

Wiggins, David. 1996. Objective and Subjective in Ethics. With Two Postscripts About Truth. In Hooker 1996, 35–50.

Williams. Bernard. 1972. *Problems of the Self. Philosophical Papers 1956–1972.* London: Cambridge UP.

———. 1996. *Truth and Truthfulness. An Essay in Genealogy.* Princeton: Princeton UP.

Williams, Donald C. 1953/1966. On the Elements of Being. In D. C. Williams, *Principles of Empirical Realism.* Springfield: Thomas, 74–109.

Williamson, Timothy. 2000. *Knowledge and Its Limits.* Oxford: Oxford UP.

Wittgenstein, Ludwig. *Investigations. Philosophical Investigations. The German Text, with a Revised English Translation.* Translated by G. E. M. Anscombe. Oxford: Blackwell, 2001.

———. *Tractatus. Tractatus Logico-Philosophicus.* Translated by David F. Pears and Brian F. McGuinness. New York: Routledge, 1962.

———. 1914–1916/1961. *Notebooks 1914–1916.* Edited by G. H. von Wright and G. E. M. Anscombe. With an English translation by G. E. M. Anscombe. Edition. New York: Harper & Brothers. German provided on facing pages.

Wolff, Christian. 1728/1996. *Discursus praeliminaris de philosophia in genere* (1728). German translation: *Einleitende Abhandlung über Philosophie im allgemeinen.* Historical-critical edition. Translated, introduced, and edited by G. Gawlick und L. Kreimendahl. Stuttgart: Frommann-Holzboog.

———. 1737. *Cosmologia Generalis, Methodo Scientifica Pertractata.* Frankfurt am Main, Leipzig: Renger.

Wright, Crispin. 1992. *Truth and Objectivity.* Cambridge (MA), London: Harvard UP.

———. 1996. Truth in Ethics. In Hooker 1996, 1–18.

Index

Aaron, R. I., 109
absolute, the, 155, 415, 443, 458, 459
absolutely necessary, 443
absoluteness, 327
abstract particular, 210
accomplishment of theoreticity, epistemic
    dimension as, 99–120
acceptability (of theories; van Fraassen), 126
action, ethical, 290–305
    context of (Habermas), 150n41
actualism, 320, 431
actuality, 389, 431, 433f
    and existence, 433–36
adequacy, empirical, 126, 422
    theoretical, 40
aesthetic theory, 321ff
aestheticity, 27, 308f
    as dimension of presentation of the
        world, 308f
aesthetics, philosophical, 315
agreement, 313f, 322, 440
Albert, H., 55f
ἀλήθεια, 142f
all-is-contingent thesis, 444f
Almeder R., 59–62
'ämät, 142f
'ämunä, 142f
Anacker, U., 335ff
anthropic principle, 329
anthropology, philosophical, 263–90
antirealism, 126
Apel, K.-O., 56
application, intended, 134
approximation, 129, 134, 140n32
architectonic, 20, 467
argument from knowledge, 287

Aristotle, 2n1, 7, 90, 111, 131, 157, 164, 165, 261,
    265, 276, 291f, 309, 334, 420n32, 439
art, 318–21
    abstract, 322
    and the semantics/ontology of possible
        worlds, 320f
    cognitive and moral aspects of, 320
    expressive theory of, 319
    formalistic theory of, 319
    mimetic theory of, 319
    priority of, 318
    semiotic theory of, 319
art, theory of, 322
artistic beauty, 31
artwork(s), 312, 319–21
    abstract, 322
    and theory, 321
    as aesthetic presentations of the world, 322
    as aesthetic presentations of transforma-
        tions of the world, 319f
    ontological status of presentations of, 321
assertion, 152f
atomism, 479
attitude, aesthetic, 311f
    ontological import of the aesthetic, 309
    practical, 311f
    theoretical, 311f
attributes (properties and relations), 160
Austin J. L., 4, 364n5
axiomatics, 45–47
    formal Hilbertian, 46
    informal Hilbertian, 46
axiomatization, 45–47
    and formalization, 125
    by definition, 46f
    Euclidean, 45

informal set-theoretical, 46f
axioms, 45–47, 128, 133

Bacon, J., 222n37
Baldwin, T., 232n43
Ballarin, R., 222n36, 442n49
Balzer, W., 75, 127n26
bare particular, 190, 213
Barwise, J., 235n45
Baumgartner, H.-M., 335ff
Beall, C., 142n33
Beaufret, J., 37n14
beauty, 321ff
    and the domains of being, 318
    as fundamental aesthetic concept, 314–18
    as harmony/consonance of the immanent
        characteristics of being, 315f
    concept of, 314f
    decreased significance of in contemporary
        aesthetic theories, 322f
    universal, 19
Bedeutung (Frege), 15, 158, 160ff
being, 16, 25, 32, 37ff, 155, 248, 414, 415ff
    absolute, 17, 416
    absolute and contingent dimensions of, 331
    absolute uniqueness of, 436
    absolutely necessary, 441–60
    additional steps in the explication of,
        451–60
    absolutely necessary minded
        (personal), 452
    and beings, 37, 417ff, 430n41
    and existence, 414f
    and God, 420f
    and structure, 246
    and the dimension of being, 415ff
    and the natural world, 257f
    and thinking, 37
    as a whole, 25, 302f, 306, 311, 350, 430, 441
    as problematic English expression, 420
        note d
    as such and as a whole, 17, 416
    as structurality, 38
    as such, 25
    as two-dimensional, 444, 446
    as universal class, 430
    beauty as immanent structural moment
        of, 440
    comprehensive systematic view of, 417
    contingent, 17, 446, 458

dimension of, 16, 415, 417ff
goodness as immanent structural moment
    of, 440
Heidegger's philosophy (metaphysics) of, 2
Heidegger's conception of, 417ff
in the objective sense, as counterpole to
    subjectivity, 415
in the primary sense, 436
in the secondary (derivative) sense, 436
inner structurality of, 436–40
itself, as the structure of all structures, 439
mental constitution of absolutely
    necessary, 458
most universal immanent characteristics
    of, 436–40
non-reductive objectivist conception
    of, 416
primordial comprehensive, 435f
primordial difference with respect to as
    the difference between the absolutely
    necessary and the contingent dimen-
    sion of, 443–46
reductionist-objectivist conception of, 417
relation between absolutely necessary and
    contingent, 452–55
self-articulation of, 18
(self-)explication of, 437
subsisting by itself, 416, 458
    and God, 447n52
the actual world and the plurality of pos-
    sible worlds, 431–36
theory of, 7, 13, 315, 318, 331, 369, 370f, 413ff
    and ontology, 370f
    structural-systematic, 19
transcendental-dualistic view of, 416f
two-dimensionality of
    as the difference between the absolutely
        necessary and the contingent
        dimensions of, 451
    indirect proof of, 444f
universal coherence as immanent struc-
    tural moment of, 439
universal intelligibility as immanent struc-
    tural moment of, 438f
being-possible, 444
belief, 126
    idealized, 119f
    Kant on, 109f
    normal, 119f
Bell, J. L., 179f

Belnap, N., 26n3, 227
Bentham, J., 84
Bien, G., 292n30
big-bang singularity, 325f
big-bang theory, 325ff
Bigelow, J., 426n38, 428, 443n50, 451
biology, 455
Blau, U., 86
body-soul dualism, 84
Boethius, 265n13
Boole, G., 30n9
Bourbaki, N., 167n6, 175–78
Bradley, F. H., 232
Brandom, R., 100f, 151n42
Brink, D. O., 298
Buddhism, 330
bundle theories, 191f, 265, 275n18

Camp, J. L., 227
Campbell, K., 192, 210
Candlish, S., 232
Cantor, G., 266
Carnap, R., 4, 9, 22f, 47, 66f, 75, 79, 82, 95, 158,
    397n21
Carter, B., 329
Cartwright, R., 423n33
categorical imperative, 293
category/categories, 15, 158, 164ff, 169
    and conceptual scheme(s), 167
    and language(s), 166
    ontological, 208f
    semantic, 183
    syntactic, 183
    theories of, 183
        mathematical, 177
causality, 327
Chalmers, D., 287
Chisholm, R., 166
Christianity, 329f, 334, 345, 351, 447
    and genuine theology, 332
    openness of to philosophy, 332
circularity Ketchum, 51
classes (mathematics), and sets, 267f
coextensionality, intentional, 276
cognition, 108
cognition, faculties of (Kant), 360
coherence, 40, 352, 410, 463, 464, 466, 467, 469
    and consistency, 464
    as systematicity, 439
    degrees of, 466

higher degrees of, 475f
    relatively maximal, 69
    systematic, 464
    universal, 19
coherence criterion, 463f
coherence methodology, 42–44
coherentist grounding, 51f, 64, 66
collective, 340f
    humanity as third-order ontological,
        339
collective configurator, 339, 341
communication, 77
    and presentation, 372f, 385
    languages of, 150
completeness, 40, 70
compositionality, principle of, 15, 147f, 158, 185,
    186ff, 194, 195, 235n45, 379
compositionality, truth-functional, 200f
comprehension, principle of, 266f
comprehensive architectonic, 8
comprehensive systematics, 17, 139, 357–60
    as structural metaphysics, 357ff
    as universal philosophical theory, 411ff
    chief obstacle to the development of a
        philosophical, 359–69
    philosophical status of, 357–413
comprehensive theory, 7
    Christian-theological, 356
compresence, 192
concept(s), 15, 158–61, 164
    and function, 2
    and structure(s), 314
    as abbreviations for structures, 167
    of the concept, 159, 160ff
conceptual clarification, 141
conceptual scheme(s), 84, 191, 338, 398ff, 410
    of the sciences (Quine), 168f
conceptualizing, 156–58
condition, 261
configuration ontology, vs. substance ontol-
    ogy, 410
configuration(s), 15f, 29, 147f, 202, 203n24,
    206f, 208f, 213f, 263f
    adequate formal articulation of the con-
        cept of, 265–69
    and mereology, 268
    and propositional-logical conjunction,
        268f
    and set(s), 265–68
    as bundle(s), 265

as ontological structure adequate for human being, 270–82
concept of, 264–69
forms of
and first-order predicate logic, 218–22
and propositional logic, 216ff
as ontological structures, 215–22
of primary entities, 29
of primary facts, 263f
conglomerates, 266
conjunction(s)
logical-structural, 181ff
of primary sentences, primary propositions, and primary facts, 208
consensus, 114n16
consequence, logical, 180n14
constitution, ontological, 264
constraints, 129
constructivism, 379n15
context principle, 15, 147f, 195, 199–203
incompatibility of with principle of compositionality, 200ff
strong version of, 199f, 201ff
context operator, 150
contingency, 327, 443
continuum, 384, 387, 391f, 408f
convergentism, 367f
conviction, 109, 126
correspondence, 7
correspondence rules (in the received view of theories), 123
Corry, L., 27n5, 29, 167n6, 175–78, 426n39
cosmology
in the natural-scientific sense, 324–29
in the philosophical sense, 324
physical, 343ff
Cover, J. A., 193n18
creation, 458
and evolution, 459
ex nihilo, 325–28
thought of, 327
creative absolute, 459
creator, 325f, 327
creator God, 325f
Cresswell, M. J., 443
culture, world of, 480
cut or gap
metaphysical, putative, 17f, 362, 402, 411f, 434, 444
ontological, putative, 367
problem, 182, 337, 359ff, 367, 375, 413, 421

starting point for a systematic clarification of, 369ff

Daly, C., 192
Danto, A. C., 323n46
data basis, 253f
datum (data), 10f, 16, 24f, 32f, 42, 43f, 68, 270–72, 475f
in the general sense as the subject matter for theories, 24
in the specific sense as concept correlative to the (dimension of) structure, 24
Davidson, D., 77, 84, 164, 167n5, 200, 235n45, 283, 398
death, 351–56
as separation of soul and body, 353
as separation of soul and body, 355ff
life after, 315–56
deep structure, 408f
definite description, Quine's elimination of, 197f
definition, 26, 387
deflationism, 231f
demand, mode of, 305
demands, and practical sentences, 93
Denkel, A., 190n17
denotation function, 184
derivability, 179
Descartes, R., 5, 100, 131, 191
descriptions, definite, Quine's elimination of, 197f
designation function, 184
detachment, law of, 179
detachment, rule of, 179
developmental factors of world history, 341
diagonalization, Cantor's procedure of, 379, 392
Diederich, W., 124n20
dilution (Koslow), 181
dimension
formal, 32
primordial and comprehensive, 412f
Dipert, R. R., 222n37
discourse
aesthetic, 89
practical/pragmatic, 89
theoretical, 89
disquotation, 234
disquotation theory of truth, 224
DN-explanation, 123
Dodd, J., 232n43, 233n44, 235n45

Dorschel, A., 56
Dowty, D. R., 187
dualism, body-soul, 84
Dummett, M., 3–7, 117, 199f

Ebbinghaus, H. D., 29n7, 211n30, 266
element, property of being an, 426
eliminativism, 283
empiricism, constructive, 422
end in itself, 303
end point of the research process (Peirce), 399
Engels, F., 334ff, 343, 350f
entanglement, 355n57
entity/entities, 163, 422
    and being(s), 419f
    mathematical, 451
    physical, 355
    uncountably many, 392ff
    with first- vs. second-order ontological
        status, 304f
epistemology, 55, 56f, 101
Ereignis (Heidegger), 82, 418f
"es gibt," literal and idiomatic meanings of,
    419 note c
"es verhält sich so dass," and "It is the case
    that", 404 note a
Erlebnis, 310 note b
eschatology, 356
esse per se subsistens, 416, 458
    and God, 447n52
ethics, 290–93
    ambiguity of normative, 291ff
    and morality, 290
    as philosophical theory, 291ff
    normative, 94, 293f
    truth in, 296ff
    with cognitivist status, 290
event(s), 164, 193, 261
evolution
    complete and adequate explanation
        of, 455f
    theory of, 455–58
existence, 444
    and being, 414f
    and the dimension of being, 444
    as indexical concept (D. Lewis), 433
existential quantifier, 381, 415
experience, 271f
    lived aesthetic, 310, 310 note b
explanandum, 123
explanans, 123

explanation of evolution, complete and
    adequate, 455f
expressa
    logical, 182f
    of logical/mathematical sentences, 171f
    of practical sentences, 92
    of practical-deontic sentences, 305
    of primary sentences, 146, 202f
    of subject-predicate sentences, 202f
    of theoretical sentences, 92, 163, 171, 174,
        178f, 202f, 204, 227, 229
    semantic, 178
    syntactic, 178f
    valuative, 309
expressibility, 78, 97f
    as basic structural moment of being and
        beings, 371
    in aesthetics, 310
    ontological, 376
    universal/immanent, 17f, 138, 205, 363f, 369,
        377, 388f, 395, 412, 439, 454f
extension, 164, 189, 218–22, 268
    of concept, 169
extensionality, law of, 212

fact ontology, 221
fact(s), 15, 158, 163, 220ff, 233f, 235n45, 261
    as true thought (Frege), 233
    Frege on, 233
    logical, 183
    logical/mathematical, 240
    primary. See primary fact(s)
faculties of cognition (Kant), 312ff
fallacy, naturalistic, 305
falsity
    as truth value, 237
    necessary, 238
    starting points for a theory of, 237ff
family resemblance, 205, 214
Feuerbach, L., 481
Feyerabend, P., 124
Feynman, R., 327
Fichte, J. G., 35, 279
field(s), 261, 269
Findlay, J. N., 232n42
fixed point, 348, 349
folk psychology, 273, 283
Forge, J., 130
Forrest P., 352n54, 453n55, 457n58
foundationalist grounding, 51f, 64, 66
Frank, M., 278ff, 280n23

freedom, 335, 342
  in the strong sense, 304
  of the creative absolute, 459
  of the will, 298
Frege, G., 4ff, 16, 79, 145, 153, 158, 159–63, 188f,
    199, 223, 232, 302, 373
Fukuyama, F., 341f
function(s), 160
  and object(s), 161
functionality, 200
function-value model, 200
fundamental fact about language, 148f

Gabbay, D., 149n40, 222
Gadamer, H.-G., 2, 338f, 364n4, 395
galaxy, 354
Gallop, D., 438n43
gap or cut
  metaphysical, putative, 17f, 362, 402, 411f,
    434, 444
  ontological, putative, 367
  problem, 337, 359ff, 367, 375, 413, 421
    starting point for a systematic
      clarification of, 369ff
Gardner, S., 310, 315, 319
geometry, 45ff
Gettier, E., 100, 104–6, 108, 118
Giere, R., 124, 126
Gilson E., 414
global systematics, 14, 22–98
God, 37, 331f, 339, 346f, 349, 415, 416, 420f,
    443n51, 445–51, 458, 459
  and being, 420f
  and being as a whole, 447
  proofs of the existence of, 447–51
  theory concerning, 447ff
  Thomas Aquinas's third proof of existence
    of, 449ff
Gödel, K., 117, 241n51
Gödelization, 427f
God's eye view, 337, 360, 362
Goldblatt, R., 29, 222n37, 426n39
good, the, 301
Goodman N., 4, 362, 400f
goodness, 321
  universal, 19
gravity, force of, 355
Greimann, D., 302
Grim, P., 19, 424ff, 428f
grounding, 9, 51–73
  and justification, 57–64

and proof, 53f
concept of
  and the problem of its definition, 57–64
  in philosophy, 54–73
  metasystematic, 482
  non-systematic, 55–64
  systematic, 64–73
  definition of, 57–64
  foundationalist, 68
  idealized concept of systematic, 71ff
  incipiently systematic, 64–67, 71, 72f
  innersystematic, 64, 67f, 71f
  metasystematic, 64, 68–73
  pragmatic, 52f, 72
  ultimate, 9, 56, 482
groups, mathematical, 46f
Grover, D. L., 227
Grünbaum, A., 326n48
Guenthner, F., 149n40, 222
Gupta, A., 26n3
Guyer, P., 108n11

Habermas, J., 100f, 114n16, 150n41, 297f, 481
Hales, S. D., 241n52
Hanna, J. F., 82
Hanson, N. R., 124
Hartle, J., 327
Hawking, S., 325ff, 329, 344
heaps, 266, 267
Hegel G. W. F., 35, 81, 88n3, 114n14, 157f, 160,
    165, 232, 251, 281n24, 304f, 318, 333, 335,
    342, 364n4, 438, 472f, 481
Heidegger, M., 2, 37, 75, 81f, 98n7, 121, 143,
    262n10, 281, 371, 414, 415, 417ff, 421,
    430n41, 438, 458, 473
Hempel, C. G., 75, 123, 125, 133
Henrich, D., 279ff
hermeneutics, 2, 338f
Hilbert D., 46, 461
Hinduism, 330
history, 37, 155
  effective- (Gadamer), 338
  end of, 342
  science of, and philosophy of world his-
    tory, 333f
history configurator, 341
history of philosophy, 3
history/histories of the universe, 327
  multiplicity of, 329
  sum of all, 327
Hodges, W., 149n40, 180n14

Hoffman, J., 166, 191
homomorphism, 187
Hooker, B., 291
Hornsby, J., 232n43
Horwich, P., 146, 231f
H-O-Schema of scientific explanation, 123
Hughes, G. E., 443
Hugly, P., 78, 380–84, 386f, 390ff
human being(s)
    as intentionally coextensive with the universe (being as a whole), 303
    place of within the universe, 302f, 330
humanity
    as collective, 339
    history of, 459
Husserl, E., 5, 162, 364n5, 415
hylomorphism, 169

I, the, 265, 269, 275
I-saying, 275, 277, 282, 340
ideal, structural-systematic method as regulative, 467
idealism, 375
    absolute, 156
    transcendental, 156
idealization
    artistic, 320
    theoretical, 320n44
identity, 211
    as limiting case of correspondence, 234
    conditions of, 206f, 277
        for primary propositions and primary facts, 203–7
        semantic and ontological, 204–7
    function, 211
    relation, 211f
        and identity function, 211
    theory of truth, 232f
    identity thesis (in truth theory), 16, 231–36
        as identity of true primary proposition and primary fact, 232ff
        Frege's, 233
immortality, 353
    problem of, 351f
    in analytic philosophy, 351f
imperative sentences, 293f
imperative, categorical, 293
implication relation, 181
implication-structure, 181ff
incompleteness theorem (Gödel), 20, 117, 243, 424, 427f

"indeed" (Tarski), 226
indeterminacy of translation (Quine), 204
individual(s), 263f
    and classes, 211f
individuation
    principle of, 264
    problem of, 277
inference to the best explanation, 352, 453
inference to the best systematization, 43n17
inscrutability of reference (Quine), 252f
integrality
    of the human being, 355f
    of the mind, 311f, 314, 321
intelligence, 276, 278
intelligibility, 40, 67, 69f, 349, 352, 364, 413, 441, 453f, 462–67
    absolute, 466
    as immanent to theoretical framework, 465f
    degrees of, 466
    higher degrees of, 475f
    of a subject matter, 373
    potential for, 244, 413, 423, 441, 444
    relatively maximal, 469
    relativity of to theoretical framework, 465f
    universal, 19, 259
intension, 160
intentionality, 288, 354
    and knowledge, 288ff
    and self-consciousness, 278–82
    and truth, 288ff
interface between knowers and the world, putative, 365f
interpretation function, 149, 174, 184
interpretation, semantics of, 148f, 174
intersubjectivity, 32
Islam, 330
it (particle), 404f, 413
"It is the case that," and "Es verhält sich so dass," 404 note a

Jackson, F., 287
James, W., 100f, 364n5
Judaism, 330
judgment, 306
    aesthetic, 306
justification, 226n40. See also grounding

Kant, I., 4, 5, 17, 33, 54, 101, 108–10, 111, 113n12, 131, 138, 156ff, 162, 165f, 243, 247, 256n7,

261n8, 270, 279, 280, 288, 308f, 312ff, 317, 336ff, 341, 359f, 376, 411, 443n51 467
Keller, A., 414
Kelley, J. L., 425, 430
Kelley-Morse, 425, 430
Ketchum, R. J., 55, 57–59, 62f
kind, 190f
Kleene, S. C., 28f, 167f, 177, 209n29
Kneale, W., and M., 30n9
knowledge, 108
    according to Kant, 108–10
    ambiguity of, 101f
    as dimension of accomplishment of theo-
        reticity, 99–120
    as philosophical problem, 102–8
    definition of, 106
    Gettier's definition of, 103ff
    non-analytic explanation of, 106f
    Oxford English Dictionary on, 103
    relativity of historical, 337f
König, G., 121
Koslow, A., 181ff
Kripke, S., 442n49
Kuhn, T., 124, 127
Kuipers, T. H. F., 400n24
Kutschera, F. v., 8n4, 293, 295, 596n31, 449ff

Landmesser, C., 142n35
language(s), 30f, 126
    actual/possible, 394ff
    adequate concept of philosophical, 372–97
    analysis of, 4ff
    and communication, 77, 372f, 381f
    and concept, 160ff
    and knowledge, 97f, 99
    and structure, 136f
    and theoretical frameworks, 386
    and tokening systems, 380–84, 390ff
    and ultimate metalanguage, 89
    and world, 37, 38, 49, 144f, 204f, 218f, 234,
        387–90, 401
    artificial or technical, 77
    as means of expressibility, 363f, 369,
        377, 396
    as semiotic system with uncountably
        many expressions, 19, 78, 98, 374–92,
        394ff
    as system of communication, 380
    basis for the assumption of a philosophi-
        cal, consisting of uncountably many
        expressions, 389f

cardinality of, 388f
centrality of for philosophy, 95
communication and dimension of presen-
    tation, 77–99
communication and presentation, 372f
consisting of uncountably many expres-
    sion, possibility in principle of, 378ff
determination of
    lifeworldly-contextual, 152f
    linguistic-pragmatic, 153
    semantic, 154
    the three levels of, 149–52
extent and expressive strength of philo-
    sophical, 374f
formal, 49, 79, 128
fully determined status of, 148f, 150, 224
fundamental fact about, 148f
fundamentality of the semantic dimension
    of, 152f
games, 89, 365
ideal, 79, 391
indeterminacy of, 137
indeterminacy vs. determinacy of, 148f
injective/surjective/bijective mapping
    between world and, 388
learnability of, 77f
limits of, putative, 31
logical and ontological dimension of, 95ff
maximal (absolutely universal), 396
normal vs. philosophical, 77–89, 367
ontological consequences of the plurality
    of theoretical, 398–411
ontological dimension of the fully deter-
    mined semantic status of, 231f
ontological interpretation and conse-
    quences of plurality of, 397–411
ordinary/natural, 11, 14, 38, 77–89, 143,
    160, 185, 193, 198n20, 202, 226f, 365, 370,
    371ff, 374f, 379, 385f, 391, 412
    unavoidability of, putative, 86, 88f
originary, 98f
philosophical, 6, 11, 80f, 86f, 89–99, 412
    segmental character of, 387–92
philosophical and scientific, as purely
    human products? 394–401
philosophy of, 55
pure semiotic (syntactic), 396
religious, 142f
representation of, 382f
segmental character of philosophical,
    387–92

segments of a, 385f
self-determination of, 151f
semantically-ontologically structured, 396f
status of in logic and mathematics, 174
structures of philosophical, 373f
syntactic, semantic, and pragmatic dimensions of, 95ff
systematic-philosophical, 94–96
theoretical, 79
tokening systems for theoretical, 385ff
uncoupling of logic from, 182
universal, 412f
use, 95
lattices, 269
law of detachment, 179
Legenhausen, G., 218–22
Leibniz, G. W., 131, 206, 452
Lewis, D., 7n4, 248, 268, 284, 285, 431–36
liar paradox, 360f, 424
life
    philosophies of, 2
    meaning of, 345
lifeworld, 95, 374, 480
light, speed of, 289
linguistic framework(s), 9, 66f, 70f, 79, 192
linguistic situation, ideal, 114n16
linguisticality, 31, 395
links, 129
logic, 9, 15, 27f, 32, 37, 44, 53, 55, 64, 86, 95,
    115n16, 118, 131, 148f, 162, 166, 169f, 172–
    75, 178–83, 185, 193f, 215–22, 237, 239ff,
    248, 268f, 296f, 347f, 353n55, 396, 407,
    419, 421, 423, 431, 441f, 474, 477
    ancient, 165
    and ontology, 215–22
    autonomy of, 172
    classical, 222
    deontic, 222
    dynamic, 222
    elementary, 216ff, 441
    expansions of classical, 222
    formal, 86, 88, 121, 125, 149n40, 169, 359,
        473, 474
    free, 222
    in relation to mathematics and philosophy, 172–75
    in relation to semantics and ontology, 95
    intuitionistic, 222
    intuitive (informal), 86
    many-valued, 222
    mathematical, 2, 122

modal, 222, 441f, 444, 474
    interpretations of, 441f
    ontological (metaphysical) neutrality
        of, 219
    ontological status of formal, 169
    predicate, 210f
    predicate (or language), first-order, 83, 85,
        164, 185, 193f, 195f, 210f, 216, 218, 253,
        380f, 398
    as regimentation of scientific language,
        195–99, 253
    ontological configurations and, 218–22
    non-standard semantics for, 218–22
    standard interpretation of, 193f
    standard semantics of, 218f
    propositional, 138, 230, 268f, 474
    ontological configurations and, 216f
    pure, 239ff
    putative ontological (metaphysical) neutrality of (Legenhausen), 221
    three-valued, 86
logicism, 174
Loux, M., 190f
Lowe, E. J., 191
Löwenheim-Skolem paradox, 137
Ludwig, G., 137n31

Machover, M., 179f
Mackie, J. L., 302, 321
MacLane, S., 426n39
mapping, principles of, 137n31
Marchal, J. H., 28n6, 36
Marx, K., 481
materialism, 7, 282ff, 415
    dialectical, 343, 351
    historical, 343
mathematics, 15, 27f, 32, 37
    and ontology, 215
    autonomy of, 172
    in relation to logic and philosophy,
        172–75
    pure, 239ff
matter, 415
maximally consistent subset, 465f
McCarthy, T., 115n16
McDowell, J., 36, 271, 365, 368f, 373
McGee, V., 423n33
McGinn, C., 296n32, 304n34
meaning, 158, 161
    concept of, 342f
    of life, 345

theory of, 6
ultimate (Nagel), 346–50
"Meaning is use", 365
Meinong, A., 232n42
Mellor, D. H., 236n45
metadiscipline(s), 461
metaethics, 293
metaframework, theoretical, 243f, 472
    external, 473f
metalanguage(s), 361
metalinguistics, 461
metalogic, 461
metamathematics, 79, 461
metaphilosophy
    as theory of philosophy, 461
    in the broader sense, 462
    in the narrower sense, 462
metaphysica generalis, 324, 477
metaphysica specialis, 324, 359, 477
metaphysics, 17, 55, 357ff, 367, 461
    general, 324, 477
    materialistic, 334ff
    ontological values as queer according to
        materialistic-physicalistic, 302ff
    special, 324, 359, 477
metasemantics, 461
metasystematics, 19f, 139, 243f,
        461–84
    and metaphilosophy, 461ff
    external, 19f, 462f, 469–81
    external intratheoretical, 469–76
    external intratheoretical interphilosophi-
        cal, 469–75
    external intratheoretical philosophical-
        nonphilosophical, 477–80
    extratheoretical, 480f
    immanent, 19f, 462f, 467ff
    status of, 461–67
method
    as confirmation, 481f
    axiomatic, 45–47, 131
    four-staged, 138–40
    holistic network or coherentist, 42, 45
    idealized four-stage, 41–52
    philosophical, 4ff
    structural-systematic, 138–48, 467ff
    therapeutic philosophical, 365
methodology, 4
Michel, D., 142n35, 400n24
mind and minding, 480
    absolute, 335

as intentionally coextenxive with the uni-
    verse (being as a whole), 276, 282, 288ff,
    330f, 339, 340, 352ff, 354ff, 435, 452f, 458
as subject, 282
as universal entity/phenomenon, 355
minimalism (as form of deflationism with
    respect to truth), 231f
modalities, 19, 441, 443, 477
    as ontological/metaphysical structures, 442
model(s), 29, 47f, 122, 125, 126, 128–30, 149, 168,
        174, 179
    actual, 128
    as structure(s), 133f
    of language(s), 29
    partial-potential, 129, 134, 135
    physical-cosmological, 326f, 328
    potential, 128, 135
model-relation, 149n40
model-structure, 29, 149n40, 220f
model-theory, 149n40
Modernity, 322f
Moerdijk, I., 426n39
Montague, R., 79f
Moore, G. E., 232
Morgan, A. de, 30n9
Mormann, T., 222n37
Moser, P. K., 203n25
Moulines, C. U., 127ff, 134f, 140n32, 400n24
M-theory, 344ff
Münchhausen trilemma, 55f
"myth of the given," 271

Nagel, T., 287, 345ff, 349f, 351ff
Narlikar, B. J., 326
narrative connections in historiography, 336
natura naturans, 250
natura naturata, 250
natural beauty, 312, 313, 318
natural law(s), 327f
natural science(s), 2, 12, 13, 34, 47, 69, 94, 142,
        250f, 255, 257–60, 273, 324ff, 328, 343ff,
        355, 375, 376 455f, 477f, 479. *See also*
        science(s)
natural world
    and the plurality of domains of
        being, 262
    categorial-structural constitution of, 261
    philosophy of the, 250–62
        and the natural sciences, 260f
        major tasks and theses of, 260–62
naturalism, 58f, 251f, 255f, 353n55, 415

and Quine's global ontological
  structuralism, 251–57
evolutionary-theoretical, 353n55
naturalization, 402
nature, 155, 250, 415
  animate, 262
  inanimate, 262
necessity, 443
  ontological, 451f
Nelson, L., 56f
Newman B., 323n46
Nietzsche, F., 298, 342
nihilum absolutum, 445f
Niiniluoto, I., 300n23
"No entity without identity" (Quine),
  170, 203
non-contradiction, principle of, 245
non-statement view of theories, 36, 51
normativity, 292f
notation, canonical, 83, 168f
nothing, the, 436, 458
nothingness, absolute, 445f
noumenon, 248
nuclear theory, 193
null predicate, 210f
null set, 168, 209
null structure, 168, 209f
number theory, 122
numbers, real, 381

object, 15, 158, 163f, 190, 194, 197, 216, 256, 261,
  311, 312ff
object language, 361
objectivity, 37, 163
  dimension of, 118–20
observation sentence, 196, 253
O'Leary-Hawthorne, J., 193n18
ontological difference, 262, 371, 430n41
ontological status, first- and second-order,
  304f, 321
ontologization, 370n8, 402
ontology/ontologies, 7, 23, 369, 422
  analytic, 422
  and metaphysics, 8n4
  and theory/theories of being, 370f
  compositional, 163, 186–90
  contextual (non-compositional), 15f, 146f,
    208f, 218, 410, 483
  critique of compositional, 190–99
  formal, 8n4
  general, 370

interchangeability of, according to Quine,
  254
methodology of, 256f
of ethical sentences, 299f
of objects (substances), 299
of primary facts, as the ontology appropriate to the truth concept of the
  structural-systematic philosophy, 235f
physicalistic, 284ff, 478
Quine's global structuralistic, 255
reductive reinterpretation of, 252
specific, 370
substance, 7. *See also* substance
unicategorial, 192
onto-theo-logic, 81f
onto-theology, 458
operator(s), 309
  aesthetic, 90–94, 307, 316, 389
  factive mental state (FMSO; Williamson),
    107, 119
  linguistic pragmatic, 151
  logical, 181ff, 201, 268f
  ontological status of logical, 182f
  primary propositions and primary facts as
    arguments of, 218
  practical, 90–94, 307, 309
  practical-deontic, 294ff
  prosentence-forming, 227f
  theoretical, 14, 18, 90–94, 294ff, 306–9, 316,
    405f
  theoretical-deontic, 295f
  theoretical-valuative, 295f
  transcendental, 403
  universal theoretical, 403f
opinion (Kant), 109
Oppenheim, P., 123
ordering structure, 269
origin
  of the universe (cosmos), 13, 325ff, 343ff
  philosophical/metaphysical concept, 13
Otto, R., 330
ought sentences, 235n45
overlapping, 268

pantheism, 447
Papineau, D., 254
Parsons, C., 240
part, 268
part-whole model, 200
Peacock, C., 159n2
Peano, G., 122

Peirce, C. S., 100f, 367f, 400f
Penrose, R., 325, 355n57
Peres, C., 315, 317, 322
perfection (*perfectio*), 300, 316, 440
permutation (Koslow), 181
perproposition, 228–31
Perry, J., 235n45
persentence, 228–31
person, 254–90
  as not materialistically-physicalistically
    explicable, 282–90
  ontological constitution of, 353
  the individual human being as, 254–90
perspective change from subjectivity to being
    as a whole, 402–4
Peters, S., 187
phenomenology, 2
philosophy, 1, 22
  ambiguity of practical, 291ff
  ambiguity of the theoretical and practical
    parts of, 308
  analytic, 1f, 6f, 54, 79, 82f, 84, 86, 87, 97,
    100, 106f, 108, 131, 142, 161, 163f, 168, 185,
    199, 218, 224, 235f, 238, 247f, 251, 253,
    261, 270f, 283, 302, 318, 334, 337, 343, 345,
    351, 359, 363, 371, 374f, 379, 398, 402, 410,
    414, 473, 474, 477, 478
  fragmentary character of, 1, 3ff
  non-systematic (fragmentary) character
    of, 1, 3ff
  and the non-philosophical sciences, 12f,
    34f, 39, 255, 257–60, 271
  as theoretical activity and discipline, 480
  as theory, 480f
  as universal science, 10, 17, 356, 477, 481ff
  global-systematic, 14
  history of, 3
  logic and mathematics and, 172–75
  non-analytic, 2, 81f
  of language, 4, 6
  of mind, 13. *See also* anthropology,
    philosophical
  practical and theoretical, 291
  strictly theoretical character of, 131
  structural-systematic, the, 11, 19, 22, 24f,
    26ff, 41, 49, 50, 64, 169, 173, 185, 282, 306,
    315, 318, 320f, 330, 459, 482ff
  basic idea of, 483
  central domain of, 483
  incompleteness of the theoretical
    framework of, 427f
  intermediate domain of, 483
  metasystematic self-determination of,
    462–67
  metasystematics of, 481–84
  non-absoluteness of the theoretical
    framework of, 427f
  overcoming of, 484
  peripheral level of, 483
  openness and incompleteness of, 243f
  purely theoretical sentences as the
    appropriate form of presentation
    for, 369f
  quasi-definition of, 26
  self-determination of, 481–84
  (self-)grounding of, 51–73, 481–84
  systematic, 1f, 5f, 20, 22, 26ff
  tradition of, 2f
  universality of, 462f
physicalism, 282ff
  an argument against, 287–90
  metaphysical, 283ff
  non-reductive, 283–87
  ontology of, 284ff
  reductive, 283
physics, 2, 17, 258, 261, 283, 289, 303f, 329, 339f,
    343ff, 478, 479
  contemporary, 479
Plantinga, A., 53n21, 104, 425n34
platitudes (C. Wright), truth-theoretical, 297
pleasure, 317
point, fixed, 348, 349
Pompa, L., 337
Popper, K. R., 124, 373, 400f
positivism, logical, 359
possibilism, 431f
possibility, 443, 445f
possible, being-, 444
post-structuralism, 27n4
potentialities, 457, 459, 465
  non-physical, 457
  ontological, 457
practicity, 27
pragmatics, transcendental, 56
predicate(s), 161, 164, 199, 202, 210f, 218ff
  as appearing in abbreviations of primary
    sentences, 202
predication, 164, 194f, 201
  and the relation of being-an-element, 216
presentation, theoretical/practical/aesthetic,
    311
prima philosophia, 58f

primary, 15, 29
primary entities, 29
  complex, 29
  simple, 29
primary fact(s), 15f, 29, 93, 147, 164, 167, 168,
    182, 203n24, 206–14, 215f, 232ff, 236, 262,
    263f, 267ff, 273–78, 282, 299, 310, 314,
    316, 321, 322, 340f, 373, 388f, 392, 393f,
    410
  aesthetically valuational, 314, 316
  as arguments of connectives, 268f
  as ontological primary structures, 236
  consituting human beings
    absolutely essential, 274
    contingent, 274
    relatively (historically) essential, 274
  kinds of, 263f
  simple, 29, 168, 208, 209–14
primary proposition(s), 15f, 29, 91, 147, 167,
    168, 179, 203–8, 211, 217f, 225, 229, 232f,
    234, 237, 239, 263, 268f, 272, 299, 307f,
    310, 314, 316, 388f
  aesthetic, 306f
  aesthetically valuative, 314
  and the truth concept, 208
  as arguments of connectives, 268f
  as semantic primary structures, 207f
  fully determined status of, 225
primary sentence(s), 15f, 91, 146f, 153, 202–5,
    206f, 208, 210f, 213f, 221f, 229, 232f, 237,
    263, 273, 370, 373, 388ff, 391
  identity conditions for, 204
primary states of affairs, 147, 168. See also
    primary propositions
primary structure(s), 15f, 29f, 147, 268f, 273
  as arguments of connectives, 268f
  as configurations, 263f
  definition of ontological, 208
  ontological, 182, 208–22, 236, 263f
  semantic, 207f, 235
  simple primary facts as simple ontological,
    208–22
principle of ontological rank (POR), 454f
  and the theory of evolution, 455–58
principle of sufficient reason, 245
principle of the identity of the indiscernible,
    206
principle of the indiscernibility of the identi-
    cal, 206
principle of the unity of the entirety of actu-
    ality, 245

principle of universal coherence, 245
process, 164, 193, 261
projection, 181
proof, 72
property (attribute), 158, 163f, 219
proposition(s), 15, 92f, 160f, 189
  aesthetic, 307
  and states of affairs, 147, 162, 163, 188, 220
  as arguments of connectives
    (Quine), 216ff
  as state(s) of affairs, 220
  contradictory, 238
  ethical-deontic, 305
  false, 237ff
  fully determined, 224, 234
  logical, 183
  practical-deontic, 305
  primary. See primary propositions
  semantic, 179
  syntactic, 179
  the ontological import of logical/math-
    ematical, 239ff
prosentence, 227
prosentential theory of truth, 227
proxy function (Quine), 252
pseudopropostion(s), 238
psychology, folk, 273, 283
Pulte, H., 121
Puntel, L. B., 2n2, 3n3, 23, 35, 81, 97, 142,
    147n38, 157, 163, 190, 200n22, 202,
    203n25, 226n40, 235n45, 270n15, 290n27,
    458n60
purposiveness, 312ff
Putnam, H., 17, 39, 122, 137, 241n51, 256n7,
    360–67

quantification, 381, 386
  and totality, 423f
  plural, 423f
quantifier, 198
quantifier shift, 450f
quantum gravitation, theory of, 344n51
quantum logic, 222
quantum physics, 360
quantum theory, 327
quasi-definition of the structural-systematic
    philosophy, 26f
Quine, W. V., 4, 36, 58f, 83f, 95, 97, 117, 146, 149,
    160, 168f, 195–99, 200, 204f, 211f, 216ff,
    222, 224ff, 231, 240f, 251–57, 273, 277f,
    381, 387, 402, 414, 441n48, 472n5, 483n8

Ramsey sentences, 124n20
rationality, 114, 297f
    communicative, 101
    ideal, 114n16, 298
    theories of, 118
    universal, 115n16
Rayo, A., 423n33
ready-made world, 11, 362, 363
ready-structured world, 39, 362, 363
realism, 126
    internal (Putnam), 362–65, 375
    internal, 362–65
    metaphysical, 362, 375, 377
    modal, 248, 431
    moral, 298
    pragmatic/direct/natural, 364f, 366f
    structural, 262
realism-antirealism debate, 374f
realism-antirealism problem, 371, 374f
reference, 160
    inscrutability of (Quine), 257
    theories of, 218f
reflexivity, 181
Reichenbach, H., 133
reification, 196f, 255
relation, 158, 163f
relationality, 478f
    complete and adequate grasp of
        holistic, 78f
relativism, 370
relativity, 52, 478f
    ontological, 137, 253, 257
    theory of (physics), 289, 327
    to theoretical framework, 9, 241f, 465f
relativization of philosophy to the subject, 18
religion(s), 16, 256n7, 262, 324, 330ff, 346ff,
        351f, 356, 363, 447, 459
    high, 330, 332
    plurality of, 330ff
    starting point for a philosophical interpre-
        tation of, 330ff
religious, the, 330ff
representation, 278
Rescher, N., 7n4, 11n6, 33, 42ff, 64, 166, 167n5,
        298, 367f, 392f, 394f, 398f, 401n25, 410,
        464n2
resemblance relation, 192
resemblance, family, 205, 214
revelation, history of, 459f
Richards, T., 427
Ricken, F., 296

Ritter, J., 461n1
Rosen, G., 432
Rosenberg, G., 287
Rosenkrantz, G. S., 166, 191
Rosenkranz, K., 322
rule(s), 41, 53
    as postulates (or bridge rules), 75
    for formal languages, 384
    linguistic, 22
    logical, 15, 45ff, 75, 122, 174, 179, 353n55,
        473, 474
    methodological, 41
    of detachment, 179
    of the proxy function (Quine), 252
    practical, 114
    semantic, 187f, 218, 319
    syntactic (rules of formation), 187f, 319
Russell, B., 198, 232, 265n14, 273, 275n18
Russell's antinomy, 266f
Ryle, G., 4

salvation, history of, 459f
Sartre, J.-P., 279n22
Sartwell, C., 78, 106, 380–84, 386f, 390ff
Schelling, F. W. J., 35, 251
Schlapkohl, C., 265n13
Schneider, C., 222n37, 261, 425n36
science(s), 341, 477. See also natural Science(s)
    particular, 479f
    practical and theoretical, 292f
Searle, J., 235n45, 285ff, 478n7, 479
Seibt, J., 164, 261n9
self and world, understanding of as articu-
        lated in ordinary language, 272f
self-consciousness
    according to the model of reflection,
        279
    and intentionality, 278–82
    as relationless primordial phenomenon
        (D. Henrich), 279ff
self-grounding, 9, 54
    of the structural-systematic philosophy, as
        a continuing process, 482ff
self-identity, 211, 214
Sellars, W., 11n5, 224, 271, 369, 373
semantics
    and ontology, 6, 16, 95, 148f, 159f, 171, 175,
        185ff, 196, 223, 374, 387, 442n49, 469,
        475f
    and pragmatics, 101, 151
    and theories of being and beings, 370f

basic features of compositional, 186–90
compositional, 15, 146f, 163, 186–95, 288
contextual (of this book), 15f, 146f, 190–99,
    218, 263, 288, 442, 483
critique of compositional, 190–99
formal, 190, 218f
in the comprehensive sense and the nar-
    rower sense, 171f
of possible worlds, 242
on the basis of a strong version of the con-
    text principle, 199–208
ontologically oriented, 185ff
propositional-conditional, 188
truth-conditional, 188
valuation, 148f, 174
sense, and *Bedeutung,* 160ff, 189, 233n44
sentence(s), 222
    aesthetic, 90, 306f, 309f, 316, 318
    and proposition(s), 216ff
    ethical, 291f, 298
    false, 237
    fully determined, 224, 225, 234
    fully determined status of true ethical, 298
    indicative (theoretical), 77, 80, 90, 92, 183,
        185, 186, 188f, 203f, 206f, 229, 309, 405f
    moral, 291n29
    normative, 292f
    objectively practical, 293
    observation (Quine), 196, 253
    of the subject-predicate form, 15, 91, 146f,
        165, 185, 195–99, 202f, 206f, 218, 235n45,
        263, 391, 396, 404, 419, 442
        as abbreviations of complex primary
        sentences, 206, 221
    practical, 89f, 291
    primarily practical, 293ff
    primary. *See* primary sentence(s)
    semantic primacy of, 84
    the three logical-semantic forms of aes-
        thetic, 306–14
    theoretical-deontic, 293–96, 298
    theoretical-ethical, 291–96
    theoretical-practical, 293, 295f
    theoretical-valuative, 293–96, 298
    truth status of theoretical-ethical, 298
    valuative, 295
    without subject-predicate structure, 210f,
        273
sentence form, 90f
sentence operator(s), 90–94
    aesthetic, 90–94, 306–7, 309f, 316, 318

practical, 90–94, 294f, 309
theoretical, 90–94, 295f, 306f, 308, 316
sentence types and sentence tokens, 395
separation, axiom of, 267
Septuagint, 142
set theory, 128
set(s)
    and classes, 267f
    concept of, 176ff, 266f
    uncountable, 243
Shalkowski, S. A., 442n49
Shapiro, S., 175, 177n12
sheaves, 269
Sher, G., 142n33
Shoenfield, J. R., 122, 380f
Shope, R. K., 109
similarity relation, 192
Simons, P., 192, 268
simplification (Koslow), 181
singularity, 325f
skeptics, 63f
slingshot argument, 235n45
Smith, N. K., 108
Smith, Q., 326n48
Sneed, J., 75, 122, 127n26, 128, 137n31, 400n24
society, 37, 340f
space of reasons, 373
speakers and subjects, 174
Spinoza, B., 4, 5, 131, 191, 250
spontaneity, 280
Stegmüller, W., 45–47, 75, 122, 127ff, 137n31,
    174n8, 180n14, 184f, 303f, 400n24
Stern, R., 232n43
Stoljar, D., 283ff
string theories, 344n51
structural core, 128ff
structural framework, 169f
structural methodology, 42
structuralism, 27n4
    Bourbaki's mathematical, 175–78
    mathematical, 175
    Quine's global ontological, 251–57
structurality, 17
structuration of the world (of being), 399,
    407f
structure(s), 11f, 15, 25, 125, 126, 135, 149n40,
    270ff
    absolutely universal, 257, 477
    abstract (pure), 28f, 49f, 167, 209f, 211, 215
    and being, 36–41, 246
    and data, 24ff

and language, 136f
and the universal datum, 469
and theories (Quine), 254ff
and truth, 222f
Bourbaki's formal concept of, 175–78
compositional-semantic, 186–90
concept of, 27–30
concrete, 28f, 167, 209f, 215
configuration(s) as, 213f
contentual (semantic and ontological),
    169f
contextual-semantic, 207f
fine-grained/coarse-grained, 408ff
formal (logical/mathematical), 12, 155, 169f,
    172–83, 269
fully determined interrelationships among
    fundamental, 222–45
fundamental, 12, 15, 155
immanent ontological import of, 469
interconnections among formal, semantic,
    and ontological, 412
logical, 178–83
logical/mathematical, 451f
mathematical, 47, 128, 175–78
multiplicity of, 222
ontological, 12, 15, 38–41, 155, 167, 180n14,
    186, 208–22, 262, 267, 270, 315, 322, 339,
    341, 364, 396f, 398, 399, 411, 452n54
ontological import of the truth of logical/
    mathematical, 239ff
ontological status of logical, 182f
particular (domain-specific), 34ff, 258
philosophically expanded concept of, 167
relation of formal and semantic, 171f
relation of logical/mathematical and onto-
    logical, 215f
relatively universal, 258, 477
semantic, 12, 155, 183–208
systematic-architectonic philosophical
    place-value of expanded concept of,
    155, 167
the three levels of fundamental, 172–222
topological, 269
universal (most general), 33ff, 169f, 257
subject(ivity), 14, 32, 37, 99ff, 265, 270f, 279ff,
    314, 338, 360, 411f
according to the transcendental perspec-
    tive, 18
and knowledge, systematically considered,
    110–17
and perspective changes, 117–20

as intentional, 100
as intentionally coextensive with being as
    a whole, 111
as taking positions, 100
foundationalistically understood, 100
particularistic, 111
pragmatic, 100f
structuration of, 110
transcendental-constitutive, 101
transcendental, 415
universal, 403
subject-operator, 111
particularistic, 111–20, 405f
universal, 111–20
subsets, axiom of, 267
substance, 69f, 164, 190, 265
and attributes, 164
and independence, 191
category of, 261
with the designation "object" 261
substance ontology, 15, 158, 164, 185, 192, 194,
    203, 221, 235f, 393, 419
critique of, 190–95
the fundamental problem of every, 193ff
vs. configuration ontology, 410
substance-schema, 70
substratum, 69f, 193, 213, 261, 275
sum/Fusion, 268
Summa, 2
supervenience, 283ff
supervenience physicalism, 283ff
Suppe, F., 75, 122–26, 128, 179
surface structure, 408
Swinburne, R., 447
System of Pragmatic Idealism (Rescher), 7n4
system(s), 1f, 17, 22f, 28, 35, 42, 360
and the observer, 360
and the theoretician, 360, 362
and thinking/mind/language, 363
formal, 24, 46, 49f, 51, 117, 121f, 148f, 174,
    218f, 380, 382
open, 20
philosophical, 2, 3f, 7n4, 10, 20, 35f, 41, 64,
    462f, 464, 482
semantic, 46
semiotic, 19, 38, 78, 97ff, 376, 376, 378ff,
    381f, 384–87, 389ff, 394, 396f, 412
"systematic," 1f
systematicity, 43
as articulated theory (Dummett), 4ff
cognitive, 64n19

ontological, 64n19
systematics of structure, 15, 155–245
    and world, 249
    concept of, 155–72
    program for a philosophical, 168ff
    status of language and semantics within,
        170ff

Tappolet, C., 142n33
Tarski, A., 87, 95, 143, 145, 152, 224ff, 360, 400
teleology, 314
temporal logic, 222
term(s)
    general, 195ff
    singular, 194, 202f
        Quine's procedure for the elimination of,
            84, 195–99
        non-referentiality of many, 196f
theology
    Christian, 142f, 447
    in the genuine (theoretical) sense, 447
theorem, of Kelley-Morse, 425, 430
theoretical framework(s), 9ff, 13, 16, 18, 20,
        22f, 41, 42, 51f, 55, 56, 70–73, 94, 99, 242,
        243f, 338, 441, 466
    absolute, 20, 430, 482
    adequacy of, 348ff
    and grounding, 62ff, 72f, 90f
    and languages with uncountably many
        expressions, 389f
    and why questions, 347ff
    as ontological arena(s), 403, 405, 407f
    comparisons between, 471f
    everyday/lifeworldly, 405f
    external metasystematic, 469f
    greater intelligibility and coherence of,
        20, 482
    metasystematic, 474f
    non-absolute, 429
    of philosophy and of specific sciences,
        477
    of specific sciences, 478
    of the structural-systematic philosophy,
        22ff, 232, 469–75
    ontological adequacy of, 405–11
    physical-cosmological, 326f
    plurality of, 117, 242, 404, 482
    relatively universal, 478
    semantic-ontological, 193
    special-scientific, 480
    structural-systematic, 148, 263, 290, 467

universal, 477
theoretical network(s), 42, 50f, 128
theoreticity, 27, 42, 74ff, 90ff, 93, 116
    as dimension of presentation, 74–76
    criterion for, 14
    full determination of, 141
    language, as medium of presentation for,
        76–89
    the linguistic criterion of, 89–96
    systematics of, 74–154
    truth as the fully determined status of, 222f
theory/theories, 4, 14, 26f, 29, 35f, 42f, 45,
        65, 99
    and structure (Quine), 254
    and the truth concept, 127
    and the world / being as a whole, 401
    articulated (Dummett), 4ff
    axiomatic, 45, 139
    constructive-empiricist conception of,
        124–27
    essential components of the structural
        concept of, 136–38
    ideal, 362
    in metalogic/metamathematics, and the
        philosophy of science, 121–30
    in the narrower sense, 14, 121–40
    logical concept of, 121f, 132
    network-structural (coherentist), 139
    non-statement view of, 75, 127
    of everything, 140n32, 324ff
    philosophical, 6, 9, 10, 15, 35f, 41, 42f, 45, 65,
        69, 85, 131, 136f, 138f, 141, 172, 291, 318,
        320f, 324, 334, 337, 343, 345, 350, 352f,
        412, 426, 441, 480
    received view of, 75, 122ff, 127, 132
    scientific concept of, 122–30
    semantic view of, 124–27
    statement view of, 75, 122
    structural concept of, 130–38
    structuralist conception of, 75, 127–30,
        140n32
    underdetermination of, 253
    Weltanschauung view of, 123
theory and practice, putative
    unity of, 481
theory concept, 121–40
    as regulative concept in philosophy, 138–40
theory elements, 128, 130
theory form(s), 45, 227
    axiomatic, 45ff
    network-structural (coherentist), 48

theory material, 42
theory of being as such and as a whole, basic
    features of, 413–60
theory of meaning, 160
theory of reference, 160
theory-dependence, 75
theory-ladenness, 68f, 254
thing in itself, 248
    and appearances, 376ff
    unknowability of (Kant), 256n7
thinkability, argument from, 287
thinking, 161f, 270f
third realm (Frege), 373
third world (Popper), 373
Thomas Aquinas, 131, 316, 366, 416, 438n45,
    440, 443n50, 443n51, 447n52, 448–51, 458
Thomason, R., 79
thought(s), 16, 158, 161ff
    Frege on, 162f, 188f
Tichý, P., 400n24
time, 325, 327
    imaginary, 327
token, 396
token physicalism, 283ff
tokening system (tokenability), 382–87
totality, 8n4, 336, 343, 419
    and Cantor's diagonal argument, 424ff
    and ontology, 427
    and quantification, 423f, 426
    and set theory, 425
    and the distinction between sets and
        classes, 425
    and thing-ontology, 427
    as set, 424ff
    as universal class, 425ff, 430
    logical/mathematical problematic of talk
        about, 19, 423ff
    maximal, 429f
    of being and of beings, 419
    of beings, 430
    of truths, 424ff
    semantic problematic of talk about, 421ff
    submaximal, 429
Toulmin, S., 124
tradition, process of (Gadamer), 338
transcendental philosophy, 112, 368
trope, 192
trope theories, 192, 210
true, taking something to be (Kant), 109f
truth, 187, 305
    absolute, 242–45

    and correctness, 297f
    and model(s), 149n40
    and objectivity, 297
    and ontological import, 7, 225, 231f, 236,
        288, 291, 435
    and ordinary/natural language, 143–45
    and structure, 149n40, 222f
    as predicate and as operator, 145–47
    as unconcealment (Heidegger), 143
    basic idea of, 148–54
    cataphoric theory of, 228
    completely defined concept of, 234
    comprehensive and subtheories of,
        147f
    concept of, 104n9, 108. See also
        truth-concept
    correspondence theory of, 144f, 231f, 234,
        235n45, 297, 362
    deflationistic theories of, 144
    ethical, 298
    for Kant, 110
    for us, 367f
    in itself, 367f
    informal-intuitive formulation of the fun-
        damental idea of, 153f
    intuitive understanding of, 143, 225f
    more precise characterization of the basic
        idea of, 223–26
    ontological dimension of ethical, 296ff
    ontological implications of, 289, 290f
    ontological import of, 7, 225, 231f, 236, 288,
        291, 435
    relative to all theoretical frameworks, 242
    relativism about, 241–45, 401
        a moderate form of, 241–45
        antinomy of, 241f
    relativity of, 241–45
    radical, 241
truth bearer(s), 147, 226f
    as structures, 227
truth candidates, 11n6, 43f
truth concept, 14, 93, 134, 222f, 438f
    and primary propositions, 208
    as composition of three functions, 227–36
    Hebraic (Biblical), 142
    in ethics, 297
truth conditions, 188
truth operator, 151, 224, 234
    anaphoric, 227f
    as composition of three functions, 229–31
    cataphoric, 228–31

truth paradox, 36f, 424
truth predicate, 226
truth theory/theories, 7, 16, 141–54, 222–45
  analytically oriented, 142
  anaphoric-deflationistic, 227f
  as relative to theoretical frameworks, 242
  cataphoric, 227–31
  criteriological, 148
  definitional, 148
  deflationistic, 144f
  deflationistic, 224ff
  substantialistic, 144f
truth value(s), 161, 149n44, 181, 233n44
  as objects (Frege), 163, 188f, 223
truth-approximation, 400f
truth-schema, 224, 360
T-schema, 225f
T-Theoreticity, 75
Tugendhat, E., 143n36, 275n19
two-dimensionality of being, 331

ugliness, 322f
unboundedness of being, 369, 371
unboundedness of the conceptual, 369, 371,
  373
unconditioned, the, 443n51
uncountability, 78
underdetermination thesis, 472n5
understanding, 338
understanding of world and self as articulated
  in ordinary language, 272f
unifying point
  and the universe, 282
  as the factor configuring human being,
    275–78
  systematic, 282, 340, 354
universal class, 425
universal quantification, 429
universal(s), 69f, 190, 192, 194, 213f, 219, 264
  and subject (substratum), 194
  problem of, 8n4
universe of discourse, 1, 18, 25, 30–33, 38, 39f,
    42, 136ff, 158, 257f, 311, 371, 395, 469, 477
  restricted, 30
  unrestricted, 10f, 12, 30–33
universe, the, 7, 11, 155, 339, 325–29, 349f, 360,
    380
  and modal concepts, 328f
  as self-contained (Hawking), 327ff
  as set, 428ff
  as such, 328

  as the absolute, 329
  contingent, 328
  physical, 355
  structured, 329
  submaximal, 428ff
urelement, 209, 215

Vallicella, W. F., 235n45
valuation semantics, 148f, 174
value(s), 158, 299f
  and the good, 301
  as primary facts, 263
  as state of affairs, 299f
  concepts of, 293, 295f
  distinction between basal-ontological and
    moral-ontological, 299f
  domain of, 301
  ethical, 290–305
  general-metaphysical perspective with
    respect to basal ontological, 300ff
  metaphysical-anthropological perspective
    with respect to basal-ontological,
    302f
  moral-ontological, 309
  ontological, 296ff
  ontological status of basal-ontological,
    300–303
  ontological status of moral-ontological,
    303ff
  practical basal-ontological, 309
  semantic, 15, 160ff, 187f, 200f, 295f
  the peculiarity of moral-ontological, 321
  the putative peculiarity of moral-
    ontological, 302ff
value-facts
  moral, 299f
  ontological, 311
van Cleve, J., 270n15
van Fraassen, B., 124–27, 132f, 137, 422
Varga, M., 174n8, 180n14, 184f
variable(s)
  and singular terms, 197f
  assignments of values to, 184, 207, 475
  values of, 197
vocabulary
  pragmatic, 150f
  semantic, 150, 151f

Wall, R. E., 187
warrant, 53n21
Weingartner, P., 165, 173

Whitehead, A. N., 10, 261n9
wholeness, 420n32
why questions, 346–50
    and theoretical frameworks, 347ff
Wiggins, D., 296
Wigner, E., 360
will, 278, 305
    absolute, 459
Williams, B., 351n53
Williams, D. C., 192
Williamson, T., 106, 108, 111, 119
Wittgenstein L., 4, 14, 16, 31, 79, 82f, 90, 95, 164,
        205, 362, 364f, 366, 420
Wolff, C., 324
Wood, A. W., 108n11
world(s), 7, 11, 16, 31, 37, 155, 306, 374, 422
    actual, 248, 431, 433f
    aesthetic, 305–23
    and being as a whole, 247ff
    and self, understanding of as articulated in
        ordinary language, 272f
    and universe of discourse, 247ff
    art as presentation of transformed, 319f
    as relative or contingent dimension of
        being, 389
    as totality of appearances, 314
    contingent, 451f
    human, 262–305
    ideal, 262
    logical/mathematical, 240f
    mathematical and physical, 240
    ontology of possible, 238
    plurality of, 400
    possible, 242, 248, 431–36, 451f
    primordial comprehensive unity of all, 435
    (self-)presentation of, 407f, 412
    structural levels of, 407f
    structuration of, 40f, 249, 363, 407f
    subdimensions of the actual, 249

    submaximal, 428ff, 451
    van Fraassen on, 422
    See also universe, being (as a whole)
world concept(s), 247–50
    in analytic philosophy, 248
    transcendental, 247
world history, 332–56, 459f
    analytically oriented conceptions of, 337
    and theory/theories of being, 334
    as third-order ontological configuration,
        350
    bases for a comprehensively systematic
        theory of, 345–50
    comprehensive theory of, 350
    comprehensive view of, 324–56
    comprehensively systematic theories
        of, 343ff
    hermeneutical conception of, 338f
    inner structure of, 340ff
    materialistic theory of, 351
    meaning of, 342–56
    metaphysical theories of, 334ff
    non-metaphysical theories of, 334ff
    ontological determination of, 334–40
    ontology of, 334–40
    presuppositions for a comprehensively
        systematic theory of the meaning of,
        350–56
    restricted theories of, 343, 345, 350
    starting points for a cosmological-
        materialistic theory of, 334
    structure of, 340ff
    transcendental view of, 336ff, 337n50
    versions of, 400
world in itself and world as appearance,
        376
world spirit, 335, 339
world systematics, 16, 246–56
Wright, C., 142n33, 152, 291, 297

CPSIA information can be obtained at www.ICGtesting.com
Printed in the USA
BVOW031157070513

320098BV00002B/187/P